Researches in Selected Plant Physiology

Researches in Selected Plant Physiology

Edited by **Clive Koelling**

R **C**ALLISTO
EFERENCE

New York

Published by Callisto Reference,
106 Park Avenue, Suite 200,
New York, NY 10016, USA
www.callistoreference.com

Researches in Selected Plant Physiology
Edited by Clive Koelling

International Standard Book Number: 978-1-63239-554-2 (Hardback)

Printed in the United States of America.

Contents

Preface

The main aim of this book is to educate learners and enhance their research focus by presenting diverse topics covering this vast field. This is an advanced book which compiles significant studies by distinguished experts in the area of analysis. This book addresses successive solutions to the challenges arising in the area of application, along with it; the book provides scope for future developments.

The aim of this book is to provide research-focused information regarding selected plant physiology. This book provides fundamental concepts and latest insights on plant physiology covering abiotic stress, plant water relations, mineral nutrition and reproduction. Plant reactions to insufficient water availability and other inorganic strains (like metals) have been studied through changes in water consumption, transport structure and through molecular and genetic methods. A comparatively new feature of fruit nourishment is presented in order to offer the basis for the development of a few fruit quality traits. Compiling researches of experts around the globe, it can serve as a source of methodologies, theories, pathways and ideas for students and researchers in different areas of plant physiology.

It was a great honour to edit this book, though there were challenges, as it involved a lot of communication and networking between me and the editorial team. However, the end result was this all-inclusive book covering diverse themes in the field.

Finally, it is important to acknowledge the efforts of the contributors for their excellent chapters, through which a wide variety of issues have been addressed. I would also like to thank my colleagues for their valuable feedback during the making of this book.

Editor

Section 1

Abiotic Stress

Abiotic Stress Responses in Plants: A Focus on the SRO Family

Rebecca S. Lamb

Department of Molecular Genetics, The Ohio State University, Columbus, OH
USA

1. Introduction

Plants are sessile organisms and as such must have mechanisms to deal with both abiotic and biotic stresses to ensure survival. The term "abiotic stress" includes many stresses caused by environmental conditions such as drought, salinity, UV and extreme temperatures. Due to global climate change it is predicted that abiotic stresses will increase in the near future and have substantial impacts on crop yields (Intergovernmental Panel of Climate Change; http://www.ipcc.ch). Therefore, understanding abiotic stress responses and the connection between such responses and agronomically important traits is one of the most important topics in plant science. Often plants will experience more than one abiotic stress at a time, making it difficult to determine the effect of a single stress under field conditions. Therefore, much of the progress in understanding plant defence signaling and response has come from laboratory studies, especially those using the model plant species *Arabidopsis thaliana*, which belongs to the family Brassicaceae.

1.1 Responses to different abiotic stresses share common components

An understanding of abiotic stress responses depends on an understanding of the molecular processes underlying those responses. Plant defences against different abiotic stresses have both common and unique elements. Common elements include increases in reactive oxygen species (ROS) and cytosolic Ca^{2+} as well as activation of kinase casades. In addition, stresses can lead to increased concentrations of hormones such as salicylic acid (SA), jasmonic acid (JA), abscisic acid (ABA) and ethylene, all of which have been implicated in response to environmental conditions (reviewed in (Hirayama & Shinozaki, 2010)).

The increase in ROS is an especially important common connection between different stresses. ROS are continuously produced in the plant through cellular metabolism and plants have many antioxidants and scavenging enzymes to maintain homeostasis. However, under stress conditions ROS accumulates. Although these molecules can damage cells (Moller et al., 2007), they are also known to have signalling functions (Foyer & Noctor, 2009). In fact, while excess ROS is toxic, a certain level of ROS production is necessary for a successful response to stress, including salt (Kaye et al., 2011). In addition, ROS accumulation has been shown to have a role in priming plants for enhanced stress resistance (reviewed in (Conrath, 2011)). However, excess ROS can lead to cell death (Kangasjarvi et al., 2005; Overmyer et al., 2005) and perturb development (Tognetti et al., 2011).

1.2 Abiotic stress causes largescale changes in gene expression

Plant defences are characterized by large reprogramming of gene expression, much of it through regulation of transcription. Research over the last two decades has lead to the identification of many stress-inducible genes, especially since the publication of the Arabidopsis genome (Arabidopsis Genome Initiative, 2000), which allowed global gene expression experiments. Since 2000, several other plant species have had their genomes sequenced, allowing expansion of this type of analysis. Functional analysis has confirmed the importance of many of these genes in stress tolerance. More recently, genes whose expression is downregulated under stress conditions have received attention (Bustos et al., 2010). It is now understood that transcriptional repression responses are an integral part of adaptive responses to stress.

To mount an effective defence, ultimately a transcription factor needs to bind and activate or repress its target genes. Since there are both common and unique effects from different stresses, comparison of the transcriptional profiles of such stresses has revealed both common and unique gene activation and repression patterns and lead to the development of models of transcriptional regulation of abiotic stress responses. The transcriptional control of stress can be divided into several temporal phases, most likely due to varying dependency on different signaling molecules or protein synthesis (Yamaguchi-Shinozaki & Shinozaki, 2006). Changes can begin within 15-30 minutes of exposure and last for several days (Kilian et al., 2007). The common stress transcriptome represent a shared response and is likely responsible for the widely observed cross-protection where exposure to a given stress increases the resistance of the plant to a second.

Many transcription factors involved in stress responses have been identified. Often the expression of genes encoding these transcription factors responds rapidly to abiotic stress treatments (Gadjev et al., 2006; Kilian et al., 2007). During domestification of crops, selection for stress tolerance has acted on such transcription factors (Lata et al., 2011), underlining their importance. These proteins have also been targets for development of abiotic stress tolerant transgenic plants (Hussain et al., 2011). Transcription factors that regulate stress responses belong to many different families. However, there are certain families that include a relatively large number of members that have been implicated in environmental response. These include the DREB1/CBF family of AP2 transcription factors (Lata & Prasad, 2011) as well as other AP2-type factors (Dietz et al., 2010), Class I homeodomain-leucine zipper proteins (Elhiti & Stasolla, 2009) and the WRKY family (Rushton et al., 2011). Interestingly, the families mentioned here are all plant-specific (Riechmann et al., 2000), suggesting that they may have evolved to help plants deal with the stress of life on land. However, members of transcription factor families that are found outside of plants have also been implicated in control of stress-inducible gene expression.

The activity of these transcription factors is also controlled at posttranscriptional levels. Of particular note, they can be regulated through protein-protein interactions and/or posttranslational modifications. For example, AtMEKK1 can phosphorylate WRKY53 and regulate its activity during senescence (Miao et al., 2007). DREB2A, which when constitutively active confers salt and high temperature tolerance (Sakuma et al., 2006b), interacts with the Med25 subunit of the Mediator complex to regulate gene expression (Elfving et al., 2011), while heterodimers of bZIP1 and bZIP53 act together to activate transcription during low energy stress (Dietrich et al., 2011). Thus, the protein complexes in which transcription factors are found and the modifications they have are essential to determine their activity.

1.3 Epigenetic control of abiotic stress response

As discussed above, upon stress plants reprogram their transcriptome. Although transcription factors are important for this reprogramming, it is thought that alteration of chromatin structure is also critical (Arnholdt-Schmitt, 2004). Genomic DNA is packaged around nucleosomes into chromatin, the confirmation of which can restrict access of proteins to the DNA. Therefore, transcription is heavily influenced by dynamic changes in chromatin structure (Kwon & Wagner, 2007). Chromatin structure is regulated by several mechanisms, including histone and DNA modifications, chromatin remodelling, which uses ATP hydrolysis to alter histone-DNA contacts, and histone variants (JM. Kim et al., 2010; Pfluger & Wagner, 2007). Alterations in chromatin structure are known to impact stress tolerance (JM. Kim et al., 2010).

Posttranslational modifications of histones are one of the best-studied aspects of chromatin regulation. Over 25 sites of histone modification have been identified in Arabidopsis (Zhang et al., 2007) and the pattern of modification is known to alter upon stress (JM. Kim et al., 2008). For example, a decrease in trimethylation of histone H3 Lys27 (H3K27me3), which is a maker of less transcriptionally active genes, is seen at cold-responsive loci upon exposure to cold (Kwon et al., 2009). Some of the proteins responsible for histone modifications have been implicated in abiotic stress response as well. The histone deacetylase HDA6 is involved in ABA signalling and salt stress response and required for jasmonate-induced gene expression in addition to a role in flowering time control (LT. Chen et al., 2010; K. Wu et al., 2008; Yu et al., 2011). It is also necessary for freezing tolerance (To et al., 2011a). Mutations in *HOS15*, which encodes a WD-repeat protein, cause hypersensitivity to freezing and HOS15 increases deacetylation of histone H4 (Chinnusamy et al., 2008; J. Zhu et al., 2008). The histone acetylase AtGCN5 has roles in gene expression in response to cold and light (Benhamed et al., 2006; Stockinger et al., 2001). Many more such connections are being discovered.

Another important level at which gene expression is epigenetically controlled is degree of nucleosome coverage of a gene. Generally, nucleosome density is decreased and chromatin structure relaxed when transcription is activated (Lieb & Clarke, 2005). Chromatin remodelling factors are necessary for the rearrangement of nucleosomes on DNA and several of these have been implicated in stress response. For example, the SWI/SNF family member AtCHR12 has been shown to mediate the transient growth arrest seen under adverse environmental conditions (Mlynarova et al., 2007). Another member of this family, SPLAYED (SYD), also regulates stress pathways (Walley et al., 2008). DEAD-box helicases, which unwind duplex DNA or RNA, can also affect chromatin structure and several have been implicated in various stress responses (Vashisht & Tuteja, 2006). Interestingly, in Arabidopsis nucleosomal DNA is more highly methylated than flanking DNA and nucleosomes are enriched on exons (Chodavarapu et al., 2010). Genes whose coding regions are methylated tend to be longer and more functionally important and include many stress-regulated genes (Takuno & Gaut, 2011). In plants DNA methylation status is dynamic, regulated by DNA methylation and demethylation reactions and influenced by histone modifications (reviewed in (He et al., 2011)). High DNA methylation is associated with silenced transposable elements. However, this modification also functions in gene regulation and transcribed genes will also contain methylated bases. Although the involvement of DNA methylation in abiotic stress response has not been extensively examined, it is involved in defence against gemini viruses (Raja et al., 2008, 2010) and important in the vernalization response (DH. Kim et al., 2009). In addition, the histone deactylase HDA6, discussed above, has been shown to regulate silencing in cooperation with the DNA methyltransferase MET1 (To et al., 2011b), providing a link from DNA methylation to ABA and jasmonate signalling.

1.4 Costs of defense responses

Plants have developed many sophisticated defence pathways to allow them to thrive even in the presence of suboptimal environmental conditions. Phenotypes involved in tolerance or defence against environmental stress can be inducible or constitutive. The evolution of induced responses is thought be the result of the high cost of maintaining the response in the absence of stress. This is because of the reallocation of energy and resources to defence from growth and reproduction (Walters & Heil, 2007). Research has begun to measure the benefits and costs of adaptation to stressful conditions, for example during cold acclimation (Zhen et al., 2011) and tolerance (Jackson et al., 2004). In addition, analysis of mutant and transgenic plants with derepressed stress responses to both biotic and abiotic stresses often have developmental abnormalities and reduced seed set. For example, *CONSTITUTIVE EXPRESSION OF PR GENES5 (CPR5)* was originally identified in a mutant screen for constitutive expression of systemic acquired resistance; the *cpr5* mutant has chlorotic lesions, reduced trichome development and stunted growth (Bowling et al., 1997). *CPR5* encodes a transmembrane protein that represses leaf senescence and pathogen-defence responses in Arabidopsis (Kirik et al., 2001; Yoshida et al., 2002). An altered cellular redox state is present in *cpr5* mutants, which underlies the chlorotic lesions and maybe the other developmental defects as well (Jing et al., 2008) and CPR5 has been hypothesized to act as a repressor of ROS accumulation (Jing & Dijkwel, 2008).

The cost of stress response is reflected in a phenotype observed in plants exposed to chronic, sublethal abiotic stress, the so-called stress-induced morphogenetic response (SIMR; (Potters et al., 2007; Tognetti et al., 2011)). SIMR is characterized by reduced cell elongation, blockage of cell division in primary meristems and activation of secondary meristems (Potters et al., 2009). Plants displaying SIMR often show accumulation of antioxidants and other compounds that act as modulators of stress responses. It is thought that these changes allow the redistribution of resources to stress response pathways, permitting plants to acclimate to their environment. Another aspect of the SIMR response is accelerated flowering, a response that has been associated with many abiotic stresses, including nutrient deficiency (Wada et al., 2010; Wada & Takeno, 2010) and salinity (Ryu et al., 2011) and is thought to guarantee reproduction before any potential lethality caused by stress. SIMR has been hypothesized to be mediated by accumulation of ROS caused by the stressful conditions and subsequent alterations in auxin accumulation and signaling (Potters et al., 2007; Tognetti et al., 2011). In Arabidopsis, SIMR has been shown to be induced under several different abiotic stress conditions (Potters et al., 2007; 2009), including salt stress (Zolla et al., 2009) and exposure to the nonprotein amino acid amino-butyric acid (CC. Wu et al., 2010).

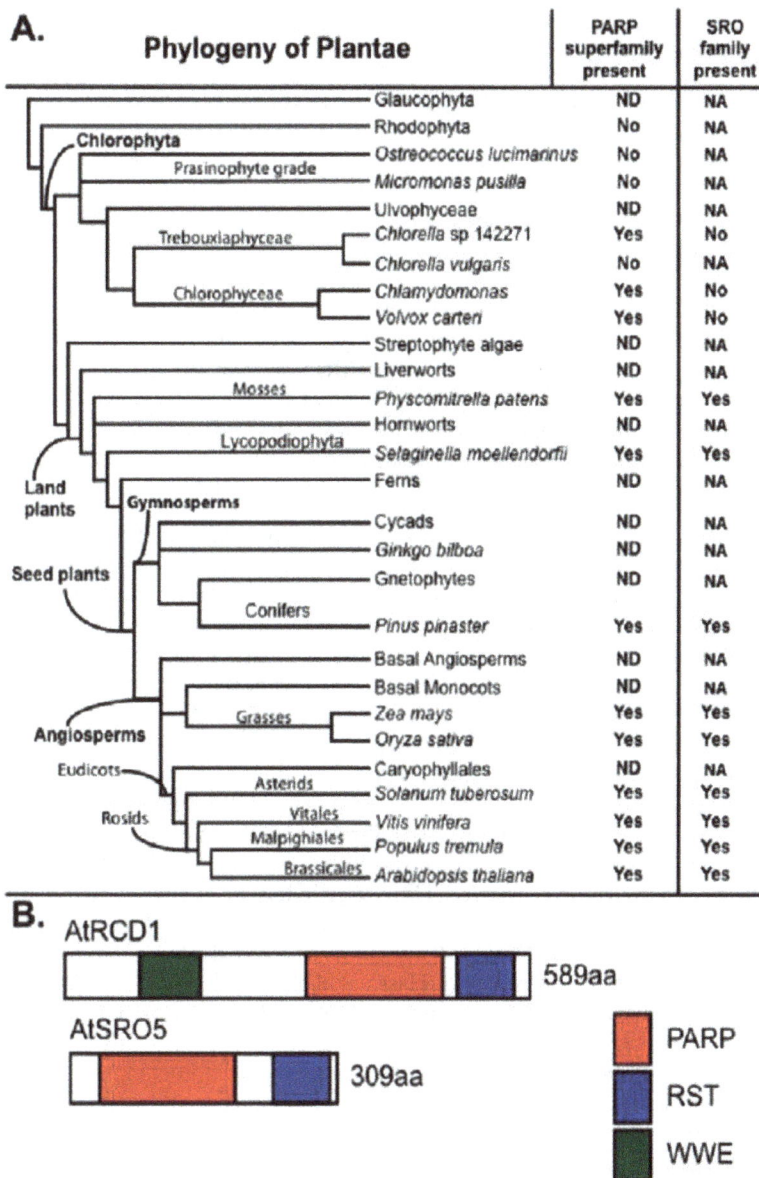

Fig. 1. The SRO family of PARP-like proteins is plant specific. A. Simplified phylogeny of Plantae. Branch lengths do not reflect genetic distance. Presence or absence of PARP superfamily members and SRO subfamily members are indicated, based on ((Citarelli et al., 2010) and searches of EuroPineDB (for *Pinus pinaster*; (Fernandez-Pozo et al., 2011)) and the potato genome (Potato Genome Sequencing Consortium, 2011)). B. Schematic representaion of domains found in two representative Arabidopsis SRO family members. Protein domains are illustrated by colored boxes and defined according to Pfam 25.0 (Finn. et al., 2010).

2. The SRO family: A novel group of poly(ADP-ribose) polymerase-like proteins found only in land plants

The poly(ADP-ribose) polymerase (PARP) superfamily is distributed across the breadth of the eukaryotes (Citarelli et al., 2010) and was first identified as enzymes that catalyze the posttranslational modification of proteins by multiple ADP-ribose moieties (poly(ADP-ribosyl)ation; (Chambon et al., 1963)). It is now recognized that there are many types of PARPs and PARP-like proteins; they are characterized by a shared PARP catalytic domain but differ outside of this domain. The functions of these proteins have also expanded and some members of this family do not act in poly(ADP-ribosyl)ation. *Bona fide* PARPs attach ADP-ribose subunits from nicotinamide adenine dinucleotide (NAD$^+$) to target proteins (MY. Kim, 2005). However, other members of the PARP superfamily have been shown to have either mono(ADP-ribose) transferase (mART) activity (Kleine et al., 2008) or to be enzymatically inactive (Aguiar et al., 2005; Jaspers et al., 2010b; Kleine et al., 2008; Till et al., 2008). Biologically, PARP superfamily members are involved in a broad range of functions, including DNA damage repair, cell death pathways, transcription and chromatin modification/remodeling (reviewed in (Hassa & Hottiger, 2008)).

Although non-enzymatically active PARP superfamily members have not been as well studied as those with known poly(ADP-ribosyl)ation activity, some information is available. Human PARP9 (HsPARP9), which does not have enzymatic activity, is inducible by interferon and is able to increase the expression of inteferon-stimulated genes (Juszczynski et al., 2006), suggesting a role in host defense against viruses. Another enzymatically inactive PARP, HsPARP13, interacts with viral RNA from select viruses and recruits factors to degrade that RNA (G. Chen et al., 2009; Gao et al., 2002; Y. Zhu & Gao, 2008). HsPARP13 is also able to induce type I interferon genes by associating with the RIG-I viral RNA receptor in a ligand dependent maner, promoting oligomerization of this protein. This stimulates ATPase activity of RIG-I and enhancement of NF-KB signaling (Hayakawa et al., 2011). Even those PARPs for which poly(ADP-ribosyl)ation activity has been demonstrated have functions that do not depend on such activity. For example, HsPARP1 was originally isolated based on its catlytic activity. However, it has been shown to function in gene expression non-enzymatically, both as a transcription factor/coregulator and at the chromatin level. For example, HsPARP1 functions as a coactivator of NF-KB but enzymatic activity is not required for this function (Hassa et al., 2003; Oliver et al., 1999). HsPARP1 can bind directly to regulatory sequences, impacting transcriptional activity, as has been shown for the *CXCL1* promoter (Nirodi et al., 2001) or bind to other proteins that mediate the DNA binding, as has been shown for the *COX-2* promoter region (Lin et al., 2011). In addition, it can bind to nucleosomes and promote compaction of chromatin by bringing together neighboring nucleosomes in the absence of NAD$^+$ or enzymatic activity (MY. Kim et al., 2004; Wacker et al., 2007). Clearly, the functions of PARP proteins extends beyond poly(ADP-ribosyl)ation.

2.1 The SRO family

Compared to mammals, in which the PARP superfamily has been greatly amplified, both in numbers and types (Hassa & Hottiger, 2008), plants have relatively few such proteins (Citarelli et al., 2010). The red and green algae do not encode members of this family or encode only one or two representatives (Fig. 1A; (Citarelli et al., 2010)). Land plants, however, have several types of PARPs and PARP-like proteins, including a novel group of PARP proteins, the SRO family (Fig. 1A, B; (Citarelli et al., 2010; Jaspers et al., 2010b)). Although first identified in *Arabidopsis thaliana* (Belles-Boix et al., 2000), these proteins are found throughout land plants and consist of two subgroups (Citarelli et al., 2010; Jaspers et al., 2010b). The first is found in all examined groups of land plants and consists of relatively long proteins with a WWE protein-protein interaction domain (Aravind, 2001) in the N-terminus and a C-terminal extension past the PARP catalytic domain (Fig. 1B). This extension contains an RST domain (Jaspers et al., 2010a). The second subgroup is confined to the eudicot group of flowering plants. These proteins appear to be truncated relative to the other subgroup and likely arose from a partial gene duplication. They have lost the N-terminal region, including the WWE domain, and retain only the catalytic domain and the RST domain (Fig. 1B). The SRO family is characterized by changes in their putative PARP catalytic domains that suggest that they may not act enzymatically. *Arabidopsis thaliana* RADICAL-INDUCED CELL DEATH1 (RCD1), the first member of the SRO family identified, has been shown to be inactive and not even bind NAD$^+$ (Jaspers et al., 2010b). However, the catalytic domains within this group show variability and this observation may not be applicable to all SRO family members (Citarelli et al., 2010).

Arabidopsis thaliana gene	Locus ID	Selected plant orthologs[a]	Expression pattern[b]	Enzyme activity	Associated with stress?
AtRCD1	At1g32230	OsQ0DLN4 OsQ336N3 OsQ0J949 OsQ654Q5 VvA7PC35 VvA5BDE5 PtB9MU68 PtB9GZJ6	Expressed in all organs	No (Jaspers et al., 2010b)	Yes
AtSRO1	At2g35510	See AtRCD1[c]	Expressed in all organs	ND	Yes
AtSRO2	At1g23550	PtB9INI8 PtB9HDP9 PtB9HDP8 PtB9HDP5	Expressed in all organs	ND	Yes
AtSRO3	At1g70440	See AtSRO2[c]	ND	ND	Yes
AtSRO4	At3g47720	VvA5BFU2 PtB9I3A2 PtB9IES0	ND	ND	ND
AtSRO5	At5g62520	See AtSRO4[c]	Expressed in all organs		Yes

Table 1. SRO family members found in *Arabidopsis thaliana*. [a] Orthologs as found in (Citarelli et al. 2010). [b]Genevestigator (Zimmermann et al., 2005). Those genes with no data are not represented on ATH1 GeneChip (Affymetrix). [c]Represent paralogs in *Arabidopsis thaliana*. NA, not applicable; ND, no data; *Mt*, *Medicago truncatula*; *Os*, Oryza sativa; *Pp*, *Physcomitrella patens*; *Pt*, *Populus trichocarpa*; *Sm*, *Selaginella moellendorffii*; *Vv*, *Vitis vinifera*; *Zm*, *Zea mays*.

3. The SRO family and abiotic stress response

Although the SRO family is found in all examined land plants, almost all of the work on this family has been carried out using Arabidopsis. In this plant there are nine genes encoding members of the SRO family (Table 1; (Belles-Boix et al., 2000; Ahlfors et al., 2004)). Two paralogous genes, *RCD1* and *SIMILAR TO RCDONE1* (*SRO1*), encode members of the ubiquitous SRO subfamily, which contains the long N-terminal region containing a WWE protein-protein interaction domain (Fig. 1B). Consistent with their paralogous natures, *RCD1* and *SRO1* are partially redundant (Jaspers et al., 2009; Teotia & Lamb, 2009). The other four genes, *SRO2-5*, encode members of the eudicot-specific subfamily encoding truncated proteins.

3.1 Loss of *RCD1* and/or *SRO1* alters abiotic stress response

The SRO family was originally discovered based on the ability of one member, Arabidopsis RCD1/CEO1, to rescue oxidative stress response defects in mutant yeast (Belles-Boix et al., 2000). Mutants in this gene were discovered based on their hypersensitivity to ozone (Overmyer et al., 2000) and resistance to methyl viologen (Fujibe et al., 2004). *rcd1* mutants are also hypersensitive to other sources of apoplastic ROS, such as H_2O_2 (Overmyer et al., 2005; Teotia & Lamb, 2009) as well as salt (Katiyar-Agarwal et al., 2006; Teotia & Lamb, 2009). Conversely, *rcd* mutants are resistant to UV-B and the herbicide paraquat, which generate reactive oxygen species in the plastid (Ahlfors et al., 2004; Fujibe et al., 2004; Teotia & Lamb, 2009). In contrast, *sro1-1* plants are not resistant to the chloroplastic ROS induced by paraquat but are resistant to apoplastic ROS and high salt levels (Teotia & Lamb, 2009). Loss of either *RCD1* or *SRO1* confers resistance to osmotic stress (Teotia & Lamb, 2009). These results suggest that the relationship between *RCD1* and *SRO1* and their contribution to abiotic stress is complex and that the two genes may have some independent functions. In addition, loss of *RCD1* or *SRO1* alters responses to a number of different abiotic stresses, suggesting that these genes have broad functions. The stress responses of *rcd1; sro1* double mutant plants are technically difficult to access. Most *rcd1-3; sro1-1* plants die as embryos (Teotia & Lamb, 2009) and of those that germinate (approximately 40%), only 10-15% will produce more than 2-3 true leaves (Jaspers et al., 2009; Teotia & Lamb, 2009). However, these double mutant seedlings do display some photobleaching under normal light conditions, suggesting they are under photooxidative stress (Fig. 2A; (Teotia & Lamb, 2009)).

Consistent with the response changes upon exposure to multiple abiotic stresses, *rcd1* single mutants have been shown to accumulate ROS (Overmyer et al., 2000) and nitric oxide (Ahlfors et al., 2009) under non-stress conditions. In addition, expression of a number of stress-regulated genes is altered in this background (Ahlfors et al., 2004; Jaspers et al., 2009). For example, expression of *AOX1A*, encoding a mitochondrial alternative oxidase, is increased in *rcd1-1*. Cold and ABA regulated genes have reduced basal expression when *RCD1* is reduced. However, for the majority of genes whose expression was examined, loss of *SRO1* does not change expression levels (Jaspers et al., 2009), presumably due to the greater role *RCD1* plays in stress response (Jaspers et al., 2009; Teotia & Lamb, 2009). An exception is *tAPX*, encoding a plastid localized ascorbate peroxidase thought to be involved in defense against H_2O_2 (Kangasjarvi et al., 2008), whose expression is lower in *sro1-1* plants. *rcd1-3; sro1-1* double mutant plants exhibit increased expression of stress response genes and accumulation of SUMOylated proteins (known to accumulate during stress; (Kurepa et al., 2003)) under nonstress conditions (Teotia et al., 2010). Taken together, these data suggest that RCD1 and SRO1 may function as inhibitors of some stress responses, perhaps through regulation of ROS accumulation, consistent with their function in responses to a broad range of abiotic stresses.

Fig. 2. Loss of *RCD1* or *RCD1* and *SRO1* leads to developmental defects. (a) *rcd1-3; sro1-1* seedling. White arrow points to potential photobleaching. (b) *rcd1-3* plant grown under short day conditions (8 hours light/16 hours dark). Red arrow points to an aerial rosette. (c) Adult Arabidopsis plants. From left to right: wild type, *sro1-1*, *rcd1-3*, *rcd1-3; sro1-1*.

3.2 Other SRO family members in Arabidopsis also contribute to stress responses

In contrast to the work on *RCD1* and *SRO1*, relatively little work has been done on *SRO2-5*. No functional data exists on *SRO3* or *SRO4* and they are not represented on the Affymetrix ATH1 genechip and, therefore, not in publically available expression databases (Table 1).

However, *SRO3* expression is significantly reduced under light stress and induced by salt stress and ozone (Jaspers et al., 2010b). *SRO2* has been shown to be upregulated in response to high light in chloroplastic ascorbate peroxidase mutants (Kangasjarvi et al., 2008). *SRO5* expression is relatively low under normal conditions but its expression has been shown to be induced by salt treatment (Borsani et al., 2005) and repressed by high light (Khandelwal et al., 2008). *sro5* plants were more sensitive to H_2O_2-mediated oxidative stress and to salt stress (Borsani et al., 2005). *SRO5* has also been implicated in regulation of proline metabolism under salt stress both at the small RNA level and by couteracting ROS accumulation caused by proline accumulation (Borsani et al., 2005). Inhibiting ROS accumulation may be a core function of the SRO family.

3.3 Loss of *RCD1* and *SRO1* leads to a SIMR-like phenotype

As discussed above, chronic exposure to abiotic stress can lead to a developmental syndrome termed SIMR (Potters et al., 2007; Tognetti et al., 2011). Single *rcd1* mutants display some phenotypes that resemble those of SIMR, including reduced height (Fig. 2C; (Ahlfors et al., 2004; Teotia & Lamb, 2009)) and shorter primary roots accompanied by a greater number of lateral roots (Teotia & Lamb, 2009). In addition, loss of *RCD1* leads to accelerated flowering under long day conditions (Teotia & Lamb, 2009). This coorlelates with accumulation of ROS and NO (Ahlfors et al., 2009; Overmyer et al., 2000), as well as changes in expression of stress-induced genes (Ahlfors et al., 2004; Jaspers et al., 2009). *sro1* plants display some subtle developmental defects, consistent with it playing a minor role compared to *RCD1* (Teotia & Lamb, 2009).

The *rcd1-3; sro1-1* double mutants are severely defective. The majority of *rcd1-3; sro1-1* individuals die during embryogenesis (Teotia & Lamb, 2009). *rcd1-3; sro1-1* plants are very small and pale green as seedlings (Fig. 2A); at least some of this decrease in size is caused by a decrease in cell elongation (Teotia & Lamb, 2009). However, double mutant plants also make fewer cells (Teotia & Lamb, 2011). In the roots of *rcd1-3; sro1-1* plants, the meristems are smaller with fewer mitotic cells and cell differentiation is disrupted. The specialized cell walls of several cell types such as lateral root cap cells and the conducting cells of the xylem, are often defective (Teotia & Lamb, 2011). These phenotypes resemble extreme SIMR phenotypes and are accompanied by molecular signs of chronic stress (Teotia et al., 2010). A resonable hypthothesis based on the available data is that RCD1 and SRO1 function to inhibit stress responses, particularly accumulation of ROS, and that in their absence, there is a derepression of these pathways, leading both to altered stress responses and developmental defects (Fig. 3A).

4. Molecular functions of the SRO family

Although the SRO family is a subgroup of the PARP superfamily, it does not appear likely that they act in poly(ADP-ribosyl)ation (Jaspers et al., 2010b). Therefore, the molecular function of these proteins remains to be elucidated. RCD1 and SRO1 accumulate in the nucleus in Arabidopsis (Jaspers et al., 2009), although there is one report that RCD1 may also be found at the plasma membrane (Katiyar-Agarwal et al., 2006). SRO5 has been reported in the mitochondria (Borsani et al., 2005) but also in other subcellular locations (Jaspers et al., 2010b). RCD1, SRO1 and SRO5 have all been shown to interact with transcription factors in yeast two-hybrid assays (Belles-Boix et al., 2000; Jaspers et al., 2009, 2010b). These interactions are mediated by the RST domain characteristic of the SRO family (Fig. 3B), which is also found in the transcription initiation complex component TAF4 (Jaspers et al., 2010a). Based on localization and binding to transcription factors, members of the SRO family may act in gene expression regulation.

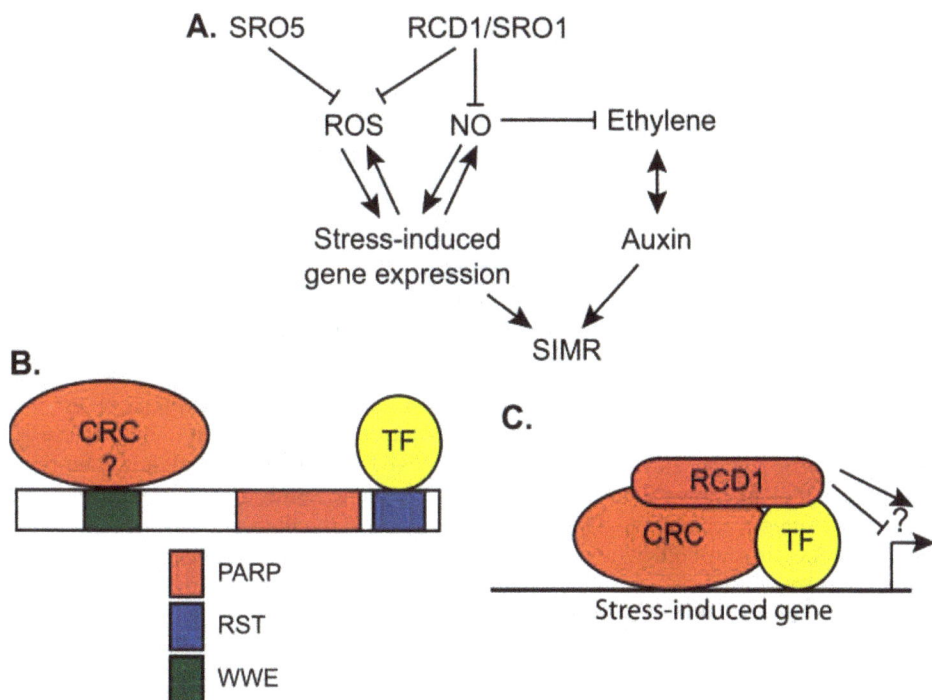

Fig. 3. Model of how SRO family members regulate abiotic stress. A. SRO family members inhibit accumulation of reactive oxygen species, which contributes both to altered abiotic stress responses and stress-induced morphogenetic response phenotypes. B. SRO family members act as scaffolds bringing together transcription factors bound to their RST domains with other proteins. Members that contain WWE domains may recruit chromatin remodeling complexes through their WWE domains. Domains shown as in Fig. 1B. C. SRO family containing complexes function to regulate gene expression.

The type of transcription factors bound by the SRO family members are diverse, including members of the bZIP, WRKY, bHLH, HSF and AP2/ERF families. A number of the identified transcription factors have been shown to be involved in abiotic stress responses. For example, SRO5 binds to a heat shock factor, HSFA1E (Jaspers et al., 2010b), which is necessary to induce expression of *HsfA2*, encoding a key regulator of the HSF network under salt and high light stress (Nishizawa-Yokoi et al., 2011). RCD1, SRO1 and SRO5 all bind to DREB2A (Jaspers et al., 2010b), an AP2/ERF transcription factor involved in cold acclimation (Sakuma et al., 2006a). Therefore, it is reasonable to hypothesize that the changes in stress-inducible gene expression seen in mutants of SRO family members arise from changes in activity of the transcription factors they bind, although this has not been demonstrated.

It is not yet clear how the binding of SRO family members to transcription factors affects the function of these proteins. Other types of PARP superfamily proteins have roles in transcriptional regulation and epigenetic control of gene expression; these roles are not always dependent on poly(ADP-ribosyl)ation activity as discussed above. HsPARP13 is not enzymatically active and has been shown to be part of multicomponent complexes in which it appears to act as a scaffold, bringing different molecules together (G. Chen et al., 2009; Gao et al., 2002; Hayakawa et al., 2011; Y. Zhu & Gao, 2008). Therefore, we hypothesize that members of the SRO family act to regulate gene expression within complexes that they anchor (Fig. 3C). Since SRO family members do not appear to have any DNA binding domains, they must be recruited to chromosomes via other proteins. These SRO-containing complexes may act directly to induce or repress transcription or act via epigenetic modification of chromatin structure to influence gene expression. The RST domain binds to transcription factors and could recruit these proteins (Fig. 3B, C). In full length SRO family members that contain WWE domains, such as RCD1, this region could be available to recruit addtional factors to the complex, such as chromatin remodeling factors (Fig. 3B, C).

Although we have been discussing the role of SRO family members in abiotic stress response, it is likely that they may also function to control gene expression in other pathways. For example, *RCD1* may have a role in control of phase change in Arabidopsis. In short days, *rcd1-3* plants cannot maintain reproductive fate; rather they bolt and then revert to vegetative fate, making aerial rosettes (Fig. 2B; (Teotia & Lamb, 2009)). The formation of the aerial rossettes is accompanied by ectopic expression of the floral repressor *FLOWERING LOCUS C (FLC)* in the bolt, where it should not be expressed. The expression of *FLC* is controlled at several levels, including epigenetic marking of histones (reviewed in (Y. He, 2009)) and by transcriptional activators (Yun et al., 2011). Therefore, the SRO family may help control gene expression beyond that involved in abiotic stress response.

5. Conclusions

The SRO family is a plant specific subfamily of PARP-like proteins that have roles in response to a number of abiotic stresses. It is interesting to note that the emergence of this family at the base of the land plants coincides with the need for protection from new stresses such as drought and increased light. Although the SRO proteins do not appear to have enzymatic activity, a possible mechanism by which they function is as part of multiprotein complexes that regulate gene expression. We hypothesize that the SRO family functions to prevent inappropriate gene expression in the absence of stress and, in their absence, ROS and other defence molecules accumulate at the expense of proper growth and development. Much work remains to test these hypotheses and clarify the contributions of individual SRO family members to stress responses as well as to move research of this important family into plants other than Arabidopsis, particularly crop plants.

6. Acknowledgments

This work was supported by funds from The Ohio State University. We apologize to colleagues whose work could not be cited due to space limitations.

7. References

Aguiar, RC., Takeyama, K., He, C., Kreinbrink, K. & Shipp, MA. (2005) B-aggressive lymphoma family proteins have unique domains that modulate transcription and exhibit poly(ADP-ribose) polymerase activity. *Journal of Biological Chemistry* Vol. 280, No. 40, pp. 33756-33765, ISSN 0021-9258.

Ahlfors, R., Lang, S., Overmyer, K., Jaspers, P., Brosche, M., Tauriainen, A., Kollist, H., Tuominen, H., Belles-Boix, E., Piippo, M., Inze, D., Palva, ET. & Kangasjarvi, J. (2004) Arabidopsis RADICAL-INDUCED CELL DEATH1 belongs to the WWE protein-protein interaction domain protein family and modulates abscisic acid, ethylene and methyl jasmonate responses. *Plant Cell* Vol. 16, No. 7, pp. 1925-1937, ISSN 1040-4651.

Ahlfors, R., Brosche, M., Kollist, H. & Kangasjarvi, J. (2009) Nitric oxide modulates ozone-induced cell death, hormone biosynthesis and gene expression in *Arabidopsis thaliana*. *The Plant Journal* Vol. 58, No. 1, pp. 1-12, doi 10.111/j.1365-313X.2008.03756.

Arabidopsis Genome Initiative. (2000) Analysis of the genome sequence of the flowering plant *Arabidopsis thaliana*. *Nature* Vol. 408, No. 6814, pp. 796-815, Accession No. 11130711.

Aravind, L. (2001) The WWE domain: a common interaction module in protein ubiquitination and ADP ribosylation. *Trends in Biochemical Science* Vol. 26, No. 5, pp. 273-275, Accession No. 11343911.

Arnholdt-Schmitt, B. (2004) Stress-induced cell reprogramming. A role for global genome regulation? *Plant Physiology* Vol. 136, No. 1, pp. 2579-2586, ISSN 0032-0889.

Belle-Boix, E., Babiychuk, E., Van Montagu, M., Inze, D., & Kushnir, S. (2000) CEO1, a new protein from *Arabidopsis thaliana*, protects yeast against oxidative damage. *FEBS Letters* Vol. 482, No. 1-2, pp. 19-24, Accession No. 11018516.

Benhamed, M., Bertrand, C., Servet, C. & Zhou, DX. (2006) Arabidopsis GCN5, HD1, and TAF1/HAF2 interact to regulate histone acetylation required for light-responsive gene expression. *The Plant Cell* Vol. 18, No. 11, pp. 2893-2903, ISSN 1040-4651.

Borsani, O., Zhu, J., Verslues, PE., Sunkar, R. & Zhu, JK. (2005) Endogenous siRNAs derived from a pair of natural cis-antisense transcripts regulate salt tolerance in Arabidopsis. *Cell* Vol. 123, No. 7, pp. 1279-1291, ISSN 0092-8674.

Bowling, SA., Clarke, JD., Liu, Y., Klessig, DF. & Dong, X. (1997) The *cpr5* mutant of Arabidopsis expresses both NPR1-dependent and NPR1-independent resistance. *The Plant Cell* Vol. 9, No. 9, pp. 1573-1584, ISSN 1040-4651.

Bustos, R., Castrillo, G., Linhares, F., Puga, MI., Rubio, V., Perez-Perez, J., Solano, R., Leyva, A. & Paz-Ares, J. (2010) A central regulatory system largely controls transcriptional activation and repression responses to phosphate starvation in Arabidopsis. *PLoS Genetics* Vol. 6, No. 9, ISSN 1553-7404.

Chambon, P., Weil, JD. & Mandel, P. (1963) Nicotinamide mononucleotide activation of new DNA-dependent polyadenylic acid synthesizing nuclear enzyme. *Biochem Biophys Res Commun* Vol. 11, pp. 39-43, Accession No. 14019961.

Chen, G., Guo, X., Lv, F., Xu, Y. & Gao, G. (2008) p72 DEAD box RNA helicase is required for optimal function of the zinc-finger antiviral protein. *Protocols of the National Academy of Sciences USA* Vol. 105, No. 11, pp. 4352-4357, ISSN 1091-6490.

Chen, LT., Luo, M., Wang, YY. & Wu, K. (2010) Involvement of Arabidopsis histone deacetylase HDA6 in ABA and salt stress response. *Journal of Experimental Botany* Vol. 61, No. 12, pp. 3345-3353, ISSN 1460-2431.

Chinnusamy, V., Gong, Z. & Zhu, JK. (2008) Abscisic acid-mediated epigenetic processes in plant development and stress responses. *Journal of Intergrative Plant Biology* Vol. 50, No. 10, pp. 1187-1195, ISSN 1744-7909.

Chodavarapu, RK., Feng, S., Bernatavichute, YV., Chen, PY., Stroud, H., Yu, Y., Hetzel, JA., Kuo, F., Kim, J., Cokus, SJ., Casero, D., Huijser, P., Clark, AT., Kramer, U., Merchant, SS., Zhang, X., Jacobsen, SE. & Pellegrini, M. (2010) Relationship between nucleosome positioning and DNA methylation. *Nature* Vol. 466, No. 7304, pp. 388-392, ISSN 1476-4687.

Citarelli, M., Teotia, S. & Lamb, RS. (2010) Evolutionary history of the poly(ADP-ribose) polymerase gene family in eukaryotes. *BMC Evolutionary Biology* Vol. 10, p. 308, ISSN 1471-2148.

Conrath, U. (2011) Molecular aspects of defence priming. *Trends in Plant Science*, Accession number 21782492.

Dietrich, K., Weltmeier, F., Ehler, A., Weiste, C., Stahl, M., Harter, K. & Droge-Laser, W. (2011) Heterodimers of the Arabidopsis transcription factors bZIP1 and bZIP53 reprogram amino acid metabolism during low energy stress. *The Plant Cell* Vol. 23, No. 1, pp. 381-395, ISSN 1532-298X.

Dietz, KJ., Vogel, MO. & Viehhauser, A. (2010) AP2/EREBP transcription factors are part of gene regulatory networks and integrate metabolic, hormonal and environmental signals in stress acclimation and retrograde signalling. *Protoplasma* Vol. 245, No. 1-4, pp. 3-14, ISSN 1615-6102.

Elfving, N., Davoine, C., Benlloch, R., Blomberg, J., Brannstrom, K., Muller, D., Nilsson, A., Ulfstedt, M., Ronne, H, Wingsle, G., Nilsson, O. & Bjorklund, S. (2011) The *Arabidopsis thaliana* Med25 mediator subunit integrates environmental cues to control plant development. *Protocols of the National Academy of Sciences USA* Vol. 108, No. 20, ISSN 1091-6490.

Elhiti, M. & Stasolla, C. (2009) Structure and function of homodomain-leucine zipper (HD-Zip) proteins. *Plant Signalling & Behavior* Vol. 4, No. 2, pp. 86-88, ISSN 1559-2324.

Fernandez-Pozo, N., Canales, J., Guerrero-Fernandez, D., Villalobos, DP., Diaz-Moreno, SM., Bautista, R., Flores-Monterroso, A., Guevara, MA., Perdiguero, P., Collada, C., Cervera, MT., Soto, A., Ordas, R., Canton, FR., Avila, C., Canovas, FM. & Claros, MG. (2011) EuroPineDB: a high-coverage Web database for maritime pine transcriptome. *BMC Genomics* Vol. 12, No. 1, p. 366, ISSN 1471-2164.

Finn, RD., Mistry, J., Tate, J., Coggill, P., Heger, A., Pollington, JE., Gavin, OL., Gunasekaran, P., Ceric, G., Forslund, K., Holm, L., Sonnhammer, EL., Eddy, SR. & Bateman, A. (2010) The Pfam protein families database. *Nucleic Acids Research* Vol. 38, pp. D211-222, ISSN 1362-4962.

Foyer, CH. & Noctor, G. (2009) Redox regulation in photosynthetic organisms: signaling, acclimation, and practical implications. *Antioxidants & Redox Signaling* Vol. 11, No. 4, pp. 861-905, ISSN 1557-7716.

Fujibe, T., Saji, H., Arakawa, K., Yabe, N., Takeuchi, Y. & Yamamoto, KT. (2004) A methyl viologen-resistant mutant of Arabidopsis, which is allelic to ozone-sensitive *rcd1*, is tolerant to supplemental ultraviolet-B irradiation. *Plant Physiology* Vol. 134, No. 1, pp. 275-285, Accession No. 14657410.

Gadjev, I., Vanderauwera, S., Gechev, TS., Laloi, C., Minkov, IN., Shulaev, V., Apel, K., Inze, D., Mittler, R. & Van Breusegem, F. (2006) Transcriptomic footprints disclose specificity of reactive oxygen species signaling in Arabidopsis. *Plant Physiology* Vol. 141, No. 2, pp. 436-445, ISSN 0032-0889.

Gao, G., Guo, X. & Goff, SP. (2002) Inhibition of retroviral RNA production by ZAP, a CCCH-type zinc finger protein. *Science* Vol. 297, No. 5587, pp. 1703-1706, ISSN 1095-9203.

Hassa, PO., Buerki, C., Lombardi, C., Imhof, R. & Hottiger, MO. (2003) Transcriptional coactivation of nuclear factor-kappaB-dependent gene expression by p300 is regulated by poly(ADP)-ribose polymerase-1. *Journal of Biological Chemistry* Vol. 278, No. 46, pp. 45145-45153, Accession No. 12960163.

Hassa, PO. & Hottiger, MO. (2008) The diverse biological roles of mammalian PARPs, a small but powerful family of poly-ADP-ribose polymerases. *Frontiers in Bioscience* Vol. 13, pp. 3046-3082, ISSN 1093-4715.

Hayakawa, S., Shiratori, S., Yamato, H., Kameyama, T., Kitatsuji, C., Kashigi, F., Goto, S., Kameoka, S., Fujikura, D., Yamada, T., Mizutani, T., Kazumata, M., Sato, M., Tanaka, J., Asaka, M., Ohba, Y., Miyazaki, T., Imamura, M. & Takaoka, A. (2011) ZAPS is a potent stimulator of signaling mediated by the RNA helicase RIG-I during antiviral responses. *Nature Immunology* Vol. 12, No. 1, pp. 37-44, ISSN 1529-2916.

He, XJ., Chen, T. & Zhu, JK. (2011) Regulation and function of DNA methylation in plants and animals. *Cell Research* Vol. 21, No. 3 (March 2011), pp. 442-465, ISSN 1748-7838.

Hirayama, T. & Shinozaki, K. (2010) Research on plant abiotic stress responses in the post-genome era: past, present and future. *The Plant Journal* Vol. 61, No. 6, pp. 1041-1052, ISSN 1365-313X.

Hussain, SS., Kayani, MA. & Amjad, M. (2011) Transcription factors as tools to engineer enhanced drought stress tolerance in plants. *Biotechnology Progress* Vol. 27, No. 2, pp. 297-306.

Jackson, MW., Stinchcombe, JR., Korves, TM. & Schmitt, J. (2004) Costs and benefits of cold tolerance in transgenic *Arabidopsis thaliana*. *Molecular Ecology* Vol. 13, No. 11, pp. 3609-3615, ISSN 0962-1083.

Jaspers, P., Blomster, T., Brosche, M., Salojarvi, J., Ahlfors, R., Vainonen, JP., Reddy, RA., Immink, R., Angenent, G., Turck, F., Overmyer, K. & Kangasjarvi, J. (2009) Unequally redundant *RCD1* and *SRO1* mediate stress and developmental responses and interact with transcription factors. *The Plant Journal* Vol. 60, No. 2, pp. 268-279, ISSN 1365-313X.

Jaspers, P., Brosche, M., Overmyer, K. & Kangasjarvi, J. (2010a) The transcription factor interacting proteins RCD1 contains a novel conserved domain. *Plant Signaling & Behavior* Vol. 5, No. 1, pp. 78-80, ISSN 1559-2324.

Jaspers, P., Overmyer, K., Wrzaczek, M., Vainonen, JP. Blomster, T., Salojarvi, J., Reddy, RA. & Kangasjarvi, J. (2010b) The RST and PARP-like domain containing SRO protein family: analysis of protein structure, function and conservation in land plants. *BMC Genomics* Vol. 11, p. 170, ISSN 1471-2164.

Jaspers, P. & Kangasjarvi, J. (2010) Reactive oxygen species in abiotic stress signaling. *Physiologia Plantarum* Vol. 138, No. 4, pp. 405-413, ISSN 1399-3054.

Jing, HC. & Dijkwel, PP. (2008) CPR5: A Jack of all trades in plants. *Plant Signaling & Behavior* Vol. 3, No. 8, pp. 562-563, ISSN 1559-2324.

Jing, HC., Hebeler, R., Oeljeklaus, S., Sitek, B., Stuhler, K., Meyer, HE., Sturre, MJ., Hille, J., Warscheid, B. & Dijkwel, PP. (2008) Early leaf senescence is associated with an altered cellular redox balance in Arabidopsis *cpr5/old1* mutants. *Plant Biology (Stuttgart)*, Vol. 10, Suppl. 1, pp. 85-98, ISSN 1435-8603.

Juszczynski, P., Kutok, JL., Li, C., Mitra, J., Aguiar, RC. & Shipp, MA. (2006) BAL1 and BBAP are regulated by a gamma interferon-responsive bidirectional promoter and are overexpressed in diffuse large B-cell lymphomas with a prominent inflammatory infiltrate. *Molecular Cell Biology* Vol. 26, No. 14, pp. 5348-5359, ISSN 0270-7306.

Kangasjarvi, J., Jaspers, P. & Kollist, H. (2005) Signalling and cell death in ozone-exposed plants. *Plant, Cell and Environment* Vol. 28, pp. 1021-1036.

Kangasjarvi, S., Lepisto, A., Hannikainen, K., Piippo, M., Luomala, EM., Aro, EM. & Rintamaki, E. (2008) Diverse roles for chloroplast stromal and thylakoid-bound ascorbate peroxidases in plant stress responses. *Biochemical Journal* Vol. 412, No. 2, pp. 275-285, ISSN 1470-8728.

Katiyar-Agarwal, S., Zhu, J., Kim, K., Agarwal, M., Fu, X., Huang, A. & Zhu, JK. (2006) The plasma membrane Na^+/H^+ antiporter SOS1 interacts with RCD1 and functions in oxidative stress tolerance in Arabidopsis. *Protocols of the National Academy of Sciences USA* Vol. 103, No. 49, pp. 18816-18821, Accession No. 17023541.

Kaye, Y., Golani, Y., Singer, Y., Leshem, Y., Cohen, G., Ercetin, M., Gillaspy, G. & Levine, A. (2011) Inositol polyphosphate 5-phosphatase7 regulates production of reactive oxygen species and salt tolerance in Arabidopsis. *Plant Physiology* Preview, DOI: 10.1104/pp.111.176883.

Khandelwal, A., Elvitigala, T., Ghosh, B. & Quatrano, RS. (2008) Arabidopsis transcriptome reveals control circuits regulating redox homeostasis and the role of an AP2 transcription factor. *Plant Physiology* Vol. 148, No. 4, pp. 2050-2058, ISSN 0032-0889.

Kilian, J., Whitehead, D., Horak, J., Wanke, D., Weinl, S., Batistic, O., D'Angelo, C., Bornberg-Bauer, E., Kudla, J. & Harter, K. (2007) The AtGenExpress global stress expression data set: protocols, evaluation and model data analysis of UV-B light, drought and cold stress responses. *The Plant Journal* Vol. 50, No. 2, pp. 347-363, ISSN 0960-7412.

Kim, DH., Doyle, MR., Sung, S. & Amasino, RM. (2009) Vernalization: winter and timing of flowering in plants. *Annual Review of Cell & Developmental Biology* Vol. 25, pp. 277-299, ISSN 1530-8995.

Kim, JM., To, TK., Ishida, J., Morowawa, T., Kawashima, M., Matsui, A., Toyoda, T., Kimura, H., Shinozaki, K. & Seki, M. (2008) Alterations of lysine modifications on the histone H3 N-tail under drought stress conditions in *Arabidopsis thaliana*. *Plant and Cell Physiology* Vol. 49, No. 10, pp. 1580-1588, ISSN 1471-9053.

Kim, JM., To, TK., Nishioka, T. & Seki, M. (2010) Chromatin regulation functions in plant abiotic stress responses. *Plant and Cell Physiology* Vol. 33, No. 4, pp. 604-611, ISSN 1365-3040.

Kim, MY., Mauro, S., Gevry, N., Lis, JT. & Kraus, WL. (2004) NAD^+-dependent modulation of chromatin structure and transcription by nucleosome binding properties of PARP-1. *Cell* Vol. 119, No. 6, pp. 803-814, ISSN 0092-8674.

Kirik, V., Bouyer, D., Schobinger, U., Bechtold, N., Herzog, M., Bonneville, JM. & Hulskamp, M. (2001) *CPR5* is involved in cell proliferation and cell death control and encodes a novel transmembrane protein. *Current Biology* Vol. 11, No. 23, pp. 1891-1895.

Kleine, H., Poreba, E., Lesniewicz, K., Hassa, PO., Hottiger, MO., Litchfield, DW., Shilton, BH. & Luscher, B. (2008) Substrate-assisted catalysis by PARP10 limits its activity to mono-ADP-ribosylation. *Molecular Cell* Vol. 32, No. 1, pp. 57-69, ISSN 1097-4164.

Kurepa, J., Walker, JM., Smalle, J., Gosink, MM., Davis, SJ., Durham, TL., Sung, DY. & Vierstra, RD. (2003) The small ubiquitin-like modifier (SUMO) protein modification system in Arabidopsis. Accumulation of SUMO1 and -2 conjugates is increased by stress. *Journal of Biological Chemistry* Vol. 279, No. 9, pp. 6862-6872, Accession No. 12482876.

Kwon, CS. & Wagner, D. (2007) Unwinding chromatin for development and growth : a few genes at a time. *Trends in Genetics* Vol. 23, No. 8, pp. 403-412, ISSN 0168-9525.

Kwon, CS., Lee, D., Choi, G. & Chung, WI. (2009) Histone occupancy-dependent and – independent removal of H3K27 trimethylation at cold-responsive genes in Arabidopsis. *The Plant Journal* Vol. 60, No. 1, pp. 112-121, ISSN 1365-313X.

Lata, C., Bhutty, S., Bahadur, RP., Majee, M. & Prasad, M. (2011) Association of an SNP in a novel DREB2-like gene *SiDREB2* with stress tolerance in foxtail millet [*Sertaria italica* (L.)]. *Journal of Experimental Botany* Vol. 62, No. 10, pp. 3387-3401, ISSN 1460-2431.

Lata, C. & Prasad, M. (2011) Role of DREBs in regulation of abiotic stress responses in plants. *Journal of Experimental Botany* Preview, ISSN 1460-2431.

Lieb, JD. & Clarke, ND. (2005) Control of transcription through intragenic patterns of nucleosome composition. *Cell* Vol. 123, No. 7, pp. 1187-1190, ISSN 0092-8674.

Lin, Y., Tang, X., Zhu, Y., Shu, T. & Han, X. (2011) Identification of PARP-1 as one of the transcription factors binding to the repressor element in the promoter region of COX-2. *Archives of Biochemistry and Biophysics* Vol. 505, No. 1, pp. 123-129.

Miao, Y., Laun, TM., Smkowski, A. & Zentgraf, U. (2007) Arabidopsis MEKK1 can take a short cut : it can directly interact with senescence-related WRKY53 transcription factor on the protein level and can bind to its promoter. *Plant Molecular Biology* Vol. 65, No. 1-2, pp. 63-76, ISSN 0167-4412.

Mlynarova, L., Nap, JP. & Bisseling, T. (2007) The SWI/SNF chromatin-remodeling gene *AtCHR12* mediates temporary growth arrest in *Arabidopsis thaliana* upon perceiving environmental stress. *The Plant Journal* Vol. 51, No. 5, pp. 874-885, Accession No. 17605754.

Moller, IM., Jensen, PE. & Hansson, A. (2007) Oxidative modifications to cellular components in plants. *Annual Review of Plant Biology* Vol. 58, pp. 459-481, ISSN 1543-5008.

Nirodi, C., NagDas, S., Gygi, SP., Olson, G., Aebersold, R. & Richmond, A. (2001) A role for poly(ADP-ribose) polymerase in the transcriptional regulation of the melanoma growth stimulatory activity (CXCL1) gene expression. *Journal of Biological Chemistry* Vol. 275, No. 12, pp. 9366-9374, ISSN 0021-9258.

Nishizawa-Yokoi, A., Nosaka, R., Hayashi, H., Tainaka, H., Maruta, M., Tamoi, M., Ikeda, M., Ohme-Takagi, M., Yoshimura, K., Yabuta, Y. & Shigeoka, S. (2011) HsfA1d and HsfA1e are involved in the transcriptional regulation of *HsfA2* function as key regulators for the Hsf signaling network in response to environmental stress. *Plant and Cell Physiology* Vol. 52, No. 5, pp. 933-945, ISSN 1471-9053.

Oliver, FJ., Menissier-de Murcia, J., Nacci, C., Decker, P., Andriantsitohaina, R., Muller, S., de la Rubia, G., Stoclet & de Murcia, G. (1999) Resistance to endotoxic shock as a consequence of defective NF-kappaB activation in poly (ADP-ribose) polymerase-1 deficient mice. *EMBO Journal* Vol. 18, No. 16, pp. 4446-4454; ISSN 021-4189.

Overmyer, K., Tuominen, H., Kettunen, R., Betz, C., Langebartels, C., Sandermann, H. Jr. & Kangasjarvi, J. (2000) Ozone-sensitive Arabidopsis *rcd1* mutant reveals opposite roles for ethylene and jasmonate signaling pathways in regulating superoxide-dependent cell death. *The Plant Cell* Vol. 12, No. 10, pp. 1849-1862, ISSN 1040-4651.

Overmyer, K., Brosche, M., Pellinen, R., Kuittinen, T., Tuominen, H., Ahlfors, R., Keinanen, M., Scheel, D. & Kangasjarvi, J. (2005) Ozone-induced programmed cell death in the Arabidopsis *radical-induced cell death1* mutant. *Plant Physiology* Vol. 137, No. 3, pp. 1029-1104, Accession No. 15728341.

Pfluger, J. & Wagner, D. (2007) Histone modifications and dynamic regulation of genome accessibility in plants. *Current Opinion in Plant Biology* Vol. 10, No. 6, pp. 645-652.

Potato Genome Sequencing Consortium. (2011) Genome sequence and analysis of the tuber crop potato. *Nature* Vol. 475, No. 7355, pp. 189-195, ISSN 1476-4687.

Potters, G., Pasternak, TP., Guisez, Y., Palme, KJ. & Jansen, MA. (2007) Stress-induced morphogenic responses: growing out of trouble? *Trends in Plant Science* Vol. 12, No. 3, pp. 98-105, ISSN 1360-1385.

Potters, G., Pasternak, TP., Guisez, Y. & Jansen, MA. (2009) Different stresses, similar morphogenic responses: integrating a plethora of pathways. *Plant, Cell and Environment* Vol. 32, No. 2, pp. 158-169, ISSN 1365-3040.

Raja, P., Sanville, BC., Buchmann, RC. & Bisaro, DM. (2008) Viral genome methylation as an epigenetic defense against geminiviruses. *Journal of Virology* Vol. 82, No. 18, pp. 8997-9007, ISSN 1098-5514.

Raja, P., Wolf, JN. & Bisaro, DM. (2010) RNA silencing directed against geminiviruses : post-transcriptional and epigenetic components. *Biochimica et Biophysica Acta* Vol. 1799, No. 3-4, pp.337-351, ISSN 0006-3002.

Riechmann, JL., Heard, J., Martin, G., Reuber, L., Jiang, C., Keddie, J., Adam, L., Pineda, O., Ratcliffe, OJ., Samaha, RR., Creelman, R., Pilgrim, M., Broun, P., Zhang, JZ., Ghandehari, D., Sherman, BK. & Yu, G. (2000) Arabidopsis transcription factors : genome-wide comparative analysis among eukaryotes. *Science* Vol. 290, No. 5499, pp. 2105-2110, Accession No. 11118137.

Rushton, DL., Tripathi, P., Rabara, RC., Lin, J., Ringler, P., Boken, AK., Langum, TJ., Smidt, L., Boomsma, DD., Emme, NJ., Chen, X., Finer, JJ., Shen, QJ. & Rushton, PJ. (2011) WRKY transcription factors: key components in abscisic acid signalling. *Plant Biotechnology Journal*, ISSN 1467-7652.

Ryu, JY., Park, CM. & Seo, PJ. (2011) The floral repressor *BROTHER OF FT AND TFL1 (BFT)* modulates flowering initiation under high salinity in Arabidopsis. *Molecules and Cells*, ISSN 0219-1032.

Sakuma, Y., Maruyama, K., Osakabe, Y., Qin, F., Seki, M., Shinozaki, K. & Yamaguchi-Shinozaki, K. (2006a) Functional analysis of an Arabidopsis transcription factor, DREB2A, involved in drought-responsive gene expression. *The Plant Cell* Vol. 18, No. 5, pp. 1292-1309, ISSN 1040-4651.

Sakuma, Y., Maruyama, K., Qin, F., Osakabe, Y., Shinozaki, K. & Yamaguchi-Shinozaki, K. (2006b) Dual function of an Arabidopsis transcription factor DREB2A in water-stress-responsive and heat-stress-responsive gene expression. *Protocols of the National Academy of Sciences USA* Vol. 103, No. 49, pp. 18822-18827, ISSN 0027-8424.

Stockinger, EJ., Mao, Y., Regier, MK., Triezenberg, SJ. & Thomashow, MF. (2001) Transcriptional adaptor and histone acetyltransferase proteins in Arabidopsis and their interactions with CBF1, a transcriptional activator involved in cold-regulated gene expression. *Nucleic Acids Research* Vol. 29, No. 7, pp. 1524-1533, ISSN 1362-4962.

Takuno, S. & Gaut, BS. (2011) Body-methylated genes in *Arabidopsis thaliana* are functionally important and evolve slowly. *Molecular Biology and Evolution* Preview, ISSN 1537-1719.

Teotia, S. & Lamb, RS. (2009) The paralogous genes *RADICAL-INDUCED CELL DEATH1* and *SIMILAR TO RCD ONE1* have partially redundant functions during Arabidopsis development. *Plant Physiology* Vol. 151, No. 1, pp. 180-198, ISSN 0032-0889.

Teotia, S., Muthuswamy, S. & Lamb, RS. (2010) *RADICAL-INDUCED CELL DEATH1* and *SIMILAR TO RCD ONE1* and the stress-induced morphogenetic response. *Plant Signaling & Behavior* Vol. 5, No. 2, pp. 143-145, ISSN 1559-2324.

Teotia, S. & Lamb, RS. (2011) RCD1 and SRO1 are necessary to maintain meristematic fate in *Arabidopsis thaliana*. *Journal of Experimental Botany* Vol. 62, No. 3, pp. 1271-1284, ISSN 1460-2431.

Till, S., Diamantara, K. & Ladurner, AG. (2008) PARP : a transferase by any other name. *Nature Structural & Molecular Biology* Vol. 15, No. 12, pp. 1243-1244, ISSN 1545-9985.

To, TK., Nakaminami, K., Kim, JM., Morosawa, T., Ishida, J., Tanaka, M., Yokoyama, S., Shinozaki, K. & Seki, M. (2011a) Arabidopsis HDA6 is required for freezing tolerance. *Biochem Biophys Res Commun* Vol. 406, No. 3, pp. 414-419, ISSN 1090-2104.

To, TK., Kim, JM., Matsui, A., Kurihara, Y., Morosawa, T., Ishida, J., Tanaka, M., Endo, T., Kakutani, T., Toyoda, T., Kimura, H., Yokoyama, S., Shinozaki, K. & Seki, M. (2011b) Arabidopsis HDA6 regulates locus-directed heterochromatin silencing in cooperation with MET1. *PLoS Genetics* Vol. 7, No. 4, p. e1002055, ISSN 1553-7404.

Tognetti, VB., Muhlenbock, P. & V. A. N. Breusegem, F. (2011) Stress homeostasis - the redox and auxin perspective. *Plant, Cell & Environment*, ISSN 1365-3040.

Vashisht, AA. & Tuteja, N. (2006) Stress responsive DEAD-box helicases: a new pathway to engineer plant stress tolerance. *Journal of Photochemistry and Photobiology B : Biology* Vol. 84, No. 2, pp. 150-160, ISSN 1011-1344.

Wacker, DA., Ruhl, DD., Balagamwala, EH., Hope, KM., Zhang, T. & Kraus, WL. (2007) The DNA binding and catalytic domains of poly(ADP-ribose) polymerase 1 cooperate in the regulation of chromatin structure and transcription. *Molecular Cell Biology* Vol. 27, No. 21, pp. 7475-7485, ISSN 0270-7306.

Wada, KC., Yamada, M., Shiraya, T. & Takeno, K. (2010) Salicylic acid and the flowering gene *FLOWERING LOCUS T* homolog are involved in poor-nutrition stress-induced flowering of *Pharbitis nil*. *Journal of Plant Physiology* Vol. 167, No. 6, pp. 447-452, ISSN 1618-1328.

Wada, KC. & Takeno, K. (2010) Stress-induced flowering. *Plant Signaling & Behavior* Vol. 5, No. 8, pp. 944-947, ISSN 1559-2324.

Walley, JW., Rowe, HC., Xiao, Y., Chehav, EW., Kliebenstein, DJ., Wagner, D. & Dehesh, K. (2008) The chromatin remodeler SPLAYED regulates specific stress signaling pathways. *PLoS Pathology* Vol. 4, No. 12, p. e1000237, ISSN 1553-7374.

Walters, D. & Heil, M. (2007) Costs and trade-offs associated with induced resistance. *Physiological and Molecular Plant Pathology* Vol. 71, pp. 3-17.

Wu, CC., Singh, P., Chen, MC. & Zimmerli, L. (2010) L-Glutamine inhibits beta-aminobutyric acid-induced stress resistance and priming in Arabidosis. *Journal of Experimental Botany* Vol. 61, No. 4, pp. 995-1002, ISSN 1460-2431.

Wu, K., Zhang, L, Zhou, C., Yu, CS. & Chaikam, V. (2008) HDA6 is required for jasmonate response, senescence and flowering in Arabidopsis. *Journal of Experimental Botany* Vol. 59, pp. 226-234.

Yamaguchi-Shinozaki, K. & Shinozaki, K. (2006) Transcriptional regulatory networks in cellular responses and tolerance to dehydration and cold stresses. *Annual Review of Plant Biology* Vol. 57, pp. 781-803, ISSN 1543-5008.

Yoshida, S., Ito, M., Nishida, I. & Watanabe, A. (2002) Identification of a novel gene HYS1/CPR5 that has a repressive role in the induction of leaf senescence and pathogen-defence responses in *Arabidopsis thaliana*. *The Plant Journal* Vol. 29, No. 4, pp. 427-437, ISSN 0960-7412.

Yu, CW., Liu, X., Luo, M., Chen, C., Lin, X., Tian, G., Lu, Q., Cui, Y. & Wu, K. (2011) HISTONE DEACETYLASE6 interacts with FLOWERING LOCUS D and regulates flowering in Arabidopsis. *Plant Physiology* Vol. 156, No. 1, pp. 173-184, ISSN 1532-2548.

Yun, H., Hyun, Y., Kang, MJ., Noh, YS., Noh, B. & Choi, Y. (2011) Identification of regulators required for the reactivation of *FLOWERING LOCUS C* during Arabidopsis reproduction. *Planta* Preview, DOI : 10.1007/s00425-011-1484-y.

Zhang, K., Sridhar, VV., Zhu, J., Kapoor, A. & Zhu, JK. (2007) Distinctive core histone post-translational modification patterns in *Arabidopsis thaliana*. *PLoS One* Vol. 2, No. 11, p. e1210, ISSN 1932-6203.

Zhen, Y., Dhakal, P. & Ungerer, MC. (2011) Fitness benefits and costs of cold acclimation in *Arabidopsis thaliana*. *American Naturalist* Vol. 178, No. 1, pp. 44-52, ISSN 1537-5323.

Zhu, J., Jeong, JC., Zhu, Y., Sokolchik, I., Miyazaki, S., Zhu, JK., Hasegawa, PM., Bohnert, HJ., Shi, H., Yun, DJ. & Bressan, RA. (2008) Involvement of Arabidopsis HOS15 in histone deacetylation and cold tolerance. *Protocols of the National Academy of Sciences USA* Vol. 105, No. 12, pp. 4945-4950, ISSN 1091-6490.

Zhu, Y. & Gao, G. (2008) ZAP-mediated mRNA degradation. *RNA Biology* Vol. 5, No. 2, pp. 65-67, ISSN 1555-8584.

Zimmermann, P., Hennig, L. & Gruissem, W. (2005) Gene-expression analysis and network discovery using Genevestigator. *Trends in Plant Science* Vol. 10, No. 9, pp. 407-409, Accession No. 16081312.

Zolla, G., Heimer, YM. & Barak, S. (2009) Mild salinity stimulates a stress-induced morphogenic response in *Arabidopsis thaliana* roots. *Journal of Experimental Botany* Vol. 61, No. 1, pp. 211-24, Accession No. 19783843.

2

Characterization of Plant Antioxidative System in Response to Abiotic Stresses: A Focus on Heavy Metal Toxicity

Miguel Mourato, Rafaela Reis and Luisa Louro Martins
UIQA, Instituto Superior de Agronomia, Technical University of Lisbon, Lisbon
Portugal

1. Introduction

During their life span, plants can be subjected to a number of abiotic stresses, like drought, temperature (both high and low), radiation, salinity, soil pH, heavy metals, lack of essential nutrients, air pollutants, etc. When affected by one, or a combination of abiotic stresses, a response is induced by changes in the plant metabolism, growth and general development.

Reactive Oxygen Species (ROS) are a natural consequence of the aerobic metabolism, and plants have mechanisms to deal with them in normal conditions, controlling the formation and removal rates. Under stress conditions, cell homeostasis is disrupted and ROS production can increase a lot putting a heavy burden on the those antioxidative mechanisms, some of which are activated in order to eliminate the excess ROS (Mittler et al., 2004).

Trace element contamination cause abiotic stress in plants and it can affect crop production and quality. Certain metals, like copper, are essential for plants, but at high concentrations (depending on plant species) can be considered toxic. Other elements like cadmium and arsenic (a metalloid), while not essential elements for plants, are widespread pollutants that are present in nature due to both natural and manmade activities.

Plants have developed different strategies to cope with these stresses. Some use an avoidance strategy to reduce trace element assimilation while others use internal defence mechanisms to cope with the increasing levels of the toxic species. Phytotoxic amounts of trace elements are known to affect several physiological processes and can cause oxidative stress. Plants have developed several trace element defence mechanisms, that allow them to grow despite the presence of variable concentrations of trace elements, but the threshold concentrations as well as the different response mechanisms strongly depend on plant species and on the type of metal. Metal toxicity can cause a redox imbalance and induce the increase of ROS concentration, activating the antioxidant defence mechanisms of plants (Sharma & Dietz, 2009). These mechanisms are very dependent on the metal and the plant but usually include the involvement of the ascorbate-glutathione cycle enzymes which is a major antioxidative defence mechanism, and of other antioxidant enzymes like catalase, peroxidases, and superoxide dismutase. Other non-enzymatic substances with reported antioxidant properties can also be involved in plant defence mechanisms, like ascorbate, glutathione, alkaloids, phenolic compounds, non-proteic amino-acids and carotenoids.

2. Oxidative stress and ROS production

ROS are produced by all aerobic organisms and are usually kept in balance by the antioxidative mechanisms that exist in all living beings. Because ROS have an important signalling role in plants (Foyer & Noctor, 2003; Vranova et al., 2002), their concentration must be carefully controlled through adequate pathways (Mittler, 2002). ROS can be formed during normal aerobic metabolic processes like photosynthesis and respiration and thus, the majority of ROS are produced in the mitochondria, chloroplast, peroxisomes, plasma membrane and apoplast (Ahmad et al., 2008; Moller, 2001). Other sources of ROS production are NADPH oxidases, amine oxidases and cell-wall peroxidases (Mittler, 2002).

Under certain stress conditions (like excess light, cold, heat, drought, heavy metals etc.) the production of ROS can exceed the capacity of the plant's defence mechanisms, an imbalance in intracellular ROS content is established and this results in oxidative stress (Gill & Tuteja, 2010). Thus, oxidative stress can be defined as the physiological changes resulting from the formation of excess quantities of reactive oxygen species (ROS) (Vangronsveld & Clijsters, 1994). This increase in ROS levels induces a metabolic response in the plant in order to eliminate them. This metabolic response is highly dependent on the plant species, plant growing state and the type and duration of the stress.

Heavy metals[1] are natural elements that are present at different concentrations throughout nature, but whose levels can increase and overtake the toxicity threshold of living beings due to both natural and anthropogenic causes (Sánchez, 2008). As plants must adapt (or die) to the conditions where they grow, the presence of heavy metals can induce oxidative stress and the activation of several defence factors in the plants (Prasad, 2004). It is important to understand how some plants can cope with high concentration of metals in order to produce crops able to grow on contaminated soils (Schröder et al., 2008), to help in environmental cleanup via phytoremediation (Adriano et al., 2004) and to breed plants with higher contents of essential nutrients (Zhao & McGrath, 2009).

2.1 Types of ROS

Molecular oxygen (O_2) is in itself a bi-radical[2], as it has two unpaired electrons that have parallel spins (Halliwell, 2006). The ground state of the oxygen molecule is the triplet oxygen ($^3O_2$ or $\cdot O\text{-}O\cdot$), because this is an energetically more favourable state. Due to this electron configuration, it doesn't react easily with organic molecules that have paired electrons with opposite spins. Thus, in order for oxygen to react, it must be activated (Garg & Manchanda, 2009). If the oxygen molecule in its ground state absorbs sufficient energy, the spin of one of the unpaired electrons can be reversed forming singlet oxygen, that can readily react with organic molecules. This can happen during photosynthesis when an excess of light energy cannot be readily dissipated by the photosynthetic machinery (Foyer et al., 1994).

---

[1] Although the term "heavy metal" can be misleading because it includes a whole range of substances some of which do not conform to the more usual chemical definitions of the term, we use it in this text in its environmental context, where it includes metals that are a cause of pollution concern. Other authors sometimes use the expression "trace elements" in this context, although this definition also has problems on its own.
[2] Radicals are usually highly unstable, and thus very reactive, species that have one or more unpaired electrons.

Another form of activation is by partial reduction adding one, two or three electrons giving rise to the superoxide radical, hydrogen peroxide and hydroxyl radical, respectively (Mittler, 2002). The complete reduction of oxygen (adding four electrons) results in water, which is the normal reduction of oxygen that occurs in the mitochondrial electron transport chain, catalyzed by cytochrome oxidase. As such, this type of activation can occur in metabolic pathways that involve an electron transport chain and can thus occur in several cell locations (Alscher et al., 2002).

In Table 1 we present the most important types of ROS. They can be free radicals or non-radicals.

Name	Structure	Type	Relative Reactivity
Singlet oxygen	$^1O_2$ (O-O:)	Radical	High
Superoxide	$O_2^{\bullet-}$ ($\cdot$O-O:)	Radical	Medium
Hydrogen peroxide	H_2O_2 (H:O-O:H)	Non-radical	Low
Hydroxyl radical	$HO^{\bullet}$ (H:O$\cdot$)	Radical	Very high

Table 1. Most important types of ROS

Singlet oxygen is mainly produced in the chloroplasts at photosystem II (Asada, 2006) but may also result from lipoxygenase activity and is a highly reactive species that can last for nearly 4 µs in water (Foyer et al., 1994). $^1O_2$ reactivity has as preferred target the conjugated double bonds present on polyunsaturated fatty acids (PUFAs) leaving a specific footprint in the cell (Moller et al. 2007) that can be followed by the detection of several aldehydes like malondialdehyde (MDA) formed by PUFA peroxidation.

The superoxide radical is mainly produced both in the chloroplasts (photosystems I and II) and mitochondria as sub products and in peroxisomes (del Rio et al., 2006; Moller et al., 2007; Rhoads et al., 2006), has a half-life of 2-4 µs and cannot cross phospholipid membranes (Garg & Manchanda, 2009) and so it is important that the cell has adequate *in situ* mechanism to scavenge this ROS. Superoxide dismutase can catalyse the conversion of this species into hydrogen peroxide. Superoxide radical can also be produced by NADPH oxidase in the plasma membrane (Moller et al. 2007).

Hydrogen peroxide is mainly produced in peroxisomes (del Rio et al., 2006) and also in mitochondria (Rhoads et al., 2006), and also results from the dismutation of superoxide. It is not a radical and can easily cross membranes diffusing across the cell and has a half-life of around 1 ms (Garg & Manchanda, 2009).

The hydroxyl radical, the most reactive of the species listed in Table 1, can be formed from hydrogen peroxide via Fenton and Fenton-like reactions (catalyzed by iron or other transition metals) and, unlike the previous two ROS mentioned, there are no known enzymatic systems able to degrade it (Freinbichler et al., 2011).

Although the superoxide radical and hydrogen peroxide are not as reactive as other species they are produced in large amounts in the cell and can initiate other reactions that lead to more dangerous species (Noctor & Foyer, 1998). In fact, superoxide radical can be converted by specific enzymes into hydrogen peroxide, and this can also be a problem as it cause the occurrence of Fenton reactions (Moller et al., 2007).

Heavy metals are known to induce oxidative stress increasing the ROS concentration. As an example, in figure 1 we present experimental results of work performed by the authors, of the effect of Cd and Cu in hydrogen peroxide content in roots of tobacco plants. As can be seen, there is a good correlation between Cu levels and H_2O_2 content, but the effect of Cd showed only a small non-significant increase. These metals, an essential and a non-essential, do seem to provoke different responses in the plant. The increase in hydrogen peroxide levels with metals is a frequently reported stress indicator (Khatun et al., 2008; Mobin & Khan, 2007).

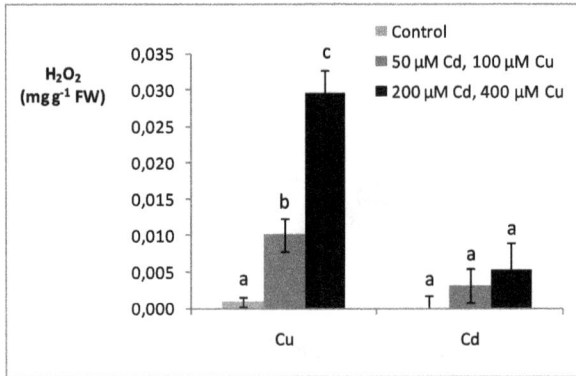

Fig. 1. Hydrogen peroxide content in *Nicotiana tabacum* L. plants grown in nutrient solution, with excess copper and cadmium.

2.2 ROS effect in different cellular components (lipids, DNA, proteins, carbohydrates)

When cell homeostasis is affected by a given stress, ROS production increases to the point where it can damage cellular components and ultimately lead to cell death. ROS can affect lipids, proteins, carbohydrates and DNA and the detailed mechanisms are well detailed in Moller et al. (2007).

Unsaturated fatty acids from lipid membranes are particularly susceptible to ROS oxidation, increasing membrane leakage. Lipid peroxidation occurs through a series of chain reactions that start when a ROS like the hydroxyl radical removes one hydrogen from a carbon from the fatty acid molecule (mainly at the unsaturation). An oxygen can then easily bond to that location forming a lipid peroxyl radical, that can continue and propagate the same kind of reactions (Gill & Tuteja, 2010).

Proteins can also suffer oxidation by ROS, causing certain enzymes to lose its catalytic function. One of the more susceptible targets in proteins are thiol groups the oxidation of which can lead to protein denaturation and loss of functional conformation (Moller et al., 2007). Also, protein oxidation leads to the production of carbonyl groups and to increased rate of proteolysis as the damaged proteins are targeted by proteolytic enzymes (Palma et al., 2002).

Changes in protein content and in protein profile can be found as a consequence of the stress induced by toxic metals. In figure 2, we show the effect of excess copper (50 µM) in the protein content of *Lupinus luteus* leaves, evidencing a significantly higher protein content after 11 days of excess copper, compared to control. In this work, lupin plants were grown in nutrient solution with the indicated Cu concentration. This protein increase could represent the positive balance from the inactivation of some proteins whereas other proteins are formed in relation to the defense response.

In figure 3, we present the protein profile of *Lupinus luteus* leaves after 11 days of exposition to different Cu concentrations. The protein profile showed some changes that can be related to the Cu concentration in nutrient solution. In fact, we found that some protein bands showed higher intensities (56.5 and 17.7 Da) while new protein bands were detected that were not present in control samples (28.5 and 14 Da). These new proteins could be related to the Cu defence mechanism of these plants.

DNA can also be attacked by ROS damaging nucleotide bases, causing mutations and genetic defects (Tuteja et al., 2001).

Both free carbohydrates and wall polysaccharides can react easily with the hydroxyl radical, and this can also be a defence mechanism if the radical reacts with these carbohydrates before damaging more biologically important molecules (Moller et al., 2007).

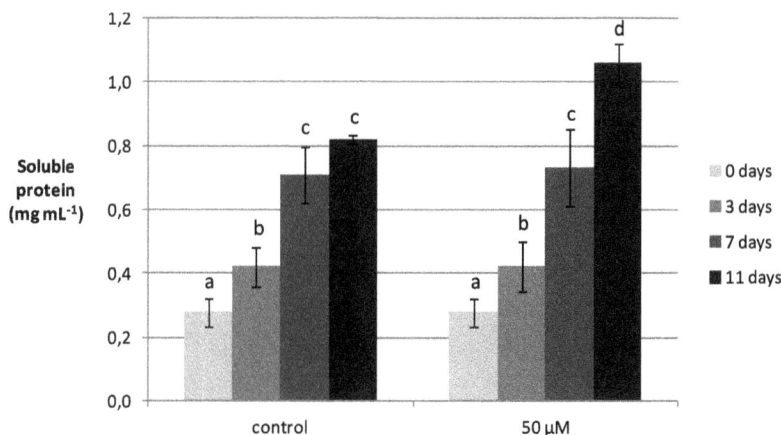

Fig. 2. Protein content of *Lupinus luteus* leaves grown in nutrient solution with 0.1 μM of Cu (control) and excess copper (50 μM) for up to 11 days.

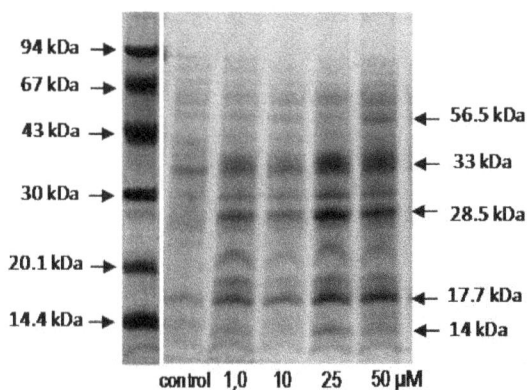

Fig. 3. SDS-PAGE protein profile of *Lupinus luteus* leaves after 11 days at different concentrations of copper. Electrophoresis was performed in a 12 % polyacrylamide gel stained with Commassie blue.

3. Antioxidant defence mechanisms

3.1 Enzymatic mechanisms

As was said before, enzymatic mechanisms and enzymes involved in specific metabolic pathways are one of the major antioxidative defence strategy of plant defence against excess ROS.

Superoxide dismutase (SOD, EC 1.15.1.1), catalyses the dismutation of superoxide molecules into hydrogen peroxide and oxygen (Alscher et al., 2002).

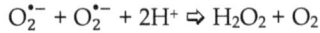

$$O_2^{\bullet-} + O_2^{\bullet-} + 2H^+ \Rightarrow H_2O_2 + O_2$$

SOD has a metal cofactor and depending on the metal can be classified in three different groups, localized in different cell compartments: FeSOD (chloroplasts), MnSOD (mitochondria and peroxisomes), Cu/ZnSOD (chloroplast and cytosol). As SOD produces hydrogen peroxide that is subsequently converted to water by peroxidases and catalases, the activity of all these enzymes must be carefully balanced.

Catalase (CAT, EC 1.11.1.6) exists mainly in the peroxisomes and as during stress the number of these organelles increase, CAT can have an important role in H_2O_2 detoxification that can diffuse into the peroxisome from other cell locations where it is produced (Mittler, 2002). CAT catalyses the hydrogen peroxide breakdown to water:

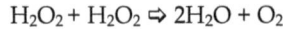

$$H_2O_2 + H_2O_2 \Rightarrow 2H_2O + O_2$$

Peroxidases (EC 1.11.1) are a member of a large family of enzymes that are ubiquitous in the cell and have numerous roles in plant metabolism (Passardi et al., 2005), namely to remove hydrogen peroxide formed due to induced stress using different reductants. They have the general reaction:

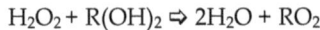

$$H_2O_2 + R(OH)_2 \Rightarrow 2H_2O + RO_2$$

$R(OH)_2$ represents different electron donors: guaiacol peroxidase (GPOD, EC 1.11.1.7) uses mainly phenolic donors, ascorbate peroxidase (APX, EC 1.11.1.11) uses ascorbic acid and glutathione peroxidase (GPX, EC 1.11.1.9) uses glutathione.

Besides their role as a scavenger of hydrogen peroxide, cell-wall peroxidases are also involved in ROS formation, both as a defence against biotic stresses and as a signalling process against several stresses, leading to the activation of other defence mechanisms (Mika et al., 2004).

APX has a much higher affinity to H_2O_2 than CAT suggesting that they have different roles in the scavenging of this ROS, with APX being responsible for maintaining the low levels of hydrogen peroxide while CAT is responsible for the removal of its excess (Mittler, 2002).

The water-water cycle (Figure 4) occurs in chloroplasts and is a fundamental mechanism to avoid photooxidative damage (Rizhsky et al., 2003), using SOD and APX to scavenge the superoxide radical and hydrogen peroxide in the location where they are produced avoiding the deleterious effects of their reactivity with other cellular components (Asada, 1999; Shigeoka et al., 2002).

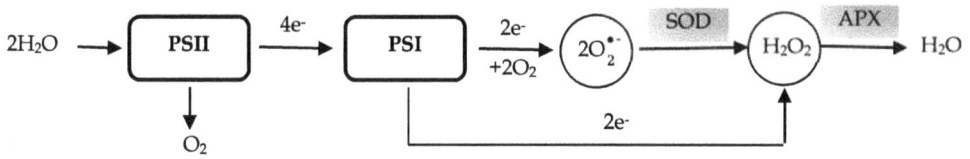

Fig. 4. The water-water cycle. PSI and PSII - Photosystems I and II, SOD - Superoxide dismutase, APX - ascorbate peroxidase.

The ascorbate-glutathione cycle (Figure 5) is an important group of reactions involved in ROS detoxification, as it converts hydrogen peroxide (formed as a consequence of an induced stress or via SOD action) and occurs in several cell compartments, like chloroplasts, cytosol, mitochondria, peroxisomes and apoplast. It uses APX and GPX as well as other enzymes like monodehydroascorbate reductase (MDHAR, EC 1.6.5.4) and dehydroascorbate reductase (DHAR, EC 1.8.5.1) that have a role in the regeneration of the reduced form of ascorbate and glutathione reductase (GR, EC 1.6.4.2), important to maintain the pool of reduced glutathione.

Other enzymes like heme oxygenase (HO, EC 1.14.99.3), that catalyzes the stereo specific cleavage of heme to biliverdin (Balestrasse et al., 2005) have been reported to have a role in plant defence mechanisms against oxidative stress.

3.2 Enzymatic responses to heavy metal stress

There has been extensive studies on the activities of the enzymes involved in ROS defence on plants subjected to heavy metal stress (Sharma & Dietz, 2009). Although different metals can cause oxidative stress, their mode of action is different. For example while copper, an essential element toxic at high concentrations, is involved in redox reactions, cadmium, which has no known biological function, cannot. However, they both can induce oxidative stress, through different mechanisms (Cuypers et al., 2010).

Of all the main enzymes involved in oxidative stress defence, like SOD, APX, CAT and peroxidases, published reports both describe an increase in its activity and a decrease (or no change), depending on plant species, plant organ, type of metal, duration of the treatment, plant age, and growing media (Gratão et al., 2005).

In table 2 we list some representative publications of this kind of studies, regarding determinations made at a single time of growth (time series studies show also changes along the time, further complicating the analysis). As can be seen, there is a huge variation between the behaviour of the enzymes involved in oxidative stress. In some situations the activity increases for lower concentrations of the metal and then decreases as the defence mechanism breaks down due to excessive concentration. This also shows that different components of the antioxidative defence system described above are active under different conditions. The increase in activity in a given enzyme can be a signal that that enzyme has been activated or its expression upregulated. On the other hand sometimes the effect of the metal can be so drastic that enzymes structures are being affected with a consequent decrease in activity. Several enzymes have metal cofactors so there could be a link between these enzymes expression and metal availability (Cohu & Pilon, 2007).

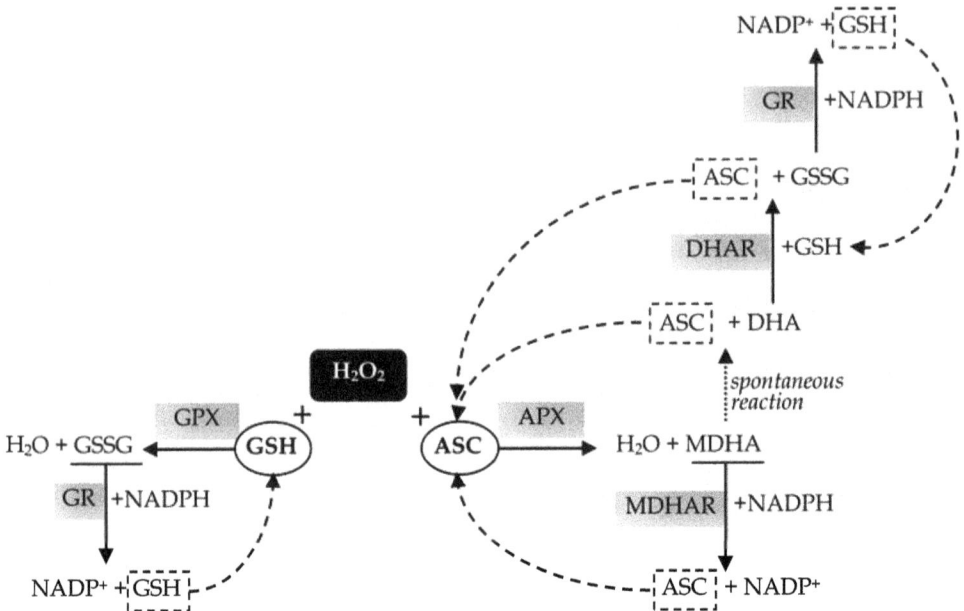

Fig. 5. The ascorbate-glutathione cycle. Non-enzymatic compounds: ASC - ascorbate, MDHA - monodehydroascorbate, DHA - dehydroascorbate, GSH - glutathione (reduced), GSSG - glutathione (oxidized). Enzymes (grey box): APX - ascorbate peroxidase, GPX - glutathione peroxidase, GR - glutathione reductase, MDHAR - monodehydroascorbate reductase, DHAR - dehydroascorbate reductase.

Consequently, enzymatic response can be complex to analyse. For example, in figure 6A we present the activity of guaiacol peroxidase in tomato roots growing for 3 days with 50 μM Cu in nutrient solution. As can be seen by the relative activity, no significant changes in POD activity were detected compared to control. However, the isoperoxidase profile (figure 6B) showed the appearance of new isoforms in tomato roots (C, D), while other isoforms showed less intensity (A, B). These results indicate that although enzymatic activity can be similar in control and stressed plants it is possible that some isoforms can be activated in response to excess copper.

Polyphenol oxidase (PPO, EC 1.10.3.1) is an oxidoreductase that catalyzes the oxidation of phenols to quinones and its activity has been shown to increase under heavy metal stress and has thus been associated to some form of defence mechanism (Ali et al., 2006; L. L. Martins & Mourato, 2006). Kováčik et al. (2009) observed an increase in root PPO activity with Cu and Cd and concluded that the formation of polymerized phenols could be used to complex free metal ions. On the other hand PPO has also been associated to a catalase-like activity (Gerdemann et al., 2001), and could thus have a role in direct hydrogen peroxide removal.

Peroxiredoxins (PRX, EC 1.11.1.15) are ubiquitous antioxidant enzymes that participate in cellular redox homeostasis and reduce hydrogen peroxide to water. PRX levels have been shown to increase under several abiotic stresses suggesting a role in the defence mechanisms (Barranco-Medina et al., 2007).

Plant Species	Metal	Concentration	Organ	Enzyme	References
Arabidopsis thaliana	Cd	5, 10, 20 µM	leaves	APX ↑↓, GPOD ↑↓, SOD =, CAT =, GR ↓	(Smeets et al., 2008)
Brassica juncea	Cd	10, 25, 50 µM	leaves	APX ↓, GPOD ↓, CAT ↑, GR =	(Nouairi et al., 2009)
"	Cd	10, 30, 50, 100 µM	leaves	APX ↑↓, CAT ↓, GR ↑↓, GPX ↓, MDHAR ↑↓, DHAR ↑↓	(Markovska et al., 2009)
"	As	5, 25 µM	leaves	APX ↑, SOD ↑, CAT =	(Khan et al., 2009)
"	Cd	5, 15, 35 mg.kg⁻¹	leaves	GPOD ↑, APX ↑↓, CAT ↑↓	(Pinto et al., 2009)
Brassica napus	Cd	10, 25, 50 µM	leaves	APX ↓, GPOD ↑, CAT ↓, GR ↑↓	(Nouairi et al., 2009)
Cannabis sativa	Cd	25, 50, 100 mg.kg⁻¹	seedlings	GPOD ↑, SOD ↑, CAT =	(Shi et al., 2009)
Daucus carota	Cu	100, 200, 400, 800 mg.kg⁻¹	leaves	APX ↓↑, SOD ↑↓, CAT ↑↓	(Ke et al., 2007)
Elsholtzia splendens	Cu	25, 50, 100, 500 µM	leaves	APX ↑, GPOD ↓↑, SOD ↑, CAT ↑	(Peng et al., 2006)
Matricaria chamomilla	Cd	3, 60, 120 µM	leaves, roots	CAT ↑, GR ↑	(Kovácik & Backor, 2008)
"	Cu	3, 60, 120 µM	leaves, roots	CAT ↑, GR ↑	(Kovácik & Backor, 2008)
Nicotiana tabacum	Cd	10, 25, 50, 100 µM	young leaves old leaves	GPOD ↑, SOD ↓ GPOD ↑, SOD ↑	(Martins et al., 2011)
"	Cd	5, 15, 35 mg.kg⁻¹	leaves	GPOD =, APX ↓↑, CAT ↑	(Pinto et al., 2009)
Solanum nigrum	Cd	5, 15, 35 mg.kg⁻¹	leaves	GPOD ↑, APX ↑, CAT ↑	(Pinto et al., 2009)
Typha angustifolia	Cd	1 mM	leaves	GPOD ↑, APX =, GPX =, SOD ↑, CAT =	(Bah et al., 2011)
"	Cr	2 mM	leaves	GPOD ↑, APX =, GPX =, SOD ↑, CAT =	(Bah et al., 2011)
"	Pb	1 mM	leaves	GPOD ↑, APX =, GPX =, SOD =, CAT =	(Bah et al., 2011)

Plant Species	Metal	Concentration	Organ	Enzyme	References
Withania somnifera	Cu	10, 25, 50, 100, 200 µM	leaves	APX ↑↓, GPOD ↑, SOD ↓, CAT ↓, GR ↓, MDHAR ↑, DHAR ↓↑	(Khatun et al., 2008)
Zea mays	Cd	300, 600, 900 µM	leaves	APX ↑↓, GPOD ↑↓, SOD ↑, GR ↑↓	(Ekmekci et al., 2008)

Table 2. Changes in enzyme activities for several plants and metals. The symbols after the enzymes indicate if its activity increased (↑), decreased (↓), remained the same (=) or increased for lower concentrations and decreased for higher concentrations or vice-versa(↑↓ or ↓↑).

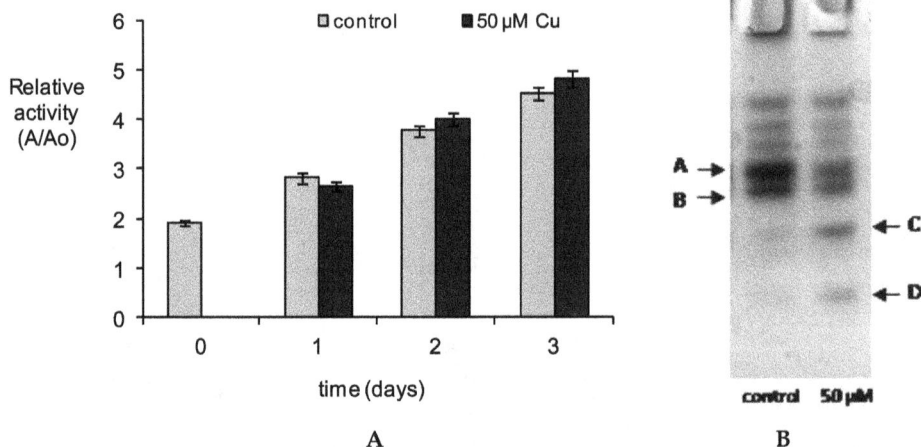

Fig. 6. Guaiacol peroxidase activity of tomato roots for 3 days of 50 µM Cu in nutrient solution (A) and the correspondent isoperoxidase profile for plants collected after 3 days (B). Cu concentration in control plants was 0.1 µM.

3.3 Non enzymatic mechanisms

Besides the enzymatic mechanisms described in the two previous chapters, there is a whole range of other substances that have been reported to be involved in antioxidative defence in plants. Some are well known and have been extensively studied (like the role of ascorbate and glutathione) while others are thought to be part of defence mechanisms but its role remains to be fully understood. While many studies report the increase in the concentration of some substance in relation to the induction of a stress, this type of correlation is not, by itself, conclusive enough to ascertain the effect of that substance in the plant metabolism.

Of course, when the levels of the compounds described in this chapter increase, most of the enzymes involved in the respective biosynthesis are also induced.

Several organic molecules have been reported to be able to form complexes with heavy metals, like phytochelatins (see section 3.4), organic acids and amino acids like proline. But exactly what kind of molecules complex the metals seem to be highly dependent not only on metal type and plant species but also on plant organ and compartment, as different complexes can be formed in the cytoplasm, the xylem and phloem, for example (Sharma & Dietz, 2006).

Ascorbate and glutathione are antioxidants that exist in relatively high concentration in cell compartments (Potters et al., 2002), and are involved in the ascorbate-glutathione cycle as described above being also a substrate for APX an enzyme important in H_2O_2 removal. Ascorbate (ASC) is an electron donor that can be oxidized to the radical monodehydroascorbate (MDHA) and this compound can then form dehydroascorbate (DHA) (Figure 7):

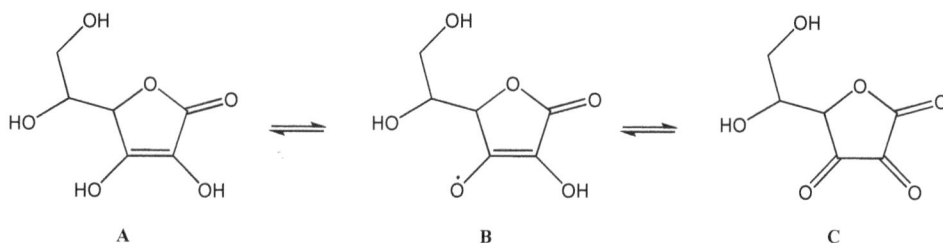

A   B   C

Fig. 7. Structure of ascorbate (A), monodehydroascorbate (B) and dehydroascorbate (C)

Besides its role as an enzyme substrate, ascorbate also reacts directly with singlet oxygen and superoxide and is important in the regeneration of α-tocopherol and certain carotenoids (Potters et al., 2002).

Glutathione (Figure 8) is a tripeptide (containing glutamate, cysteine and glycine) that can exist in two predominant forms: the reduced form (usually represented by GSH) and the oxidized form (usually represented by GSSG) (Noctor & Foyer, 1998). It is involved in the sulphur metabolism and in defence reactions against oxidative stress (Potters et al., 2002). It can also lead to the synthesis of phytochelatins that are important sequesters for certain heavy metals (Cobbett & Goldsbrough, 2002).

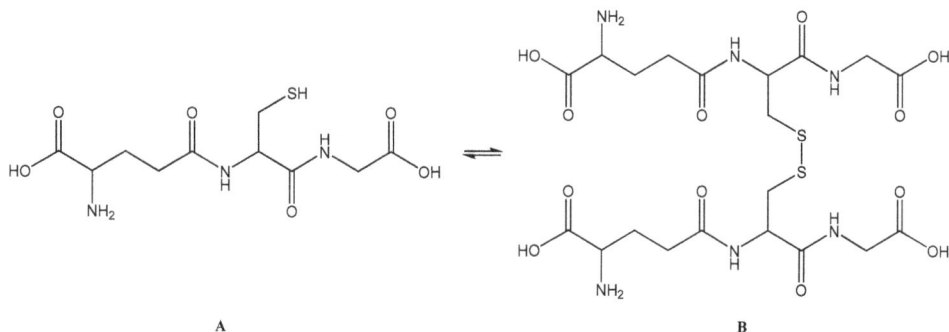

A   B

Fig. 8. Structure of glutathione in its reduced (A) and oxidized forms (B)

Heat shock proteins (HSP) are proteins that not only showed increased expression under heat stress but are also molecular chaperones that protect other proteins from stress induced damage (Feder & Hofmann, 1999). HSP help proteins maintain or recover their native conformation, and remove potentially harmful polypeptides. HSP expression has been associated to high temperature, cold, drought, light, heavy metals, salt and ozone stresses (Timperio et al., 2008). Kochhar & Kochhar (2005) detected the induction of both high and low molecular weight HSPs in response to combined heat and cadmium stress. ROS also have a signaling role during a stress, in order to induce HSP production (Timperio et al., 2008).

Carotenoids have an important protective role during photosynthesis as these molecules can quench the excited states of chlorophyll in order to avoid the production of singlet oxygen. As a consequence, the carotenoid molecules become themselves excited but this is not a big problem as they don't have enough energy to form this ROS species (Taiz & Zeiger, 2002).

Terpenoids are a large class of organic compounds derived from the isoprene unit that could also have an antioxidative role in plants, although that is not yet clear (Grassmann et al., 2002).

Flavonoids are organic molecules with a structure similar to flavone (Figure 9), that have been shown to have a protective role against several stresses (Jaakola et al., 2004), both by themselves and in conjugation with peroxidases (Mika et al., 2004). Anthocyanins, a type of flavonoids (they are glucosides of anthocyanidins, Figure 9) present in the vacuoles, have an antioxidative capacity (Kahkonen & Heinonen, 2003) but its location in the cell prevents them to contact directly with ROS production sites, although its levels have been reported to increase under Cd stress (Mobin & Khan, 2007).

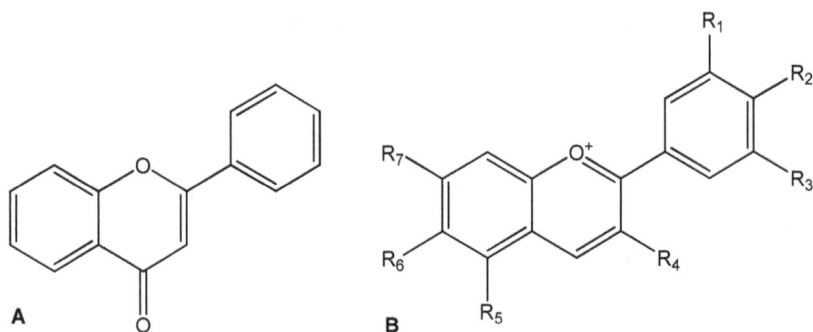

Fig. 9. Structure of flavone (A) and of anthocyanidins (B, where R_1 toR_7 can be H, OH or OCH_3 according to the exact type of anthocyanidin)

Thiols can play an important antioxidative role, protecting membrane lipids. Lipoic acid (Figure 10), both in its reduced and oxidized form, is reported to have antioxidative properties due to its direct scavenging of ROS. It is also able to chelate several metal ions that induces oxidative stress (Navari-Izzo et al., 2002) and thus can have an important role in cell protection.

Tocopherols are a class of compounds synthesized by photosynthetic organisms that have vitamin E activity. α-Tocopherol (Figure 11) is the most common form in leaves while γ-tocopherol is more common in roots (Abbasi et al., 2007). They have a role in ROS protection as they can quench singlet oxygen (Gill & Tuteja, 2010) and can act as an antioxidant and terminate chain reactions occurring during lipid peroxidation (Garg & Manchanda, 2009).

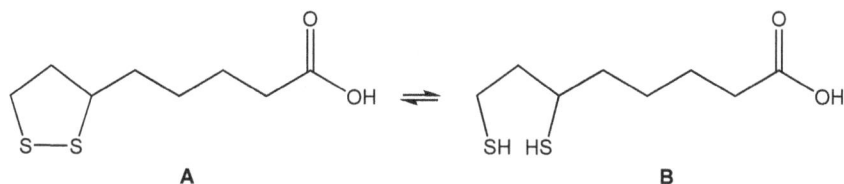

Fig. 10. Structure of lipoic acid in its oxidized (A) and reduced (B) forms.

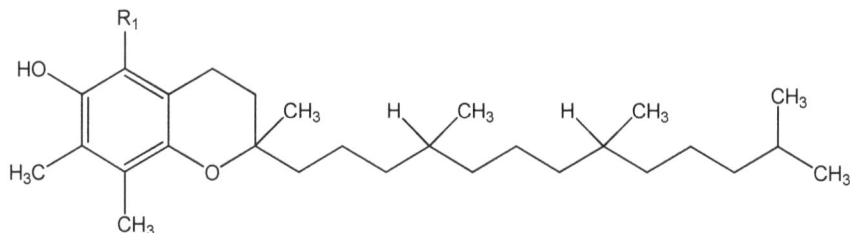

Fig. 11. Common structure of α-tocopherol (with $R_1=CH_3$) and of γ-tocopherol (with $R_1=H$).

Proline (an amino acid, shown in figure 12A), is a compatible solute that participate in the osmotic adjustment of plant cells being able to balance water stress. Proline has been reported to improve plant resistance to oxidative stress by scavenging ROS (namely by quenching singlet oxygen and hydroxyl radicals), increasing the activity of antioxidative enzymes and protecting them and maintaining redox homeostasis (Matysik et al., 2002), and they could also participate in signalling pathways that regulate stress related genes (Khedr et al., 2003). Although the concentration of proline has been shown to increase under heavy metal stress in certain plants (Martins et al., 2011), its exact role in heavy metal detoxification is unclear as, under certain conditions, it could be only an indirect response due to heavy-metal induced disturbances in plant water balance (Schat et al., 1997).

Histidine (figure 12B) is another amino acid that has been mainly linked to Ni hyperaccumulator plants (Sharma & Dietz, 2006).

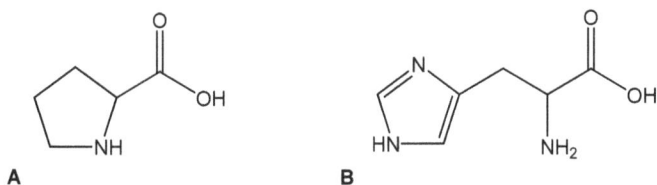

Fig. 12. Structure of proline (A) and histidine (B).

Glycine betaine (2-trimethylammonioacetate, figure 13A) is another compatible solute that is involved in plant resistance against abiotic stresses, namely salt stress (Banu et al., 2009). It helps not only in controlling water balance but can also help to maintain protein and membrane structure (F.-L. Zhang et al., 2008). Nicotianamine (figure 13B) is an amino acid that has a known role in heavy metal transport in plants (Stephan & Scholz, 1993), but recent findings have suggested an important role also in heavy metal tolerance namely in relation to hyper accumulating species (Sharma & Dietz, 2006). Mugineic acid (figure 13C) is a siderophore (that is, an iron-chelating compound) that promotes iron acquisition from the rhizosphere, but that may also participate in the distribution of other metals in the plant (Haydon & Cobbett, 2007). Meda et al. (2007) reports that phytosiderophores can alleviate Cd toxicity in maize.

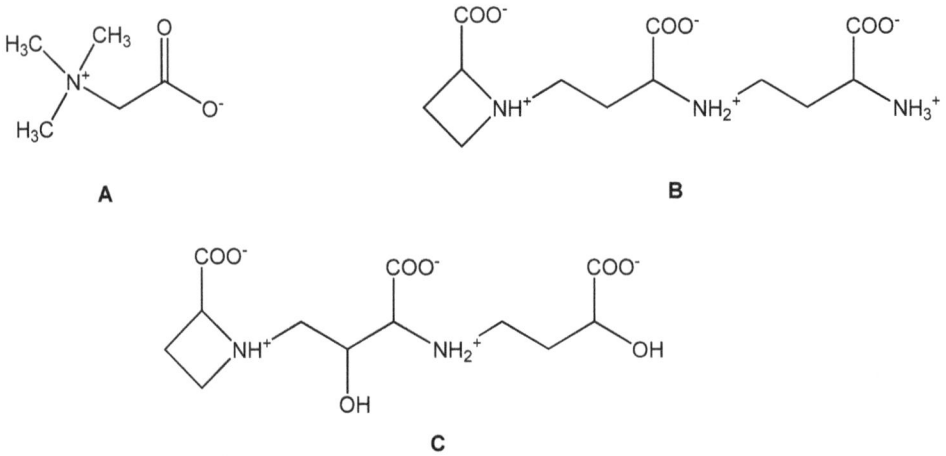

Fig. 13. Structure of glycine betaine (A), nicotianamine (B) and mugineic acid (C).

Polyamines are low molecular weight amines (figure 14) that have a role in plant growth and developmental processes (Kakkar & Sawhney, 2002). The concentration of some polyamines has been reported to increase under abiotic stress. Mascher (2002) showed that the concentration of putrescine, spermidine and spermine increased when red clover plants were subjected to As toxicity. However it is still not clear the exact role these compounds play in heavy metal defence, but a participation in the stabilization and protection of the membrane systems has been proposed (Sharma & Dietz, 2006).

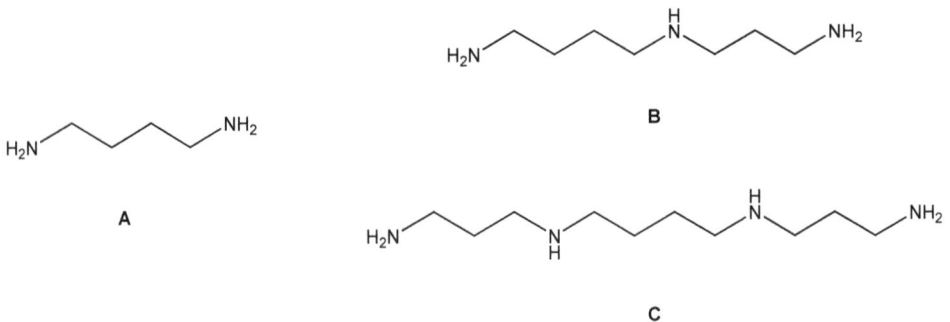

Fig. 14. Structure of several polyamines. A - putrescine (a di-amine); B - spermidine (a tri-amine); C - spermine (a tetra-amine)

Although soluble sugars have been linked to metabolic pathways that produce ROS, they can also have an important role in ROS scavenging mechanisms. Increased glucose levels can increase the production of NADPH (via the pentose-phosphate pathway), that is an important intermediate in the ascorbate-glutathione cycle (Couée et al., 2006) as NADPH is the primary electron donor that assures a intracellular reduction status. Both glucose and sucrose levels have been shown to increase in some plant species under certain abiotic stresses but it is not obvious that this happens due to a putative defence mechanism induction (Couée et al., 2006), although these sugars also participate in signalling mechanisms. Van den Ende and Valluru (2009) suggest that sucrose might have a protective role against stress due to its capacity to scavenge ROS. Other sugars like raffinose (Figure 15A) and fructans (which are fructose polymers) are also reported to have a protective role of membranes against several stresses, namely freezing and drought stress (Van den Ende & Valluru, 2009).

Trehalose (Figure 15B) is a non-reducing disaccharide that can also participate in the stabilization of membranes and protection of proteins under abiotic stresses (Luo et al., 2008). Although trehalose exists in numerous organisms (like bacteria, fungi and nematodes) it is not found widespread in plants, and when it is found is usually in very low concentrations (Wingler, 2002), but several studies have correlated the availability of this disaccharide (or the expression of the genes related to its synthesis) with stress responses. Almeida et al. (2005) found that transgenic tobacco plants over-expressing trehalose-6-phosphate synthase had higher resistance against different abiotic stresses (like drought and temperature). Nery et al. (2008) stated that trehalose participated in protecting cells against hydrogen peroxide, by preventing oxidation of both membranes and proteins.

Brassinosteroids are a family of polyhydroxysteroids that have been reported to modify the activity of antioxidant enzymes and the level of non-enzymatic compounds (like ascorbic acid and tocopherols), when plants are subjected to different abiotic stresses, but its effect is still poorly understood (Bajguz & Hayat, 2009). ROS also have a signalling role in hormone responses, like ABA, auxin and ethylene (Kwak et al., 2006).

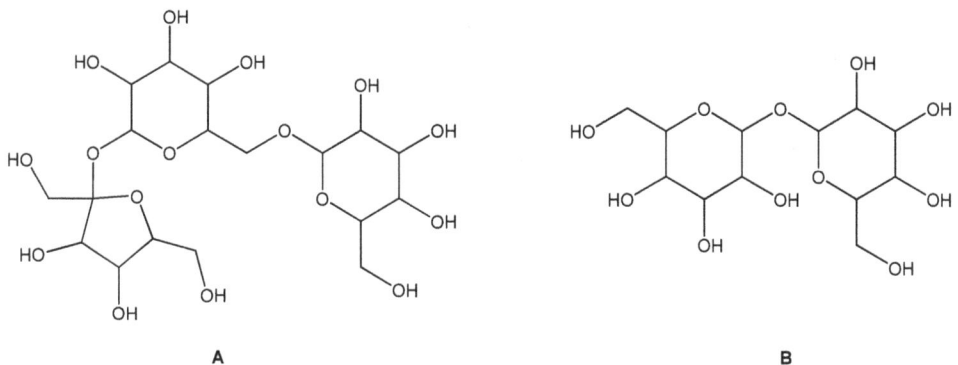

A

B

Fig. 15. Structure of raffinose (A) and trehalose (B)

3.4 Other non enzymatic substances involved in heavy metal tolerance

Several other substances, not described above, have been reported to be involved in the tolerance mechanisms against heavy metal tolerance. Metals are rarely available in the free ionic form in plants but are bound to different types of organic molecules. These include organic acids (like citric and maleic acid) and some amino acids (Haydon & Cobbett, 2007). In fact, organic acids, among other organic solutes of several types, can accumulate in leaves in stress conditions, such as water stress, with an important role on subcellular structures protection acting as osmolytes (Pinheiro et al., 2004). One possible effect of heavy metal stress is that plants restrict water uptake, thus indirectly inducing water stress.

Any changes in the concentrations of the intermediate organic acids can reflect an influence of a heavy metal in metabolic pathways as they are involved in primary plant metabolism such as cell respiration and the formation of ATP. Organic acids can have a detoxification role complexing metals, and reducing their availability to the plant, but this role is not yet fully understood (Hall, 2002).

The metabolism of several phenols has been reported to change under Cu and Cd stress indicating a putative role of these compounds in heavy metal detoxification by decreasing the presence of the free metal ions (Kováčik et al., 2009). Lignin accumulation has also been described as a consequence of Cu toxicity (Lequeux et al., 2010) but whether this is related to a heavy metal defence mechanism is still not clear.

Zorrig et al. (2010) suggested a transport role for citrate to translocate Cd from the roots to the shoots of lettuce plants while Panfili et al. (2009) also studied the effect of citrate in the uptake of Cd. Citrate and malate have been reported to be major ligands for Zn and Ni in several studies (Haydon & Cobbett, 2007).

Phytochelatins (PC) are small metal-binding peptides with the structure $(\gamma\text{-Glu-Cys})_n\text{-Gly}$ (figure 16) where n ranges between 2 and 11 (Cobbett & Goldsbrough, 2002). Its synthesis is catalyzed by phytochelatin synthase using glutathione as a substrate (Clemens, 2006; Grill et al., 1989). They form complexes with metals like Cd (due to the high affinity metals have to the thiol group present in cysteine) and sequester them to the vacuoles and are thus an important mechanism that plants use to avoid heavy metal toxicity. Other peptides with a structure similar to phytochelatins have been identified, with the terminal Gly being replaced by other amino acids like serine, glutamic acid and glutamine in the case of iso-phytochelatins or alanine in homophytochelatins (Oven et al., 2002). All of these peptides have been reported to participate in the detoxification mechanisms of various plant species against several metals or metalloids, besides Cd, like As (Vázquez et al., 2009) and Pb (Z. C. Zhang et al., 2008).

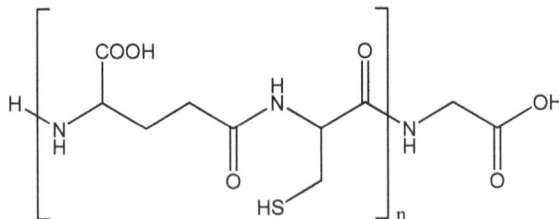

Fig. 16. General structure of phytochelatins (n ranges between 2 and 11).

Metallothioneins (MT) are other small peptides and in plants they contain typically between 60 to 85 amino acids (Freisinger, 2009), also containing cysteine in various amounts according to the type of MT. They have a different synthesis pathway from phytochelatins (Cobbett & Goldsbrough, 2002), and are thought to be important in metal complexation , but can also have other roles like ROS scavenging (Hassinen et al., 2011). MTs are reported to be much more important in Cu tolerance of certain plants than PCs (Mijovilovich et al., 2009).

4. Acknowledgments

The authors acknowledge the financial support of *Fundação para a Ciência e Tecnologia*, through its project PTDC/AGR-AAM/102821/2008 and the Research Unit Environmental Chemistry.

5. References

Abbasi, A.-R., Hajirezaei, M., Hofius, D., Sonnewald, U., & Voll, L. M. (2007). Specific Roles of {alpha}- and {gamma}-Tocopherol in Abiotic Stress Responses of Transgenic Tobacco. *Plant Physiology*, Vol.143, No.4, 1720-1738.

Adriano, D. C., Wenzel, W. W., Vangronsveld, J., & Bolan, N. S. (2004). Role of assisted natural remediation in environmental cleanup. *Geoderma*, Vol.122, 121-142.

Ahmad, P., Sarwat, M., & Sharma, S. (2008). Reactive oxygen species, antioxidants and signaling in plants. *Journal of Plant Biology*, Vol.51, No.3, 167-173.

Almeida, A. M., Villalobos, E., Araújo, S. S., Leyman, B. V. D. P., Alfaro-Cardoso, L., Fevereiro, P. S., Torné, J. M., & Santos, D. M. (2005). Transformation of tobacco with an Arabidopsis thaliana gene involved in trehalose biosynthesis increases tolerance to several abiotic stresses. *Euphytica*, Vol.146, 165-176.

Alscher, R. G., Erturk, N., & Heath, L. S. (2002). Role of superoxide dismutases (SODs) in controlling oxidative stress in plants. *Journal of Experimental Botany*, Vol.53, No.372, 1331-1341.

Asada, K. (1999). The Water-Water Cycle in Chloroplasts: Scavenging of Active Oxygens and Dissipation of Excess Photons. *Annual Review of Plant Physiology & Plant Molecular Biology*, Vol.50, No.1, 601.

Asada, K. (2006). Production and scavenging of reactive oxygen species in chloroplasts and their functions. *Plant Physiology*, Vol.141, No.2, 391-396.

Bah, A. M., Dai, H., Zhao, J., Sun, H., Cao, F., Zhang, G., & Wu, F. (2011). Effects of Cadmium, Chromium and Lead on Growth, Metal Uptake and Antioxidative Capacity in Typha angustifolia. *Biological Trace Element Research*, Vol.142, No.1, 77-92.

Bajguz, A., & Hayat, S. (2009). Effects of brassinosteroids on the plant responses to environmental stresses. *Plant Physiology and Biochemistry*, Vol.47, No.1, 1-8.

Balestrasse, K. B., Noriega, G. O., Batlle, A., & Tomaro, M. L. (2005). Involvement of heme oxygenase as antioxidant defense in soybean nodules. *Free Radical Research*, Vol.39, No.2, 145-151.

Banu, M. N. A., Hoque, M. A., Watanabe-Sugimoto, M., Matsuoka, K., Nakamura, Y., Shimoishi, Y., & Murata, Y. (2009). Proline and glycinebetaine induce antioxidant defense gene expression and suppress cell death in cultured tobacco cells under salt stress. *Journal of Plant Physiology*, Vol.166, No.2, 146-156.

Barranco-Medina, S., Krell, T., Finkemeier, I., Sevilla, F., Lazaro, J.-J., & Dietz, K.-J. (2007). Biochemical and molecular characterization of the mitochondrial peroxiredoxin

PsPrxII F from Pisum sativum. *Plant Physiology and Biochemistry*, Vol.45, No.10-11, 729-739.

Clemens, S. (2006). Evolution and function of phytochelatin synthases. *Journal of Plant Physiology*, Vol.163, No.3, 319-332.

Cobbett, C., & Goldsbrough, P. (2002). Phytochelatins and metallothioneins: Roles in heavy metal detoxification and homeostasis. *Annual Review of Plant Biology*, Vol.53, 159-182.

Cohu, C. M., & Pilon, M. (2007). Regulation of superoxide dismutase expression by copper availability. *Physiologia Plantarum*, Vol.129, No.4, 747-755.

Couée, I., Sulmon, C., Gouesbet, G., & El Amrani, A. (2006). Involvement of soluble sugars in reactive oxygen species balance and responses to oxidative stress in plants. *Journal of Experimental Botany*, Vol.57, No.3, 449-459.

Cuypers, A., Plusquin, M., Remans, T., Jozefczak, M., Keunen, E., Gielen, H., Opdenakker, K., Nair, A., Munters, E., Artois, T., Nawrot, T., Vangronsveld, J., & Smeets, K. (2010). Cadmium stress: an oxidative challenge. *BioMetals*, Vol.23, No.5, 927-940.

del Rio, L. A., Sandalio, L. M., Corpas, F. J., Palma, J. M., & Barroso, J. B. (2006). Reactive oxygen species and reactive nitrogen species in peroxisomes. Production, scavenging, and role in cell signaling. *Plant Physiology*, Vol.141, No.2, 330-335.

Ekmekci, Y., Tanyolac, D., & Ayhan, B. (2008). Effects of cadmium on antioxidant enzyme and photosynthetic activities in leaves of two maize cultivars. *Journal of Plant Physiology*, Vol.165, No.6, 600-611.

Feder, M. E., & Hofmann, G. E. (1999). Heat-shock proteins, molecular chaperones, and the stress response: Evolutionary and ecological physiology. *Annual Review of Physiology*, Vol.61, 243-282.

Foyer, C. H., Lelandais, M., & Kunert, K. J. (1994). Photooxidative Stress in Plants. *Physiologia Plantarum*, Vol.92, No.4, 696-717.

Foyer, C. H., & Noctor, G. (2003). Redox sensing and signalling associated with reactive oxygen in chloroplasts, peroxisomes and mitochondria. *Physiologia Plantarum*, Vol.119, No.3, 355-364.

Freinbichler, W., Colivicchi, M. A., Stefanini, C., Bianchi, L., Ballini, C., Misini, B., Weinberger, P., Linert, W., Vareslija, D., Tipton, K. F., & Della Corte, L. (2011). Highly reactive oxygen species: detection, formation, and possible functions. *Cellular and Molecular Life Sciences*, Vol.68, No.12, 2067-2079.

Freisinger, E. (2009). Metallothioneins in Plants. In A. Sigel, H. Sigel & R. K. O. Sigel (Eds.), *Metallothioneins and Related Chelators* (Vol. 5, pp. 107-153). Cambridge: The Royal Society of Chemistry.

Garg, N., & Manchanda, G. (2009). ROS generation in plants: Boon or bane? *Plant Biosystems*, Vol.143, No.1, 81 - 96.

Gill, S. S., & Tuteja, N. (2010). Reactive oxygen species and antioxidant machinery in abiotic stress tolerance in crop plants. *Plant Physiology and Biochemistry*, Vol.48, No.12, 909-930.

Grassmann, J., Hippeli, S., & Elstner, E. F. (2002). Plant's defence and its benefits for animals and medicine: role of phenolics and terpenoids in avoiding oxygen stress. *Plant Physiology and Biochemistry*, Vol.40, No.6-8, 471-478.

Gratão, P. L., Polle, A., Lea, P. J., & Azevedo, R. A. (2005). Making the life of heavy metal stressed plants a little easier. *Functional Plant Biology*, Vol.32, 481-494.

Grill, E., Loffler, S., Winnacker, E. L., & Zenk, M. H. (1989). Phytochelatins, the Heavy-Metal-Binding Peptides of Plants, Are Synthesized from Glutathione by a Specific Gamma-

Glutamylcysteine Dipeptidyl Transpeptidase (Phytochelatin Synthase). *Proceedings of the National Academy of Sciences of the United States of America,* Vol.86, No.18, 6838-6842.

Hall, J. L. (2002). Cellular mechanisms for heavy metal detoxification and tolerance. *Journal of Experimental Botany,* Vol.53, No.366, 1-11.

Halliwell, B. (2006). Reactive species and antioxidants. Redox biology is a fundamental theme of aerobic life. *Plant Physiology,* Vol.141, No.2, 312-322.

Hassinen, V. H., Tervahauta, A. I., Schat, H., & Karenlampi, S. O. (2011). Plant metallothioneins - metal chelators with ROS scavenging activity? *Plant Biology,* Vol.13, No.2, 225-232.

Haydon, M. J., & Cobbett, C. S. (2007). Transporters of ligands for essential metal ions in plants. *New Phytologist,* Vol.174, No.3, 499-506.

Jaakola, L., Määttä-Riihinen, K., Kärenlampi, S., & Hohtola, A. (2004). Activation of flavonoid biosynthesis by solar radiation in bilberry (Vaccinium myrtillus L.) leaves. *Planta,* Vol.218, No.5, 721-728.

Kahkonen, M. P., & Heinonen, M. (2003). Antioxidant activity of anthocyanins and their aglycons. *Journal of Agricultural and Food Chemistry,* Vol.51, No.3, 628-633.

Kakkar, R. K., & Sawhney, V. K. (2002). Polyamine research in plants - a changing perspective. *Physiologia Plantarum,* Vol.116, No.3, 281-292.

Ke, W. S., Xiong, Z. T., Xie, M. J., & Luo, Q. (2007). Accumulation, subcellular localization and ecophysiological responses to copper stress in two Daucus carota L. populations. *Plant and Soil,* Vol.292, No.1-2, 291-304.

Khan, I., Ahmad, A., & Iqbal, M. (2009). Modulation of antioxidant defence system for arsenic detoxification in Indian mustard. *Ecotoxicology and Environmental Safety,* Vol.72, No.2, 626-634.

Khatun, S., Ali, M. B., Hahn, E.-J., & Paek, K.-Y. (2008). Copper toxicity in Withania somnifera: Growth and antioxidant enzymes responses of in vitro grown plants. *Environmental and Experimental Botany,* Vol.64, No.3, 279-285.

Khedr, A. H. A., Abbas, M. A., Wahid, A. A. A., Quick, W. P., & Abogadallah, G. M. (2003). Proline induces the expression of salt stress responsive proteins and may improve the adaptation of Pancratium maritimum L. to salt stress. *Journal of Experimental Botany,* Vol.54, No.392, 2553-2562.

Kochhar, S., & Kochhar, V. K. (2005). Expression of antioxidant enzymes and heat shock proteins in relation to combined stress of cadmium and heat in Vigna mungo seedlings. *Plant Science,* Vol.168, No.4, 921-929.

Kovácik, J., & Backor, M. (2008). Oxidative status of Matricaria chamomilla plants related to cadmium and copper uptake. *Ecotoxicology,* Vol.17, No.6, 471-479.

Kováčik, J., Klejdus, B., Hedbavny, J., Štork, F., & Bačkor, M. (2009). Comparison of cadmium and copper effect on phenolic metabolism, mineral nutrients and stress-related parameters in Matricaria chamomilla plants. *Plant and Soil,* Vol.320, No.1, 231-242.

Kwak, J. M., Nguyen, V., & Schroeder, J. I. (2006). The role of reactive oxygen species in hormonal responses. *Plant Physiology,* Vol.141, No.2, 323-329.

Lequeux, H., Hermans, C., Lutts, S., & Verbruggen, N. (2010). Response to copper excess in Arabidopsis thaliana: Impact on the root system architecture, hormone distribution, lignin accumulation and mineral profile. *Plant Physiology and Biochemistry,* Vol.48, No.8, 673-682.

Luo, Y., Li, W.-M., & Wang, W. (2008). Trehalose: Protector of antioxidant enzymes or reactive oxygen species scavenger under heat stress? *Environmental and Experimental Botany*, Vol.63, No.1-3, 378-384.

Markovska, Y. K., Gorinova, N. I., Nedkovska, M. P., & Miteva, K. M. (2009). Cadmium-induced oxidative damage and antioxidant responses in Brassica juncea plants. *Biologia Plantarum*, Vol.53, No.1, 151-154.

Martins, L. L., Mourato, M. P., Cardoso, A. I., Pinto, A. P., Mota, A. M., Goncalves, M. d. L. S., & de Varennes, A. (2011). Oxidative stress induced by cadmium in Nicotiana tabacum L.: effects on growth parameters, oxidative damage and antioxidant responses in different plant parts. *Acta Physiologiae Plantarum*, Vol.33, No.4, 1375-1383.

Mascher, R., Lippmann, B., Holzinger, S., & Bergmann, H. (2002). Arsenate toxicity: effects on oxidative stress response molecules and enzymes in red clover plants. *Plant Science*, Vol.163, No.5, 961-969.

Matysik, J., Alia, Bhalu, B., & Mohanty, P. (2002). Molecular mechanisms of quenching of reactive oxygen species by proline under stress in plants. *Current Science*, Vol.82, No.5, 525-532.

Meda, A. R., Scheuermann, E. B., Prechsl, U. E., Erenoglu, B., Schaaf, G., Hayen, H., Weber, G., & von Wiren, N. (2007). Iron Acquisition by Phytosiderophores Contributes to Cadmium Tolerance. *Plant Physiology*, Vol.143, No.4, 1761-1773.

Mijovilovich, A., Leitenmaier, B., Meyer-Klaucke, W., Kroneck, P. M. H., Gotz, B., & Kupper, H. (2009). Complexation and Toxicity of Copper in Higher Plants. II. Different Mechanisms for Copper versus Cadmium Detoxification in the Copper-Sensitive Cadmium/Zinc Hyperaccumulator Thlaspi caerulescens (Ganges Ecotype). *Plant Physiology*, Vol.151, No.2, 715-731.

Mika, A., Minibayeva, F., Beckett, R., & Lüthje, S. (2004). Possible functions of extracellular peroxidases in stress-induced generation and detoxification of active oxygen species. *Phytochemistry Reviews*, Vol.3, No.1, 173-193.

Mittler, R. (2002). Oxidative stress, antioxidants and stress tolerance. *Trends in Plant Science*, Vol.7, No.9, 405-410.

Mittler, R., Vanderauwera, S., Gollery, M., & Van Breusegem, F. (2004). Reactive oxygen gene network of plants. *Trends in Plant Science*, Vol.9, No.10, 490-498.

Mobin, M., & Khan, N. A. (2007). Photosynthetic activity, pigment composition and antioxidative response of two mustard (Brassica juncea) cultivars differing in photosynthetic capacity subjected to cadmium stress. *Journal of Plant Physiology*, Vol.164, No.5, 601-610.

Moller, I. M. (2001). Plant mitochondria and oxidative stress: Electron transport, NADPH turnover, and metabolism of reactive oxygen species. *Annual Review of Plant Physiology and Plant Molecular Biology*, Vol.52, 561-591.

Moller, I. M., Jensen, P. E., & Hansson, A. (2007). Oxidative modifications to cellular components in plants. *Annual Review of Plant Biology*, Vol.58, 459-481.

Navari-Izzo, F., Quartacci, M. F., & Sgherri, C. (2002). Lipoic acid: a unique antioxidant in the detoxification of activated oxygen species. *Plant Physiology and Biochemistry*, Vol.40, No.6-8, 463-470.

Nery, D. d. C. M., da Silva, C. G., Mariani, D., Fernandes, P. N., Pereira, M. D., Panek, A. D., & Eleutherio, E. C. A. (2008). The role of trehalose and its transporter in protection against reactive oxygen species. *Biochim Biophys Acta*, Vol.1780, No.12, 1408-1411.

Noctor, G., & Foyer, C. H. (1998). Ascorbate and glutathione: Keeping active oxygen under control. *Annual Review of Plant Physiology and Plant Molecular Biology*, Vol.49, 249-279.

Nouairi, I., Ben Ammar, W., Ben Youssef, N., Ben Miled, D. D., Ghorbal, M., & Zarrouk, M. (2009). Antioxidant defense system in leaves of Indian mustard (Brassica juncea) and rape (Brassica napus) under cadmium stress. *Acta Physiologiae Plantarum*, Vol.31, No.2, 237-247.

Oven, M., Page, J. E., Zenk, M. H., & Kutchan, T. M. (2002). Molecular characterization of the homo-phytochelatin synthase of soybean Glycine max - Relation to phytochelatin synthase. *Journal of Biological Chemistry*, Vol.277, No.7, 4747-4754.

Palma, J. M., Sandalio, L. M., Corpas, F. J., Romero-Puertas, M. C., McCarthy, I., & del Rio, L. A. (2002). Plant proteases, protein degradation, and oxidative stress: role of peroxisomes. *Plant Physiology and Biochemistry*, Vol.40, No.6-8, 521-530.

Panfili, F., Schneider, A., Vives, A., Perrot, F., Hubert, P., & Pellerin, S. (2009). Cadmium uptake by durum wheat in presence of citrate. *Plant and Soil*, Vol.316, No.1, 299-309.

Passardi, F., Cosio, C., Penel, C., & Dunand, C. (2005). Peroxidases have more functions than a Swiss army knife. *Plant Cell Reports*, Vol.24, No.5, 255-265.

Peng, H. Y., Yang, X. E., Yang, M. J., & Tian, S. K. (2006). Responses of antioxidant enzyme system to copper toxicity and copper detoxification in the leaves of Elsholtzia splenden. *Journal of Plant Nutrition*, Vol.29, No.9, 1619-1635.

Pinheiro, C., Passarinho, J. A., & Ricardo, C. P. (2004). Effect of drought and rewatering on the metabolism of Lupinus albus organs. *Journal of Plant Physiology*, Vol.161, No.11, 1203-1210.

Pinto, A. P., Alves, A. S., Candeias, A. J., Cardoso, A. I., de Varennes, A., Martins, L. L., Mourato, M. P., Goncalves, M. L. S., & Mota, A. M. (2009). Cadmium accumulation and antioxidative defences in Brassica juncea L. Czern, Nicotiana tabacum L. and Solanum nigrum L. *International Journal of Environmental Analytical Chemistry*, Vol.89, No.8-12, 661-676.

Potters, G., De Gara, L., Asard, H., & Horemans, N. (2002). Ascorbate and glutathione: guardians of the cell cycle, partners in crime? *Plant Physiology and Biochemistry*, Vol.40, No.6-8, 537-548.

Prasad, M. N. V. (2004). *Heavy Metal Stress in Plants*: Springer.

Rhoads, D. M., Umbach, A. L., Subbaiah, C. C., & Siedow, J. N. (2006). Mitochondrial reactive oxygen species. Contribution to oxidative stress and interorganellar signaling. *Plant Physiology*, Vol.141, No.2, 357-366.

Rizhsky, L., Liang, H. J., & Mittler, R. (2003). The water-water cycle is essential for chloroplast protection in the absence of stress. *Journal of Biological Chemistry*, Vol.278, No.40, 38921-38925.

Sánchez, M. L. (2008). *Causes And Effects Of Heavy Metal Pollution*. New York: Nova Science Publishers, Inc.

Schat, H., Sharma, S. S., & Vooijs, R. (1997). Heavy metal-induced accumulation of free proline in a metal-tolerant and a nontolerant ecotype of Silene vulgaris. *Physiologia Plantarum*, Vol.101, No.3, 477-482.

Schröder, P., Herzig, R., Bojinov, B., Ruttens, A., Nehnevajova, E., Stamatiadis, S., Memon, A., Vassilev, A., Caviezel, M., & Vangronsveld, J. (2008). Bioenergy to save the world. *Environmental Science and Pollution Research*, Vol.15, No.3, 196-204.

Sharma, S. S., & Dietz, K.-J. (2009). The relationship between metal toxicity and cellular redox imbalance. *Trends in Plant Science*, Vol.14, No.1, 43-50.

Sharma, S. S., & Dietz, K. J. (2006). The significance of amino acids and amino acid-derived molecules in plant responses and adaptation to heavy metal stress. *Journal of Experimental Botany*, Vol.57, No.4, 711-726.

Shi, G. R., Cai, Q. S., Liu, Q. Q., & Wu, L. (2009). Salicylic acid-mediated alleviation of cadmium toxicity in hemp plants in relation to cadmium uptake, photosynthesis, and antioxidant enzymes. *Acta Physiologiae Plantarum*, Vol.31, No.5, 969-977.

Shigeoka, S., Ishikawa, T., Tamoi, M., Miyagawa, Y., Takeda, T., Yabuta, Y., & Yoshimura, K. (2002). Regulation and function of ascorbate peroxidase isoenzymes. *Journal of Experimental Botany*, Vol.53, No.372, 1305-1319.

Smeets, K., Ruytinx, J., Semane, B., Van Belleghem, F., Remans, T., Van Sanden, S., Vangronsveld, J., & Cuypers, A. (2008). Cadmium-induced transcriptional and enzymatic alterations related to oxidative stress. *Environmental and Experimental Botany*, Vol.63, No.1-3, 1-8.

Stephan, U. W., & Scholz, G. (1993). Nicotianamine - Mediator of Transport of Iron and Heavy-Metals in the Phloem. *Physiologia Plantarum*, Vol.88, No.3, 522-529.

Taiz, L., & Zeiger, E. (2002). *Plant Physiology* (3rd ed.). Sunderland, MA: Sinuaer Associates, Inc.

Timperio, A. M., Egidi, M. G., & Zolla, L. (2008). Proteomics applied on plant abiotic stresses: Role of heat shock proteins (HSP). *Journal of Proteomics*, Vol.71, No.4, 391-411.

Tuteja, N., Singh, M. B., Misra, M. K., Bhalla, P. L., & Tuteja, R. (2001). Molecular mechanisms of DNA damage and repair: Progress in plants. *Critical Reviews in Biochemistry and Molecular Biology*, Vol.36, No.4, 337-397.

Van den Ende, W., & Valluru, R. (2009). Sucrose, sucrosyl oligosaccharides, and oxidative stress: scavenging and salvaging? *Journal of Experimental Botany*, Vol.60, No.1, 9-18.

Vangronsveld, J., & Clijsters, H. (1994). Toxic effects of metals. In M. E. Farago (Ed.), *Plants and the chemical elements. Biochemistry, uptake, tolerance and toxicity* (pp. 149-177). Weinheim: VCH Verlagsgesellschaft.

Vázquez, S., Goldsbrough, P., & Carpena, R. O. (2009). Comparative analysis of the contribution of phytochelatins to cadmium and arsenic tolerance in soybean and white lupin. *Plant Physiology and Biochemistry*, Vol.47, No.1, 63-67.

Vranova, E., Inze, D., & Van Breusegem, F. (2002). Signal transduction during oxidative stress. *Journal of Experimental Botany*, Vol.53, No.372, 1227-1236.

Wingler, A. (2002). The function of trehalose biosynthesis in plants. *Phytochemistry*, Vol.60, No.5, 437-440.

Zhang, F.-L., Niu, B., Wang, Y.-C., Chen, F., Wang, S.-H., Xu, Y., Jiang, L.-D., Gao, S., Wu, J., Tang, L., & Jia, Y.-J. (2008). A novel betaine aldehyde dehydrogenase gene from Jatropha curcas, encoding an enzyme implicated in adaptation to environmental stress. *Plant Science*, Vol.174, No.5, 510-518.

Zhang, Z. C., Gao, X., & Qiu, B. S. (2008). Detection of phytochelatins in the hyperaccumulator Sedum alfredii exposed to cadmium and lead. *Phytochemistry*, Vol.69, No.4, 911-918.

Zhao, F.-J., & McGrath, S. P. (2009). Biofortification and phytoremediation. *Current Opinion in Plant Biology*, Vol.12, No.3, 373-380.

Zorrig, W., Rouached, A., Shahzad, Z., Abdelly, C., Davidian, J.-C., & Berthomieu, P. (2010). Identification of three relationships linking cadmium accumulation to cadmium tolerance and zinc and citrate accumulation in lettuce. *Journal of Plant Physiology*, Vol.167, No.15, 1239-1247.

3

Plant-Heavy Metal Interaction: Phytoremediation, Biofortification and Nanoparticles

Elena Masarovičová and Katarína Kráľová
Comenius University in Bratislava
Slovak Republic

1. Introduction

Contamination of ecosystems and action of toxic metals to plants is one of the major problems of all over the world. Most of these metals are present in the environment as a consequence of geological and/or anthropogenic activities. Metal contamination in agricultural environments can originate from atmospheric pollution, pesticide applications, contamination by chemical fertilizers, and irrigation with wastewater of poor quality. Although, some of metals are bioelements (macro- and micronutrients) at normal concentration, they can cause harmful effects on the plants in excess. Moreover, these metals have strong impact on human health through the food chain. Toxic metals or metals, as bioelements in higher than normal concentrations, are group of substances belonging to the xenobiotics. Frequently is also used term heavy metals, that are metals with specific weight higher than 5 gcm^{-3}, e.g. Cd, Hg, Pb, Cr, Ag and Sn (di Toppi & Gabbrielli, 1999). Knowledge of dominant fluxes of metals in the soil-root-shoot continuum can also help agronomic strategy to address the problem of crop growth under metal-excess, biomass production and food quality. In addition to the highly toxic heavy metals, light metals (e.g. Mg, Al) and metalloids (e.g. As and Se) are of great environmental and health significance. Extraordinarily dangerous for both humans and plants are mainly metals that attack activity of the enzymes containing –SH group (e.g. Cd, Hg and Pb). These metals can initiate acute or more dangerous and frequently occurring chronic diseases. Therefore, new and environmental-friendly technologies, such phytoremediation appeared to remove the harmful metals from the environment. Based on their strategy, plants growing on soil containing metal can be classified as accumulators and excluders. Later the following groups of plants were suggested: metal excluders and metal non-excluders (indicators, hyperaccumulators) (in detail see Masarovičová et al., 2010).

Medicinal plants have an increasing economic importance and the food, pharmaceutical and cosmetic industry produce important goods with these plants. Products based on natural substances enjoy an increasing value. However, anthropogenic activity and its effects on environment recently showed that medicinal plants have also responded on the changing environmental conditions. Some medicinal plants produce specific secondary metabolites, which can detoxify some of toxic metals. Thus, medicinal plants can also have non-traditional use e.g. in phytoremediation technologies. In the last years the practical use of alternative medicine in healing processes showed continually increasing tendency. Several species of medicinal plants can be used as supplementary nutrition due to their ability to accumulate some essential nutrition elements (e.g. Se, Zn, and Fe) in the edible parts of these plants. Such fortification of plants with essential nutrients (phytofortification) in an easily assimilated form can help to feed the rapidly increasing world population and improve human health through balanced mineral nutrition. In general, data related to toxic metal contents (e.g. Cd) in pharmaceutically utilized parts of the medicinal plants are also considered from the aspect of „food safety". Therefore it should be stressed that cultivation and use of medicinal plants have to respect the potential hazard connected with environmental contamination, mainly with toxic metals (Masarovičová & Kráľová, 2007).

2. Crops

While the green revolution resulted in the development of new cultivars of crops suited to high input of fertilizer and water, many regions of the world rely on contaminated soils utilized for food production. The continuous application of large amounts of fertilizers and other soil amendments to agricultural land has raised concern regarding the possible accumulation of elevated levels of trace element constituents and potential harm to the environment. Furthermore, increasing amounts of urban and industrial wastes which may contain also significant quantities of toxic metals are being disposed on the agricultural lands (Raven and Leoppert, 1997). Mining, processing metal ore, wastewater irrigation, solid waste disposal, sludge application, vehicular exhaust and many industrial activities are the major and harmful sources of soil contamination with toxic metals. Therefore, an increased uptake of toxic metals by food crops grown on such soils together with human health risks are often recorded. Excessive accumulation of toxic metals in agricultural soils and the aforesaid elevated toxic metal uptake by crops affect both food quality and safety. Numerous regions in the world are known where cultivation of food and feed crops is irresponsible due to the presence of excessive amounts of plant-available toxic metals, leading to economic losses and negative effect for food chain (Meers et al., 2010). Severe toxic metal contamination of soil may cause a variety of problems, including the reduction of crop yield, serious damage of plants and intoxication of animals and humans.

2.1 „Second" green revolution

According to Lynch (2007) at present we need a "second" green revolution, to improve the yield of crops grown in infertile (or low fertile) soils by farmers with little access to fertiliser. Just as the green revolution was based on crops responsive to high soil fertility, the "second" green revolution will be based on crops tolerant to low soil fertility. Root architecture is critically important for both soil exploration and nutrients and water uptake. Architectural traits that enhance topsoil foraging are important for acquisition of phosphorus from infertile soils. Genetic variation in the length and density of root hairs is important for the acquisition of immobile nutrients such as phosphorus and potassium. Genetic variation in root cortical aerenchyma formation is important in reducing the metabolic costs of root growth and soil exploration. Genetic variation in rhizosphere modification through the efflux of protons, organic acids and enzymes is important for the mobilisation of nutrients such as phosphorus and transition metals, and the avoidance of aluminum toxicity. Manipulation of ion transporters may be useful for improving the acquisition of nitrate and for enhancing salt tolerance. Genetic variation in these traits is associated with substantial yield gains in low-fertility soils. In breeding crops for low-fertility soils, selection for specific root traits through direct phenotypic evaluation or molecular markers is likely to be more productive than conventional field screening. Crop genotypes with greater yield in infertile soils will substantially improve the productivity and sustainability of low-input agroecosystems, and in high-input agroecosystems will reduce the environmental impacts of intensive fertilisation. Above mentioned author stressed that population growth, ongoing soil degradation and increasing costs of chemical fertiliser will make the „second" green revolution a priority for plant biology in the 21st century.

2.2 Effect of metals on physiological processes

In general, the metals play an important role in the metabolic pathway during the growth and development of plants, when available in appreciable concentration. Bioelements are essential for many proteins in plants, although they are toxic in excess. Plants differ greatly in metal uptake and accumulation characteristics. Thus, uptake and distribution of metals in plant body depends upon availability and concentration of metals as well as on plant species.

Kranner & Colville (2011) published review paper providing information of metal impact on seed of plants (including the crops and medicinal plants) from the aspect of biochemical and molecular implications and their significance for seed germination. The authors noticed that metals generally affect seed variability and germination in a strictly concentration-dependent manner when exogenously applied in the micro- to mili-molar range. The seeds of metal tolerant plants and hyperaccumulators may have a substantially higher threshold for toxicity than non-tolerant ones. However, knowledge of the effects of metals on seeds is only starting to emerge, in particular on the mechanisms of damage. In addition, more studies are needed that investigate the effects of metals on seed germination *in situ*. It was found, that the inhibition of seed germination by certain metals, e.g. Cd, may be reversible by rinsing the seeds in water. There is some evidence that Cd and Cu can inhibit seed water uptake, which has the potential to impact on germination. On the other hand, seed dormancy status can also be affected by low metal concentrations, but the precise mechanisms of action are far from understood. Moreover, it is largely unknown where metals are deposited in developing seeds and which levels are toxic to the embryo as compared to the endosperm or cotyledons. More work is needed to understand how seed longevity is affected by metals. According to the authors, regarding the mechanism of damage and seed stress response, surprising few studies have concentrated on the detoxification of metals by phytochelatins in seed, although some papers have reported on metallothioneins. Importantly, for a more comprehensive understanding of the effects of metals on seed viability and dormancy, the signalling networks need to be explored including the interaction of seed hormones with metals. Similarly, Shanker et al. (2005) published review article devoted to the chromium toxicity in plants. It was stressed that toxicity of Cr to plants depends on its valence state: Cr(VI) is highly toxic and mobile whereas Cr(III) is less toxic. Since plants lack a specific transport system for this metal, it is taken up by varied transporters of essential ions such as sulphate or iron. Toxic effects of Cr on plant growth and development include alterations in the germination process as well as the growth of roots, stems and leaves, which may affect total dry mass production and yield.

Cr also causes deleterious effects on plant physiological processes such are photosynthesis, water relations and mineral nutritions. Metabolic alterations due to Cr exposure have also been described in plants by a direct effect on enzymes or other metabolites or by its ability to generate reactive oxygen species which may cause oxidative stress. These authors noticed that Cr, in contrast to other toxic metals like Cd, Pb, Cu or Al, has received little attention from plants scientists. Its complex electronic chemistry has been major hurdle in unravelling its toxicity mechanism in plants.

We have already mentioned that Cd and Cu were very often used for investigation of their negative effects on plants. Thus, Hattab et al. (2009) studied physiological effects of Cd and Cu on growth and photosynthesis of pea (*Pisum sativum* L.). It was found, that root and shoot lengths, the concentration of photosynthetic pigments and the rate of photosynthesis were affected by the high metal concentrations. The analysis of metal accumulation showed that leaves significantly accumulate Cd for all tested concentrations. However, Cu was significantly accumulated only with the highest tested dose. These findings can explain the higher inhibitory effects of Cd on growth and photosynthesis in pea plants. These results are valuable for understanding the biological consequences of toxic metals contamination particularly in soils devoted to organic agriculture. Similarly, Shukla et al. (2003) examined the influence of cadmium on the wheat (*Triticum aestivum* L.) plant. The root, shoot-leaf length and the root, shoot-leaf biomass progressively decreased with increasing Cd^{2+} concentration in the nutrient medium. Cd^{2+} uptake and accumulation was found to be maximum during the initial growth period. Cd^{2+} also interfered with the nutrients uptake, especially Ca^{2+}, Mg^{2+}, K^+, Fe^{2+}, Zn^{2+}, and Mn^{2+} from the growth medium. Growth reduction and altered levels of major biochemical constituents such as chlorophyll, protein, free amino acids, starch, and soluble sugars, that play a major role in plant metabolism, were observed in response to varying concentrations of Cd^{2+} in the nutrient medium. In this paper the effects of Cd^{2+} on growth, biomass productivity, mineral nutrients, chlorophyll biosynthesis, protein, free amino acid, starch, and soluble sugars content in wheat plants were estimated to establish an overall picture of the Cd^{2+} toxicity at structural and functional levels.

According to the Gupta & Gupta (1998) nutrient toxicities in crops are more frequent for manganese and boron than for other nutrients. Manganese toxicity was found on acid soils in many parts of the world, boron toxicities occurred in irrigated regions where irrigation waters were exceptionally high in B. Most other nutrient toxicities occurred when large amounts of nutrients in question have been added in waste, e.g. sewage sludge. Crops grown near mines and smelters were prone to nutrient toxicities. In general, the symptoms of toxicity in crops occurred as burning, chlorosis and yellowing of leaves. Toxicities can result in decreased yield and/or impaired crop quality. Use of crop species and genotypes less sensitive to toxicity were recommended where toxicity is suspected.

Tudoreanu & Phillips (2004) in their paper concluded that there is currently only a limited understanding and quantification of key parameters which would allow a comprehensive mechanistic model of Cd uptake by different plant genotypes to be constructed, and also that there is a limited number of empirical observations of key endpoints for an empirical model. Further work on these aspects is essential to facilitate the construction of effective models to control excessive Cd accumulation in the food chain.

Metal toxicity is influenced not only by metal concentration, but also by mineral composition and organic substances in the soil, pH, redox-potential and presence of other metals in the soil. The most important factor which modifies metal toxicity is relationship to the mineral substances in the plants grown in contaminated areas. Toxic metals can induce excessive input or output of elements from plants and thus also modify metal content in the plants. It was found that, e.g. Cd acts negatively on uptake of many nutrients through deterioration of plasmatic membrane of the roots (lipid peroxidation, protein degradation), by effect on ATP-ses and other carriers, as well as by respiration decrease of root with following decrease of active uptake and transport of nutrients. Decline of essential elements in the plants can be also consequence of their washing out from the damaged roots and immobilisation of elements in the roots (Siedlecka, 1995). This author found that Cd also induced Fe deficiency despite the fact that Fe was in medium in sufficient amount. As a consequence of negative effect of Cd there were observed leaf –rolling, –yellowing, –darkening or leaf wilting (e.g. Hagemeyer et al., 1986). Cd decreased water potential and transpiration rate as a consequence of decrease of stomatal conductance (Barceló & Poschenrieder, 1990). This metal seriously damaged photosynthetic apparatus, induced inhibition of both photosystems (PS I and PS II), non-cyclic transport of electrons (Siedlecka & Krupa, 1996) as well as Calvin cycle including activity of Rubisco and PEP-carboxylase (e.g. Stiborová et al., 1986). The other authors (e.g. Šeršeň & Kráľová, 2001) confirmed direct interaction Cd^{2+} with photosynthetic apparatus, namely with light phase of photosynthesis. Cadmium ions can interact with tyrosine radicals located on 161st position of D1 or D2 proteins as well as with reaction centre of PS I. These ions can release Mn^{2+} from manganese cluster which is component of water splitting complex. As a mechanism of action it is supposed formation of complexes of Cd^{2+} with amino acids of proteins in photosynthetic centres. Effect of Cd on respiration rate was found at low Cd concentration stimulative and at high Cd concentration inhibitive (influence of Krebs cycle) (Burzynski, 1990). In our recent review paper (Masarovičová et al., 2011) we presented comprehensive outline related to the effect of metals on cytology, anatomy, physiology and production of crops used as a food, fodder as well as technical plants.

2.3 Defense reactions of plants

Plants respond to stress induced by cadmium using some defense reactions such are immobilization, exclusion, accumulation of cadmium ions in vacuole, synthesis of phytochelatins and stress proteins as well as production of ethylene (e.g. di Toppi & Gabbrielli, 1999). Immobilization in both, cell wall and extra cellular polysaccharides (mucilage, callose) is the first barrier against Cd^{2+} transport into the root. Exclusion represents prevention of Cd^{2+} enter into the cytosol through plasmatic membrane. However, in the case of higher Cd^{2+} concentration the plasmatic membrane can be seriously damaged (in detail see Budíková & Mistrík, 1999). One of the frequent mechanism of plant detoxification is synthesis of phytochelatins which form with their–SH group and cadmium ions various types of complexes (Grill et al., 1985). Formation of stress proteins is induced by any stress factors including toxic metals. In root cells contaminated by cadmium was found production of specific m-RNA transcripts which control synthesis of stress proteins (in detail see Tamás et al., 1997). Plants under Cd stress produce more ethylene, whereby higher amount of this gas is formed in roots than in the shoots (Rodecap et al. 1981).

Subsequently, the further reactions are initiated such are lignine formation (Ievinsh et al., 1995), induction of ascorbate peroxidase activity (Mehlhorn, 1990) and change in gene expression (Whitelaw et al., 1997).

2.4 Food – Quality, sufficiency, safety, and human health risks

Since food production lag behind the rate of growth in population, it is necessary to develop not only high-yielding varieties of food crops but also to develop strategies for integrated nutrient management, integrated pest management, and efficient utilization of water and soil resources. The identification of the origin, quality and authenticity of food, including ingredients and food sources is of prime importance for the protection of consumers. Moreover, traceability of substances in food plays important role for both food safety and human health risks. It means the ability to trace the substances in food "through all stages of production, processing and distribution": primary production, storage, transport, sale, importation, manufacture, distribution, supply (Regulation EC/178/2002 – General Food Low).

World Health Organisation (WHO) has stated its commitment on food safety as an essential public health issue and jointly with the Food and Agriculture Organisation (FAO) sponsor the Codex Alimentarius Commission where international standards on guidance are prepared for member and non-member states. In Europe, the European Food Safety Authority (EFSA) was also constituted for need to guarantee a high level of food safety.

Means and opportunities by which to satisfy the health and nutritional needs (using also GM crops) of impoverished nations and communities differ significantly from those who enjoy greater affluence. The planet´s resources and scientific ingenuity are sufficient to satisfy everyone´s need, but not everyone´s greed. Present and predictable world-wide demand for bioscientists and bioengineers exceeds best estimates of supply. Systematically planned, long-term investments by governments and bioindustries to generate adequate qualified people are urgently needed. (Hulse, 2002).

Food chain contamination is one of the important pathways for entry toxic metals as well as excessive and so harmful concentration of essential nutrients into the human body. The consumption of plants produced on contaminated agricultural soils together with inhalation of contaminated particles are two principal factors contributing to human exposure to metals. Wastewater irrigation, solid waste disposal, sludge applications, vehicular exhaust and various industrial activities also contribute to both soil and food crops contamination with toxic metals. Especially, wastewater-irrigated plants were contaminated with Cr, Cu, Ni, Pb and Cd and exceeded the permissible limits for vegetables set by SEPA (State Environmental Protection Administration, China) as well as WHO. Both, adults and children consuming food crops grown in wasterwater-irrigated soil ingest significant amount of the metals studied (in detail see Khan et al., 2008).

Potential health risks of humans and animals from consumption of food and fodder can be due to metal uptake from contaminated soils through plant roots or direct deposition of contaminants from the atmosphere onto plant surfaces. In the absence of a basic understanding of metal behavior in each specific situation, a more precautionary approach to toxic metal additions to soils is warranted (McBride, 2003). Therefore information about toxic metal concentrations in food products and their dietary intake is very important for assessing their risk to human health (Zhuang et al., 2009).

It can be concluded that cultivation of crops for human (food) or livestock (fodder) consumption on soils contaminated by toxic metals can lead to the uptake and accumulation of these metals in the edible parts of plants with potential risk to both, animal and human health. However, different metals can behave entirely differently in the same soil, as can a particular metal in different soils. Detailed knowledge of the soil at the application site, especially pH, buffering capacity, organic matter and clay content, is so essential (McBride, 2003).

3. Medicinal plants

Chemical studies concerning therapeutical properties of medicinal plants are mostly directed to the determination of the structure of organic substances, especially those which might have to do with their medicinal uses. The importance of their mineral constituents is often overlooked in spite of evidence that many of them are essential to the medicinal plants themselves, and in most cases the synthesis of these organic compounds requires the action of enzymes containing metallic ions at their active centres. Furthermore, one has to consider that these same elements are important to animal and human beings and the knowledge of their concentration could help to explain the therapeutical properties of medicinal plants. Elements such are manganese, iron, copper or zinc operate as activators or enzymes important for healing process – synthesis of extracellular substances, cellular division, digestion of necrotic tissue etc. Considering that the elements are complexed with organic substances they can help to concentrate organic substances responsible for the action or the action can be due to the whole complex (in detail see Pereira & Felcman, 1998). Moreover, it could be stressed that toxic substances including toxic metals (or essential bioelements in higher than physiological concentration) are also very serious factor which negatively influences not only growth, production, structure and function of these important group of plants but also diminish their phytotherapeutical action.

3.1 Response of medicinal plants to metal presence

As has already been mentioned, medicinal plants have also responded on the changing environmental conditions. In our laboratory we focused on three medicinal species: *Hypericum perforatum* L., *Matricaria recutita* L. and *Salvia officinalis* L. which are in general the most frequent medicinal plants used in phytotherapy. Major constituents of *Hypericum perforatum* L. extracts include several classes of compounds exemplified by flavonols, flavonol glycosides, biflavones, naphthodianthrones, phloroglucinols, tannins, coumarins, essential oils, xanthophylls and others (Nahrstedt & Butterweck, 1997). These compounds are very important for this medicinal plant to preserve it against toxic metal stress. Thus, we studied tolerance of *H. perforatum* to toxic effect of copper and cadmium with respect to metal accumulation in individual plant organs (Kráľová & Masarovičová, 2004). It was confirmed that as the most sensitive parameter to Cu treatment was found to be the root dry mass. The length of shoot as well as shoot dry mass was not significantly affected. Lower values of root dry mass could be explained with significant reduction of lateral roots and root hairs by Cu treatment. The roots of *H. perforatum* accumulated markedly higher concentrations of Cu than the shoots. The metal accumulation in both plant organs showed an increase with increasing metal concentration. Bioaccumulation factors (BAF), i.e. quotients obtained by dividing the concentration of the metal in dry mass of individual plant tissues (root and shoot, respectively) by its concentration in the external exposure medium, were also calculated. Taking into account the actual dry mass of individual plant organs (root and shoot), Cu portion in shoot was in the investigated concentration range approximately 20 % from the total uptaken metal content by the whole plant. With respect to relatively high Cd content in the shoot dry mass (1087 µg g^{-1}) H. perforatum could be classified as Cd hyperaccumulator. In this paper was firstly discussed possible contribution of the formation of metal complexes with secondary metabolites of H. perforatum to the plant metal tolerance. Later we stated (Masarovičová & Kráľová, 2007) that for plants producing specific secondary metabolites (medicinal plants) the further, additive mechanism of tolerance arose. This additive mechanism is connected with chelatation of metal ions by some specific secondary metabolites such are hypericin and pseudohypericin (Fig. 1) produced by H. perforatum. The toxic ionic form of metal is thus changed into non-toxic metal chelate. This assumption was based on the findings of Falk and Schmitzberger (1992) and Falk and Mayr (1997) who stated that the pronounced acidity of the bay-region hydroxyl groups of hypericin make salt formation a definite possibility: hypericin is present in the plant material mainly as its potassium salt. In addition, in structurally similar bay-hydroxylated fringelites salt formation with divalent ions, such as Ca^{2+} yields polymeric systems, which because of their extreme insolubility are highly stabile in fossils.

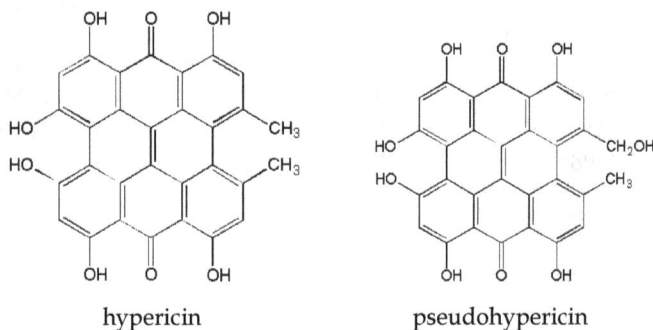

hypericin   pseudohypericin

Fig. 1. Structures of hypericin and pseudohypericin.

The peri-hydroxyl groups situated in the neighbourhood of the carbonyl groups display the best prerequisites to form chelates with transition metal ions. Such coordination complexes could be characterized in the case of fringelite D and Zn^{2+} (Falk & Mayr, 1997).

Afterwards Palivan et al. (2001) investigated the formation of copper complexes with hypericin in solutions using EPR spectroscopy and found that hypericin forms a four-coordinated copper species where the solvent participates to the coordination sphere of the metal. Taking into account the above mentioned results concerning complex formation between hypericin and copper it could be assumed that this secondary metabolite of H. perforatum as well as structurally similar pseudohypericin will form similar complexes with further transition metal ions (Fig. 2). Due to such interaction the concentration of free metal ions will decrease and their toxic effect will diminished. Thus, formation of complexes between heavy metals and the above mentioned secondary metabolites could be regarded as a further additive mechanisms contributing to enhanced tolerance of *H. perforatum* against divalent metals such as copper and cadmium.

The above-described chelatation of cadmium ions with hypericin could contribute to enhanced tolerance of *H. perforatum* plants to cadmium stress and to their cadmium hyperaccumulating ability (in detail see Masarovičová et al., 2011). Classification of this medicinal plant species as a Cd hyperaccumulator firstly was confirmed by Marquard and Schneider (1998) and consequently also in our research (Masarovičová et al., 1999; Kráľová et al., 2000).

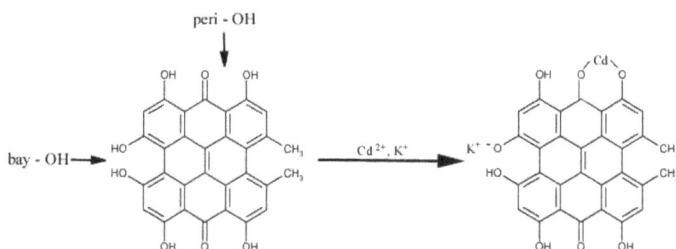

Fig. 2. Complex of hypericin with metal ions Cd^{2+} and K^+.

According to Murch et al. (2003) metal contamination can change the chemical composition of *H .perforatum*, thereby, seriously impacting the quality, safety and efficacy of natural plant substances produced by medicinal plant species. The seedlings of *H. perforatum* lost completely the capacity to produce or accumulate hyperforin and demonstrated a 15-20-fold decrease in the concentration of pseudohypericin and hypericin. Several authors confirmed that *M. recutita* species tolerate Cd concentrations corresponding to the middle-strong contaminated soils (Grejtovský & Prič, 2000; Masarovičová et al., 2003) and high Cd concentration in shoots assigns also this medicinal plant to Cd hyperaccumulators.

In our further paper (Kráľová et al., 2000) the effect of cadmium (12 µM $Cd(NO_3)_2$; pH = 5.5) on growth, plant biomass (root and shoot) and root dark respiration rate of *H. perforatum* (cultivated hydroponically) as well as cadmium accumulation in all plant organs was investigated. On the basis of found results it could be concluded that Cd enhanced the permeability of membranes in both root and shoots. Consequently, above mentioned metal ions were transported into the leaves where their higher content was estimated. Cd administration did not affect the growth and dry biomass of the shoot and root and the root:shoot ratio. However, the root dark respiration rate of the Cd-treated plants was faster than those of control plants.

It has been already mentioned that *M. recutita* L. belongs to the most favoured medicinal plants not only in Slovakia but also over the world (Masarovičová & Kráľová, 2007). This species produces a variety of volatile secondary metabolites, e.g. chamomillol, gossonorol, cubenol, α-cadinol, chamazulene, β-farnesene, (-)-α-bisabolol, (-)-α-bisabololoxide A, (-)-α-bisabololoxide B, 1-azulenethanol acetate and (-)-α-bisabolol acetate, herniarin, etc. Traditionally, in Eastern Slovakia large regions are used for commercial chamomile cultivation. As chamomile species are long-term cultivated in field conditions, it is important to know how many of Cd is taken up from the soil, transported and accumulated in individual parts of plants. Therefore Cd content in the pharmaceutical important plant part - anthodium was also estimated (Šalamon et al., 2007).

Marquard & Schneider (1998) were the first to confirm that chamomile plants had the potential to accumulate high levels of cadmium from the soil. In our paper (Pavlovič et al., 2006) two tetraploid cultivars of *Matricaria recutita* L. (cv. Goral and cv. Lutea) were investigated in response to Cd application. Significant inhibition of root growth was observed in both chamomile cultivars after Cd-treatment. We did not found any differences in Cd accumulation in root between cultivars, but cv. Lutea accumulated slightly higher amount of Cd in the shoot. In root test we observed fragility, browning and twisting of roots. In shoots leaf roll, chlorosis and leaf growth inhibition occurred. During the root test chamomile plants cv. Goral formed the anthodia in all concentrations except control, despite the fact that the plants were only 3 weeks old. According to our observation, the plants started blossoming when they are 8 – 12 weeks old, however Cd treatment resulted in reduced size of flowers. The measurements confirmed higher inhibition of photosynthesis in cv. Lutea, although these plants accumulated less Cd than cv. Goral. Similar decrease of shoot dry weight in both cultivars was also detected. Decrease of net photosynthetic rate could be due to structural and functional disorders in many different levels. Shoot and root respiration rates were not changed significantly in both chamomile cultivars. We confirmed that chamomile belongs to the group of Cd accumulator species. If we take into account high content of Cd in chamomile shoot (over 300 µg g^{-1} at 12 µmol dm^{-3} Cd in solution), only small extent of damages occurred in Cd treated plants. Therefore this medicinal plant species exhibited high tolerance to Cd treatment. This fact was also confirmed by Masarovičová et al. (2003) when the effect of cadmium and zinc separately (10 µmol dm^{-3} for Cd and 50 µmol dm^{-3} for Zn), as well as combined application of these ions on physiological processes (photosynthetic rate and dark respiration rates of leaves and roots, chlorophyll concentration) and production parameters (shoot and root biomass, shoot:root ratio, length of shoots and roots) of young plants of *H. perforatum* and *M. recutita* was investigated. As the applied metal concentrations did not significantly affect studied parameters (except of root respiration rate) we can conclude that both investigated medicinal plants could be used in phytoextraction and subsequent remediation of soils contaminated by cadmium and zinc. Jakovljevic et al. (2000) investigated the influence of the different doses of sodium selenate (0, 100 and 500 g Se per hectare) applied by foliar spraying on the yield and quality of chamomile. The applied doses of Se did not influence the formation of dry chamomile flowers yield and the content of essential oil. However, the applied Se caused the significant increase of the content of bisabolol oxide A and B, followed by the decrease of the chamazulene content in the chamomile essential oil. Significant increase of Se content in the chamomile flowers (12.9 to 53.6 ppm) has also been observed.

Salvia officinalis L. is in general also one of the most important medicinal and aromatic plants with the great spectrum of application in phytotherapy, cosmetic and food industry. The genus *Salvia* includes more than 400 species. *S. officinalis* as a perennial plant originates from the Mediterranean region. In regard to the analysis of sage essential oil the major compounds are thujone, cineole, camphor and caryophyllene. These secondary metabolites are biologically active compounds in *herba salviae* having application in phytotherapy. In food industry this aromatic plant species is recommended as a spice or additive substance (cf. Langer et al., 1996; Perry et al., 1999). From all above-mentioned aspects it is important to have information of toxic metal effects on growth and metal accumulation into the different plant organs of this species. Since Marquard & Schneider (1998) characterised *S. officinalis* as excluder of cadmium we studied effect of large external concentration range of cadmium (30 – 480 µmol dm^{-3} Cd(NO$_3$)$_2$) on production characteristics (length of roots and shoots as well as dry mass of roots and shoots) of this species. We tested two cultivars: cv. Krajova (Slovakian provenance) and cv. Primorska (Yugoslavian provenance). Two months old plants were cultivated hydroponically seven days under controlled conditions in Hoagland solution without and in the presence of Cd(NO$_3$)$_2$ (Masarovičová et al., 2004). There were found differences in phenology and production parameters between two tested cultivars of different provenance. Cv. Krajova was already sensitive to the concentration of 60 µmol dm^{-3} Cd(NO$_3$)$_2$ when the oldest leaves dried. At concentration of 120 µmol dm^{-3} Cd(NO$_3$)$_2$ all older leaves were dried and younger leaves were wilt. At the concentration 240 µmol dm^{-3} Cd(NO$_3$)$_2$ the brown spots were observed on the leaves and at the both applied highest metal concentrations (360 and 480 µmol dm^{-3} Cd(NO$_3$)$_2$) the all leaves were dried and on apical side of the leaves depigmentation was observed. Cultivar Primorska seems to be more tolerant to metal treatment. Visual changes occurred at the concentration of 120 µmol dm^{-3} Cd(NO$_3$)$_2$ when dried only some of older leaves of the plant. At the concentration of 240 µmol dm $^{-3}$ Cd(NO$_3$)$_2$ the leaves were dried but they were green coloured. This fact confirms the disturbance of water regime and indicated strong water stress. At highest tested Cd concentrations (360 and 480 µmol dm^{-3} Cd(NO$_3$)$_2$) all leaves were already dried and the damage of leaf pigmentation as the brown coloured spots was observed. In spite of high concentration of Cd (360 – 480 µmol dm^{-3}) length of the roots in both cultivars was almost not influenced. Also for the shoots of both cultivars only a slight reduction of length was found. On the other hand, dry mass of the shoots decreased at all applied Cd concentrations more expressively than the dry mass of the roots. The negative effect of the high Cd concentrations on the shoot dry mass was manifested mainly in cv. Krajova. Both cultivars uptaken the greatest portion of Cd into the roots, but cv. Primorska accumulated in the shoot app. two-times more Cd than cv. Krajova. However, differences were found in translocation of cadmium from roots into the shoots. Cv. Krajova did not allocate Cd from roots into the shoots already at 240 µmol dm^{-3} Cd(NO$_3$)$_2$ which confirm existence of some barriers in the roots.

4. Classification of metallophytes and plant strategies for metal uptake

A metallophyte is a plant that can tolerate high levels of toxic (heavy) metals. Such plants range between "obligate metallophytes" (which can only survive in the presence of these metals), and "facultative metallophytes" which can tolerate such conditions but are not confined to them. Metallophytes commonly exist as specialised flora found on spoil heaps of mines. Primarily these plants have potential for use for phytoremediation of contaminated ground. In our earlier paper (Masarovičová et al., 2010) classification of metallophytes and actual theory of the strategies in the response of plants to toxic metals were presented, and for hyper/accumulators and excluders the bioaccumulation and translocation factors were discussed. Based on this review the following outline could be described. Strategies of the plants grown on the metal containing soil, classified as accumulators and excluders, were first time published by Baker in 1981 (Baker, 1981) based on the ratio between leaf: root metal concentration. Later this conception was improved suggesting the following two groups of plants: metal excluders and metal non-excluders (indicators, hyperaccumulators). The plant strategy of hyperaccumulators was originally introduced to define plants containing > 0.1 % (1000 μg g^{-1} on dry matter –DM-basis) of Ni in dried plant tissues (Brooks et al., 1977). For other metals such as Zn and Mn the threshold is 10 000 μg g^{-1} (1%) of metal in aerial dry mass. Baker et al. (1994) determined threshold for Cd 100 μg g^{-1} (0.01%). Nowadays the accepted concentration defining hyperaccumulation for Cd is still 0.01 % of this metal in the shoot. Hyperaccumulator plants can be regarded as one subset of a larger category of metal-tolerant plants. However, the exact relationship between metal tolerance and metal hyperaccumulation has not yet been fully resolved (Pollard et al., 2002). Recently, conditions for above-mentioned classification of plant strategies were improved by two further characteristics: bioaccumulation factor (BF or BAF) and translocation factor (TF). Both factors have to be considered for evaluation whether a particular plant is a metal hyperaccumulator (Ma et al., 2001). The term BF, defined as the ratio of metal concentrations in plant dry mass (μg.g^{-1} DM) to those in soils (μg g^{-1} soil), has been used to determine the effectiveness of plants in removing metals from soils (Tu and Ma, 2002). The term TF, defined as the metal concentration in plant shoot to this in the roots, has been used to determine the effectiveness of plants in translocating metal from the root to the shoot (Stoltz and Greger, 2002; Tu and Ma, 2002; Deng et al., 2004). However, fraction of accumulated metal allocated in shoots related to the total amount of metal accumulated by plants respects also actual plant biomass. Consequently, relative high value of this fraction could be also reached when TF value is lower, however actual shoot biomass is high. In the course of hyperaccumulation the following processes are usually observed: (a) higher metal uptake connected with high effectiveness of metal translocation from the root into the shoot, (b) preference of biomass allocation into the root where also high metal concentration occurs (c) development of larger root system in comparison with shoot biomass, which is favourable for total ion uptake by the plant (Cosio, 2004). These features are currently included in BAF. It should be stressed that complex study of plant species is needed to assort it to group of metal hyper/accumulators, metal excluders or plants tolerant to the tested metal. Moreover, all important parameters (e.g. bioaccumulation factor, translocation factor) should be estimated before recommendation if tested plant species is utilizable in phytoremediation technology.

4.1 Hyperaccumulating plants – How and why do they do it?

Recently appeared paper (Rascio & Navari-Izzo, 2011) describing three basic hallmarks to distinquish hyperaccumulators from related non-hyperaccumulating species: a strongly enhanced rate of toxic metal uptake, a faster root-to-shoot translocation and greater ability to detoxify and sequester metals in the leaves. An interesting breakthrough that has emerged from comparative physiological and molecular analyses of hyperaccumulators and related non-hyperaccumulators is that most key steps of hyperaccumulation rely on different regulation and expression of genes found in both kinds of plants. In particular, a determinant role in driving the uptake, translocation into the leaves, sequestration in vacuoles or cell walls of great amounts of toxic metals, is played in hyperaccumulators by constitutive overexpression of genes encoding transmembrane transporters. Among the hypotheses proposed to explain the function of hyperaccumulators, most evidence has supported the "elemental defence" hypothesis, which states that plants hyperaccumulate toxic metals as defence mechanisms against natural enemies, such as herbivores. According to the most recent hypothesis of "joint effects", toxic metals can operate in concert with organic defensive compounds leading to enhanced plant defence overall. Above-mentioned authors stressed, that more elements and a larger number of hyperaccumulators need to be examined to validate the hypothesis of defensive effects of toxic metals. Moreover, the investigations need to move from laboratory to field conditions to provide realistic information about defence mechanism of the plants under natural stand. It should be emphasized the interest in the potential exploiting of hyperaccumulators as a rich genetic resource to develop engineered plants with enhanced nutritional value for improving public health or for contending with widespread mineral deficiencies in human vegetarian diets (Palmgren et al., 2008). This aspect concerning biofortification is discussed in the section 6 of this chapter.

5. Phytoremediation: Environmental-friendly, cost-effective and natural green biotechnology

Rapid expansion and increasing sophistication of various industries in the last century has remarkably increased the amount and complexity of toxic or hazardous substances, such are heavy metals, radionuclides, organic and inorganic wastes, pesticides, etc. which may be bioremediated by suitable organisms (plants and microbes). This technology was termed as bioremediation or phytoremediation (in detail see Singh & Tripathi, 2007). Phytoremediation is environmental-friendly, cost-effective and natural green biotechnology for the removing xenobiotics, including toxic metals, from the environment using some species of the plants. It seems unbelievable, that these pioneering ecological approaches appeared over the last 20 years. However, problem of toxic metals contamination of environment is still continuously worsening due to intensive, various and mostly negative human activities. This unfavourable state leads to intensification of research dealing with phytotoxicity of the metals and with mechanisms used by plants to face against their harmful effects. Great interest has been gained by the behaviour of hyperaccumulators growing on metalliferous soils, which accumulated toxic metals in the leaves at concentration much higher than other plant species. Aims of studying these hyperaccumulators has been to highlight physiological and molecular mechanisms underlying the hyperaccumulation ability, to discover adaptive functions performed by hyperaccumulation in these plants and to explore the possibility of using them as tools to remove metals from contaminated or natural metal-rich soils. However, in spite of important progress made in recent years by the numerous studies accomplished, the complexity of hyperaccumulation is far being understood and several aspects of this remarkable feature still await explanation (c.f. Rascio & Navari-Izzo, 2011).

5.1 Phytoremediation classification and principles

There are several types of phytoremediation technologies currently and quite successfully available for clean-up of both contaminated soils and water. The most important of them could be characterised as follows: reduction of metal concentration in the soil by cultivating plants with a high capacity for metal accumulation in the shoots **(phytoextraction)**, adsorption or precipitation of metals onto roots or absorption by the roots of metal-tolerant aquatic plants **(rhizofiltration)**, immobilization of metals in soils by adsorption onto roots or precipitation in the rhizosphere (**phytostabilization**), absorption of large amounts of water by fast growing plants and thus prevent expansion of contaminants into adjacent uncontaminated areas (**hydraulic control**), decomposition of organic pollutants by rhizosphere microorganisms (**rhizodegradation**), and re-vegetation of barren area by fast grown plants that cover soils and thus prevent the spreading of pollutants into environment (**phytorestauration**) (in detail see Masarovičová et al., 2009).

It was found that the most effective but also the most technically difficult phytoremediation technology is phytoextraction. This technology is based on hyperaccumulation of the metals into the whole plants. This approach involves cultivation of metal-tolerant plants (hyperaccumulators) that concentrate metals in the aboveground organs of the plant. At the end of the growing season, plant biomass is harvested, dried, and the contaminant-enriched mass is deposited in a special dump or added into a smelter. The energy gained from burning of the biomass could support the profitability of this technology, if the resultant fumes can be cleaned appropriately. For phytoextraction to be effective, the dry biomass or the ash derived from aboveground tissues of a phytoremediator crop should contain substantially higher concentrations of the contaminant than the polluted soil (Krämer, 2005).

Metal-tolerant species (including some of energetic crops such are *Hordeum vulgare, Triticum aestivum, Brassica napus, Brassica juncea, Helianthus annuus,*) can accumulate high concentration of some toxic metals in their aboveground biomass. As has already been mentioned, one group of metallophytes are hyperaccumulators (metal extractors). However, besides hyperaccumulators the fast-growing (high-biomass-producing) plants (e.g. *Salix* spp., *Populus* spp.) can also be used in phytoremediation technology. In spite of lower shoot metal-bioaccumulating capacity of these species, the efficient clean-up of contaminated substrates is connected with their high biomass production. Aronsson et al. (2002) have already recognized that it is both environmentally and economically appropriate to use vegetation filters of short rotation willow to purify waters and soils.

Capability of the plants to reduce the amount of heavy metals in contaminated soils depends on plant biomass production and their metal bioaccumulation factor, which is the ratio of metal concentration in the shoot tissue to the soil. The bioaccumulation factor is determined by the ability and capacity of the roots to take up metals and load them into the xylem, by the mass flow in the xylem stream as driven by transpiration, and by the ability to accumulate, store and detoxify metals while maintaining metabolism, growth and biomass production (Guerinot & Salt, 2001). With the exception of hyperaccumulators, most plants have metal bioconcentration factors less than 1, which means that it takes longer than a human lifespan to reduce soil contamination by 50%. To achieve a significant reduction of contaminants within one or two decades, it is therefore necessary to use plants that excel in either of these two factors, e.g. to cultivate crops with a metal bioconcentration factor of 20 and a biomass production of 10 tonnes per hectare (Mg/ha), or with a metal bioconcentration factor of 10 and a biomass production of 20 Mg/ha (Peuke & Rennenberg, 2005).

It was concluded that the most frequent practical application has phytoextraction which has been growing rapidly in popularity worldwide for the last twenty years. In general, this process has been tried more often for extraction of toxic metals than for organic substances. Phytoextraction as an environment friendly approach could be used for cleaning up sites that are contaminated with toxic metals. However, the method has been questioned because it produces a biomass-rich secondary waste containing the extracted metals. Therefore, further treatment of this biomass is necessary - gasification (i.e. pyrolysis) which could help make phytoextraction more cost-effective. Hence, processing of biomass to produce energy and valuable ash in a form which can be used as ore or disposed safely at low cost is advantageous. Recovery of energy by biomass burn or pyrolysis could help to make the phytoextraction a cost-effective technology (Li et al., 2003).

5.2 Phytoattenuation

In general, term „ attenuation" is used in physics (in some context also called extinction) as a gradual loss in intensity of any kind of flux through a medium. Recently (Meers et al., 2010) appeared the first time term „phytoattenuation" for risk-reduction of metals in the produced biomass while allowing maximum economic valorisation of marginal land as main objectives. Before this notice, Meers et al. (2005) observed that of four biomass producing crops – energy plants (*Brassica rapa, Cannabis sativa, Helianthus annuus and Zea mays*) maize exhibited the highest biomass potential on moderately metal contaminated land, with the lowest metal accumulation in the harvestable plant parts. This fits well within the intended scope of phytoattenuation, namely risk-reduction of metals in the produced biomass (by using an excluder species) while allowing maximum economic valorisation of marginal land as main objectives. Otherwise, the primary objective in the current context is allowing an optimal economic use of marginal/contaminated land with risk-reduction of the metals in the produced biomass, whereas the second objective is the gradual "attenuation" of the metals from the soil.

It was recognized, that worldwide there are numerous regions where conventional agriculture is affected by the presence of elevated amounts of plant-available trace elements, causing economic losses and endanger or diminish food and feed quality and safety. Phytoremediation as a soil remediation technology only appears feasible if the produced biomass might be valorised in some manner. It was proposed the use of energy crops (such are maize, sunflower, or sorghum) aiming at risk-reduction and generation of an alternative income for agriculture, yet in the long run also a gradual reduction of the pollution levels. Since the remediation aspect is demoted to a secondary objective with sustainable risk-based land use as first objective, Meers et al. (2010) suggested and introduced the term "phytoattenuation". This concept is in analogy with "natural attenuation" of organic pollutants in soils where also no direct intended remediation measures but a risk-based management approach is implemented. In the current field experiment, cultivation of energy maize resulted in 33,000-46,000 kW h of renewable energy (electrical and thermal) per hectare per year which by substitution of fossil energy would imply a reduction of up to 21 tons ha^{-1} year^{-1} CO_2 if used to substitute a coal fed power plant. Metal removal was very low for Cd and Pb but more significant for Zn with an annual reduction of 0.4-0.7 mg kg^{-1} in the top soil layer. Above mentioned authors stressed that removal efficiency could be further enhanced by introducing winter crops for bio-energy purposes in crop rotation. The use of whole plant for industrial purposes is currently the only realistic scenario to combine phytoremediation with risk-based soil management. Metal concentrations in the green maize shoot were too high for use as fodder, but still were acceptable use as feedstock for anaerobic digestion. However, climatic conditions also play important role. Thus for warmer more southern regions in

Europe other crops such are maize, sunflower or sorghum may potentially serve as alternative energy crops for similar purpose. There is also an urgent need to find and characterize other hyperaccumulators, to cultivate them and better assess agronomic practices and management to enhance plant growth and metal uptake by selective breeding and gene manipulation. Even then, metal uptake might pose environmental risks, unless the biomass produced during the phytoremediation process could be rendered economically by burning it to produce bio-ore and converting it into bioenergy. However, according to the authors Rascio & Navari-Izzo (2011) it is only matter of time before the commercialization of phytoextraction using high-biomass hyperaccumulator plants becomes widespread, considering that not only will it remediate contaminated sites but will generate income from agricultural lands otherwise not utilized.

6. Biofortification a phytofortification

Biofortification is the process of increasing the bioavailable concentrations of essential nutrients in edible portions of food crops through agronomic intervention or genetic selection (White & Broadley, 2005). Phytofortification as a part of biofortification is the fortification of plants with essential nutrients, vitamins and metabolites during their growth and development, there by making these additives more readily available for human/animal consumption (in detail see Kráľová & Masarovičová, 2006). The idea of fortifying food crops with the essential minerals required for a healthy diet is thus relatively new. As many of the metals that can be hyperaccumulated are also essential nutrients, it is easy to see that food fortification and phytoremediation are two sides of the same coin (Guerinot & Salt, 2001).

Plants are at the beginning of food chain, therefore improving the nutrients uptake from soil and enhancing their movement and bioavailability in the edible parts (see below) of crops will provide benefits for both, animal and humans. It can be stated that biofortification provides a truly feasible means of reaching malnourished populations in relatively remote rural areas, delivering naturally-fortified foods to population groups with limited access to commercially-marketed fortified foods. According to Palmgren at al. (2008), there are two main challenges ahead: (i) to develop crops that have an increased content of essential nutrients in the edible parts of plant but that at the same time exclude toxic elements that exhibit similar chemical properties; and (ii) to avoid sequestration of nutrients in the inedible parts of plants, for example in the roots. A breeding approach to produce nutritionally improved food crops relies on genetic diversity in natural populations that can be crossbred to introduce traits/genes from one variety or line into a new genetic background.

Phytofortification is divided into agronomic and genetic phytofortification. The first one uses soil and spray fertilizers enriched by individual essential elements (e.g. Fe, Zn and Se). This approach has been adopted with success in Finland for enrichment of crops by Se. Agronomic biofortification could be used as a cost-effective method to produce high-Se wheat products that contain most Se in the desirable selenomethionine form. Increasing Se content in wheat is a food systems strategy that could increase the Se intake of whole populations. Genetic phytofortification presents the possibility to enrich food crops by selecting or breeding crop varieties, which enhanced Se accumulation characteristics

(Broadley et al., 2006). A strategy of genetic phytofortification offers a sustainable, cost-effective alternative to conventional supplementation and fortification programs. Genc et al. (2005) suggested that a combined strategy utilising (a) plant breeding for higher micronutrient density, (b) maximising the effects of nutritional promoters (e.g. inulin, vitamin C) by promoting favourable dietary combinations, as well as by plant breeding; and (c) agronomic biofortification (e.g. applying selenate to cereal crops by spraying or adding to fertiliser) is likely to be the most effective way to improve the nutrition of populations. Because selenium as an essential micronutrient for humans and animals is deficient in at least a milliard people worldwide, selenium-accumulating plants are a source of genetic material that can be used to alter selenium metabolism and tolerance to help develop food crops that have enhanced levels of anticarcinogenic selenium compounds (Ellis and Salt, 2003). Wheat (*Triticum aestivum* L.) is a major dietary source of Se. Agronomic biofortification (e.g. application of selenate on soil) could be used by food companies as a cost-effective method to produce high-Se wheat products that contain most Se in the desirable selenomethionine form. Increasing the Se content in wheat is a food systems strategy that could increase the Se intake of whole human population (Lyons et al., 2003, 2005). Finland, for example, fortifies with Se, although there is not strong evidence of selenium deficiency-related public health problems (Aro et al., 1998).

Finally, it has to be emphasized the interest in the potential exploiting of hyperaccumulators as a rich genetic resource to develop engineered plants with enhanced nutritional value for improving public health or for contending with widespread mineral deficiencies in human vegetarian diets (Palmgren et al., 2008). However, the strategies of food crop biofortification are still in infancy, even though that importance of biofortification for the humans makes this an exciting line of future research in the field of hyperaccumulation of essential bioelements.

7. Nanoparticles and plants

Nanoparticles (nano-scale particles = NSPs) are atomic or molecular aggregates with dimension between 1 and 100 nm that can drastically modify their physico-chemical properties compared to the bulk material. NSPs can be made from variety of bulk materials and they can act depending on chemical composition, size or shape of the particles. According to the Ruffini-Castiglione & Cremonini (2009) there were identified three types of NSPs: natural (e.g. volcanic or lunar dust, mineral composites), incidental (resulting from anthropogenic activity, e.g. diesel exhaust, coal combustion, welding fumes) and engineered. To the last type of NSPs belong also metal based materials – quantum dots, nanogold, nanozinc, nanoaluminium, TiO_2, ZnO and Al_2O_3 (Li & Xing, 2007). There is now an extensive discussion about the risks of the anthropogenic or engineered NSPs into environment, plants as well as human health. Handy et al. (2008) in their recent paper summarized information concerning the current status, knowledge gaps, challenges and future needs of NSPs ecotoxicology. These authors concluded that NSPs can be toxic to bacteria, algae, invertebrates and fish species, as well as mammals. However, much of the ecotoxicological data is limited to species used in regulatory testing and freshwater organism. Data on bacteria, terrestrial species, marine species and higher plants is particularly lacking. Till now only some studies have focused on the effects and mechanisms of nanomaterials on higher (vascular) plants. The results of these studies have been reported by Ruffini-Castiglione & Cremonini (2009) with the aim to provide further insight into connections between plants and NSPs.

Stampoulis et al. (2009) studied effect of Ag, Cu, ZnO, and Si nanoparticles and their corresponding bulk counterparts on seed germination, root elongation, and biomass production of *Cucurbita pepo* plants which were grown in hydroponic solutions. It was found that seed germination was unaffected by any of the treatments, but Cu nanoparticles reduced emerging root length by 77% and 64% relative to untreated controls and seeds exposed to bulk Cu powder, respectively. Biomass of plants exposed to Ag nanoparticles was reduced by 75%. Although bulk Cu powder reduced biomass by 69%, Cu nanoparticle exposure resulted in 90% reduction of biomass relative to control plants. For Ag and Cu metals it was found, that half of the observed phytotoxicity originated from nanoparticles. Similarly, Li & Xing (2007) investigated effects of some metal nanoparticles (Al, Zn, ZnO) on seed germination and root growth of six plant species (radish, rape, rye-grass, lettuce, corn, and cucumber). Seed germination was not affected except for the inhibition of nanoscale zinc on ryegrass and zinc oxide on corn at 2000 mg/L. Inhibition on root growth varied greatly among nanoparticles and plants. Suspensions of 2000 mg/L nano-Zn or nano-ZnO practically terminated root elongation of the tested plant species. Fifty percent inhibitory concentrations (IC_{50}) of nano-Zn and nano-ZnO were estimated to be near 50 mg/L for radish, and about 20 mg/L for rape and ryegrass. The inhibition occurred during the seed incubation process rather than seed soaking stage.

Saison et al. (2010) studied toxic effect of core–shell copper oxide nanoparticles on the green alga *Chlamydomonas reinhardtii* with regards to the change of algal cellular population structure, primary photochemistry of photosystem II and reactive oxygen species formation. Algal cultures were exposed to 0.004, 0.01 and 0.02 g/L of core–shell copper oxide nanoparticles for 6 h. It was found that core–shell copper oxide nanoparticles induced cellular aggregation processes and had a deteriorative effect on chlorophyll by inducing the photoinhibition of photosystem II. The inhibition of photosynthetic electron transport induced a strong energy dissipation process via non-photochemical pathways. The deterioration of photosynthesis was interpreted as being caused by the formation of reactive oxygen species induced by core–shell copper oxide nanoparticles. However, no formation of reactive oxygen species was observed when *C. reinhardtii* was exposed to the core without the shell or to the shell only.

7.1 Biosynthesis of metal nanoparticles

In modern nanotechnology one of the most exciting area of research is the formation of nanoparticles (biosynthesis of metal nanoparticles) through the biological interventions. It is known, that plants also have inherent capacity to reduce metal through their specific metabolic pathway. Shekhawat & Arya (2009) used seedlings of *Brassica juncea* prepared *in vitro* to produce silver nanoparticles. Two weeks old seedlings were transferred into nutrient solution augmented with silver nitrate. After seven days (in hydroponics cultivated) plants were harvested and analyzed through UV-VIS spectrophotometer and by Transmission Electron Microscopy (TEM) that confirmed the nanoscale silver nanoparticles. Beattie & Haverkamp (2011) investigated sites for the reduction of metal ions Ag^+ and Au^{3+} to Ag^0 and Au^0 metal nanoparticles in *Brassica juncea* plants. Harvested plants were sectioned and studied by transmission electron microscopy, total metal content was analysed by atomic absorption spectroscopy and chemical state of the both metals was determined using X-ray absorption spectroscopy. Nanoparticles of Ag^0 and Au^0 were found in leaves, stem, roots and cell walls of the plants at a concentration of 0.40% Ag and 0.44% Au in the leaves. It is interesting, that the sites of the most abundant reduction of metal salts to nanoparticles were chloroplasts, regions of high reducing sugar (glucose and fructose) content. Above mentioned authors proposed that these sugars are responsible for the reduction of these metals. At present these gold nanoparticles are also named „green-gold". Similarly, Ankamwar (2010) reported synthesis of gold nanoparticles in aqueous medium using *Terminalia catappa* leaf extract as the reducing and stabilizing agent. On treating chloroauric acid with this leaf extract rapid reduction of chloroaurate ions was observed leading to the formation of highly stable gold nanopaticles in solution. Gold nanoparticles ranged in size from 10 to 35 nm with mean size of 21.9 nm. It was stressed by Kumar & Yadav (2009) that biosynthetic processes for nanoparticles would be more useful if nanoparticles were produced extracellularly using plants or their extracts and in a controlled manner according to their size, dispersity and shape. Plant use can also be suitably scaled up for large-scale synthesis of nanoparticles. These authors summarized plant species which can form silver and gold nanoparticles, to which belong well known species such are *Triticum aestivum, Avena sativa, Medicago sativa, Cicer arietinum, Pelargonium graveolens, Aloe vera* or many other species, e.g. *Cymbopogon flexuosus, Cinnamommum camphora, Azadirachta indica, Tamarindus indica, Emblica officinalis,* etc.

While many studies were aimed to metal uptake by plants, particularly with regard to phytoremediation and hyperaccumulation, only few have distinguished between metal deposition and metal salt accumulation. Therefore Haverkamp & Marshall (2009) described the uptake of $AgNO_3$, $Na_3Ag(S_2O_3)_2$, and $Ag(NH_3)_2NO_3$ solutions by hydroponically cultivated plants of *Brassica juncea* and the quantitative measurement of the conversion of these salts to silver metal nanoparticles. It was found that there is a limit on the amount of metal nanoparticles that may be deposited, of about 0.35 wt.% Ag on a dry plant basis, and that higher levels of silver are obtained only by the concentration of metal salts within the plant, not by deposition of metal. The limit on metal nanoparticle accumulation, across a range of metals, is proposed to be controlled by the total reducing capacity of the plant for the reduction of the metal species and limited to reactions occurring at an electrochemical potential greater than zero volts. Metal nanoparticles in plants were observed for gold, silver and copper. However, silver is very often used as a model compound because this metal not only can form metal nanoparticles in plants, but high levels of silver have been `achieved in plants, silver nanoparticles exhibit good catalytic properties, this metal has a high electrochemical reduction potential and also other useful properties.

It should be noticed, that not only vascular plants but also microorganisms such as bacteria, yeasts, algae, fungi and actinomycetes can be used for biosynthesis of nanoparticles (in detail see Sastry et al., 2003). However, Navarro et al. (2008) stressed, that there is a remarkable lack of information on some key aspects, which prevents a better understanding and assessment of the toxicity and ecotoxicity of nanoparticles, especially engineered nanoparticles to living organisms - vascular as well as non-vascular plants. Therefore, collaboration between ecotoxicologists, toxicologists, biologists, chemists, biophysicists, and analytical researchers is needed.

8. Utilization of metallomics for better understanding of metal-induced stress in plants

In the future for elucidation of the physiological roles and functions of biomolecules binding with metallic ions in the biological systems increasingly widespread use of metallomics could be expected. The term "metallome" was first coined by Williams (2001) who referred to it as an element distribution, equilibrium concentrations of free metal ions, or as a free element content in a cellular compartment, cell, or organism and the ensemble of research activities related to metal ions in biological systems has been recently referred to as "metallomics" (Haraguchi, 2004). Later it was proposed that the term "metallome" should be extended to the entirety of metal and metalloid species present in a cell or tissue type (Szpunar, 2004). In the study of metallomics, elucidation of the physiological roles and functions of biomolecules binding with metallic ions in the biological systems should be the most important research target. According to Haraguchi (2004), metallomics may be called, in another words, "metal-assisted function biochemistry". The wider use of molecular biology methods is expected to complement the *in vivo* bioanalytical data with *in vitro* molecular genetic data and lead to an understanding of metal functions at the molecular level (Mounicou and Lobinski, 2008). Lombi et al. (2011) summarized in detail the main techniques used to investigate metal(loids) *in situ* in plants, including histochemical techniques, autoradiography, laser ablation inductively coupled plasma mass spectrometry ((LA)-ICP-MS), secondary ion mass spectrometry (SIMS), scanning electron microscopy (SEM), proton/particle induced X-ray emission (PIXE), synchrotron techniques (utilising high energy X-rays), X-ray fluorescence spectroscopy, tomography — differential absorption, tomography — fluorescence, as well as X-ray absorption spectroscopy (XAS). However, it could be stressed that sample preparation will continue to be a critical step for the majority of *in situ* techniques and in this regard the physiological aspects, in relation to sample preparation, should be at least as important as the methodological needs. Comprehensive review papers related to metallomics were published by Mounicou et al. (2009), Lobinski et al. (2010) and Arruda &Azevedo (2009).

9. Some aspects of bioethics

It should be stressed that negative effects of toxic metals on crops and medicinal plants is manifested not only in undesirable decrease of yield but also in endangered food safety what is serious action and intervention into the whole human population (cf. Nasreddine & Parent-Massin, 2002; Prasad 2004; Reeves & Chaney, 2008). From global aspect it can be stated that crops have dominant use as a food and fodder. Only in advanced and highly-developed countries these plants are also used as technical plants (e.g. for alternative source of energy or environment protection). Therefore it is so stressed ethical aspect if the crops (e.g. maize, cereals, potatoes, rapeseed, and sunflower) could be used exclusively for alimentary purposes or also as an alternative energy source. Moreover, topic of the effect of toxic substances including heavy metals on physiological and production characteristics of the both, crops and medicinal plants is in general extraordinarily important.

10. Conclusion

Numerous regions in the world are known where cultivation of food crops and medicinal plants is irresponsible due to the presence of excessive amounts of plant-available toxic metals, leading to economic losses and negative effects for the human food chain as well as for human health. It was stressed that cultivation and use of crops and medicinal plants have to respect the potential hazard connected with environmental contamination, mainly with toxic metals. Although, some of metals are bioelements (macro- and micronutrients) at normal concentration, they can cause harmful effects on the plants in excess. Therefore, new and environmental-friendly, cost-effective and natural green technologies, such phytoremediation appeared to remove the harmful metals from the environment. This review has illustrated from the known behaviour of toxic metals that broad assumptions about the general behaviour and bioavailability of metal contaminants should be considered from many aspects including the metal nanoparticles – their formation as well as action on physiological processes of the plants. It was described that there were developed strategies of the plants grown on the metal containing soil, classified as metal excluders and metal non-excluders (indicators, hyperaccumulators). Since anthropogenic activity also affects medicinal plants it was mentioned that this group of plants likewise have non-traditional use in phytoremediation and biofortification technologies. Moreover, crops and medicinal plants can be used as supplementary nutrition due to their ability to accumulate some essential nutrition elements (e.g. Se, Zn, and Fe) in the edible parts of these plants. Such fortification of plants with essential nutrients (phytofortification) in an easily assimilated form can help to feed the rapidly increasing world population and improve human health through balanced mineral nutrition. It was stated that crops have dominant use as a food and fodder. However, only in advanced and highly-developed countries these plants are also used as technical plants. Therefore it was stressed ethical aspect if the crops could be utilised exclusively for alimentary purposes or also as an alternative energy source.

11. Acknowledgments

This study was in part financially supported by Association for Production and Use of Biofuels (Civil Society Organization), Bratislava, Slovak Republic.

12. References

Ankamwar, B. (2010). Biosynthesis of gold nanoparticles (green-gold) using leaf extract of *Terminalia catappa*. *E-Journal of Chemistry*, Vol. 7, No. 4, (October – December 2010), pp. 1334-1339, ISSN 0973-4945

Aro, A.; Alfthan, G.; Ekholm, P. & Varo, P. (1998). Effects of selenium supplementation of fertilizers on human nutrition and selenium status. *Environmental Chemistry of Selenium*, Book Series: *Books in Soils Plants and the Environment*, W.T. Frankenberger & R.A. Engberg, (Ed.), Vol. 64, pp. 81-97

Aronsson, P.; Heinsoo, K.; Perttu, K. & Hasselgren, K. (2002). Spatial variation in above-ground growth in unevenly wastewater-irrigated willow *Salix viminalis* plantations. *Ecological Engineering*, Vol. 19, No. 4, (October 2002), pp. 281-287, ISSN 0925-8574

Arruda, M.A.Z. & Azevedo, R.A. (2009). Metallomics and chemical speciation: towards a better understanding of metal-induced stress in plants. *Annals of Applied Biology*, Vol. 155, No. 3, pp. 301-307, ISSN 0003-4746

Baker, A.J.M. (1981). Accumulators and excluders – strategies in the response of plants to heavy metals. *Journal of Plant Nutrition*, Vol. 3, No. 1-4, pp. 643–654, ISSN 0190-4167

Baker, A.J.M.; Reeves, R.D. & Hajar A.S.M. (1994). Heavy-metal accumulation and tolerance in British populations of the metallophyte *Thlaspi caerulescens* J and C pressl (Brassicaceae). *New Phytologist*, Vol. 127, No. 1, (May 1994), pp. 61–68, ISSN 0028-646X

Barceló, J. & Poschenrieder, C. (1990). Plant water relations as affected by heavy metal stress: A review. *Journal of Plant Nutrition*, Vol. 13, No. 1, (January 1990), pp. 1-37, ISSN 0190-4167

Beattie, I.R. & Haverkamp, R.G. (2011). Silver and gold nanoparticles in plants: sites for their reduction to metal. *Metallomics*, Vol. 3, No. 6, (May 2011), pp. 628-632, ISSN 1756-5901

Broadley, M.R.; White, P.J.; Bryson, R.J.; Meacharn, M.C.; Bowen, H.C.; Johnson, S.E.; Hawkesford, M.J.; McGrath, S.P.; Zhao, F.J.; Breward N.; Harriman, M. & Tucker, M. (2006). Biofortification of UK food crops with selenium. *Proceedings of the Nutrition Society*, Vol. 65, No. 2, (May 2006), pp.169-181, ISSN 0029-6651

Brooks, R.R.; Lee, J.; Reeves, R.D. & Jaffré, T. (1977). Detection of nickeliferous rocks by analysis of herbarium specimens of indicator plants. *Journal of Geochemical Exploration*, Vol. 8, No. 3, pp. 49–57. ISSN 0375-6742

Budíková, S. & Mistrík, I. (1999). Cultivar characterization of aluminium tolerance of barley seedlings by root growth, aluminium and callose distribution. *Biologia (Bratislava)*, Vol. 54, No. 4, (August 1990), pp. 447-451. ISSN 0006-3088

Burzynski, M. (1990). Activity of some enzymes involved in NO_3^- assimilation in cucumber seedlings treated with lead or cadmium [NADH-glutamate synthase]. *Acta Physiologiae Plantarum*, Vol. 12, No. 2, pp. 105-110. ISSN 0137-5881

Codex alimentarius. Available from http://www.codexalimentarius.net/

Cosio, C. (2004). Phytoextraction of heavy metal by hyperaccumulating and non hyperaccumulating plants: comparison of cadmium uptake and storage mechanisms in the plants. PhD Thesis, École Polytechnique Fédérale de Lausanne, 119 p, Available from http://biblion.epfl.ch/EPFL/theses/2004/2937/EPFL_TH2937.pdf

Deng, H.; Ye, Z.H.; Wong, M.H. (2004). Accumulation of lead, zinc, copper and cadmium by 12 wetland plant species thriving in metal contaminated sites in China. *Environmental Pollution*, Vol. 132, No. 1, (November 2004), pp.29-40, ISSN 0269-7491

di Toppi, L.S. & Gabbrielli, R. (1999). Response to cadmium in higher plants. *Environmental and Experimental Botany*, Vol. 41, No. 2, (April 1999), pp.105-130, ISSN 0098-8472

Ellis, D.R. & Salt, D.E. (2003). Plants, selenium and human health. *Current Opinion in Plant Biology*, Vol. 6, No. 3, (June 2003), pp. 273-279, ISSN 1369-5266

Falk, H. & Schmitzberger, W. (1992). On the nature of soluble hypericin in *Hypericum* species. *Monatshefte für Chemie*, Vol. 123, No. 8-9, (August-September 1992), pp. 731-739, ISSN 0026-9247

Falk, H. & Mayr, E. (1997). Concerning bay salt and peri chelate formation of hydroxy-phenanthroperylene quinines (fringelites). *Monatshefte für Chemie*, Vol. 128, No. 4, (April 1997), pp. 353-360, ISSN 0026-9247

Food and Agriculture Organization of United Nations.Available from http://www.fao.org/

Genc, Y.; Humphries, M.J.; Lyons, G.H. & Graham, R.D. (2005). Exploiting genotypic variation in plant nutrient accumulation to alleviate micronutrient deficiency in populations. *Journal of Trace Elements in Medicine and Biology*, Vol. 18, No. 4, (June 2005), pp.319-324, ISSN 0946-672X

Grejtovský, A. & Prič, R. (2000). The effect of high cadmium concentration in soil on growth, uptake of nutrient and some heavy metals on *Chamomilla recutita* (L.) Rauschert. *Journal of Applied Botany-Angewandte Botanik*, Vol. 74, No. 5-6 (December 2000), pp. 169-174, ISSN 0949-5460

Grill, E.; Winnacker, E.L. & Zenk, M.H. (1985). Phytochelatins: the principal heavy-metal complexing peptides of higher plants. *Science*, Vol. 230, No. 4726, (November 8, 1985), pp. 674-676, ISSN 0036-8075

Guerinot, M.L. & Salt, D.E. (2001). Fortified foods and phytoremediation. Two sides of the same coin. *Plant Physiology*, Vol. 125, No. 1, (January 2001), pp. 164-167, ISSN 0032-0889

Gupta, U.C. & Gupta, S.C. (1998). Trace element toxicity relationships to crop production and livestock and human health: Implications for management. *Communications in Soil Science and Plant Analysis*, Vol. 29, No. 11-14, pp. 1491-1522, ISSN 0010-3624

Hagemeyer, J.; Kahle, H.; Breckle, S.W. & Waisel, Y. (1986). Cadmium in *Fagus sylvatica* L. trees and seedlings: leaching, uptake and interconnection with transpiration. *Water, Air and Soil Pollution*, Vol. 29, No. 4, (August 1986), pp. 347-359, ISSN 0049-6979

Handy, R.D., Owen, R. & Valsami-Jones, E. (2008). The ecotoxicology of nanoparticles and nanomaterials: current status, knowledge gaps, challenges and future needs. *Ecotoxicology*, Vol. 17, No. 5, (July 2008), pp. 315-325, ISSN 0963-9292

Haraguchi, H. (2004). Metallomics as integrated biometal science. *Journal of Analytical Atomic Spectrometry*, Vol. 19, No. 1, (January 2004), pp. 5-14, ISSN 0267-9477

Hattab, S.; Dridi, B.; Chouba, L.; Kheder, M.B. & Bousetta, H. (2009). Photosynthesis and growth response of pea *Pisum sativum* L. under heavy metals stress. *Journal of Environmental Sciences*, Vol. 21, No. 11, pp. 1552-1556, ISSN 1001-0742

Heverkamp, R.G. & Marshall, A.T. (2009). The mechanism of metal nanoparticle formation in plants: limits on accumulation. *Journal of Nanoparticle Research*, Vol 11, No. 6, (August 2009), pp. 1453-1463, ISSN 1388-0764

Hulse, J.H. (2002). Ethical issues in biotechnologies and international trade. *Journal of Chemical Technology and Biotechnology*, Vol. 77, No .5, (May 2002), pp. 607-615, ISSN 0268-2575

Ievinsh, G.; Valcina, A. & Ozola, D. (1995). Induction of ascorbate peroxidase activity in stressed pine *(Pinus sylvestris* L.) needles: a putative role for ethylene. *Plant Science*, Vol. 112, No. 2, (December 1995), pp. 167-173, ISSN 0168-9452

Jakovljevic, M.; Antic-Mladenovic, S.; Ristic, M.; Maksimovic, S. & Blagojevic, S. (2000). Influence of selenium on the yield and quality of chamomile (*Chamomilla recutita* (L.) Rausch.). *Rostlinná Výroba-Plant Production*, Vol. 46, No. 3, pp. 123-126, ISSN 0370-663X

Khan, S.; Cao, Q.; Zheng, Y.M.; Huang, Y.Z. & Zhu, Y.G. (2008). Health risks of heavy metals in contaminated soils and food crops irrigated with wastewater in Beijing, China. *Environmental Pollution*, Vol. 152, No. 3, (April 2008), pp. 686-692, ISSN 0269-7491

Kráľová, K.; Masarovičová, E. & Bumbálová, A. (2000). Toxic effect of cadmium on *Hypericum perforatum* plants and green alga *Chlorella vulgaris*. *Chemia i Inżynieria Ekologiczna*, Vol.7, No. 11, pp. 1200-1205, ISSN 1231-7098

Kráľová, K. & Masarovičová, E. (2004). Could complexes of heavy metals with secondary metabolites induce enhanced metal tolerance of *Hypericum perforatum*? In: *Macro and Trace Elements. Mengen- und Spurenelemente.* ed. M. Anke et al., 411-416. 22. Workshop, Jena, 24.-25.09, Friedrich Schiller Universität

Kráľová, K. & Masarovičová, E. (2006). Plants for the future. *Ecological Chemistry and Engineering*, Vol. 13, No. 11, pp. 1179-1207, ISSN 1231-7098

Krämer, U. (2005). Phytoremediation: novel approaches to cleaning up polluted soils. *Current Opinion in Biotechnology*, Vol. 16, No. 2, (April 2005), pp. 133-141, ISSN 0958-1669

Kranner, I. & Colville, L. (2011). Metals and seeds: Biochemical and molecular implications and their significance for seed germination. *Environmental and Experimental Botany*, Vol. 72, No. 1, (August 2011), pp. 93-105, ISSN 0098-8472

Kumar, V. & Yadav, S.K. (2009). Plant-mediated synthesis of silver and gold nanoparticles and their application. *Journal of Chemical Technology and Biotechnology*, Vol. 84, No. 2, (February 2009), pp. 151-157, ISSN 0268-2575

Langer, R.; Mechtler, C. & Jurenitsch, J. (1996). Composition of essential oils of commercial samples of *Salvia officinalis* L. and *S. fruticosa* Mill.: a comparison of oils obtained by extraction and steam distillation. *Phytochemical Analysis*, Vol. 7, No. 6, (November-December 1996), pp. 289-293, ISSN 0958-0344

Li, Y.M.; Chaney, R.; Brewer, E.; Roseberg, R.; Angle, J.S.; Bake, A.; Reeves, R. & Nelkin, J. (2003). Development of a technology for commercial phytoextraction of nickel: economic and technical considerations. *Plant and Soil*, Vol. 249, No. 1, (February 2003), pp. 107-115, ISSN 0032-079X

Lin, D. & Xing, B. (2007). Phytotoxicity of nanoprticles: inhibition of seed germination and root growth. *Environmental Pollution*, Vol.150, No. 2, (November 2007), pp. 243-250, ISSN 0269-7491

Lobinski, R.; Becker, J.S.; Haraguchi, H. & Sarkar, B. (2010). Metallomics: Guidelines for terminology and critical evaluation of analytical chemistry approaches (IUPAC Technical Report). *Pure and Applied Chemistry*, Vol. 82, No. 2, (February 2010), pp. 493–504, ISNN 0033-4545

Lombi, E.; Scheckel, K.G. & Kempson, I.M. (2011). *In situ* analysis of metal(loid)s in plants: plants: State of the art and artefacts. *Environmental and Experimental Botany*, Vol. 72, No. 1, (August 2011), pp. 3–17, ISSN 0098-8472

Lynch, J. P. (2007). Roots of the second green revolution. *Australian Journal of Botany*, Vol. 55, No. 5, pp. 493-512, ISSN 0067-1924

Lyons, G.; Stangoulis, J. & Graham, R. (2003). High-selenium wheat: biofortification for better health. *Nutrition Research Reviews*, Vol. 16, No. 1, (June 2003), pp. 45-60, ISSN 0954-4224

Ma, L.Q.; Komar, K.M.; Tu, C.; Zhang, W.H.; Cai, Y. & Kennelley, E.D. (2001). A fern that hyperaccumulates arsenic - A hardy, versatile, fast-growing plant helps to remove arsenic from contaminated soils. *Nature*, Vol. 409, No. 6820, (February 2001), p. 579, ISSN 0028-0836

Marquard, R. & Schneider, M. (1998). Zur Cadmiumproblematik im Arzneipflanzenbau. In: *Fachtagung Arznei- und Gewürzpflanzen.* R. Marquard & E. Schubert, (Ed.), 9-15, 1.-2. 10.1998, Giessen, Germany.

Masarovičová, E.; Kráľová, K.; Šeršeň, F.; Bumbálová, A. & Lux, A. (1999). Effect of toxic effects on medicinal plants. In: *Mengen- und Spurelemente*, M. Anke et al., (Ed.), 189-196, 19. Arbeitstagung, Jena, 3 - 4. 2. 1999, Leipzig, Verlag Harald Schubert

Masarovičová, E.; Kráľová, K. & Streško, V. (2003). Effect of metal ions on some medicinal plants. *Chemia i Inżynieria Ekologiczna*, Vol. 10, No. 3-4, pp. 275-279, ISSN 1231-7098

Masarovičová, E.; Kráľová, K. & Streško, V. (2004). Comparative study of uptake, accumulation and some effects of cadmium in two cultivars of *Salvia officinalis* L. *Chemia i Inżynieria Ekologiczna*, Vol. 11, No. 2-3, pp. 209-214, ISSN 1231-7098

Masarovičová, E. & Kráľová, K. (2007). Medicinal plants – past, nowadays, future. *Acta Horticulturae*, Vol. 749, pp. 19- 27, ISSN 0567-7572, ISBN 978 90 6605 530 8

Masarovičová, E.; Kráľová, K. & Peško, M. (2009). Energetic plants – Cost and benefit. *Ecological Chemistry and Engineering* S, Vol.16, No.3, pp. 263-276, ISSN 1231-7098

Masarovičová, E.; Kráľová, K. & Kummerová, M. (2010). Principles of classification of medicinal plants as hyperaccumulators or excluders. *Acta Physiologiae Plantarum*, Vol. 32, No. 5, (September 2010), pp. 823-829, ISSN 0137-5881

Masarovičová, E.; Kráľová, K. & Šeršeň, F. (2011). Plant Responses to Toxic Metal Stress. In: *Handbook of Plant and Crop Stress*, 3rd Edition, M. Pessarakli, (Ed.), 595-634, Boca Raton, CRC Press, ISBN 978-1-4398-1396-6

McBride, M.B. (2003). Toxic metals in sewage sludge-amended soils: has promotion of beneficial use discounted the risks? *Advances in Environmental Research*, Vol. 8, No. 1, (October 2003), pp. 5-9, ISSN 1093-0191

Meers, E.; Ruttens, A.; Hopgood, M.; Lesage, E. & Tack, F.M.G. (2005). Potential of *Brassica rapa, Cannabis sativa, Helianthus annuus* and *Zea mays* for phytoextraction of heavy metals from calcareous dredged sediment derived soils. *Chemosphere*, Vol. 61, No, 4, (October 2005), pp. 561-572, ISSN 0045-6535

Meers, E.; Van Slycken, S.; Adriaensen, K.; Ruttens, A.; Vangronsveld, J.; Du Laing, G.; Witters, N.; Thewys, T. & Tack, F.M.G. (2010). The use of bio-energy crops (*Zea mays*) for „phytoattenuation" of heavy metals on moderately contaminated soils: A field experiment. *Chemosphare*, Vol. 78, No. 1, (January 2010), pp. 35-41, ISSN 0045-6535

Mehlhorn, H. (1990). Ethylene-promoted ascorbate peroxidase activity protects plants against hydrogen peroxide, ozone and paraquat. *Plant Cell and Environment*, Vol. 13, No. 9, (December 1990), pp. 971-976, ISSN 0140-7791

Mounicou, S. & Lobinski, R. (2008). Challenges to metallomics and analytical chemistry solutions. *Pure and Applied Chemistry*, Vol. 80, No. 12, (December 2008), pp, 2565-2575, ISSN 0033-4545

Mounicou, S.; Szpunar, J. & Lobinski, R. (2009). Metallomics: the concept and methodology. *Chemical Society Reviews*, Vol. 38, No. 4, pp. 1119-1138, ISSN 0306-0012

Murch, S.J.; Haq, K.; Rupasinghe, H.P.V. & Saxena, P.K. (2003). Nickel contamination affects growth and secondary metabolite composition of St. John's wort (*Hypericum perforatum* L.). *Environmental and Experimental Botany*, Vol. 49, No. 3, (June 2003), pp. 251-257, ISSN 0098-8472

Nahrstedt, A. & Butterweck, V. (1997). Biologically active and other chemical constituents of the herb of *Hypericum perforatum* L. *Pharmacopsychiatry*, Vol. 30, Suppl. 2, (September 1997), pp. 129-134, ISSN 0176-3679

Nasreddine, L. & Parent-Massin, D. (2002). Food contamination by metals and pesticides in the European Union. Should we worry? *Toxicology Letters*, Vol. 127, No. 1-3, February 2002), pp. 29-41. ISSN 0378-4274

Navarro, E.; Baun, A.; Behra, R.; Hartmann, N.B.; Fisler, J.; Miao, A.J.; Quigg, A.; Santschi, P.H. & Sigg, L. (2008). Environmental behaviour and ecotoxicity of engineered nanoparticles to algae, plants and fungi. *Ecotoxicology*, Vol. 17, No. 5, (July 2008), pp. 372-386, ISSN 0963-9292

Palivan, C.G.; Gescheidt, G. & Weiner, L. (2001). The formation of copper complexes with hypericin, in solutions: An EPR study. *Journal of Inorganic Biochemistry*, Vol. 86, No. 1, (August 2001), pp. 369-369, ISSN 0162-0134

Palmgren, M.G.; Clemens, S.; Williams, L.E.; Krämer, U.; Borg, S.; Schjorring, J.K. & Sanders, D. (2008). Zinc biofortification of cereals: problems and solutions. *Trends in Plant Science*, Vol. 13, No. 9, (September 2008), pp. 464-473, ISSN 1360-1385

Pavlovič, A.; Masarovičová, E.; Kráľová, K. & Kubová, J. (2006). Response of chamomile plants (*Matricaria recutita* L.) to cadmium treatment. *Bulletin of Environmental Contamination and Toxicology*, Vol. 77, No. 5, (November 2006), pp. 763-771, ISSN 0007-4861

Pereira, C.E.D. & Felcman J. (1998). Correlation between five minerals and the healing effect of Brazilian medicinal plants. *Biological Trace Element Research*, Vol. 65, No. 3, (December 1998), pp. 251-259, ISSN 0163-4984

Perry, N.B.; Anderson, R.E.; Brennan, N.J.; Douglas, M.H.; Heaney, A.J.; McGimpsey, J.A. & Smallfield, B.M. (1999). Essential oils from dalmatian sage (*Salvia officinalis* L.): variations among individuals, plant parts, seasons, and sites. *Journal of Agricultural and Food Chemisry*, Vol. 47, No. 5, (May 1999), pp. 2048-2054, ISSN 0021-8561

Peuke, A.D. & Rennenberg, H. (2005). Phytoremediation. *EMBO Reports*, Vol. 6, No. 6, (June 2005), pp. 497–501, ISSN 1469-221X
http://www.pubmedcentral.nih.gov/articlerender.fcgi?artid=1369103

Pollard, A.; Powell, K.; Harper, F. & Smith, J. (2002). The genetic basis of metal hyperaccumulation in plants. *Critical Reviews of Plant Sciences*, Vol. 21, No. 6, pp. 539– 566, ISSN 0735-2689

Prasad, M.N.V. (2004). *Heavy metal stress in plants: from biomolecules to ecosystems*, 2nd ed., Berlin: Springer, ISBN 3-540-40131-8

Prescott, V.E. & Hogan, S.P. (2006). Genetically modified plants and food hypersensitivity diseases: Usage and implications of experimental models for risk assessment. *Pharmacology & Therapeutics*, Vol. 111, No. 2, (August 2006), pp. 374-383, ISSN 0163-7258

Rascio, N. & Navari-Izzo, F. (2011). Heavy metal hyperaccumulating plants: How and why do they do it ? And what makes them so interesting ? *Plant Science*, Vol. 180, No. 2, (February 2011), pp. 169-181, ISSN 0168-9452

Raven, K.P. & Leoppert, R.H. (1997). Trace element composition of fertilizers and soil amendments. *Journal of Environmental Quality*, Vol. 26, No. 2, (March-April 1997), pp. 551-557, ISSN 0047-2425

Reeves, P.G. & Chaney, R.L. (2008). Bioavailability as an issue in risk assessment and management of food cadmium. A review. *Science of the Total Environment*, Vol. 398, No. 1-3, (July 2008), pp. 13-19, ISSN 0048-9697

Regulation (EC) No 178/2002 of the European Parliament and of the Council of 28 January 2002, Available from http://eurlex.europa.eu/LexUriServ/LexUriServ.do?uri=CELEX:32002R0178:EN: NOT

Rodecap, K.D.; Tingey, D.T. & Tibbs, J.H. (1981). Cadmium-induced ethylene production in bean plant. *Zeitschrift für Pflanzenphysiologie*, Vol. 105, pp. 65–74, ISSN 0044-328X

Ruffini-Castiglione, M. & Cremonini, R. (2009). Nanoparticles and higher plants. *Caryologia*, Vol. 62, No. 2, (April 2009), pp. 161-165, ISSN 0008-7114

Saison, C.; Perreault, F.; Daigle J.C.; Fortin, C.; Claverie, J.; Morin, M. & Popovic, R. (2010). Effect of core-shell copper oxide nanoparticles on cell culture morphology and photosynthesis (photosystem II energy distribution) in the green alga, *Chlamydomonas reinhardtii*. *Aquatic Toxicology*, Vol. 96, No. 2, (January 2010), pp. 109-114, ISSN 0166-445X

Šalamon, I.; Kráľová, K. & Masarovičová, E. (2007). Accumulation of cadmium in chamomile plants cultivated in Eastern Slovakia regions. *Acta Horticulturae. (ISHS)*, Vol. 749, pp. 217-222, ISSN 0567-7572, ISBN 978 90 6605 530 8

Sastry, M.; Ahmad, A.; Khan, M.I. & Kumar, R. (2003). Biosynthesis of metal nanoparticles using fungi and actinomycete. *Current Science*, Vol. 85, No. 2, (July 2003), pp. 162–170, ISSN 0011-3891

Seršeň, F. & Kráľová, K. (2001). New facts about CdCl$_2$ action on the photosynthetic apparatus of spinach chloroplasts and its comparison with HgCl$_2$ action. *Photosynthetica*, Vol. 39, No. 4, pp. 575-580, ISSN 0300-3604

Shanker, A.K.; Cervantes, C.; Loza-Tavera, H. & Avudainayagam, S. (2005). Chromium toxicity in plants. *Environment International*, Vol. 31, No. 5, (July 2005), pp. 739-753, ISSN 0160-4120

Shekhawat, G. S. & Arya, V. (2009). Biological synthesis of Ag nanoparticles through *in vitro* cultures of *Brassica juncea* Czern. *Nanomaterials and Devices: Processing and Applications*. Book Series: Advanced Materials Research, Vol. 67, pp. 295-299, ISSN 1022-6680, ISBN 978-0-87849-328-9

Shukla, U.C.; Singh, J.; Joshi, P.C. & Kakkar, P. (2003). Effect of bioaccumulation of cadmium on biomass productivity, essential trace elements, chlorophyll biosynthesis, and macromolecules of wheat seedlings. *Biological Trace Element Research*, Vol. 92, No. 3, (June 2003), pp. 257-273, ISSN 0163-4984

Siedlecka, A. (1995). Some aspects of interactions between heavy metals and plant mineral nutrients. *Acta Societatis Botanicorum Poloniae*, Vol. 64, No. 3, pp. 265-272, ISSN 0001-6977

Siedlecka, A. & Krupa, A. (1996). Interaction between cadmium and iron and its effects on photosynthetic capacity of primary leaves of *Phaseolus vulgaris*. *Plant Physiology and Biochemistry*, Vol. 34, No. 6, (November-December 1996), pp. 833-842, ISSN 0981-9428

Singh, S.N. & Tripathi, R.D. (2007). Environmental bioremediation technologies. Springer Verlag, Berlin, Heidelberg, Pp. 518, ISBN 978-3-540-34790-3

Stampoulis, D.; Sinha, S.K. & White, J.C. (2009). Assay-dependent phytotoxicity of nanoparticles to plants. *Environmental Science and Technology*, Vol. 43, No. 24, (December 2009), pp. 9473-9479, ISSN 0013-936X

Stiborová, M.; Doubravová, M. & Leblová, S. (1986). A comparative study of the effect of heavy metal ions on ribulose-1,5-bisphosphate carboxylase and phosphoenolpyruvate carboxylase. *Biochemie und Physiologie der Pflanzen*, Vol. 181, No. 6, pp. 373-379, ISSN 0015-3796

Stoltz, E. & Greger, M. (2002). Accumulation properties of As, Cd, Cu, Pb and Zn by four wetland plant species growing on submerged mine tailings. *Environmental and Experimental Botany*, Vol. 47, No. 3, (May 2002), pp. 271-280, ISSN 0098-8472

Szpunar, J. (2004). Metallomics: a new frontier in analytical chemistry. *Analytical and Bioanalytical. Chemistry*, Vol. 378, No. 1, (January 2004), pp. 54-56, ISSN 1618-2642

Tamás, L.; Mistrík, I. & Huttová, J. (1997). Accumulation of apoplasmic proteins in barley primary leaves induced by aluminium and pathogen treatment. *Biologia* (Bratislava), Vol. 52, No. 4, (August 1997), p. 585-589, ISSN 0006-3088

Tu, C. & Ma, L.Q. (2002). Effects of arsenic concentrations and forms on arsenic uptake by the hyperaccumulator ladder brake. *Journal of Environmental Quality*, Vol. 31, No. 2, (April 2002), pp. 641–647, ISSN 0047-2425

Tudoreanu, L. & Phillips, C.J.C. (2004). Modeling cadmium uptake and accumulation in plants. *Advances in Agronomy*, Vol. 84, Book Series: Advances in Agronomy, pp. 121-157, ISSN 0065-2113

White, P.J. & Broadley M.R. (2005). Biofortifying crops with essential mineral elements. *Trends in Plant Science*, Vol. 10, No. 12, (December 2005), pp. 586-593, ISSN 1360-1385

Whitelaw, C.A.; Le Huquet, J.A.; Thurman, D.A. & Tomsett, A.B. (1997). The isolation and characterisation of type II metallothionein-like genes from tomato (*Lycopersicon esculentum* L.). *Plant Molecular Biology*, Vol. 33, No. 3, (February 1997), p. 503-511. ISSN 0167-4412

Williams, R.J.P. (2001). Chemical selection of elements by cells. *Coordination Chemistry Reviews*, Vol. 216–217, (June-July 2001), pp. 583-595, ISSN 0010-8545

Zhuang, P., McBridge, M.B., Xia, H., Li, N. & Li, Z. (2009). Health risk from heavy metals via consumption of food crops in the vicinity of Dabaoshan, South China. *Science of the Total Environment*, Vol. 407, No. 5, (February 2009), pp. 1551-1561, ISSN 0048-9697

4

Genetic and Molecular Aspects of Plant Response to Drought in Annual Crop Species

Anna M. De Leonardis, Maria Petrarulo,
Pasquale De Vita and Anna M. Mastrangelo
CRA-Cereal Research Centre, Foggia
Italy

1. Introduction

Stress is defined as any soil and climatic conditions or combination of both that hinders the full realization of genetic potential of a plant, limiting their growth, development and reproduction. These effects in plants of agricultural interest have a major impact on productivity and quality and thus represent, together with biotic stress, the cause of the gap between yield potential and actual production (Ciais et al., 2005). Stressful environmental conditions are extreme air temperature, drought, excessive presence of salts, anoxia and hypoxia, ozone and heavy metals. Among these factors, heavy damages on agricultural production in Mediterranean environments are exerted by drought, salt stress and early spring low temperatures. The changes in climate forecasted for the near future are expected to exacerbate the onset and magnitude of events of stress due to increased drought and erratic rainfall and rise of evapotranspiration rates due to growing temperatures.

Responses to drought are species specific and often genotype specific (De Leonardis et al., 2007). Moreover, the nature of drought response of plants is influenced by the duration and severity of water loss (Pinheiro & Chaves, 2011), the age and stage of development at the point of drought exposure (De Leonardis et al., 2007), as well as the organ and cell type experiencing water deficits (Pastori & Foyer, 2002).

Plants use various mechanisms to cope with drought stress including their morphology, physiology and metabolism at organ and cellular levels (Levitt, 1972). The Figure 1 shows the drought response strategies which include i) escape, ii) avoidance, and iii) tolerance. Escaping strategy, via a short life cycle or developmental plasticity (Araus et al., 2002), allows the plant to complete its life cycle during the period of sufficient water supply before the onset of drought. The drought avoidance mechanism, via enhanced water uptake and reduced water loss (Chaves et al., 2002), involves strategies which help the plant to maintain high water status during periods of stress, either by a more efficient water absorption from roots or by reducing evapotranspiration from aerial parts. Drought tolerance, via osmotic adjustment, enhanced antioxidative capacity and physical desiccation tolerance of the organs, allows to withstand water deficit with low tissue water potential (Ingram & Bartels, 1996). The osmotic compounds synthesized include proteins and aminoacids (like proline, aspartic acid and glutamic acid), methylated quaternary ammonium compounds (e.g. glycine betaine, alanine betaine), hydrophilic proteins (e.g. late embryogenesis abundant (LEA), carbohydrates (like fructan and sucrose) and cyclitols (e.g. D-pinitol, mannitol).

DROGHT RESPONSE STRATEGIES

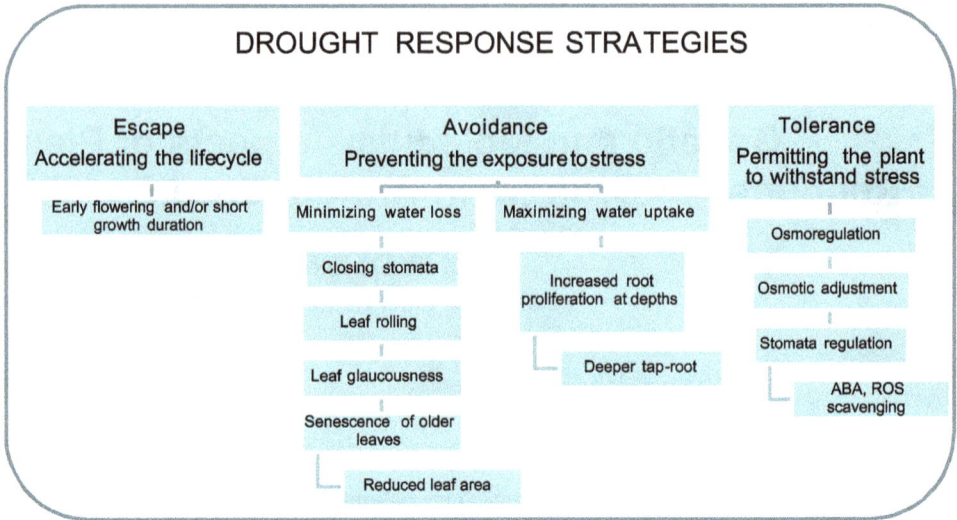

Fig. 1. Plant drought response mechanisms and main related traits.

Recent research has uncovered physiological-, biochemical- and molecular-based mechanisms involved in the drought response in plants (Amudha & Balasubramani, 2011).

More research into how plants respond to drought conditions is needed and will become more important in the future based on climate change predictions of an increase in arid areas (Petit et al., 1999). Understanding plant responses to drought is of great importance in order to select plants more tolerant to stress (Reddy et al., 2004). Advances in the understanding of these processes may lead to genetically modified drought tolerant crop plants.

This chapter focuses on the most recent findings on water stress response in plants. Both morpho-physiological traits and molecular changes contribute to promote stress resistance. In particular, the future perspectives of breeding for drought tolerance are viewed as resulting from the integration of genomic approaches based on the identification of genomic regions involved in the control of stress-related traits and a deep knowledge of the molecular mechanisms acting at cellular level in response to drought stress.

2. Morpho-physiological traits involved in the response to water stress

As the damage exerted by water stress is translated into important loss in amount and quality of crop yield, the improvement of drought tolerance represented and still represents one of the major objectives of plant breeding. At this purpose, a very important task consists of the identification of the main phenotypic features for plant to cope with drought, and therefore the formulation of the drought-tolerant ideotype. Physiological traits relevant for the responses to water deficits and/or modified by water deficits span a wide range of vital processes.

Morphological traits as early plant vigour, wider leaves and a more prostate growth habit can sustain a rapid ground cover thus avoiding loss of water by soil evaporation and suppressing weed competition for water, with a clear advantage on maintaining a favourable plant water status in order to sustain transpiration and yield (Mastrangelo et al., 2011a).

Plant phenology (escape) represents an important aspect for selecting drought tolerant crops, as it allows the alignment of plant life cycle to the features of the target drought environment. In this regard, the genetic improvement of crops has to take into account the modality of drought stress occurrence in the various environments, and in particular the stress timing, frequency and intensity. As an example, earliness is an effective breeding strategy for enhancing yield stability in Mediterranean environments where crops are exposed to terminal drought stress, even if an extreme earliness leads to yield penalty (Cattivelli et al., 1994). Nevertheless, in the case of cereal species in environments in which the drought stress is experienced in early season during the initial vegetative stage, late flowering, followed by a short grain-filling period, can lead to higher yield (van Ginkel et al., 1998). However, early-flowering varieties will escape terminal drought, but they are not necessarily considered drought-resistant.

One basic mechanism for reducing the impact of drought is early stomatal closure at the beginning of a period of water deficit. Stomatal closure reduces water loss, but also reduces the gas exchange between the plant and the ambient air. The reduced CO_2 intake then results in reduced photosynthesis (Chaves et al., 2002). Nevertheless high yield requires high stomatal conductance to sustain a great CO_2 fixation. Some leaf traits, such as stomatal number/density and leaf mesophyll structure can be important in increasing the water use efficiency. In particular, studies carried out in wheat suggest that the high-yielding modern varieties are "opportunistic", that is they have high rates of stomatal conductance with optimal soil moisture, but markedly reduce stomatal conductance when soil moisture is limiting (Siddique et al., 1990; Rizza et al., 2011). Maximal rates of photosynthesis were also positively correlated with increased yields of advanced varieties, while leaf temperatures were negatively correlated (Fischer et al., 1998). Leaf permeability is another crucial trait, as leaves can lose water through cuticle, increasing crop transpiration without an associated benefit in CO_2 fixation. Glaucousness, which is caused by the presence of epicuticular wax, can prevent these losses (Kerstiens, 2006).

At cellular level, osmotic adjustment is an adaptive mechanism in which the accumulation of solutes helps to maintain a favourable gradient of water potential in the soil-plant-air system. It allows to maintain a sufficient water absorption from a relatively dry soil for sustaining photosynthetic and transpiration activity, and cell expansion for root growth (Mastrangelo et al., 2011a; Dichio et al., 2006). Regarding the importance of this trait in improving grain yield in water stressed environment, a positive correlation between osmotic adjustment and yield increases has been found in particular in conditions of severe water stress (Serraj & Sinclair, 2002).

Besides above-ground traits of plant, deep rooted cultivars have demonstrated a clear yield advantage under water stress conditions. An increased root development in presence of water stress represents a complementary strategy to stomatal closure regulation. The influence of root architecture on yield and other agronomic traits, especially under stress conditions, has been widely reported in all major crops (Tuberosa et al., 2002a; de Dorlodot et al., 2007).

A deep and expanded root system should permit to explore a greater soil volume and extract more water. The information available on the genetic control of root traits in the field and their relationships with yield is limited, mainly due to difficulty of measuring root characteristics in a large number of genotypes. Moreover, field studies on roots often require destructive approaches and are complicated by heterogeneity in soil profile, structure, and composition. The acquisition and analysis of root parameters such as total root length are tedious, time-consuming, and often inaccurate (Zoon & Van Tienderen, 1990). Furthermore, environmental effects on root development have been documented by a number of researcher. Many of the root characteristics, such as length, average diameter, surface area, and mass have been used to asses the quantity of roots and the functional fraction of the root system. Total root mass is usually viewed as easier to measure than root length or surface area and has frequently been used to compare root systems. However, total root mass alone cannot describe many root functions adequately involved in plant-soil relationship.

In the case of annual crops capture of water, at sowing for establishment and late in the season for grain filling, may be the most important target for root system traits. The location and the timing of these water sources within the soil profile depend on the soil type and its water holding capacity, the preceding crop and its water use, the soil water content at sowing and the pattern of rainfall during and after the crop growing season. Modelling can estimate when and where valuable water is likely to be present in the soil profile for targeting root traits (Lilley & Kirkegaard, 2007; Sadras & Rodriguez, 2007). Late-season water, for example, may be located mid-profile, or at the bottom of the root zone. This suggests that it would be beneficial to combine root vigour with other root characteristics to favour resource capture.

The other characteristics may include weak root gravitropism to promote a more wide-spreading root system for shallower water uptake, or a strong gravitropism to promote deeper root penetration and deep-water uptake (Ho et al., 2005; Manschadi et al., 2006), faster extension towards moisture (hydrotropism) (Eapen et al., 2005) and more or less nodal and seminal root axes (Hochholdinger et al., 2004). Root growth in soil can be limited by physical, chemical, and biological properties of the soil. Despite the intense work carried out on these topics, there is still insufficient understanding upon the soil factors which limit root growth, and the influence of time period and weather conditions on them. Without this information, it is difficult to manage soil to maximize crop production. In terms of physical limitations to root growth, water stress (too little water for root growth), hypoxia or anoxia (too little oxygen), and mechanical impedance (soil that is too hard for roots to penetrate rapidly) are the major causes of poor root system growth and development. In particular, there is a strong interplay between the strength and water content of soil. As soils dry, capillary forces make matric potential more negative, often causing strength to increase rapidly (Whitmore & Whalley, 2009). A review from Bengough et al. (2011) describes selectively both old and new literature on root elongation in drying soil and the role of water stress, mechanical impedance, and their likely interactions.

3. Breeding for drought tolerance improvement

Drought tolerance has been historically one of the major targets of genetic improvement of crops, and some relevant results have been obtained during the last century despite the low heritability, due to a high genotype x environment (G x E) interaction, of this trait. Consistent genetic gains (from 10 to 50 kg ha^{-1} yr^{-1}) have been registered for cereals and legumes over the last century in all countries, including those characterized by vast drought-prone regions (Calderini & Slafer, 1998; Abeledo et al., 2002). Many studies suggest that cultivars selected for high yield in stress free environments are also adapted to stress prone environments (Cattivelli et al., 2008). In different field experiments modern durum wheat genotypes outperformed the old ones in all test environments including those with moderate drought stress and showed a stronger responsiveness to improved fertility (De Vita et al., 2010). Moderate drought stress is defined physiologically as reduced cell turgor that generally results in reduced stomatal conductance (reduced water loss from the leaf), and lower cellular water potential, which allows the tissue to hold onto the water that is in the leaf more tenaciously (Levitt 1972). This suggests that some of the traits selected to improve potential yield can still sustain yield at least in mild to moderate drought conditions ensuring yield stability (Slafer et al., 2005; Tambussi et al., 2005). A possible explanation is that the main targets of selection (high harvest index in wheat and barley, stay green in maize and sorghum, resistance to pests and diseases, nitrogen use efficiency) are equally useful under dry and wet conditions and, often, the best performances for these traits were overriding the differences in drought adaptability (Mastrangelo et al., 2011a).

In some cases adaptive traits were shown to contribute significantly to performance under drought. Retrospective studies on maize showed that most of the genetic yield improvement for hybrids bred in the second half of the last century could be attributed to traits related to tolerance to stress, like high plant population density, weed interference, low night temperatures during the grain-filling period, low soil moisture, and low soil N (Cattivelli et al., 2008; Tollenaar & Wu, 1999; Tollenaar & Lee, 2002).

Because of the complex nature of drought tolerance, conventional breeding has obtained little success in this regard. Successful cases of genetic improvement for yield in drought-prone environments have been obtained by selecting for secondary traits related to drought tolerance. In maize the silk-tassel interval was identified as a highly indicative secondary trait for drought-resistant breeding (Bolanos & Edmeades, 1996). Spikelet fertility can be visually estimated under field conditions and has been used as an indirect index for drought screening in rice (Garrity & O'Toole, 1994). Another example is based on the use of carbon isotope discrimination (Δ) as a surrogate for water use efficiency to select wheat lines with high water use efficiency in drought-prone environments (Rebetzke et al., 2002). During photosynthesis plants discriminate against the heavy isotope of carbon (^{13}C) and, as a result, in several C3 species, Δ is positively correlated with the ratio of internal leaf CO_2 concentration to ambient CO_2 concentration (Ci/Ca) and negatively associated with transpiration efficiency. Thus, a high Ci/Ca leads to a higher Δ and a lower transpiration efficiency (Farquhar & Richards, 1984).

In the last years a great effort has been devoted to the identification of genomic regions involved in the control of traits related to drought stress tolerance. Once the region has been mapped, closely linked molecular markers are identified, which can be used in breeding programs based on MAS (Marked Assisted Selection). The wide range of physiological and biochemical mechanisms involved in dehydration response explains the complexity of plant response to drought, for which a high number of quantitative trait loci (QTLs) widespread on many chromosomes have been found (Cattivelli et al., 2008).

As an example, Yang et al., (2007) reported several QTLs for accumulation and remobilization of water-soluble carbohydrates in wheat stems. Depending on cultivars and environments, stem water-soluble carbohydrates accumulated before flowering, and during the early periods after flowering, contributed up to 70% or more of the grain weight under terminal drought conditions (Yang et al., 2001). Major genomic regions controlling productivity and related traits (Carbon isotope ratio, osmotic potential, chlorophyll content, flag leaf rolling index) were identified on chromosomes 2B, 4A, 5A and 7B by Peleg et al., (2009) in durum wheat. QTLs for productivity were associated with QTLs for drought-adaptive traits, suggesting the involvement of several strategies in wheat adaptation to drought stress. Sixteen QTLs were identified in durum wheat by Maccaferri et al. (2008), including two major QTLs on chromosome arms 2BL and 3BS that affected grain yield and showed significant effects in multiple environments (rainfed and irrigated).

Five QTLs for anther-silking interval were identified in the maize drought tolerant line Ac7643 and transferred to the susceptible line CML247 by marker-assisted backcross. Hybrid lines were obtained that performed better than controls in well watered and mild drought condition in terms of grain yield (Ribaut & Ragot, 2007).

The identification of markers or genes associated with root growth and architecture would be particularly useful for breeding programmes to improve root traits by molecular marker-assisted selection. Few papers have described work on the identification of QTLs for root traits in wheat.

Root system architecture (RSA), the spatial configuration of a root system in the soil, is used to describe the shape and structure of root systems. Its importance in plant productivity lies in the fact that major soil resources are heterogeneously distributed in the soil, so that the spatial development of roots will determine the ability of plant to secure edaphic resources (Lynch, 1995). The search for QTLs has been a major research avenue in investigating the genetic variation of RSA, a task that is complicated by the strong responses of RSA to environmental conditions. In several instances overlap of QTLs for root features with those for productivity (yield, water use o capture) has suggested the possible role of the former in determining the latter (Tuberosa et al., 2002a; 2002b; Steele et al., 2007). Although there are few examples of QTLs that individually explained up to 30% of phenotypic variation for root traits in rice (Price & Tomos, 1997) and in maize (Giuliani et al., 2005) and for the response of RSA to environmental factors, root morphology is in most cases regulated by a suite of small-effect loci that interact with the environment (de Dorlodot et al., 2007). This is one of the constrains that limit progress from QTL discovery to the release of new varieties.

Some recent papers have reviewed in details the QTLs identified for traits related to drought stress tolerance (Maccaferri et al., 2009; Ashraf, 2010), furthermore, for many crop plants information on drought-related QTL findings have been collected in open source databases, such as GRAMENE (http://www.gramene.org/) or GRAINGENES (http://wheat.pw. usda.gov/GG2/quickquery.shtml#qtls). In particular, Courtois et al. (2009) extracted information from about sixty papers published between 1995 and 2007 and compiled a database containing QTLs for drought tolerance traits and for 29 root parameters. The data describe 2137 root and drought QTLs, out of which 675 for root traits detected in 12 mapping populations.

In rice, several QTLs for root deepness were transferred from the japonica upland cultivar "Azucena", adapted to rainfed conditions, to the lowland indica variety "IR64". MAS selected lines showed a greater root mass in low rainfall trials and higher grain yield (Steele et al., 2007). Following these studies, a highly drought tolerant variety, Birsa Vikas Dhan 111 was released in India, characterized by early maturity, high drought tolerance and high grain yield with good grain quality (Steele, 2009).

Linked molecular markers were identified for resistance to cereal cyst nematode (CCN) root disease and the root tolerance to the toxic element Al and are currently used by commercial breeding companies.

Combining, or pyramiding, a number of root characteristics for a target environment can be achieved by phenotype selection in the short term. In future, molecular markers may be available for these characteristics since a gene regulating hydrotropism has been identified in Arabidopsis (Kobayashi et al., 2007); a gene regulating specific root types, including seminal versus nodal roots, has been identified in maize (Taramino et al., 2007); and a significant QTL associated with large root system size was identified in Arabidopsis growing in agar under high osmotic stress (Fitzgerald et al., 2006).

Over the past few years there have been several mapping studies that have targeted drought tolerance and other abiotic stress tolerance loci associated with performance in low yielding environments. However, despite this substantial research effort the only markers that have found their way into practical plant breeding programmes are those for boron and aluminium tolerance (Gupta et al., 2010).

4. Molecular bases of plant response to water stress

Molecular and biochemical response of plant to water stress is a very complex task depending on multiple factors (Rizhsky et al., 2002; Bartels & Sourer, 2004). Changes in membrane integrity and modulation of lipid synthesis are key factors in the primary sensing of abiotic stress (Kader & Lindberg, 2010). Secondary, osmotic stress-induced signalling involves changes in plasma membrane H^+-ATPase and Ca^{2+}-ATPase activities that trigger concerted changes of Ca^{2+} influx, cytoplasmic pH, and apoplastic production of ROS (Beffagna et al., 2005).

Transcription factors represent the first level of regulation of mRNA metabolism, controlling the synthesis of pre-mRNA. These molecules are then subject to a splicing process that produces mature mRNA. A well studied phenomenon, with a clear role in regulation of gene expression in stress conditions, is alternative splicing, in which different mRNAs can be produced starting from the same pre-mRNA molecule (Mastrangelo et al., 2011b). The amount of mRNAs in the cell can also be controlled by mechanisms affecting their stability. Not only proteins but also small non-coding RNA molecules are involved in the regulation of these processes, and they have been recognized as important regulators of gene expression and genome integrity (Ambrosone et al., 2011). Epigenetic regulation, which comprises histone variants and post-translational modifications, DNA methylation and certain small-interfering RNA (siRNA) pathways, controls chromatin structure which can be modified in response to stress. Finally, availability of mRNAs for translation affects the synthesis of the corresponding proteins. In the last years, a new mechanism of post-transcriptional regulation of gene expression was identified in the sequestration of mRNAs in the cytoplasm to generate Stress Granules (SG). SG, produced as result of stress condition, were represented by a subset of mRNAs aggregated with specific proteins, allowing physical separation of these mRNAs from the translational machinery and resulting in transient translational repression (Anderson & Kedersha, 2009).

4.1 Transcriptional factors influencing the expression of genes in response to environmental signals

Plant transcription factors are involved in the response to environmental stresses as critical regulators of the expression of stress-related genes. More than 1,500 genes coding for transcription factors have been annotated in Arabidopsis, and they are classified into several families based on the structure of their DNA-binding domains (Ratcliffe & Riechmann, 2002 – http://datf.cbi.pku.edu.cn/). In particular, members of the MYB, MYC, ERF, bZIP, and WRKY transcription factor families have been implicated in the regulation of plant stress responses (Hussain et al., 2011).

Studies carried out in the model species Arabidopsis allowed to identify different stress signal transduction pathways leading to the activation of members of the above mentioned transcription factor families. These pathways can be either dependent or not by the plant hormone ABA (Hirayama and Shinozaki, 2010).

Among transcription factors depending on ABA, bZIPs are a large family of transcription factors with 75 members annotated in the Arabidopsis genome. Regarding water stress response, the ABRE-binding factor (ABF)/ABA-responsive-element-binding (AREB) proteins respond at the transcriptional and post-transcriptional level to drought and salt stress (Choi et al., 2000; Uno et al., 2000), increasing drought stress toleracence (Table 1).

MYC and MYB proteins have a role in late stages of stress response and are also activated following accumulation of endogenus ABA. They generally promote water stress tolerance by acting as positive regulators (Table 1), even if a different mechanism was described for the AtMYB60 and AtMYB44 genes, that are involved in stomatal movements, and function as transcriptional repressors (Cominelli et al., 2005; Jung et al., 2008).

More than 100 members of the NAC gene family have been identified in both Arabidopsis and rice (Fang et al., 2008; Ooka et al., 2003). Members of this family are involved in drought and salinity stress response, as well as in diverse processes as developmental programs, and biotic stress responses (Olsen et al., 2005). RD26, a dehydration-induced NAC protein induced by drought, high salinity, ABA, and JA treatments, represents a key factor in mediating cross-talk between ABA signalling and JA signalling during drought and wounding stress responses (Fujita et al., 2004).

An example of ABA-independent transcription factors acting in drought response are zinc finger homeodomain (ZFHD) proteins. Arabidopsis ZFHD1 binds the ZFHDR motif in the promoter of ERD1 gene, which is also regulated by NAC proteins (Hirayama & Shinozaki, 2010).

Ethylene responsive factors (ERFs) represent a class of genes which function in both ABA-dependent and independent pathways. They are a transcription factor superfamily that is unique to plants, with 124 members in Arabidopsis (Riechmann et al., 2000). ERF proteins share a conserved 58–59 amino-acid domain (the ERF domain) that binds to two similar cis-elements: the GCC box, which is found in several PR (Pathogenesis-Related) gene promoters where it confers ethylene responsiveness, and the C-repeat (CRT)/dehydration-responsive element (DRE) motif, which is involved in the expression of dehydration- and low-temperature-responsive genes.

Gene	Gene family	Species	Gene expression	Phenotype of transgenic or mutant plants	Reference
SodERF3		Sugarcane	overexpressed	Improved ABA, Salt and Woundig tolerance (Tobacco)	Trujillo et al., 2008
WXP1		Medicago	overexpressed	Improved Drought tolerance (Arabidopsis)	Zhang et al., 2007
GmERF3	ERF	Soybean	overexpressed	Improved Drought, salt and desease tolerance (Tobacco)	Zhang et al., 2009
RAP2.6		Arabidopsis	overexpressed	Hypersensitive to ABA, salt, osmotic and cold stress (Arabidopsis)	Zhu et al., 2010
DREB1C		Arabidopsis	overexpressed	Enhanced dessication tolerance (Arabidopsis)	Novillo et al., 2004
AtDREB1A		Wheat	overexpressed	Delayed wilting under drought stress (Wheat)	Pellegrineschi et al., 2004
AtDREB1A		Tobacco	overexpressed	Improved Drought and cold tolerance (Tobacco)	Kasuga et al., 2004
AtDREB1A	DRE binding	Rice	overexpressed	Improved Drought and salt tolerance (Rice)	Oh et al., 2005
AtCBF4	protein 1	Arabidopsis	overexpressed	Improved Drought and freezing tolerance (Arabidopsis)	Haake et al., 2002
OsDREB1		Rice	overexpressed	Improved Drought, Salt and freezing tolerance (Rice)	Ito et al., 2006
HvCBF4		Barley	overexpressed	Increased Drought, Salt and freezing tolerance (Rice)	Oh et al., 2007
AREB1		Arabidopsis	overexpressed	Improved Dehydration survival (Arabidopsis)	Fujita et al., 2005
ABF3/ABF4		Arabidopsis	overexpressed	Improved Drought tolerance (Arabidopsis)	Kang et al., 2002
AREB1		Arabidopsis	knock-out mutant	Reduced Drought tolerance (Arabidopsis)	Yoshida et al., 2010
ABP9		Maize	overexpressed	Improved photosynthetic capacity under drought stress (Arabidopsis)	Zhang et al., 2008
SlAREB	bZIP	Tomato	overexpressed	Improved Drought and Salt tolerance (Arabidopsis and Tomato)	Hsieh et al., 2010
OsABF1-1, OsABF1-2		Rice	mutant	More sensitive to drought and salinity treatments (Rice)	Amir Hossain et al., 2010
OsbZIP23		Rice	overexpressed	Improved Drought and Salt stress tolerance (Rice)	Xiang et al., 2008
WRKY25, WRKY 33		Arabidopsis	overexpressed	Increased sensitivity to ABA and improved salt tolerance (Arabidopsis)	Jiang & Deyholos, 2009
WRKY63	WRKY	Arabidopsis	knock out mutant	Decreased drought tolerance and hypersensitive to ABA (Arabidopsis)	Ren et al., 2010
OsWRKY45		Rice	overexpressed	Improved drought tolerance and enhanced desease resistance (Arabidopsis)	Qiu et al., 2009
AtMYB60	TF involved in stomatal movements	Arabidopsis	null mutation	Decreased wilting under water stress conditions (Arabidopsis)	Cominelli et al., 2005
AtMYB44		Arabidopsis	overexpressed	Improved drought and salt tolerance (Arabidopsis)	Jung et al., 2008
AtMYB15		Arabidopsis	overexpressed	Improved drought tolerance and enhanced sensitivity to ABA (Arabidopsis)	Ding et al., 2009
AtMYB41	R2R3 MYB	Arabidopsis	overexpressed	Negative regulation of transcriptional responses to osmotic stress (Arabidopsis)	Lippold et al., 2009
AtRD26		Arabidopsis	overexpressed	Enhanced sensitivity to ABA (Arabidopsis)	Fujita et al., 2004
ANAC019, ANAC055, ANAC072		Arabidopsis	overexpressed	Improved drought and salt tolerance (Arabidopsis)	Tran et al., 2004
ONAC045	NAC	Rice	overexpressed	Improved drought and salt tolerance (Rice)	Zheng et al., 2009
OsNAC10		Rice	overexpressed	Improved drought tolerance and grain yield (Rice)	Jeong et al., 2010

Table 1. Examples of transcription factors regulating drought tolerance in plants.

Therefore, these proteins can have a role in both biotic and abiotic stress responses, as demonstrated for soybean *GmERF3* and the the Arabidopsis ABA-responsive *RAP2.6* genes (Zhang et al., 2009; Zhu et al., 2010). In Arabidopsis, two distinct gene families of DRE/CRT binding proteins (*CBF/DREB1* and *DREB2*) were described as two distinct targets of cold and drought ABA-independent signalling transduction pathways, respectively (Shinozahi & Yamaguchi-Shinozaki, 2000). Nevertheless *CBF4*, a member of CBF/DREB1 family, was described as an ABA-dependent regulator of drought adaptation in Arabidopsis (Haake et al, 2002). *CBF/DREB1* and *DREB2* represent therefore a point of integration of different signal transduction pathways in response to abiotic stresses. The importance of *CBF/DREB* genes for tolerance to abiotic stresses has been well established in particular in cereals, with evidences at level of phenotypic evaluation of over-expressing plants, and co-segregation of CBF genes with QTLs for frost tolerance (Vàgùjfalvi et al., 2005).

Finally, WRKY proteins contain either one or two WRKY domains, 60-amino-acid regions that contain the sequence WRKYGQK, and a zinc-finger-like-motif. They are involved in the regulation of diverse plant processes including development, response to various biotic and abiotic stresses, and hormone-mediated pathways (Ramamoorthy et al., 2008). *A. thaliana* *WRKY25* and *WRKY33* genes are responsive to osmotic stress but they also are regulated by oxidative stress (Miller et al., 2008). Down-stream regulated target genes of WRKY33 include transcripts with function in ROS detoxification as peroxidases and glutathione-S-transferases (Jiang & Deyholos, 2009).

4.2 Stress related transcripts from alternative splicing events

Alternative splicing is a process which generates two or more different transcripts from the same pre-mRNA molecule by using different splice sites. The rate of plant genes subject to alternative splicing is comprised between 20 and 70%, depending on the species considered (Mastrangelo et al., 2011b). Alternative splicing events do not randomly affect mRNA of all genes, rather they seem to occur preferentially to mRNAs of certain classes of genes commonly involved in signal transduction, or encoding enzymes, receptors and transcription factors (Ner-Gaon & Fluhr, 2006; Lareau et al., 2004). Four main types of alternative splicing have been described: exon skipping, alternative 5' and 3' splice sites and intron retention. The last one is the most common alternative splicing type in plants and fungi (>50% McGuire et al., 2008).

Alternative splicing has been proposed as one of the regulatory mechanisms amplifying the number of proteins that can be produced from a single coding unit. Nevertheless, alternative transcripts containing in frame stop codons, often resulting from retained introns, can be targeted to degradation by nonsense-mediated decay. This mechanism contributes to the fine regulation of the amount of functional protein that will be produced in stressed conditions. Otherwise, truncated polypetides can be produced, which are not necessarily functionless forms of the full length protein. An example is a stress-related transcript of the *MPK13* gene, encoding a protein kinase. This transcript is translated into a truncated protein that has no protein kinase activity, but enhances the MKK6-dependent activation of the MPK13 full-length protein (Lin et al., 2010).

Many of the above described stress-related transcription factors are regulated by alternative splicing. In Arabidopsis, the *AtMYB59* and *AtMYB48* genes were found to code for alternative proteins differing for their MYB repeats and probably for their binding affinities to gene promoters (Li et al., 2006; Fig. 2). A stress-dependent alternative splicing mechanism was described for the *OsDREB2B* gene and its homologs in different species (Mastrangelo et al., 2011b). A transcript containing a shorter ORF (*OsDREB2B1*) accumulated in non stress conditions, and was rapidly converted in the full length transcript (*OsDREB2B2*) by removal of an exon carrying an in frame stop codon in response to stress exposure. This mechanism is probably aimed to finely and rapidly regulate the amount of functional protein. Moreover, this mechanism can keep the transcription of *OsDREB2B* constitutively active without affecting plant growth (Matsukura et al., 2010).

The serine/arginine proteins are a class of RNA binding proteins involved in splicing regulation. Twenty genes encoding serine/arginine proteins have been identified in Arabidopsis up to now, and most of their mRNAs undergo alternative splicing following developmental and environmental stimuli producing nearly 100 different transcripts (Palusa et al., 2007). They can promote alternative splicing of their own transcripts as well as of other gene products in response to a number of abiotic stresses (Wang & Brendel, 2006). A similar behavior has been shown for some glicyne-rich RNA-binding proteins as AtGRP7 and AtGRP8, which are able to auto-regulate their own splicing and cross-regulate with each other in a negative feed-back loop (Schoning et al., 2008). Alternative splicing regulation of genes producing transcripts that alter the splicing of other genes in turn might considerably enhance and amplify the signal-transduction cascade in response to stress stimuli.

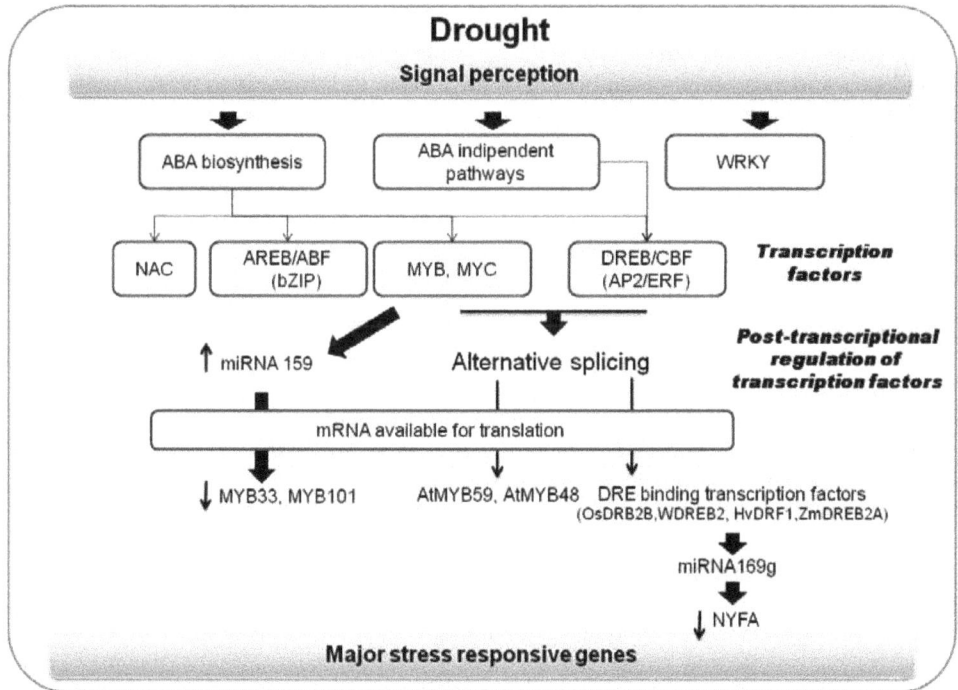

Fig. 2. ABA-dependent and independent pathways of response to drought in plants.

Finally, alternative splicing events also have been described for proteins acting in the regulation of gene expression at post-translational level. E3 ubiquitin ligases represent a very large and complex gene family involved in regulation of protein half life by spliceosome-mediated protein degradation. Alternative splicing events were described for two Arabidopsis

E3 genes (*At4g39140* and *At2g21500*) and for the durum wheat homolog *6G2*, whose mRNA retained the last 3'UTR-located intron following exposure to dehydration and cold stress (Mastrangelo et al., 2005). The same stresses induced the accumulation of an alternative transcript for the Arabidopsis SKP1-like 20 (ASK20) gene (Ogura et al., 2008).

4.3 Regulatory proteins affecting mRNA availability and activity

Transport, initiation of translation and degradation by RNA interference have been shown to regulate mRNA levels of genes in response to water stress. Many of these processes are mediated by RNA-binding proteins (RBP), a variety of heterogeneous proteins involved in diverse aspects of post-transcriptional regulation by direct interaction with single/double strand RNA molecules. The processes in which they are involved comprise mRNA maturation events such as splicing, capping, polyadenylation and export from the nucleus. At level of the cytoplasm, they can regulate mRNA localization, stability, decay and translation (Burd & Dreyfuss, 1994; Dreyfuss et al., 2002). RBPs are characterized by conserved RNA-binding motifs, such as RNA recognition, K homology, glycine-rich, arginine-rich, zinc finger (mainly CCCH type - C-x8-C-x5-C-x3-H), and double-stranded RNA-binding motifs. RRM motifs in particular are involved in RNA recognition and in protein–protein interactions, leading to the formation of heterogeneous ribonucleoprotein (RNP) complexes. More than 200 putative RBP genes have been identified in the Arabidopsis and rice genomes, and many of them seem to be unique to plants, suggesting that they might serve plant specific functions (Lorkovic, 2009; Cook et al., 2010).

Expression and/or activity of a number of RBPs were found to be regulated in response to environmental variables, including water deficit, temperature, light and low-oxygen stresses (Park et al., 2009; Sachetto-Martins et al, 2000; Sahi et al., 2007). Several RBPs resulted also to be ABA-regulated, supporting the regulatory role of ABA in the control of post-transcriptional RNA metabolism (Kuhn & Schroeder, 2003).

The role of RBPs in response to drought stress was also demonstrated by using plant mutants. The supersensitive to ABA and drought 1 (*sad1*) mutant line was isolated in Arabidopsis and exhibited enhanced responses to ABA and drought (Xiong et al., 2001). The ABA hypersensitive 1 (*abh1*) mutant showed ABA hypersensitive stomatal closing and reduced wilting during drought treatment (Hugouvieux et al., 2001). *SAD1* encodes an Sm-like protein possibly involved in RNA transport, splicing or degradation, while *ABH1* encodes a mRNA cap binding protein which can effectively control ABA signalling components at the RNA level (Covarrubiales & Reyes, 2010). These two genes have been identified as negative regulators of ABA-dependent germination and drought tolerance, together with *CBP20* (Cap-Binding Protein 20) and *HYL1* (Hyponastic Leaves 1) which codes for a double stranded RNA-binding factor necessary for the biogenesis of miRNAs and crucial for the precise and efficient cleavage of several primary-miRNAs (Vazquez et al., 2004; Szarzynska et al., 2009; Kuhn & Schroeder, 2003).

Some glycine-rich proteins, containing a dispersed CCHC-type zinc finger at the C-terminus (Karlson et al., 2002), have been identified in plants as cold shock domain protein (CSDP) (Verslues et al., 2006). Arabidopsis AtRZ-1a is a cold shock domain protein and has a negative impact on seed germination and seedling growth of Arabidopsis under salt or dehydration stress conditions (Kim et al., 2007).

Not only the sequence information, but also the secondary and tertiary structures of RNA molecules contribute to their biological activity. RNA helicases are RBPs that catalyze RNA secondary structure rearrangements, and are potentially required in any cellular process involving RNA maturation (Tanner & Linder, 2001; Rocak & Linder, 2004). The majority of RNA helicases belong to the superfamily 2 (SF2) composed of three subfamilies, termed DEAD, DEAH and DExH/D (Tanner & Linder, 2001). Amino acid sequences outside a common core (Asp-Glu-Ala-Asp) are not conserved and are believed to provide helicase specificity for target RNAs or protein–protein interactions. RNA helicases are associated with a diverse range of biotic cellular functions and are involved in cellular response to abiotic stress. Recently, a temperature-regulated RNA helicase, LOS4, has been linked with developmental processes including flowering and vernalization in Arabidopsis (Gong et al., 2002; 2005). These processes also involve ABA, to which the los4 mutants are sensitive (Gong et al., 2005).

Two DEAD-box-related helicases, DNA Helicase 47 (PDH47) and PDH45 are induced by a variety of abiotic stresses in pea (Chinnusamy et al., 2004; Sanan-Mishra et al., 2005; Vashisht et al., 2005). The expression of PDH47 in particular is regulated in a tissue specific manner: the gene is induced by cold and salinity stress in shoots and roots, and by heat and ABA treatment only in roots (Chinnusamy et al., 2004).

Finally, evidences have been reported that helicases can be regulated by the stress-induced alteration of subcellular localization, and by phosphorylation, which provides the opportunity to directly link helicase activity with environmental sensing-signal transduction phosphorylation cascades (Owttrim, 2006).

4.4 Degradation of stress related transcript by siRNAs and miRNAs

MicroRNAs (miRNAs) and siRNAs are small noncoding RNAs that have recently emerged as important regulators of mRNA degradation, translational repression, and chromatin modification.

miRNAs form an abundant class of tiny RNAs characterized by a high level of conservation across species, suggesting a common evolutionary basis. They act in regulating the expression of protein-coding genes in multicellular eukaryotes (Bartel, 2004). Plant miRNAs participate in numerous processes, including development, pattern formation, flowering time, hormone regulation, nutrient limitation, response to stress, and even self-regulation of the miRNA biogenesis pathway (Jones-Rhoades et al., 2006). Regarding their involvement in stress response, abiotic stresses like cold, dehydration, salt stress and nutrient starvation regulate the expression of different plant miRNAs (Lu & Huang, 2008). An example is the Arabidopsis miR393, that is up-regulated by cold, dehydration, high salinity, and abscisic acid (ABA) treatments (Sunkar & Zhu, 2004). In maize 21 miRNA differentially expressed under drought stress were identified (Chen et al., 2010).

In order to understand the mechanisms by which they exert a role in stress protection, it is important to characterize their target mRNAs. At this regard, an interesting feature of miRNAs is the fact that their targets are often regulatory genes (Jones-Rhoades & Bartel, 2004; Rhoades et al., 2002; Zhang et al., 2006). The level of miR159 increased in Arabidopsis seedlings water stressed. In arabidopsis transgenic plants the over-expression of miRNA159 reduced the level of MYB33 and MYB101 transcripts, and a hyposensitive phenotype to ABA was observed (Reyes & Chua, 2007; Fig. 2).

Sunkar and Zhu (2004) reported other ABA induced miRNAs (miR397b and miR402) but also cases of miRNA down-regulated by this hormone (miR389a).

Two members of the miR169 gene family, miR169a and miR169c, are repressed following drought treatments in Arabidopsis. As their target is the nuclear factor Y transcription factor *NFYA5*, the aboundance of this transcript increases and promotes stress response in mature plants (Li et al., 2008). Nevertheless, even if the same conserved miRNA family regulates homologous targets in two different plant species, the effects of this regulation can be different. Members of the miR169 family in rice, miR169g and miR169n/o are induced by salt (Zhao et al., 2009) and drought (Zhao et al., 2007) and differences in levels of induction can be observed in different tissues, being more prominent in roots than in shoots. Interestingly, miR-169g, that acts reducing the expression of NFYA, may be regulated directly by DREB transcriptional factors (Zhao et al., 2009).

Ten percent of Arabidopsis genes are in convergent overlapping gene pairs, also known as natural cis-antisense gene pairs and overlapping transcripts in antisense orientation could form double-stranded RNAs that may be processed into small RNAs (Jen et al., 2005; Wang et al., 2005). These nat-siRNAs (natural antisense transcripts-generated siRNAs) have recently emerged as important players in plant stress responses. A study in Arabidopsis demonstrated the involvement of nat-siRNA in the accumulation of proline during response to stress. As an example, the Arabidopsis *P5CDH* gene, involved in proline catabolism, is down-regulated in response to salt stress following the induction of *SRO5*, a gene of unknown function. The two genes form an antisense overlapping gene pair that generates two siRNAs (Borsani et al., 2005).

4.5 Epigenetic contribution to water stress response in plants

Epigenetic regulation is emerging as an important mechanism in response to stress. Drought induced linker histone variant H1-S was shown to be involved in the negative regulation of stomatal conductance based on the phenotypic analysis of antisense transgenic H1-S tomato plants (Scippa et al., 2004). Several hystone deacetylases (HDACs) are induced by ABA in rice (Fu et al., 2007) and Arabidopsis (Sridha et al., 2006). Transgenic Arabidopsis plants overexpressing AtHD2C exhibited enhanced expression of ABA-responsive genes and greater salt and drought tolerance than the WT plants (Sridha et al., 2006).

Besides acetylation and de-acetylation, other post-translational mechanisms can regulate the abundance and activity of histones. In particular, histone phosphorylation and ubiquitination showed a role in enhancing gene transcription (Sridhar et al., 2007; Zhang et al., 2007), while biotinylation and sumoylation repress gene expression (Nathan et al., 2006; Camporeale et al., 2007). In the desert shrub *Zygophyllum dumosum* methylation level of histone H3 was higher in presence of water than under dry growth conditions indicating post-translational regulation of gene expression activity (Granot et al., 2009).

ABA-mediated pathways also are involved in epigenetic modifications, as suggested by the ABA-dependent regulation of barley Polycomb proteins expression, with a role in histone methylation control (Kapazoglou et al., 2010).

Studies on Arabidopsis over-expressing or knock out lines for the SNF2/ BRAHMA-type chromatin remodeling gene *AtCHR12* indicated a role of this gene in regulation of growth, in particular under drought and heat stresses (Mlynarova et al., 2007). In *Pisum sativum* ABA and drought stress induced the expression of the chromatin remodelling *PsSNF5* gene. PsSNF5 protein interacts with Arabidopsis SWI3-like proteins (SWI3A and SWI3B), which in turn interact with FCA, a protein involved in the regulation of flowering (Sarnowski et al., 2005; Rios et al., 2007). This is a clear example in which stress response and plant development are co-ordinately regulated through chromatin remodeling.

Direct DNA methylation can also be involved in plant stress response. Drought and salt stresses induced a switch in photosynthesis mode from C3 to CAM in the facultative halophyte *Mesembryanthemum crystallinum* L. This metabolic change was associated with stress-induced-specific CpHpG-hypermethylation of satellite DNA (Dyachenko et al., 2006). In natural populations of mangroves DNA was hypomethylated when grown under saline conditions in contrast to populations from non-saline sites (Lira- Medeiros et al., 2010).

Although global analysis in plants such as Arabidopsis and rice suggests that the vast majority of transposons are inactive, methylated, and targeted by siRNAs (Nobuta et al., 2007; Lister et al., 2008), the induction of alternative epigenetic states not only triggers the formation of novel epialleles but also promotes the movement of DNA transposons and retroelements that are very abundant in plant genomes (Reinders et al., 2009; Mirouze et al., 2009). A lot of examples of environmentally induced transposon activities were reported (Slotkin & Martienssen, 2007), as the family of copia retrotrasposon, named Onsen, activated by heat stress in Arabidopsis (Ito et al., 2011). In natural populations, stress may play a role in transposon amplification. An example is the copy number of BARE-1 retrotransposons in barley, which varies in natural populations depending on aridity of growth environment (Vicient et al., 1999). These evidences indicate that plant populations living in stressed environments may carry inherited memories of stress adaptation and transfer this epigenetically to next generations.

5. The molecular response of plants to water stress: A complex frame resulting from integration of multiple regulation layers

The plant response mechanisms to water deficit strictly depend on plant developmental stage, stress intensity and stress duration (Bartels & Souer 2004; De Leonardis et al., 2007). A study on 325 rice transcription factors demonstrated that many of them have a tissue or developmental stage specific expression (Duan et al., 2005). In a genome wide study, Bray (2004) compared three independent array experiments dedicated to the Arabidopsis water stress response. The experiments differed for plants age, substrate of growth and stress applications. Only a small set of genes were commonly induced or repressed. Similar results were obtained in wheat, barley and rice (Ozturk et al., 2002; Lan et al., 2005; Mohammadi et al., 2007). In Arabidopsis the *Nine-Cisepoxycarotenoid Dioxygenase 3* (*NCED3*), *DREB2A* and *RD29B* genes were expressed with different levels and timing following two different kinds of stress imposition (Harb et al., 2011). These differences observed following a rapid or gradual water stress are probably due to the need of plants to optimally react to a stress event as it occurs in field conditions. Therefore, many internal and external stimuli have to be integrated into common signalling pathways.

Moreover, plants usually are exposed in field to different kinds of stress simultaneously, and the effect of the combined stresses in terms of gene expression is not simply the sum of the effects produced by the stresses applied separately (Rizhsky et al., 2002).

The superimposed complexity levels in the response to environmental changes, are therefore aimed to ensure temporally and spatially appropriate patterns of downstream stress-related gene expression.

After the translation, many post-translational mechanisms can target proteins modifying their activity, sub-cellular localization and half-life (Downes & Vierstra, 2005). Phosphorylation is one of the best known mechanism that plays a key role in many biological processes, as phosphorylation/de-phosphorylation cascades commonly translate extracellular stimuli into the activation of specific responses (Boudsocq & Laurière, 2005).

Among the polypeptides, ubiquitin and SUMO conjugations are emerging as major post-translational regulatory processes in all eukaryotes (Stone & Callis, 2007; Miura et al., 2007). The covalent binding of poly-ubiquitin usually targets proteins for proteolysis. Conversely, monoubiquitination regulate the location and activity of proteins, affecting various cellular processes from transcriptional regulation to membrane transport (Hicke, 2001). Similar effects are produced by the covalent conjugation of the SUMO (Small Ubiquitin-like Modifier) peptide (Hay et al., 2005). Both ubiquitination and sumoylation are involved in the promotion of stress tolerance in plants, and they offer a very clear example of multiple layer control of key regulators of the stress response. Along this chapter DREB/CBF proteins have been described as transcription factors with a pivotal role in plant tolerance to cold and drought stress. Their expression has shown to be modulated at transcriptional level, but also by alternative splicing (Matsukura et al., 2010). Furthermore, the HOS1 protein, correspondending to an E3 ubiquitin ligase, mediates the ubiquitination of the master regulator for the response to cold, the transcription factor Inducer of CBF Expression 1, ICE1, and repressor of MYB15 expression. This leads to its proteasome-mediated degradation during exposure to cold (Dong et al., 2006). ICE1 protein, in turn, is stabilized by sumoylation that therefore acts in this pathway with an antagonistic role with respect to ubiquitination (Ishitani et al., 1998). Finally, DREB transcriptional factors can down-regulate the expression of NFYA through activation of miRNAs (Zhao et al., 2009).

The great complexity of the pathway of regulation of gene expression in plant response to water stress makes the analysis of transcriptome in different conditions not suitable alone to draw a clear picture of tolerance mechanisms. Variations at level of proteins and ultimately of metabolites have to be investigated to achieve a more complete evaluation. In this light, recent advances in profiling of plant proteome and metabolome in water stress conditions have provided chances to integrate data from gene expression and protein activities studies. Outcomes indicate an important role of post-transcriptional and post-translational mechanisms in coordinating the plant molecular response to water stress (Mazzucotelli et al., 2008).

6. The contribution of genetic and molecular knowledge to the improvement of drought tolerance in field

A very complex network of gene interactions in response to water stress has been described in the last years, and a high number of QTLs, widespread in the genome, have been identified for tolerance, each of them controlling a low percentage of explained phenotypic variabilty. In some cases the molecular basis of resistance QTLs has been explained. Genes having a role in stress tolerance were shown to co-localize with tolerance QTLs in mapping populations. This is the case of DREB/CBF genes, for which a large gene cluster has been mapped in correspondence of QTLs for frost and drought tolerance in barley and wheat (Vàgùjfalvi et al., 2005; Francia et al., 2007). On the other side, many efforts are in course to isolate the gene(s) behind tolerance QTLs, in order to have access to the transgenic approach, or to design perfect molecular markers to pyramid different QTLs into the same genotype through MAS without the risk of losing association due to recombination. The Arabidopsis *ERECTA* gene was cloned, as the sequence beyond a QTL for transpiration efficiency (Masle et al., 2005).

Even if the molecular basis of QTLs is not known, some examples are available in which the transfer of some tolerance QTLs in MAS programs has contributed to increase grain yield in water stress conditions (see paragraph n. 3).

Anyway, the investigation of molecular mechanisms which concur in regulating the water stress response in plant allows the identification of genes/processes with a key role in determining tolerance. The expression of these genes can be altered in transgenic plants in order to obtain a tolerant phenotype. Besides the genes reported in Table 1, in some cases this approach has been successful in increasing agronomic performance of plants in the field. An example is represented by transgenic wheat constitutively expressing the barley *HVA1* gene, encoding a member of the group 3 late embryogenesis abundant (LEA) proteins. Results of nine field experiments over six cropping seasons, showed that the HVA1 protein confers a significant protection from water stress (Bahieldin et al., 2005). Aquaporins mediate most of the symplastic water transport in plants, which represents a limiting factor for plant growth and vigor in particular under unfavorable growth conditions and abiotic stress. Differential expression of genes that encode different aquaporin isoforms during plant development has been shown to be associated with various physiological processes. Such processes include stomatal closure and opening, organ movement, cell elongation, and cell division (Kaldenhoff et al., 2008). The *SlTIP2* gene coding for an aquaporin was particularly effective in improving water stress resistance of tomato plants (Sade et al., 2009). Another successful gene is *OsNAC10*, introduced in field-tested rice plants under the control of the constitutive promoter GOS2 and the root-specific promoter RCc3 (Jeong et al., 2010).

7. Perspectives

Two different but complementary approaches have been presented in this chapter for the improvement of water stress tolerance. In the first one, the phenotypic and molecular evaluation of suitable genetic materials leads to the identification of genomic regions involved in the control of tolerance. At the same time, closely linked molecular markers are found, which can be used in MAS programs to transfer useful alleles for tolerance. In the second one, the molecular study of the water stress response in plant leads to the identification of genes/processes with a key role in determining tolerance.

In the last years, strong technical advances have been realized, in the frame of the "omic" technologies, which make the study of genomes, transcriptomes, proteomes, metabolomes and phenomes more rapid and precise. Methods for a more fine phenotypic evaluation of a high number of individuals, in both controlled and field conditions, are needed for an accurate genetic analysis on segregating populations or germplasm collections.

The development in particular of new DNA sequencing technologies rapidly is producing huge amounts of sequence information with a number of applications including genome resequencing and polymorphism detection, mutation mapping, DNA methylation and histone modification studies, transcriptome sequencing, gene discovery, alternative splicing identification, small RNA profiling and DNA-protein interactions (Lister et al., 2008; Delseny et al., 2010). Thanks to these advancements, new perspectives are open for the investigation of genetic and molecular basis of water stress tolerance. Sequencing of entire genomes of crop species is expected to provide a huge opportunity to clone QTLs for drought-related traits in the near future. Moreover, sequence analysis on a genome-wide scale allows the fast and low-cost development of extremely high number of molecular markers. The availability in particular of large SNP (Single Nucleotide Polymorphism) panels for crops will accelerate the QTL discovery and transfer in MAS programs already in course for single marker-trait associations. Furthermore, it will be possible to apply a new method called genomic selection (Meuwissen et al., 2001), which predicts breeding values using data deriving from all molecular markers covering the whole genome at the same time. In this way, breeders now have an opportunity to integrate classical phenotype-based selection with selection on the basis of genotype. In particular, they will have the possibility to follow genomic variations associated to many traits of interest at the same time.

8. Acknowledgements

This study was supported by the Italian Ministry of Agriculture (MiPAAF), with the special grants AGRONANOTECH and MAPPA 5A, and by the Ministry of Education, University and Research (MIUR), with the special grant AGROGEN.

9. References

Abeledo, L.G., Calderoni, D.F. & Slafer, G.A. (2002). Genetic improvement of barley yield potential and its physiological determinants in Argentina (1944–1998), *Euphytica* Vol.130: 325-334.

Ambrosone, A., Costa, A., Leone, A. & Grillo, S. (2011). Beyond transcription: RNA-binding proteins as emerging reguators of plant response to environmental constrains, *Plant Science* doi: 10.1016/J.plantsci. 2011.02.004.

Amir Hossain, M., Lee, Y., Cho, J.I., Ahn ,C.H., Lee, S.K., Jeon, J.S., Kang, H., Lee, C.H., An, G. & Park, P.B. (2010), The bZIP transcription factor OsABF1 is an ABA responsive element binding factor that enhances abiotic stress signaling in rice, *Plant Molecular Biology* Vol.72: 557–566.

Amudha, J., & Balasubramani, G. (2011). Recent molecular advances to cambat abiotic stress tolerance in crop plants, *Biotechnology and Molecular Biology Review* Vol.6(No. 2): 31–58.

Anderson, P. & Kedersha, N. (2009). RNA granules: post-transcriptional and epigenetic modulators of gene expression, *Naural. Review Molecular Cell Biology* Vol.10: 430-436.

Araus, J.L., Slafer, G.A., Reynolds, M.P. & Royo, C. (2002). Plant breeding and drought in C-3 cereals: what should we breed for?, *Annals of Botany* Vol.89: 925–940.

Ashraf, M. (2010). Inducing drought tolerance in plants: recent advances, Biotechnol Advances Vol. 28(No.1): 169–183.

Bahieldin, A., Mahfouz, H.T., Eissa, H.F., Saleh, O.M., Ramadan, A.M., Ahmed, I.A., Dyer, W.E., El-Itriby, H.A. & Madkour, M.A. (2005). Field evaluation of transgenic wheat plants stably expressing the HVA1 gene for drought tolerance, *Physiologia Plantarum* Vol. 123: 421-427.

Bartels, D. & Sourer, E. (2004). Molecular responses of higher plants to dehydration, in H. Hirt, K. Shinozaki (ed.), *Plant Responses to Abiotic Stress*, Springer Verlag, Berlin, Heildelberg, pp. 9–38.

Bartel, D.P. (2004). MicroRNAs: Genomics, biogenesis, mechanism, and function, *Cell* Vol.116: 281–297.

Beffagna, N., Buffoli, B. & Busi, C. (2005). Modulation of reactive oxygen species production during osmotic stress in Arabidopsis thaliana cultured cells: involvement of the plasma membrane Ca^{2+}- ATPase and H^+ ATPase, *Plant Cell Physiology* Vol. 46 :1326–1339.

Bengough, A.G., McKenzie, B.M., Hallett, P.D. & Valentine, T.A. (2011). Root elongation, water stress, and mechanical impedance: a review of limiting stresses and beneficial root tip traits, *Journal of Experimental Botany* Vol.62 (No. 1). 59–68.

Bolanos, J. & Edmeades, G.O. (1996). The importance of anthesis-silking interval in breeding for drought tolerance in tropical maize, *Field Crops Research* Vol.48(No. 1): 65-80.

Borsani, O., Zhu, J., Verslues, P.E., Sunkar, R. & Zhu, J.-K. (2005). Endogenous siRNAs derived from a pair of natural cis-antisense transcripts regulate salt tolerance in Arabidopsis, *Cell* Vol.123: 1279–1291.

Boudsocq, M. & Laurière, C. (2005). Osmotic signaling in plants: multiple pathways mediated by emerging kinase families, *Plant Physiology* Vol.138: 1185–1194.

Bray, E.A. (2004). Genes commonly regulated by water-deficit stress in Arabidopsis thaliana, *Journal Experimental Botany* Vol.55: 2331–2341.

Burd, C.G. & Dreyfuss, G. (1994). Conserved structures and diversity of functions of RNA-binding proteins, *Science* Vol.265: 615–621.

Calderini, D.F. & Slafer, G.A. (1998). Changes in yield and yield stability in wheat during the 20th century, *Field Crops Reseach* Vol.57: 335-347.

Camporeale, G., Oommen, A.M., Griffin, J.B., Sarath, G. & Zempleni, J. (2007). K12-biotinylated histone H4 marks heterochromatin in human lymphoblastoma cells, *The Journal of Nutritional Biochemistry* Vol. 18: 760-768.

Cattivelli, L., Delogu, G., Terzi, V. & Stanca, A.M. (1994). Progress in barley breeding, in Slafer GA (ed.), Genetic Improvement of Field Crops, Marcel Dekker, Inc. New York, pp. 95-181.

Cattivelli, L., Rizza, F., Badeck, F.-W., Mazzucotelli, E., Mastrangelo, A.M., Francia, E., Marè, C., Tondelli, A. & Stanca, A.M. (2008). Drought tolerance improvement in crop plants: an integrated view from breeding to genomics, *Field Crops Reseach* Vol.105: 1-14.

Chaves, M.M., Pereira, J.S., Maroco, J., Rodrigues, M.L., Ricardo, C.P.P., Osòrio, M.L., Carvalho, I., Faria, T. & Pinheiro, C. (2002). How Plants Cope with Water Stress in the Field? Photosynthesis and Growth, *Annals of Botany* Vol.89(No. 7): 907–916.

Chen, X., Yang, R.-F., Li, W.-C. & Fu, F.-L. (2010). Identification of 21 microRNAs in maize and their differential expression under drought stress, *African Journal of Biotechnology* Vol. 9 (No. 30): 4741-4753.

Chinnusamy, V., Schumaker, K. & Zhu, J.K. (2004). Molecular genetic perspectives on cross-talk and specificity in abiotic stress signalling in plants, *Jornal Experimental Botany* Vol.55: 225-236.

Choi, H., Hong, J., Ha, J., Kang, J. & Kim, S.Y.(2000). ABFs, a family of ABA-responsive element binding factors, *Journal of Biological Chemistry* Vol. 275:1723-1730.

Ciais, P., Reichstein, M., Viovy, N., Granier, A., Ogèe, J. et al. (2005). *Europe-wide reduction in primary productivity caused by the heat and drought in 2003*, Nature *Vol.472: 529-533.*

Cominelli E., Galbiati M., Vavasseur A., Conti L., Sala T., Vuylsteke M., Leonhardt N., Dellaporta S. & Tonelli C. (2005). A guard-cell-specific MYB transcription factor regulates stomatal movements and plant drought tolerance, *Current Biology* Vol.15:1196–1200.

Cook, K.B., Kazan, H., Zuberi, K., Morris, Q. & Hughes, T.R. (2010). RBPDB: a database of RNA-binding specificities, *Nucleic Acids Research* Vol.39(suppl.1): 301-308.

Courtois, B., Ahmadi, N., Khowaja, F., Price, A.H., Rami, J.F., Frouin, J., Hamelin, C. & Ruiz, M. (2009). Rice Root Genetic Architecture: Meta-analysis from a Drought QTL Database, *Rice* Vol.2: 115–128.

Covarrubiales, A.A. & Reyes, J.L. (2010). Post-transcriptional gene regulation of salinity and drought responses by plant microRNAs, *Plant Cell and Environment* Vol.33: 481–489.

de Dorlodot, S., Forster, B., Pagè, L., Price, A., Tuberosa, R. & Draye, X. (2007). Root system architecture: opportunities and constraints for genetic improvement of crops, *Trends in Plant Science* Vol.12: 474–481.

De Leonardis, A. M., Marone, D., Mazzucotelli, E., Neffar, F., Rizza, F., Di Fonzo, N., Cattivelli, L. & Mastrangelo, A.M. (2007). Durum wheat genes up-regulated in the early phases of cold stress are modulated by drought in a developmental and genotype dependent manner, *Plant Science* Vol.172: 1005–1016.

De Vita, P., Mastrangelo, A.M., Matteu, L., Mazzucotelli, E., Virzì, N., Palumbo, M., Lo Storto, M., Rizza, F. & Cattivelli, L. (2010). Genetic improvement effects on yield stability in durum wheat genotypes grown in Italy, *Field Crops Research* Vol.119: 68–77.

Delseny, M., Han, B. & Hsing, Y.I. (2010). High throughput DNA sequencing: The new sequencing revolution, *Plant Science* Vol. 179: 407-422.

Dichio, B., Xiloyannis, C., Sofo, A. & Montanaro, G. (2006). Osmotic regulation in leaves and roots of olive tree (Olea europaea L.) during water deficit and rewatering. *Tree Physiology*, Vol. 26: 179–185.

Ding, Z., Li, S., An, X., Liu, X., Qin, H., & Wang, D. (2009). Transgenic expression of MYB15 confers enhanced sensitivity to abscisic acid and improved drought tolerance in Arabidopsis thaliana, *Journal of Genetics & Genomics* Vol.36: 17–29.

Dong, C.H., Agarwal, M., Zhang, Y., Xie, Q. & Zhu, J.K. (2006). The negative regulator of plant cold responses, HOS1, is a RING E3 ligase that mediates the ubiquitination and degradation of ICE1, *Proc. Natl. Acad. Sci. U.S.A.* Vol.103: 8281–8286.

Downes, B. & Vierstra, R.D. (2005). Post-translational regulation in plants employing a diverse set of polypeptide tags, *Biochemical Society Transaction* Vol.33: 393–399.

Dreyfuss, G., Kim, V.N. & Kataoka, N. (2002). Messenger-RNA-binding proteins and the messages they carry, *Nature Reviews Molecular Cell Biology* Vol.3: 195–205.

Duan, K., Luo, Y.-H., Luo, D., Xu, Z.-H. & Xue, H.-W. (2005). New insights into the complex and coordinated transcriptional regulation networks underlying rice seed development through cDNA chip-based analysis, *Plant Molecular Biology* Vol.57: 785–804.

Dyachenko, O.V., Zakharchenko, N.S., Shevchuk, T.V., Bohnert, H.J., Cushman, J.C. & Buryanov, Y.I. (2006). Effect of hypermethylation of CCWGG sequences in DNA of Mesembryanthemum crystallinum plants on their adaptation to salt stress, *Biochemistry (Moscow)* Vol.71: 461-465.

Eapen, D., Barroso, M.L., Ponce, G., Campos, M.E. & Cassab, G.I. (2005). Hydrotropism: Root growth responses to water, *Trends in Plant Science* Vol.10: 44–50.

Fang, Y., You, J., Xie, K., Xie, W. & Xiong, L. (2008). Systematic sequence analysis and identification of tissue-specific or stress-responsive genes of NAC transcription factor family in rice, *Molecular Genetics & Genomics* Vol.280: 547–563.

Farquhar, G.D. & Richards, R.A. (1984). Isotopic composition of plant carbon correlates with water-use efficiency of wheat genotypes, *Australian Journal of Plant Physiology* Vol.11: 539–552.

Fischer, R.A., Rees, D., Sayre, K.D., Lu, Z.-M., Condon, A.G. & Larque Saavedra, A. (1998). Wheat Yield Progress Associated with Higher Stomatal Conductance and Photosynthetic Rate, and Cooler Canopies, *Crop Science* Vol. 38 (No. 6): 1467-1475.

Fitzgerald, J.N., Lehti-Shui, M.D., Ingram, P.A., Deak, K.I., Biesiada, T. & Malamy, J.E. (2006). Identification of quantitative trait loci that regulate Arabidopsis root system size and plasticity, *Genetics* Vol.172: 485–498.

Francia, E., Barabaschi, D., Tondelli, A., Laidò, G. Rizza, F., Stanca, A.M., Busconi, M. & Fogher, C. (2007). Fine mapping of a HvCBF gene cluster at the frost resistance locus Fr-H2 in barley, *Theoretical and Applied Genetics* Vol.115: 1083–1091.

Fu, W., Wu, K. & Duan, J. (2007). Sequence and expression analysis of histone deacetylases in rice, Biochemical and *Biophysical Research Communication* Vol.356: 843-850.

Fujita. M., Fujita, Y., Maruyama, K., Seki, M., Hiratsu, K., Ohme-Takagi, M., Tran, L.S.P., Yamaguchi-Shinozaki, K. & Shinozaki, K. (2004). A dehydration-induced NAC protein, RD26, is involved in a novel ABA-dependent stress-signaling pathway *The Plant Journal* Vol.39: 863–876.

Fujita, Y., Fujita, M., Satoh, R., Maruyama, K., Parvez, M., Seki, M., Hiratsu, K., Ohme-Takagi, M., Shinozaki, K. & Yamaguchi-Shinozaki, K. (2005). AREB1 is a transcriptional activator of novel ABRE dependent ABA signaling that enhances drought stress tolerance in Arabidopsis, *Plant Cell* Vol.17: 3470–3488.

Garrity, D.P. & O'Toole, J.C. (1994). Screening rice for drought resistance at the reproductive phase, *Field Crops Research* Vol.39: 99-110.

Giuliani, S., Sanguineti, M.C., Tuberosa, R., Bellotti, M., Salvi, S. & Landi, P. (2005). Root-ABA1, a major constitutive QTL, affects maize root architecture and leaf ABA concentration at different water regimes, *Journal of Experimental Botany* Vol.56: 3061–3070.

Gong, Z., Dong, C.H., Lee, H., Zhu, J., Xiong, L., Gong, D., Stevenson, B. & Zhu, J.K. (2005). A DEAD box RNA helicase is essential for mRNA export and important for development and stress responses in Arabidopsis, *Plant Cell* Vol.17: 256–267.

Gong, Z., Lee, H., Xiong, L., Jagendorf, A., Stevenson, B. & Zhu, J.K. (2002). RNA helicase-like protein as an early regulator of transcription factors for plant chilling and freezing tolerance, *Proc. Natl Acad. Sci. USA* Vol.99: 11507–11512.

Granot, G., Sikron-Persi, N., Gaspan,O., Florentin, A., Talwara, S., Paul, L.K., Morgenstern, Y., Granot, Y. & Grafi, G. (2009). Histone modifications associated with drought tolerance in the desert plant Zygophyllum dumosum Boiss, *Planta* Vol.231: 27–34.

Gupta, P., Langridge, P. & Mir, R.R. (2010). Marker-assisted wheat breeding: present status and future possibilities, *Molecular Breeding* Vol.26: 145-161.

Haake, V., Cook, D., Riechmann, J.L., Pineda, O., Thomashow, M.F. & Zhang J.Z. (2002). Transcription factor CBF4 is a regulator of drought adaptation in Arabidopsis, *Plant Physiology* Vol.130: 639–648.

Harb, A., Krishnan, A., Ambavaram, M.M.R. & Pereira, A. (2011). Molecular and Physiological Analysis of Drought Stress in Arabidopsis Reveals Early Responses Leading to Acclimation in Plant Growth, *Plant Physiology* Vol.154: 1254–1271.

Hay, R.T. (2005). SUMO: a history of modification, *Molecular Cell* Vol.18: 1–12.

Hicke, L. (2001). Protein regulation by monoubiquitin, *Nature Reviews Molecular Cell Biology* Vol.2: 195–201.

Hirayama, T. & Kazuo Shinozaki, K.(2010). Research on plant abiotic stress responses in the post-genome era: past, present and future, *The Plant Journal* Vol.61: 1041–1052.

Ho, M.D., Rosas, J.C., Brown, K.M. & Lynch, J.P. (2005). Root architectural tradeoffs for water and phosphorus acquisition, *Functional Plant Biology* Vol.32: 737–748.

Hocholdinger, F., Park, W.J., Sauer, M. & Woll, K. (2004). From weeds to crops: Genetic analysis of root development in cereals, *Trends in Plant Science* Vol.9: 42–48.

Hsieh, T.H., Li, C.W., Su, R.C., Cheng, C.P., Sanjaya, T. et al., (2010). A tomato bZIP transcription factor, SlAREB, is involved in water deficit and salt stress response, *Planta* Vol.231: 1459–1473.

Hugouvieux, V., Kwak, J.M. & Schroeder, J.I. (2001). An mRNA cap binding protein, ABH1, modulates early abscisic acid signal transduction in Arabidopsis, *Cell* Vol.106: 477–487.

Hussain, S.S., Kayani, M.A. & Amjad, M. (2011). Transcription Factors as Tools to Engineer Enhanced Drought Stress Tolerance in Plants, *American Institute of Chemical Engineers February* Vol.7: 297-306.

Ingram, J. & Bartels D. (1996). The molecular basis of dehydration tolerance in plants. *Annual Reviews Plant Physiology Plant Molecular Biology* Vol.47: 377–403.

Ishitani, M., Xiong, L., Lee, H., Stevenson, B. & Zhu, J.K. (1998). HOS1, a genetic locus involved in cold-responsive gene expression in Arabidopsis, *Plant Cell* Vol.10: 1151–1162.

Ito, Y., Katsura, K., Maruyama, K., Taji, T., Kobayashi, M., Seki, M., Shinozaki, K. & Yamaguchi-Shinozaki, K. (2006). Functional analysis of rice DREB1/CBF-type transcription factors involved in cold-responsive gene expression in transgenic rice, *Plant Cell Physiology* Vol.47: 141–153.

Ito, H., Gaubert, H., Bucher, E., Mirouze, M., Vaillant, I. & Paszkowski, J. (2011). An siRNA pathway prevents transgenerational retrotransposition in plants subjected to stress, *Nature* Vol.472: 115-119.

Jen, C.-H., Michalopoulos, I., Westhead, D.R. & Meyer, P. (2005). Natural antisense transcripts with coding capacity in Arabidopsis may have a regulatory role that is not linked to double-stranded RNA degradation, *Genome Biology* Vol.6: R51.

Jeong, J.S., Kim, Y.S., Baek, K.H., Jung, H., Ha, S.H., Do Choi, Y., Kim, M., Reuzeau, C. & Kim, J.K. (2010). Root-specific expression of Os-NAC10 improves drought tolerance and grain yield in rice under field drought conditions, *Plant Physiology* Vol.153: 185–197.

Jiang, Y. & Deyholos. M.K. (2009). Functional characterization of Arabidopsis NaCl-inducible WRKY25 and WRKY33 transcription factors in abiotic stresses, *Plant Molecular Biology* Vol.69: 91–105.

Jones-Rhoades, M.W. & Bartel, D.P. (2004). Computational identification of plant microRNAs and their targets, including a stress-induced miRNA, *Molecular Cell* Vol.14: 787-799.

Jones-Rhoades, M.W., Bartel, D.P. & Bartel, B. (2006). MicroRNAs and their regulatory roles in plants, *Annual Review of Plant Biology* Vol. 57: 19–53.

Jung, C., Seo, J.S., Han, S.W., Koo, Y.J., Kim, C.H., Song, S.I., Nahm, B.H., Choi, Y.D. & Cheong, J.J. (2008). Overexpression of AtMYB44 enhances stomatal closure to confer abiotic stress tolerance in transgenic Arabidopsis, *Plant Physiology* Vol.146: 623–635.

Kader, M.A. & Lindberg, S. (2010). Cytosolic calcium and pH signaling in plants under salinity stress, *Plant Signaling & Behaviour* Vol. 5: 233–238.

Kaldenhoff, R., Ribas-Carbo, M., Flexas Sans, J., Lovisolo, C., Heckwolf, M. & Uehlein, N. (2008). Aquaporins and plant water balance, *Plant, Cell and Environment* Vol.31: 658–666.

Kang, J., Choi, H., Im, M. & Kim, S.Y. (2002). Arabidopsis basic leucine zipper proteins that mediate stress-responsive abscisic acid signalling, *Plant Cell* Vol.14: 343-357.

Kapazoglou, A., Tondelli, A., Papaefthimiou, D., Ampatzidou, H., Francia, E., Stanca, M.A., Bladenopoulos, K. & Tsaftaris, A.S. (2010). Epigenetic chromatin modifiers in barley: IV. The study of barley polycomb group (PcG) genes during seed development and in response to external ABA, *BMC Plant Biology* Vol.10: 73.

Karlson, D., Nakaminami, K., Toyomasu, T. & Imai, R. (2002). A cold-regulated nucleic acid-binding protein of winter wheat shares a domain with bacterial cold shock proteins, *The Journal of Biological chemistry* Vol.277: 35248–35256.

Kasuga, M., Miura, S. & Yamaguchi-Shinozaki, K. (2004). A combination of the Arabidopsis DREB1A gene and stress inducible rd29A promoter improved drought and low temperature stress tolerance in tobacco by gene transfer, *Plant Cell Physiology* Vol.45: 346–350.

Kerstiens, G. (2006). Water transport in plant cuticles: an update, *Journal of Experimental Botany* Vol. 57(No. 11): 2493-2499.

Kim, Y.O., Pan, S.O., Jung, C.H. & Kang, H. (2007). A zinc finger-containing glycine-rich RNA- binding protein, atRZ-1a, has a negative impact on seed germination and seedling growth of Arabidopsis thaliana under salt or drought stress conditions, *Plant Cell Physiology* Vol.48: 1170–1181.

Kobayashi, A., Takahashi, A., Kakimoto, Y., Miyazawa, Y., Fujii, N., Higashitani, A. & Takahashi, H. (2007). A gene essential for hydrotropism in roots, *Proc. Natl. Acad. Sci. U.S.A.* Vol.104: 4724– 4729.

Kuhn, J.M. & Schroeder, J.I. (2003). Impacts of altered RNA metabolism on abscisic acid signaling, *Current Opinion in Plant Biology* Vol.6: 463–469.

Lan, L., Li, M., Lai, Y., Xu, W., Kong, Z., Ying, K., Han, B. & Xue, Y. (2005). Microarray analysis reveals similarities and variations in genetic programs controlling pollination/fertilization and stress responses in rice (Oryza sativa L.), *Plant Molecular Biology* Vol.59: 151-164.

Lareau,L.F., Green, R.E., Bhatnagar, R.S. & Brenner, S.E. (2004). The evolving roles of alternative splicing, *Current Opinion in Cell Biology* Vol.14: 273–282.

Levitt (1972). Responses of plants to environmental stresses. Academic Press, New York.

Li, J., Li, X., Guo, L., Lu, F., Feng, X., He, K., Wei, L., Chen, Z., Qu, L. J. & Gu, H. (2006). A subgroup of MYB transcription factor genes undergoes highly conserved alternative splicing in Arabidopsis and rice, *Journal of Experimantal Botany* Vol. 57: 1263–1273.

Li, W.X., Oono, Y., Zhu, J., He, X.J., Wu, J.M., Iida, K., Lu, X.Y., Cui, X., Jin, H. & Zhu, J.K. (2008). The Arabidopsis NFYA5 transcription factor is regulated transcriptionally and posttranscriptionally to promote drought resistance, *The Plant Cell* Vol.20: 2238–2251.

Lilley, J.M. & Kirkegaard, J.A. (2007). Seasonal variation in the value of subsoil water to wheat: Simulation studies in southern New South Wales, *Australian Journal of Agricoltural Research* Vol.58: 1115–1128.

Lin, W.-Y., Matsuoka, D., Sasayama, D. & Nanmori, T. (2010). A splice variant of Arabidopsis mitogen-activated protein kinase and its regulatory function in the MKK6–MPK13 pathway, *Plant Science* Vol.178: 245–250.

Lippold, F., Sanchez, D.H., Musialak, M., Schlereth, A., Scheible, W.R., Hincha, D.K. &, Udvardi, M.K. (2009). AtMyb41 regulates transcriptional and metabolic responses to osmotic stress in Arabidopsis, *Plant Physiology* Vol.149: 1761–1772.

Lira-Medeiros, C.F., Parisod, C., Fernandes, R.A., Mata, C.S., Cardoso, M.A. & Ferreira, P.C. (2010). Epigenetic variation in mangrove plants occurring in contrasting natural environment, *PLoS One* Vol.5: e10326.

Lister, R., O'Malley, R.C., Tonti-Filippini, J., Gregory, B.D., Berry, C.C., et al. (2008). Highly integrated single base resolution maps of the epigenome in Arabidopsis, *Cell* Vol.133: 523–536.

Lorkovic, Z.J. (2009). Role of plant RNA-binding proteins in development, stress response and genome organization, *Trends Plant Science* Vol.14: 229–236.

Lu, X.-Y & Huang, X.-L. (2008). Plant miRNAs and abiotic stress responses, *Biochemical and Biophysical Research Communications* Vol.368: 458–462.

Lynch, J. (1995). Root architecture and plant productivity, *Plant Physiology* Vol.109: 7–13.

Maccaferri, M., Sanguineti, M.C., Corneti, S., Araus Ortega, J.L., Ben Salern, M. et al. (2008). Quantitative trait loci for grain yield and adaptation of durum wheat (Triticum durum Desf.) across a wide range of water availability, *Genetics* Vol.178: 489-511.

Maccaferri, M., Sanguineti, M.C., Giuliani, S. & Tuberosa, R. (2009). Genomics of Tolerance to Abiotic Stress in the Triticeae. In: Feuillet C, Muehlbauer G (eds) Plant Genetics and Genomics: Crops and Models, Vol. 7 Genetics and Genomics of the Triticeae, Springer, pp. 481-558.

Manschadi, A.M., Christopher, J., Devoil, P. & Hammer, G.L. (2006). The role of root architectural traits in adaptation of wheat to water-limited environments, *Functional Plant Biology* Vol.33: 823–837.

Masle, J., Gilmore, S.R. & Farquhar, G.D. (2005). The ERECTA gene regulates plant transpiration efficiency in Arabidopsis, *Nature* Vol.436: 866-870.

Mastrangelo, A.M., Belloni, S., Barilli, S., Reperti, B., Di Fonzo, N., Stanca, A.M. & Cattivelli, L. (2005). Low temperature promotes intron retention in two e-cor genes of durum wheat, *Planta* Vol.221: 705–715.

Mastrangelo, A.M., Mazzucotelli, E., Guerra, D., De Vita, P. & Cattivelli, L. (2011a). Improvement of drought resistance in crops: from conventional breeding to genomic selection In "Crop Stress and its Management: Perspectives and Strategies" Arun K. Shanker eds, DOI 10.1007/978-94-007-2220-0_7.

Mastrangelo, A.M., Marone, D., Laidò, G., De Leonardis, A.M. & De Vita, P. (2011b). Alternative splicing: Enhancing ability to cope with stress via transcriptome plasticity, *Plant Science* doi:10.1016/j.plantsci.2011.09.006

Matsukura, S., Mizoi, J., Yoshida, T., Todaka, D., Ito, Y., Maruyama, K., Shinozaki, k. & Yamaguchi-Shinozaki, K. (2010). Comprehensive analysis of rice DREB2-type genes that encode transcription factors involved in the expression of abiotic stress-responsive genes, *Molecular Genetics & Genomics* Vol.283: 185–196.

Mazzucotelli, E., Mastrangelo, A.M., Crosatti, C., Guerra, D., Stanca, A.M. & Cattivelli, L. (2008). Abiotic stress response in plants: when post-transcriptional and post-translational regulations control transcription, *Plant Science* Vol.174:420–431.

McGuire, A.M., Pearson, M.D., Neafsey, D.E. & Galagan, J.E. (2008). Cross-kingdom patterns of alternative splicing and splice recognition, *Genome Biology* Vol.9: R50.

Meuwissen, T., Hayes, B.J. & Goddard, M.E. (2001). Prediction of total genetic value using genome-wide dense marker maps, *Genetics* Vol.157: 1819–1829.

Miller, G., Shulaev, V. & Mittler, R. (2008). Reactive oxygen signaling and abiotic stress, *Physiologia Plantarum* Vol.133: 481–489.

Mirouze, M., Reinders, J., Bucher, E., Nishimura, T., Schneeberger, K., Ossowski, S., Cao, J., Weigel, D., Paszkowski, J. & Mathieu, O. (2009). Selective epigenetic control of retrotransposition in Arabidopsis, *Nature* Vol.461: 427-430.

Miura, K., Jin, J.B. & Hasegawa, P.M. (2007). Sumoylation, a post-translational regulatory process in plants, *Current Opinion in Plant Biology* Vol.10: 495–502.

Mlynarova, L., Nap, J.P. & Bisseling, T. (2007). The SWI/SNF chromatinremodeling gene AtCHR12 mediates temporary growth arrest in Arabidopsis thaliana upon perceiving environmental stress, *Plant Journal* Vol.51: 874-885.

Mohammadi, M., Nat, N.V.K. & Deyholoso, M.K. (2007). Transcriptional profiling of hexaploid wheat (Triticum aestivum L.) roots identifies novel dehydration-responsive genes, *Plant Cell Environ* Vol.30: 630-645.

Nathan D., Ingvarsdottir K., Sterner D.E., Bylebyl G.R., Dokmanovic M., Dorsey J.A., Whelan K.A., Krsmanovic M., Lane W.S., Meluh P.B. et al. (2006). Histone sumoylation is a negative regulator in Saccharomyces cerevisiae and shows dynamic interplay with positive-acting histone modifications, *Genes Development* Vol.20: 966-976.

Ner-Gaon, H. & Fluhr, R. (2006). Whole-genome microarray in Arabidopsis facilitates global analysis of retained introns, *DNA Research* Vol.13: 111–121.

Nobuta, K., Venu, R.C., Lu, C., Belo, A., Vemaraju, K., et al. (2007). An expression atlas of rice mRNAs and small RNAs, *Nature Biotechnology* Vol 25: 473–77.

Novillo, F., Alonso, J.M., Ecker, J.R. & Salinas, J. (2004). CBF2/DREB1C is a negative regulator of CBF1/DREB1B and CBF3/DREB1A expression and plays a central role in stress tolerance in Arabidopsis, *Proc. Natl. Acad. Sci. U.S.A.* Vol.101: 3985–3990.

Ogura, Y., Ihara,N., Komatsu, A., Tokioka, Y., Nishioka, M., Takase, T. & Kiyosue, T. (2008). Gene expression, localization, and protein–protein interaction of Arabidopsis SKP1-like (ASK) 20A and 20B, *Plant Science* Vol.174: 485–495.

Oh S.J., Song S.I., Kim Y.S., Jang H.J., Kim M. & Kim Y.K. (2005). Arabidopsis CBF3/DREB1A and ABF3 in transgenic rice increased tolerance to abiotic stress without stunting growth, *Plant Physiology* Vol.138: 341-351.

Oh, S.J., Kwon, C.W., Choi, D.W., Song ,S.I.K. & Kim, J.K. (2007). Expression of barley HvCBF4 enhances tolerance to abiotic stress in transgenic rice, *Journal of Plant Biotechnology* Vol.5: 646–656.

Olsen, A.N., Ernst, H.A., Leggio, L.L. & Skriver, K. (2005). NAC transcription factors: structurally distinct, functionally diverse, *Trends in Plant Sciences* Vol.10: 79-87.

Ooka, H., Satoh, K., Doi, K., Nagata, T., Otomo, Y., Marukami, K., Matsubara, K., Osato, N., Kawai, J., Carninci, P., Hayashizaki, Y., Suzuki, K., Kojima, K., Takahara, Y., Yamamoto, K. & Kikuchi, S. (2003). Comprehensive analysis of NAC family genes in Oryza sativa and Arabidopsis thaliana, *DNA Research* Vol.10: 239–247.

Owttrim, G.W. (2006). RNA helicases and abiotic stress, *Nucleic Acids Research* Vol. 34(No. 11): 3220-3230.

Ozturk, Z.N., Talame, V., Deyholos, M., Michalowski, C.B., Galbraith, D.W., Gozukirmizi, N., Tuberosa, R. & Bohnert, H.J. (2002). Monitoring large-scale changes in transcript abundance in drought- and salt stressed barley, *Plant Molecular Biology* Vol.48: 551-573.

Palusa, G.S. Ali, G.S. & Reddy, A.S.N. (2007). Alternative splicing of pre-mRNAs of Arabidopsis serine/arginine-rich proteins: regulation by hormones and stresses, *Plant Journal* Vol.49: 1091-1107.

Park, H.Y., Kang, I.S., Han, J.S., Lee, C.H., An, G. & Moon, Y.H. (2009). OsDEG10 encoding a small RNA-binding protein is involved in abiotic stress signaling, *Biochemical and Biophysical Research Communication* Vol.380: 597-602.

Pastori, G.M. & Foyer, C.H. (2002). Common components, networks and pathways of cross-tolerance to stress. The central role of 'redox' and abscisic-acid-mediated controls, *Plant Physiology* Vol.129: 460-468.

Peleg, Z., Fahima, T., Krugman, T, Abbo, S., Yakir, D., Korol, A.B. & Saranga, Y. (2009). Genomic dissection of drought resistance in durum wheatxwild emmer wheat recombinant inbreed line population. *Plant, Cell and Environment* Vol.32: 758-779.

Pellegrineschi, A., Reynolds, M., Pacheco, M., Brito, R.M., Almeraya, R., Yamaguchi-Shinozaki, K. & Hoisington, D. (2004). Stress induced expression in wheat of the Arabidopsis thaliana DREB1A gene delays water stress symptoms under greenhouse conditions. *Genome* Vol.47: 493-500.

Petit, J.R., Jouzel, J., Raynaud, D., Barkov, N.I., Barnola, J.-M., Basile, I., Bender, M., Chappellaz, J., Davisk, M. et al. (1999). Climate and atmospheric history of the past 420,000 years from the Vostok ice core, Antarctica, *Nature* Vol. 399: 429-436.

Pinheiro, C. & Chaves, M.M. (2011). Photosynthesis and drought: can we make metabolic connections from available data?, *Journal of Experimental Botany* Vol.62(No. 3): 869-882.

Price, A.H. & Tomos, A.D. (1997). Genetic dissection of root growth in rice (Oryza sativa L.). 2 mapping quantitative trait loci using molecular markers, *Theoretical and Applied Genetics* Vol.95: 143-152.

Qiu, Y. & Yu, D. (2009). Over-expression of the stress-induced OsWRKY45 enhances disease resistance and drought tolerance in Arabidopsis, *Environment and Experimental Botany* Vol.65: 35-47.

Ramamoorthy, R., Jiang, S.Y., Kumar, N., Venkatesh, P.N. & Ramachandran, S. (2008). A comprehensive transcriptional profiling of the WRKY gene family in rice under various abiotic and phytohormone treatments, *Plant Cell Physiology Vol.49*: 865-879.

Ratcliffe,O.J. & Riechmann, J.L. (2002). Arabidopsis transcription factors and the regulation of flowering time: a genomic perspective, *Current Issues in Molecular Biology* Vol.4: 77-91.

Rebetzke, G.J., Condon, A.G., Richards, R.A. & Farquhar, G.D. (2002). Selection for reduced carbon-isotope discrimination increases aerial biomass and grain yield of rainfed bread wheat, *Crop Science* Vol. 42: 739-745.

Reddy, A.R., Chaitanya, K.V. & Vivekanandan, M. (2004). Drought-induced responses of photosynthesis and antioxidant metabolism in higher plants, *Journal of Plant Physiology* Vol.161: 1189-1202.

Reinders, J., Wulff, B.B.H., Mirouze, M., Marì-Ordònez, A., Dapp, M., Rozhon, W., Bucher, E., Theiler, G. & Paszkowski, J. (2009). Compromised stability of DNA methylation and transposon immobilization in mosaic Arabidopsis epigenomes, *Genes Development* Vol.23: 939-950.

Ren, X., Chen, Z., Liu, Y., Zhang, H., Zhang, M., Liu, Q., Hong, X., Zhu, J.K. & Gong, Z. (2010). ABO3, a WRKY transcription factor, mediates plant responses to abscisic acid and drought tolerance in Arabidopsis, *Plant Journal* Vol.63: 417–429.

Reyes, J.L. & Chua, N.H. (2007). ABA induction of miR159 controls transcript levels of two MYB factors during Arabidopsis seed germination, *The Plant Journal* Vol.49: 592–606.

Rhoades, M.W., Reinhart, B.J., Lim, L.P., Burge, C.P., Bartel, B. & Bartel, D.P. (2002). Prediction of plant microRNA targets, *Cell* Vol.110: 513-520.

Ribaut, J.M. & Ragot, M. (2007). Marker-assisted selection to improve drought adaptation in maize: the backcross approach, perspectives, limitations, and alternatives, *Journal of Exp Botany* Vol.58: 351-360.

Riechmann, J.L., Heard, J., Martin, G., Reuber, L., Jiang, C., Keddie, J., Adam, L., Pineda, O., Ratcliffe, O.J. & Samaha, R.R., Creelman, R., Pilgrim, M., Broun, P., Zhang, J.Z., Ghandehari, D., Sherman, B.K. & Yu, B.K. (2000). Arabidopsis transcription factors: genome-wide comparative analysis among eukaryotes, *Science* Vol.290: 2105-2110.

Rios, G., Gagete, A.P., Castillo. J., Berbel. A., Franco, L. & Rodrigo, M.I. (2007). Abscisic acid and desiccation-dependent expression of a novel putative SNF5-type chromatin-remodeling gene in Pisum sativum, *Plant Physiology and Biochemstry* Vol.45: 427-435.

Rizhsky, L., Liang, H. & Mittler, R. (2002). The combined effect of drought stress and heat shock on gene expression in tobacco, *Plant Physiology* Vol.130: 1143–1151.

Rizza, F., Ghashghaie, J., Meyer, S., Matteu, L., Mastrangelo, A.M. & Badeck, F.-W. (2011). Constitutive differences in water use efficiency between two durum wheat cultivars, *Fields Crops Research* doi: 10.1016/J fcr.2011.09.001.

Rocak, S. & Linder, P. (2004). DEAD-box proteins: the driving forces behind RNA metabolism. *Nature Review Molecular Cell Biology* Vol.5: 232–241.

Sachetto-Martins, G., Franco, L.O. & deOliveira, D.E. (2000). Plant glycine-rich proteins: a family or just proteins with a common motif?, *Biochimica et Biophysica Acta* Vol.1492: 1–14.

Sade, N., Vinocur, B.J., Diber, A., Shatil, A., Ronen, G., Nissan, H., Wallach, R., Karchi, H. & Moshelion, M. (2009). Improving plant stress tolerance and yield production: is the tonoplast aquaporin SlTIP2;2 a key to isohydric to anisohydric conversion?, *New Phytologist* Vol.181: 651–661.

Sadras, V.O., & Rodriguez, D. (2007). The limit to wheat water use efficiency in eastern Australia. II. Influence of rainfall patterns, *Australian Journal of Agricoltural* Research Vol.58: 657–669.

Sahi, C., Agarwal, M., Singh, A. & Grover, A. (2007). Molecular characterization of a novel isoform of rice (Oryza sativa L.) glycine rich-RNA binding protein and evidence for its involvement in high temperature stress response, *Plant Science* Vol.173(No. 2): 144-155.

Sanan-Mishra, N., Pham, X.H., Sopory, S.K. & Tuteja,N. (2005). Pea DNA helicase 45 overexpression in tobacco confers high salinity tolerance without affecting yield, *Proc. Natl Acad. Sci. USA* Vol.102: 509–514.

Sarnowski, T.J., Rìos, G,. Jasik. J., Swiezewski, S., Kaczanowski, S., Li, Y., Kwiatkowska, A., Pawlikowska, K., Kozbiał, M., Kozbiał, P. et al. (2005). SWI3 subunits of putative SWI/SNF chromatin-remodeling complexes play distinct roles during Arabidopsis development, *Plant Cell* Vol.17: 2454-2472.

Schoning, J.C., Streitner, C., Meyer, I.M., Gao, Y. & Staiger, D. (2008). Reciprocal regulation of glycine-rich RNA-binding proteins via an interlocked feedback loop coupling alternative splicing to non sense-mediated decay in Arabidopsis, *Nucleic Acids Research* Vol.36: 6977–6987.

Scippa, G.S., Di Michele, M., Onelli, E., Patrignani, G., Chiatante, D. & Bray, E.A. (2004). The histone-like protein H1-S and the response of tomato leaves to water deficit, *Journal of Experimental Botany* Vol.55: 99-109.

Serraj, R. & Sinclair, T.R. (2002). Osmolyte accumulation: can it really increase crop yield under drought conditions?, *Plant Cell and Environment* Vol.25: 333-341.

Shinozahi, Z.K. & Yamaguchi-Shinozaki K. (2000). Molecular responses to dehydration and low temperature: differences and cross-talk between two stress signalling pathways, *Current Opinion in Plant Biology* Vol. 3: 217–223.

Siddique,K.H.M., Tennant, D., Perry, M.W. & Belford, R.K. (1990). Water use and water use efficiency of old and modern wheat cultivars in a Mediterranean-type environment, *Australian Journal of Agricultural Research* Vol.41(No. 3): 431 – 447.

Slafer, G.A., Araus, J.L,, Royo, C. & Del Moral, L.F.G. (2005). Promising ecophysiological traits for genetic improvement of cereal yields in Mediterranean environments, *Annals of Applied Biology* Vol.146: 61-70.

Slotkin, R.K. & Martienssen, R.A. (2007). Transposable elements and the epigenetic regulation of the genome, *Nature Reviews Genetics* Vol.8: 272-285.

Sridha, S. & Wu, K. (2006). Identification of AtHD2C as a novel regulator of abscisic acid responses in Arabidopsis, *Plant Journal* Vol.46: 124-133.

Sridhar, V.V., Kapoor, A., Zhang, K., Zhu, J., Zhou, T., Hasegawa, P.M.,Bressan, R.A. & Zhu, J.K. (2007). Control of DNA methylation and heterochromatic silencing by histone H2B deubiquitination, *Nature* Vol.447: 735-738.

Steele, K. (2009). Novel upland rice variety bred using marker-assisted selection and client-oriented breeding released in Jharkhand. India: Bangor University.

Steele, K.A., Virk, D.S., Kumar, R., Prasad, S.C. & Witcombe, J.R. (2007). Field evaluation of upland rice lines selected for QTLs controlling root traits, *Field Crops* Research Vol.101: 180–186.

Stone, S.L. & Callis, J.(2007). Ubiquitin ligases mediate growth and development by promoting protein death, *Current Opinion in Plant Biology* Vol.10: 624–632.

Sunkar, R. & Zhu, R.-K. (2004). Novel and stress-regulated microRNAs and other small RNAs from Arabidopsis, *Plant Cell* Vol.16: 2001–2019.

Szarzynska, B., Sobkowiak, L., Pant, B.D., Balazadeh, S., Scheible, W., Mueller- Roeber, B., Jarmolowski, A. & Szweykowska-Kulinska, Z. (2009). Gene structures and processing of Arabidopsis thaliana HYL 1-dependent pri-miRNAs, *Nucleic Acids Research* Vol.37: 3083–3093.

Tambussi, E.A., Nogues, S., Ferrio, P., Voltas, J. & Araus, J.L. (2005). Does higher yield potential improve barley performance in Mediterranean conditions?A case of study, *Field Crop Research* Vol.91: 149-160.

Tanner, N.K. & Linder, P. (2001). DExD/H box RNA helicases: from generic motors to specific dissociation functions, *Molecular Cell* Vol.8: 251–262.

Taramino, G., Sauer, M., Stauffer, J.L., Multani, D., Niu, X., Sakai, H. & Hoccholdinger, F. (2007). The maize (Zea mays L.) RTCS gene encodes a LOB domain protein that is a key regulator of embryonic seminal and post-embryonic shoot-borne root initiation, *Plant Journal* Vol.50: 649-659.

Tollenaar, M. & Wu, J. (1999). Yield in temperate maize is attributable to greater stress tolerance, *Crop Science* Vol.39: 1597-1604.

Tollenaar, M. & Lee, E.A. (2002), Yield stability and stress tolerance in maize, *Field Crop Research* Vol.75: 161-169.

Tran, L.S., Nakashima, K., Sakuma, Y., Simpson, S.D., Fujita, Y., Maruyama, K., Fujita, M., Seki, M., Shinozaki, K. & Yamaguchi-Shinozaki, K. (2004). Isolation and functional analysis of Arabidopsis stress inducible NAC transcription factors that bind to a drought responsive cis-element in the early responsive to dehydration stress 1 promoter, *Plant Cell* Vol.16: 2481-2498.

Trujillo, L.E., Sotolongo, M,. Menendez, C., Ochogava, M.E., Coll, Y., Hernandez, I., Borras-Hidalgo, O., Thomma, B.P.H.J., Vera, P. & Hernandez, L. (2008). SodERF3, a novel sugarcane ethylene responsive factor (ERF), enhances salt and drought tolerance when over-expressed in tobacco plants, *Plant Cell Physiology* Vol.49: 512-515.

Tuberosa, R., Sanguineti, M.C., Landi, P., Giuliani, M.M., Salvi, S. & Conti,S. (2002a). Identification of QTLs for root characteristics in maize grown in hydroponics and analysis of their overlap with QTLs for grain yield in the field at two water regimes, *Plant Molecular Biology* Vol.48: 697-712.

Tuberosa, R., Salvi, S., Sanguineti, M.C., Landi, P., Maccaferri, M. & Conti, S. (2002b). Mapping QTLs regulating morphophysiological traits and yield: case studies, shortcomings and perspectives in drought stressed maize, *Annals of Botany* Vol.89: 941-963.

Uno, Y., Furihata, T., Abe, H., Yoshida, R., Shinozaki, K. & Yamaguchi-Shinozaki, K. (2000). Arabidopsis basic leucine zipper transcription factors involved in an abscisic acid-dependent signal transduction pathway under drought and high-salinity conditions, *Proc Natl Acad Sci USA* Vol. 97: 11632-11637.

Vàgùjfalvi, A., Aprile, A., Miller, A., Dubcovsky, J., Delugu, G., Galiba, G. & Cattivelli, L. (2005). The expression of several Cbf genes at the Fr-A2locus is linked to frost resistance in wheat, *Molecular Genetics & Genomics* Vol.274: 506-514.

van Ginkel, M., Calhoun, D.S., Gebeyehu, G., Miranda, A., Tian-you, C., Pargas Lara, R., Trethowan, R.M., Sayre, K., Crossa, J. & Rajaram, S. (1998). Plant traits related to yield of wheat in early, late, or continuous drought conditions, *Euphytica* Vol.100: 109-121.

Vashisht, A.A., Pradhan, A., Tuteja, R. & Tuteja,N. (2005). Cold- and salinity stress-induced bipolar pea DNA helicase 47 is involved in protein synthesis and stimulated by phosphorylation with protein kinase C, *Plant Journal* Vol.44: 76-87.

Vazquez, F., Gasciolli, V., Crété, P. & Vaucheret, H. (2004). The nuclear dsRNA binding protein HYL1 is required for microRNA accumulation and plant development, but not post-transcriptional transgene silencing, *Current Biology* Vol.14: 346-351.

Verslues, P.E., Agarwal, M., Katiyar-Agarwal, S., Zhu, J. & Zhu, J.K. (2006). Methods and concepts in quantifying resistance to drought, salt and freezing, abiotic stresses that affect plant water status, *Plant Journal* Vol.45: 523-539.

Vicient, C.M., Suoniemi, A., Anamthawat-Jonsson, K., Tanskanen, J., Beharav, A,. et al. (1999). Retrotransposon BARE-1 and its role in genome evolution in the genus Hordeum, *Plant Cell* Vol.11: 1769-1784.

Wang, X.-J., Gaasterland, T. & Chua, N.-H. (2005). Genome-wide prediction and identification of cis-natural antisense transcripts in Arabidopsis thaliana, *Genome Biology* Vol.6: R30.

Wang, B.B. & Brendel, V. (2006). Genome wide comparative analysis of alternative splicing in plants, Proc. Natl. Acad. Sci. USA Vol. 103: 7175-7180.

Whitmore, A.P. & Whalley, W.R. (2009). Physical effects of soil drying on roots and crop growth, *Journal of Experimental Botany* Vol.60: 2845–2857.

Xiang, Y., Tang, N,, Du, H., Ye, H. & Xiong, L. (2008). Charaterization of Osb- ZIP23 as a key player of the basic leucine zipper transcription factor family for conferring abscisic acid sensitivity and salinity and drought tolerance in rice, *Plant Physiology* Vol.148: 1938– 1952.

Xiong, L., Gong, Z., Rock, C.D., Subramanian, S., Guo, Y., Xu, W., Galbraith, D. & Zhu, J.K. (2001). Modulation of abscisic acid signal transduction and biosynthesis by an Sm-like protein in Arabidopsis, *Developmental Cell* Vol.1: 771–781.

Yang, K.Y., Liu, Y. & Zhang, S. (2001). Activation of a mitogen-activated protein kinase pathway is involved in disease resistance in tobacco, *Proc. Natl. Acad. Sci. U.S.A.* Vol.98: 741–746.

Yang, D.L., Jing, R.L., Chang, X.P. & Li, W. (2007). Identification of quantitative trait loci and environmental interactions for accumulation and remobilization of water-soluble carbohydrates in wheat (Triticum aestivum L.) stems, *Genetics* Vol.176: 571–584.

Yoshida, T., Fujita, Y., Salama, H., Kidokoro, S., Maruyama, K., Mizoi, J., Shinozaki, K. & Yamaguchi-Shinozaki, K. (2010). AREB1, AREB2, and ABF3 are master transcription factors that cooperatively regulate ABRE-dependent ABA signaling involved in drought stress tolerance and require ABA for full activation, *Plant Journal* Vol.61: 672–685.

Zhang, B., Pan, X., Cannon, C.H., Cobb, G.P. & Anderson, T.A. (2006). Conservation and divergence of plant microRNA genes, *Plant Journal* Vol.46: 243–259.

Zhang, K., Sridhar, V.V., Zhu, J., Kapoor, A. & Zhu, J.K. (2007). Distinctive core histone post-translational modification patterns in Arabidopsis thaliana, *PLoS ONE* Vol.11: e1210.

Zhang, X., Wollenweber, B., Jiang, D., Liu, F. & Zhao, J. (2008). Water deficit and heat shock effects on photosynthesis of a transgenic Arabidopsis thaliana constitutively expressing ABP9, a bZIP transcription factor, *Journal of Experimental Botany* Vol. 59: 839–848.

Zhang, G., Chen, M., Li, L., Xu, Z., Chen, X., Guo, J. & Ma, Y. (2009). Overexpression of the soybean GmERF3 gene, an AP2/ERF type transcription factor for increased tolerances to salt, drought, and diseases in transgenic tobacco, *Journal of Experimental Botany* Vol.60(No. 13): 3781–3796.

Zhao, B.T., Liang, R.Q., Ge, L.F., Li, W., Xiao, H.S., Lin, H.X., Ruan, K.C. & Jin, Y.X. (2007). Identification of drought-induced microRNAs in rice, *Biochemical and Biophysical Research Communication* Vol.354: 585–590.

Zhao, B., Ge, L., Liang, R., Li, W., Ruan, K., Lin, H. & Jin, Y. (2009). Members of miR-169 family are induced by high salinity and transiently inhibit the NF-YA transcription factor, *BMC Molecular Biology* Vol.10: 29.

Zheng, X., Chen, B., Lu, G. & Han, B. (2009). Overexpression of a NAC transcription factor enhances rice drought and salt tolerance, *Biochemical and Biophysical Research Communication* Vol.379: 985–989.

Zhu, Q., Zhang, J., Gao, X., Tong, J., Xiao, L., Li, W. & Zhang, H. (2010). The Arabidopsis AP2/ERF transcription factor RAP2.6 participates in ABA, salt and osmotic stress responses, *Gene* Vol.457: 1–12.

Zoon, F.C. &. Van, Tienderen P.H. (1990). A rapid quantitative measurement of root length and root branching by microcomputer image analysis, *Plant Soil* Vol.126: 301–308.

Section 2

Plant Water Relations

5

Plant Water Relations: Absorption, Transport and Control Mechanisms

Geraldo Chavarria[1] and Henrique Pessoa dos Santos[2]
[1]The University of Passo Fundo
[2]Embrapa Grape & Wine
Brazil

1. Introduction

Although water is abundant on Earth - covering 71% of the total surface - its distribution is not uniform and can easily cause restrictions in availability to vegetal production. At global scale, these restrictions are easily observed in dry climates and can appear in other regions which do not currently experience drought, as provided by the future backdrop of climate change (IPCC, 2007).

The influences of water restriction on losses in the production and distribution of vegetation on the terrestrial surface are significantly larger than all other losses combined which are caused by biotic and abiotic factors (Boyer, 1985). This striking effect of water on plants emerges from its physiological importance, being an essential factor for successful plant growth, involving photosynthesis and several other biochemical processes such as the synthesis of energetic composites and new tissue. Therefore, in order to characterise the growth and productive behaviour of plant species it is essential to have an understanding of plant water relations, as well as the consequences of an inadequate water supply. Broadly, the water state of a plant is controlled by relative rates of loss and absorption, moreover it depends on the ability to adjust and keep an adequate water status. This will be considered throughout this chapter.

2. Absorption and water flow through plants

Independent of the species, plants require from the soil a water volume that overcomes its metabolic necessities. Through the transpiration process plants transmit to the atmosphere the majority of the water absorbed from soil (generally around 90%). From this perspective, it is noted that the plant water requirements are defined primarily by the atmosphere evapotranspirative demand, which is a predominately passive process. Figuratively, and with some caveats, we can compare a plant water flow with the principles of oil flow in the wick of an old fashion lampion (Fig. 1).

When it is fired, the oil that is burned on the upper extremity of the wick is quickly replaced by new one that is situated just below, and so on - following the physical forces of interaction between liquid and tissue - until reaching the level of the fuel reservoir, in the basal extremity of the wick. Applying this example to the plant, the burning of oil can be analogous to the process of the loss of water vapour through their leaves, i.e. the *transpiration*, which is caused by the pressure gradient of vapour between tissue saturated with water from the leaves and air, the "dry" atmosphere. The variations in this pressure gradient of the vapour will define the *evaporative demand* of the environment where the plant is. In the other extreme - where it represents the liquid reservoir of the lampion - we have water content present in the soil. In this scene, it is noted that the water flow through the plant is dependent on the energy formed by the gradient of the water content that is established between the soil and the atmosphere. However, we will see throughout this chapter that plants, unlike our lampion, can and must modulate this gradient in order to survive the wide variations of water availability between types of soils, weather and seasons.

Fig. 1. Schematic representation of water flow through the plant (arrows), by analogy with the oil flow through the wick of an old fashion lampion.

2.1 Water potential

The water content in the soil, plants and atmosphere is usually described as *water potential* (Ψw). This is based on the relation between the water content in the part of a system and pure water at the same temperature and atmospheric pressure, measured in pressure units (megapascal-MPa or bars-Bar). By definition, the potential of free pure water at atmospheric pressure and at a temperature of 25°C corresponds to 0 (zero) MPa. The contrast in the water potential between two points invariably determines the direction of water transport in a system. More precisely, the water potential represents all the water pressure in a given system and it is the sum of osmotic potential ($\Psi\pi$), matrix potential (Ψm), hydrostatic pressure or the turgor potential ($\Psi\rho$) and the gravitational potential (Ψg).

The osmotic potential ($\Psi\pi$) is the chemical potential of water in a solution due to the presence of dissolved substances (solutes). This is always negative because the water moves from one point with a lower concentration of solutes (for example, pure water) to a point with a higher concentration. So, the higher concentration of the solutes at a point which makes the system more negative will be the osmotic potential in this place. The water potential can also be influenced by a charged surface - mainly by soil components and cell walls - which compose the influence of the matrix potential (Ψm). In the soil, this influence of the matrix is so great that water potential is assumed negligible and therefore equivalent to the matrix potential. Concerning the potential of hydrostatic pressure (Ψp), it is noted that this component of the water potential can be positive or negative and it refers to the physical pressure that water exerts on a given system. For example, if we observe a turgid cell of a root cortex or a leaf mesophyll, the hydrostatic pressure is positive. However, in a xylem vessel subjected to a stressful condition - in a transpiring plant - this component of hydrostatic pressure is negative. Finally, we should emphasise that the gravitational potential (Ψg) - ignored in most cases - is very important in studies of the water potential of tree species, where plant height exerts a great influence on water flow. Considering that this gravitational component fluctuates at a rate of 0.1 MPa for every 10 meters of vertical displacement, it is suggested to consider if when plant height is 10 m or more.

2.2 Water dynamics in soil-plant-atmosphere system

From these components of water potential we return to our lampion scheme (Fig. 1) and show how the potential can vary over the continuum soil-plant-atmosphere, exposing the control points of each step of water flow from the soil to the atmosphere.

2.2.1 Soil water

The water potential in soil affects water reservoir and its availability for plants, hence it has a large impact on plant growth and production. Furthermore, the soil water content exerts a great influence on some physical and chemical properties of soil, such as the oxygen content, which interferes with root breathing, microbial activity and soil chemical status. Water potential is directly dependent on soil physical characteristics, and varies with time and space, depending on soil water balance. That balance is determined by input (rain, irrigation) and output of the soil (drainage, evaporation and root absorption). It is noteworthy that the amount of rain affecting soil water reservoir is only the *effective precipitation*. This is the amount of precipitation that is actually added and stored in the soil. For example, during drier periods less than 5 mm of daily rainfall would not be considered effective, as this amount of precipitation would likely evaporate from the surface before soaking into the ground.

It is important to emphasise that behaviour of water into soil differs from that in a pot, like the oil in the lampion reservoir (Fig. 1). That is, soil water interacts with the matrix and solutes, and it is under pressure or tension, resulting in various energy states, relative to free water (Kirkham, 2005). With regard to the physiological aspect, it is important to point out that the water content in soil is associated with three terms: *field capacity*, the *permanent wilting point* and the *available water content*.

The term "field capacity" corresponds to the maximum water content that a given soil can retain by capillarity, after saturation and gravity drainage, and it is conventionally estimated as the water content when the matrix potential is -0.03 MPa (-0.3 Bar). In spite of the great applicability of this term to irrigation management, field capacity has been recognized as an imprecise term due to theoretical advances and precise irrigation techniques. It is because the capillary soil water constantly (even slowly) decreases (due to evaporation from soil surface or drainage losses) and never stabilises (Fig. 2), it turn the soil water potential decreases while the matrix potential increases. This is most evident with medium and fine texture soils (for example, those rich in clay and organic matter), which maintain a significant drainage rate over a long time. Therefore, there is no real and unique value for accurately characterising the field capacity of a given soil. Furthermore, the continuous drainage can induce an overestimation of the water consumption of the plant. Despite these uncertainties, the term *field capacity* is still useful for a qualitative understanding - rather than a quantitative understanding - of the water behaviour of a particular soil, providing an estimate of the maximum limit of water accumulation. It is noteworthy that the inaccuracy of the field capacity determination occurs mainly when analysis takes place on samples in the laboratory, which can be contoured with evaluations directly in the soil, with specific sensors and considering together all characteristics of each site. In general, clay soils or those with higher content of organic matter (upper to 5% of organic matter) present a higher soil water holding capacity (average field capacity ranging from 35 to 40%vol). In contrast, sandy soils has a lower water holding capacity and field capacity typically ranges from 10-15%vol. It is important to observe that *field capacity* cannot be regarded as a maximum limit of the water available to plants, due the fact that plants also use free water that is in contact with the roots at the moment of soil drainage.

The *wilting point* (WP) is another important parameter in soil water dynamics as it dramatically affects plant physiology. This term is also known as the *permanent wilting point*, and can be defined as the amount of water per unit weight (or volume) of soil that is so tightly retained by the soil matrix that roots are unable to absorb causing the wilting of plant. In others words, it corresponds to the water potential of soil under which plants cannot maintain turgor pressure, even if a series of defence mechanisms have been triggered (e.g. increased ABA synthesis, stomatal closure, osmotic adjustment, leaf fall) (For more details see the Chapter by Mastrangelo et al.).

Similarly with FC, the value of water content in a soil at WP is not a unique and precise value despite it is conventionally measure at -1.5 MPa (-15 Bar) (Fig. 2). The WP is influenced by the physical and chemical characteristics of soil, but also by the plant species considered. This is because various plant species differ in their ability to deal with low soil water content due to differences in roots anatomy and depth, osmotic adjustment capacity and other defence drought mechanisms.

Conventionally, the *wilting point* is estimated as the water content when the matrix potential of the soil is -1.5 MPa (-15 bar). Nevertheless, some species of plants can absorb water soil at a potential much smaller than this limit. For example olive trees can set a water potential gradient between dry soil (-3 MPa) and leaf (-7 MPa) (Dichio et al., 2006). Similarly, *Larrea divaricata* may absorb water at -6.0 MPa soil water potential (Kirkham, 2005). Another species of the same genus of desert plant (*Larrea tridentata*) can survive with soil water potentials up to -11.5 MPa, maintaining the photosynthetic activity of leaves within the range between -5 and -8 MPa (Fitter & Hay, 2002). These examples serve to explain that the permanent wilting point does not exclusively depend on the soil but also on the plant species. At the *permanent wilting point*, the water potential of soil tends to be less than or equal to the osmotic potential of the plant, which is extremely low in plants adapted to dry environments.

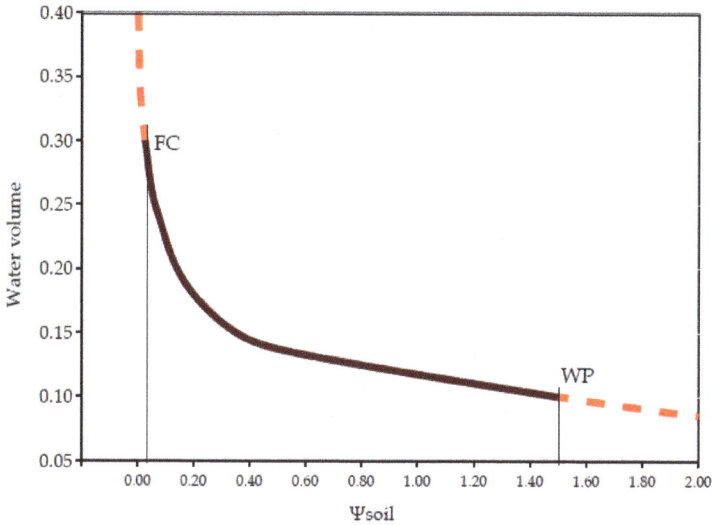

Fig. 2. Variation of the matrix of the water potential of soil (Ψsoil, negative values) in relation to water volume (cm³.cm⁻³ of soil), characterising the limits of the field capacity (FC, -0.03 MPa) and wilting point (WP, -1.5 MPa) of a given soil. The curve was generated from results obtained by Santos, H.P., in Bento Gonçalves-RS, Brazil. 2010.

The indiscriminate use of a fixed value to estimate *field capacity* (FC) and the *permanent wilting point* (WP) can generate false interpretations. However, this reference to the water content in the soil is essential for calculating the *available water content* (AWC) for the plants. The AWC is calculated considering the soil volume explored by roots and the % of water content determined as the difference between FC and WP. Due to this interval of water availability, one may assume that water could be absorbed by the roots with the same facility in the range between FC and WP. For some plants this may be true, given that the energy to extract water from the soil is small, compared to the energy needed to transport the water from the root system to the atmosphere. However, with the reduction of soil water potential, there is also a reduction in its hydraulic conductivity (i.e. water moves slowly in the soil), limiting the water absorption capacity of the roots. In this scene - and for a majority of crops - the yields are reduced if the water content in the soil approaches the *wilting point*. Thus, the available water content should be considered as a relative value and, for the same soil water potential, it may have different proportions of accessibility, depending on the ability of each species to exploit or capture available water.

2.2.2 Water absorption by the roots

As was pointed out in Figure 1, the water flow of a plant is primarily controlled by the transpiration rate. In this flow system it is essential indeed that there are no limitations on water absorption by the root system. As the roots absorb water, there is a reduction in the water potential in the soil that is in contact with the roots (rhizosphere). This process establishes a water potential gradient between the rhizosphere and a neighbouring region of the soil which presents a higher water potential and which coordinates the water movement towards the roots of a transpiring plant (Fig. 3). This water movement in the soil occurs mainly through mass flow due to the fact that the water filled micropores of the soil are interconnected. Therefore, water flows from soil to root at a rate depending on the water potential gradient between soil and plant which is affected by plant water need, hydraulic conductivity of the soil, soil type and soil water content. Sandy soils have higher conductivity due to greater porosity, but they also retain less water in relation to clay soils or soils rich in organic matter.

At *field capacity*, water is initially removed from the centre of the largest pores (spaces ≥ 50 nm, that are too large to have any significant capillary force) between the soil particles, maintaining the water next to the particles due to adhesive forces. The reduction in water content causes a drastic decrease in soil hydraulic conductivity, because the water is replaced by air in the spaces between the soil particles (Fig. 3). Thus, the water movement in the soil is limited to the periphery of soil pores, which can promote restrictions in the hydraulic conductivity to the root surface and reach the *permanent wilting point* (discussed previously).

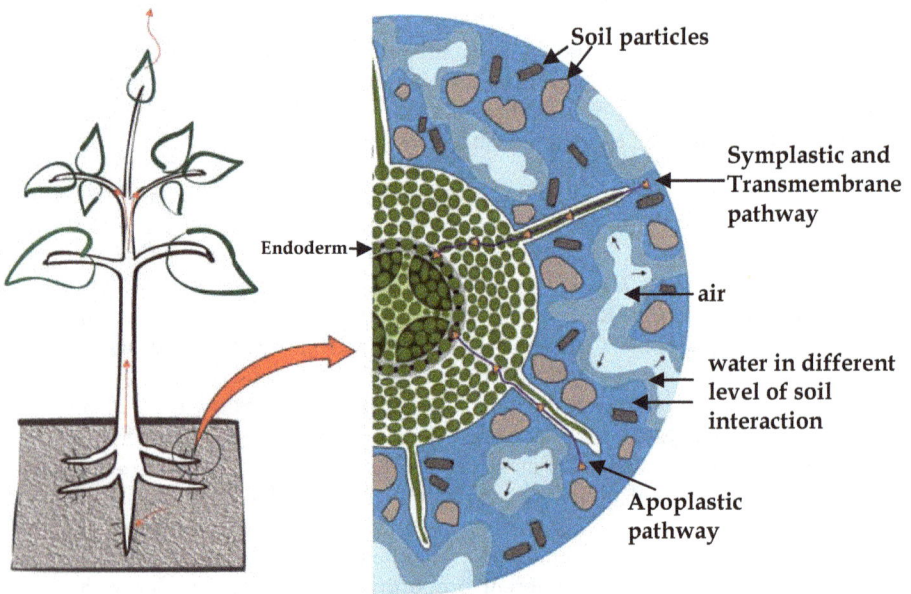

Fig. 3. Detail of the rhizosphere. Note the water adsorbed in soil particles. As the water is absorbed by the roots, the open spaces filled with air increase (small arrows). During absorption, water can flow by symplastic, apoplastic and transmembrane pathways (detailed in the text) to the endoderm cells, where the Casparian strip is present (represented by black points between the endoderm cells).

The water absorption by the roots is related to its surface directly in contact with soil. Thus, longer and younger (less suberised) roots with more root hairs are essential for increasing the contact surface and improve the water absorption capacity of the soil (Fig. 3). Moreover, the distribution and proportion of the roots is very important for meeting the water demand of a plant. In humid regions, as tropical rain forest, plants usually do not require very extensive root systems (i.e. root:shoot ratio < 0.15, Abdala et al., 1998), because a small volume of soil can meet the demands of transpiration. In addition, the water absorbed from that small soil volume is frequently (and easily) replenished by rainfall. This condition in turn induces a reduction of the root:shoot ratio. On the other hand, in dry regions, the plants invest more in their roots, increasing the root:shoot ratio such that the roots can represent upper to 90% of a plant biomass in some species of a desert climate, such as observed in some species from open areas of the Bana woodland in southern Venezuela (i.e. root:shoot > 5, Bongers et al., 1985) and from savanna in Brazil (Abdala et al., 1998). It is important to note that the use of this root:shoot relation in the classification of plants with respect to their *habitat* must be made with caution. In many species, a higher investment in roots is more related to the accumulation of reserves and not specifically to an increased root surface for water absorption (e.g. *Manihot spp.*). A higher investment in roots can also support a process called *hydraulic lift*, when the roots translocate the water from the soil positions with a greater water potential (for example, deeper) to soil positions with a less negative water potential. This process promotes a *hydraulic redistribution* (Burgess & Bleby, 2006) in the soil independently from plant transpiration, because it occurs when the stomata are closed (e.g. at night in C_3 and C_4 plants, and during the day in CAM - Crassulacean Acid Metabolism - plants).

With the water reaching the roots, the absorption process is directly dependent on the water potential gradient between the rhizosphere and the root xylem. There are two ways to establish this gradient, characterised by two absorption processes: 1) *osmotically driven absorption*, common in plants with low transpiration activity; and 2) *passive absorption*, which dominates in plants with high transpiration activity. The osmotically driven absorption occurs in plants under conditions of heat and non-limiting water availability in the soil, but with a restricted capacity of transpiration (for example, without leaves or with a limited vapour pressure deficit). In these cases, there is an accumulation of solutes in xylem vessels (for example, sucrose by degradation of starch reserves in the roots), reducing the xylem water potential in relation to the soil water potential (Kramer & Boyer, 1995). This condition results in water absorption and an increase in root pressure, which is itself responsible for the guttation that means the leaf water output through the hydathodes (pores located at the margins of the leaves) (Fig. 4). Moreover, increase in root pressure also promotes the water exudation in lesions of branches, easily observed in some species (e.g. exudation on branches of grape and kiwifruit plants after pruning in early spring).

In passive absorption and with an increasing rate of transpiration, the tension in the xylem vessels increases, indicating a predominance of the pressure potential influence upon the osmotic potential through the establishment of the water potential gradient between the root xylem and the rhizosphere. Under these conditions, the roots become a passive absorption organ, where the water is sucked into a mass flow promoted by the transpiration activity of aerial parts of the plant. A grapevine, for example, which during its annual growth and production cycle transpires between 650 to 900 mm of water, in accordance to environmental conditions where it is growing, and this volume corresponds to about 85% of all its absorbed water (Mullins *et al.*, 1992).

Fig. 4. Detail of a leaf of a wheat plant (*Triticum aestivum* L.) presenting guttation in the morning. Photo: Ana Cláudia Pedersen.

The water intake in the roots can follow three ways into the root tissue in relation to the route of the epidermis to the endoderm of the root, called *radial water transport* (Fig. 3): 1) *apoplastic*, where the water moves through the intercellular spaces and does not pass through any membranes, exclusively occupying the continuous network of the cell walls; 2) *symplastic*, where the water moves exclusively from one cell to another through plasmodesmata connections; and 3) *transmembrane*, which corresponds to a mixed path between the first two, where the water goes in one direction through the root tissue, entering (symplastic) and exiting (apoplastic) cells. The relative importance of these pathways is still a cause of much discussion, but there is some evidence for the suggestion that plants displaying low transpiration activity predominantly witness symplastic transport, while those displaying high transpiration activity witness a greater proportion of apoplastic transport (Boyer, 1985; Steudle, 2001). Another important detail in relation to these different pathways is relevant only in the outer layers of the root tissue, because in the endoderm the water apoplastic flow is limited due to the Casparian strip (Fig. 3). In this hydrophobic barrier, the radial and transverse endodermal cell walls are impregnated with lignin, suberin, structural wall proteins and wax. Note that in many plants this barrier also occurs in the epidermal cells, forming a double layered hydrophobic barrier in the roots (Enstone et al., 2003). It is important to note that the Casparian strip does not always establish a barrier that is totally impermeable to water and solutes coming from the soil. This can be observed in - for example - the development of young roots where pericycle growth can break parts of the endoderm and allow free access to water until the reconstitution of the tissue.

With regard to water absorption control in the roots, plants also present a family of membrane water transporter proteins (water-channel proteins), called *aquaporins*. These proteins have a critical role in water absorption, reducing the resistance to the water flow along the transcellular path. The number of these proteins available for the root surface is variable throughout the day, being higher during the photoperiod due to the higher demands of photo-transpiration. The *aquaporins* are controlled by many **endogenous and exogenous** factors of the roots, such as pathogens, phosphorylation, pH, solute gradient, temperature and all environment factors that interfere in hydraulic conductance along the water flow by the plant (Chaumont et al., 2005; Maurel et al., 2008).

2.2.3 Ascension of water through the plant: Vascular system

The presence of plants outside the water environment - among other factors - has been related to the evolution of the vascular system, which allows for the speedy upward movement of water to meet the demand of transpiration from the leaves. Water supply through cells by diffusion (difference in chemical gradient) alone is not able to maintain the hydration of a perspiring canopy plant. The need for a vascular system is more evident when we observe the hydraulic dynamic of a tree during a hot day, which demands a large flow of water (for example, 200 to 400 liters day^{-1}) to fit a transpiring surface that is situated along elevated positions, and in some species is higher than 100 meters (e.g. *Sequoiadendron gigantea*).

The water flows from the roots to the shoot of the plant through the xylem. The general mechanism to explain this upward movement of water is the *cohesion-tension theory*, which was proposed in the late 19th century. Basically, this theory holds that the water evaporated in leaves establishes a tensile strength in the xylem, where the hydrogen bonds provide a continuous intermolecular attraction (cohesion) between the water molecules from the leaf to the root (Fig. 1). Thus, the water column in the xylem lumen is driven out of a region with a higher water potential, i.e. from the root and the stem, to a region with a lower water potential, as the leaves, and finally toward to the air that can reach very low water potential (e.g. -100 MPa, at 50% of air relative humidity).

Recently, the cohesion theory has been questioned as a result of assessments of tension in xylem vessels, which do not present a direct relation with tension values measured on leaves through pressure chambers. Furthermore, it is assumed that the hydrophobic interaction between the internal walls of the xylem and the sap composition (lipids, proteins, polysaccharides etc.) prevents the development of a tensile strength larger than 1 MPa (Zimmerman et al., 2004), which is smaller than the estimated tension of rising water in a 30 m high tree (3 MPa). However, despite these questions, many studies argue that the fundamentals of the *cohesion-tension theory* are still valid for explaining the water flow in the continuous soil-plant-atmosphere (Richter, 2001; Steudle, 2001; Cochard, 2002; Tyree, 2003). These elements support the idea that the water column of leaves to the roots provides an auto-regulation mechanism between the process of loss and absorption of water by the plants. Therefore, although the importance of the *cohesion-tension theory* has been neglected by some critics, this mechanism is considered to be essential for the survival of plants during the transpiration process, i.e. the loss of water.

With rising water in the trunk, in addition to pressure force, there is also capillarity strength in the vessels. In a perspiring plant, the water moves continuously from the xylem bundles to the intercellular spaces in the leaves, where the water potential is lower. Due to capillarity strength, water which evaporates through leaf stomata is replaced by the water contained in the lumen of the vascular bundles. In physiological temperatures (25°C), the cohesive forces between the water molecules are sufficient to prevent the disruption of the water column. This tension and the capillarity forces present in the vascular bundles also present resistance to the water flow along the plant by two major ways: 1) the inherent properties of the xylem flow and 2) the geometric aspects of the xylem conduits (*vessel elements* and *tracheids*). In this respect, it is notable that plants with the *vessel elements* of xylem can present a significantly lower hydraulic resistance than plants with *tracheids* (Tyree & Zimmermann, 2002). As such, the xylem diameter has a great influence on the hydraulic conductivity or water flow (J_v, mm s^{-1}), according to the Hagen-Poiseuille equation which describes the transport of fluids in ideal capillaries:

$$J_v = (\pi R4 \Delta \Psi) / 8 \eta L \tag{1}$$

In this equation, $\Delta \Psi$ is the difference in the water potential (MPa) between two points of observation throughout the capillary, R is the radius of the capillary (mm) with a determined length L (mm), through which occurs a flow with a constant viscosity η (1,002 x 10^{-3} Pa second^{-1}, at 20°C) (Nobel, 2009). Accordingly, this equation shows that J_v is proportional to the fourth power of the vessel conductor diameter (the xylem). According to this logic, a trunk that presents few xylem vessels with large diameter has a greater J_v than a trunk with the same xylem area distributed in a larger number of vessels of smaller diameter. In general, xylem vessels with larger diameter are also longer. Xylem vessel diameter is also variable throughout the growth season, being larger in those vessels formed early in the growing cycle. This makes growth rings visible in transversal cuts of tree species. The vessels diameter is an important factor in preventing cavitation or embolism (formation of air bubbles by breaking the water column under high tensile strength, such as values close to -30 MPa), with thinner vessels being less susceptible to such a water column breakage. Generally, the cross-sectional area of vascular bundles is proportional to the transpiring leaf surface. This may be observed in plants adapted to arid environments, which have thinner vessels and a small transpiring surface, as reduced root:shoot ratio.

2.2.4 Leaf water and transpiration

Returning to our lampion scheme (Fig. 1), we emphasise that the leaves are the final frontier of the water flow in the continuous soil-plant-atmosphere system. In leaf mesophyll there is an extensive system of intercellular spaces - present in cell walls - which correspond to the internal surface of water contact with the air. By this interface between the cell walls and the intercellular spaces is established a water potential gradient, mobilising the water by the cell walls from the final extremities of the xylem bundles. This water flow by the cell walls occurs in an analogous way to the principles of water movement through the soil matrix, and the water interacts with cellulose microfibrils and other hydrophilic components of the cell wall (Fig. 5). Due to the high surface tension and as a result of water evaporation in the surface of the cell walls which are in contact with the air in the intercellular spaces, it is established that the tensile strength is transmitted to the xylem. Therefore, it is the tensile strength that drives the upward flow of the water column from the root and is produced in the internal evaporation process in the leaves. The maintenance of this process - as a result - depends on the output (in the atmosphere) of water vapour present in the intercellular spaces. As the leaf cuticle represents a barrier to water outlet - allowing on average only 5% of water permeability - the water vapour moves from leaf intercellular spaces to the atmosphere predominantly through stomatal diffusion. This process of the loss of water vapour by the leaves is called *transpiration* and corresponds to the majority (90%) of the volume of water absorbed by plants.

Transpiration has a number of positive effects (e.g. helps with mineral transport and leaf cooling) however it also may contribute to induce water stress when soil dry.

In the *continuum* soil-plant-atmosphere of water flow, there are two major factors determining the water potential of a plant: 1) the water potential of the soil, which characterises the water supply; and 2) transpiration, which defines the loss of water. The plant, which is an intermediate in this process, may regulate the water potential gradient between the soil and the atmosphere primarily through the regulation of stomatal conductance.

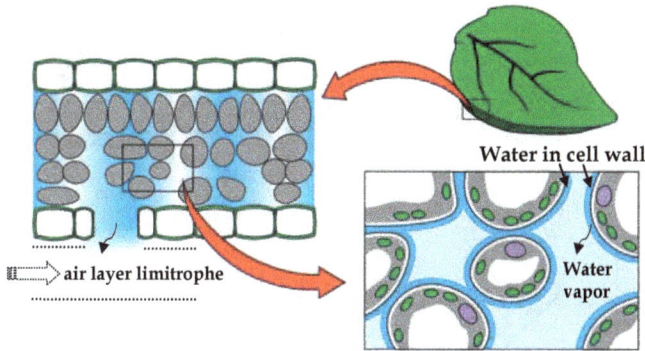

Fig. 5. Leaf water. The evaporation from the cell walls of mesophyll should be noted. The water vapour escapes through the stomatal opening and this flow is directly influenced by the boundary layer of air.

The stomata have a quick and fine control of the water relations of a plant, coordinating the control of the water potential gradient between the leaf and the air. In this interface of the leaf with the environment, it is important to note that small changes in the relative humidity of the air are reflected in major changes in the water potential gradient, which requires a stomatal control so as to maintain the water stability of the plant. A simple variation from 100% to 99% of the relative humidity already corresponds to a decrease of -1.36 MPa in the water potential of the air. This decrease becomes more evident in the water potential of the air in average (80%) and in extreme (50%) conditions of relative humidity, which respectively provide values of -14 and -93.6 MPa at 20°C (Nobel, 2009). If we compare the water potentials of the air with the average water potential of a mesophyte plant (-0.5 MPa), the high gradient always determines that the water is diffused from the leaves to the air (Fig. 4). Throughout this water route between the leaf and the air there are two components that can exert resistance to the diffusion process: 1) *stomatal resistance*, which is coordinated by the stomatal opening; and 2) *resistance of the air boundary layer*, which is located closest to the leaf surface (Fig. 5) and it is directly influenced by wind speed. The higher the speed of the wind, the greater is the frequency of air renewal in this layer surrounding the leaf, restricting diffusion resistance for the maintenance of a major gradient of the water potential. Morphological and anatomical variations among leaves can interfere with the speed of displacement of this thin layer of air, restricting the rate of transpiration in dry environments. Among these modifications, we highlight the presence of hair, the stomata located at the lower surface of the leaf, and the shape and size of the leaves. Although these changes interfere directly with the transpiration rate, they do not exercise a variable and instantaneous control in relation to the ambient conditions, as is done by the stomata.

During the day, great changes occur in the water potential along the soil-plant-atmosphere system. Initially, let us consider a mesophyte plant in a constant atmospheric condition of 75% relative humidity at 20°C (-39 MPa of air water potential) and soil without water restriction (at *field capacity*). During the nocturnal period, the transpiration is virtually nil, by stomatal closure, promoting an equilibrium between soil, root and leaf water potentials leading the potential gradient to be near zero (Fig. 6). With the first rays of the sun, at dawn, the stomata opens, allowing the water diffusion of the leaf (transpiration) and, as a consequence, reducing the leaf water potential.

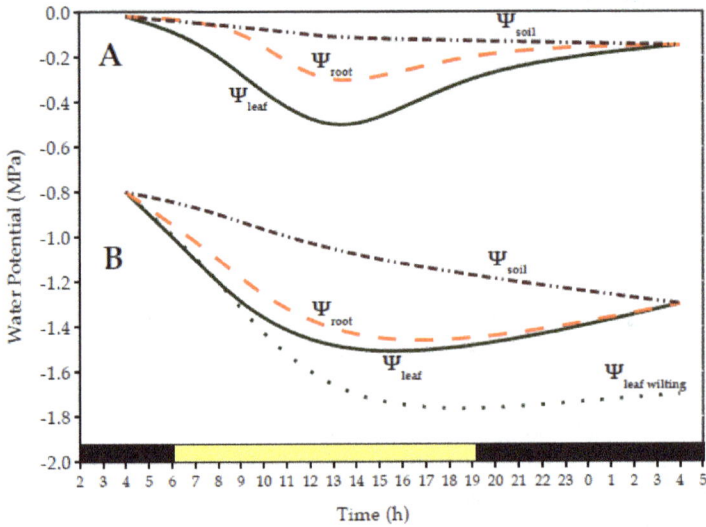

Fig. 6. Schematic daily variation of soil (Ψ_{soil}), root (Ψ_{root}) and leaf (Ψ_{leaf}) water potential of a plant well hydrated (A) and under water restriction (B).

When leaf water potential does not equilibrate with Ψ_{soil} at the end of day (case B), the permanent wilting point ($\Psi_{leaf\,wilting}$) is reached.

As a result - and with a delay that depends on the cohesion-tension forces of water and the water column size between the leaf and root - it begins the reduction of the root water potential. This reduction is slight (on average -0.3 MPa), due to high water availability of the soil. The reduction of the water potential in the plant reaches the minimum limits during the hottest times of the day, forcing the stomata to close for small intervals for extreme cases of transpiration demand (Fig. 6A). At dusk, these variations of the water potential in the leaf and the root are reversed.

In considering a plant under the same atmospheric conditions, but with severe water restriction (a soil water potential near to the permanent wilting point), the water potentials of the leaf and root necessarily reach values that are more negative for the sustainability of the water flow by the plant. At high limits of negative potential, the differences among the water potential of the leaf and root are smaller, but with great difference in relation to the soil water potential (Fig. 6B). If this condition persists, there will be a decrease in the turgor pressure of the leaves, causing temporary leaf wilting, which is recomposed during the nocturnal period. Mesophytic plants can tolerate this reduction in soil moisture up to the limit of -1.5 MPa, while some xerophyte plants can reach limits of -5.5 MPa (Nobel, 2009).

3. Physiological and biochemical aspects of water in plants

In a plant cell, the water is predominantly located in the vacuole and represents the majority of the mass of growing tissue (on average 90%). This predominance is due to the importance that this universal solvent represents in the physiological and biochemical processes of a plant cell. At the cellular level - due to its polar structure - the water acts on the dissolution and mobilisation of ions and organic metabolites, such as amino acids, proteins, carbohydrates and hormones. These water properties are also crucial for the flow between parts of the plant, acting directly in the transportation of nutrients, carbohydrates and hormones. The water acts in membrane integrity and the support of herbaceous plants through the cells turgor pressure, which varies between 1 and 5 MPa. This turgor pressure is also essential in the expansion process of cell walls and in growing tissues. This subsection will expose some physiological and biochemical aspects of the water relations in plants.

3.1 Water deficit and its effects on plant growth

When a plant is under water deficit, responses at physiological, biochemical and molecular scale are triggered (Shao et al., 2008). Physiological responses are linked to a condition of recognition of stress by the root system, turgor changes and water potential and, consequently stomatal conductance, internal CO_2 concentration and photosynthetic activity decrease. In biochemical terms, there will be a decrease in the photochemical activity of photosynthesis, rubisco enzyme activity and the accumulation of secondary metabolites linked to stress (such as glutathione and polyamines). From a molecular perspective, several genes expressed under stress conditions are activated, such as genes linked to the biosynthesis of abscisic acid and the synthesis of specific proteins.

As the cells hydration is reduced and the plant goes into a condition of water deficit, abscisic acid and solutes increase in the plant, especially in the root system (the increase in solutes occurs in a relative manner, due to water reduction). These factors will reduce stomatal conductance and, consequently, photosynthetic activity which ultimately will result in a reduction in the synthesis of proteins and cell walls, as well as a decrease in the rate of cell expansion (Taiz & Zeiger, 2004). The sum of these responses to water deficit contribute to explain the reduction of plant growth.

3.2 Elements that define water demand

Basically, according the cohesion-tension theory already described, the water demand of plants is generated by a gradient at the top of the plant which generates negative hydrostatic pressure that "suck" soil water through the plant into the atmosphere. Two main environmental factors will determine this *evaporative demand*: wind speed and solar radiation (Chavarria et al., 2009). Thus, those plants that exist in environments with high winds (e.g. 20 km h^{-1} or more) or intense solar radiation (e.g. 2500 μE m^{-2} s^{-1} of photosynthetically active radiation) will suffer a greater water loss to the atmosphere. These plants need to make use of water control mechanisms in order to tolerate these environmental conditions. Plants can suffer morphological and anatomical alterations and osmotic regulation to improve stomatal resistance and increase water absorption through the root system.

3.3 Mechanisms of water status regulation

3.3.1 Morphological and anatomical characteristics associated with water control

Plants live with a constant dilemma, namely to undergo photosynthesis while preventing water loss. Every time plants open their stomata to allow the influx of carbon dioxide, they lose water by diffusion.

When plants achieved their *"terrestrial status"* and left behind the conditions of *algae*, they required several changes to adapt to this new environment. They acquired a root system for physical support and absorption of water and nutrients, vascularity for the movement of water and photoassimilates, and a stomatal complex and a wax layer for the regulation of water loss to the atmosphere.

It is known that intense solar radiation is accompanied by elevated temperatures, which can cause severe physical damage to leaves and lead to senescence and leaf abscission or, on small scale, reduce carbon assimilation. This situation forces plants to adopt strategies to minimise such negative effects.

In order to adjust the balance between water availability and atmospheric demand, plants can reduce the size and number of leaves. The organisation of the mesophyll and leaf dimensions is modified by water restrictions, and can provide a strategy for plants to affect the stomatal conductance and CO_2 diffusion (Evans et al., 2009). Plants which need greater efficiency in water use (μmol CO_2 fixed / H_2O transpired) have a strategic demand to retain higher CO_2 concentrations inside. Characteristics of leaf anatomy related to water deficit are reduction of thickness, higher cell density and smaller intercellular spaces, all of which try to mitigate the problems of excessive water loss (Chartzoulakis et al., 2002). The reduction in cell size and, consequently, the reduction of tissues are associated with the turgor decrease of cells (Ogbonnaya et al., 1998). However, smaller cells may stay more *turgid* when compared to larger cells, having better capacity to tolerate conditions of water restriction (Burghardt et al., 2008). These changes in cell size result in an increased internal surface to of CO_2 exchange per leaf area, seeking to maintain the photosynthetic rate with reduced stomatal conductance (Ennajeh et al., 2010; Syvertsen et al., 1995).

The relation of the root with shoots is also strongly modified due to conditions of water unavailability. One way that plants use to keep their water status stable is by reducing the growth of the shoot, in order to reduce leaf area and water loss to the atmosphere. At the same time, plants can invest in the growth of the root system, in order to increase the soil volume explored and in turn water absorption. All these mechanisms occur through hormonal changes as initiated by abscisic acid signals. Nevertheless, later on it will be the effects of interactions between auxins, cytokines and gibberellins which define the relation between root and shoot.

Anatomically, the change in vascular diameter may be a response to water deficit conditions and tends to decrease under water deficit (Kutlu et al., 2009). By reducing the radius of vessels, xylem conductivity is reduced (increased resistance) according to equation 1. There are some situations of high *evaporative demand* which increase the tension in the xylem vessels, causing a disruption of the water column and the formation of air bubbles by embolism (Tyree & Sperry, 1989). The disruption of the water column can also occur under conditions where the water freezes inside the plant.

Another mechanism which protects the leaf from attack by insects and helps the plant to avoid water loss is the presence of leaf trichomes (Molina-Montenegro et al., 2006). Trichomes reduce water loss by: 1) reducing the arrival of solar radiation on the leaf surface; and 2) reducing the interference of wind on the boundary layer, which reduces the differences in water potential between the leaf and the atmosphere (Ehleringer, 1984; Vogelmann, 1993).

Leaf cuticle presents variations in anatomical position and chemical composition. Chemically, these cuticles are characterised by two specific groups of lipids: 1) cutin, which forms the support structure of the membrane; and 2) waxes deposited on the external surface - called *epicuticular waxes* - and also strongly dispersed within the matrix of the cutin (below the surface) called *cuticular waxes* (Devine et al., 1993). The most important function of epicuticular waxes is to avoid leaf water loss to the environment. In addition, they reduce leaf nutrient loss, prevent excessive solar radiation, pathogenic microorganisms, cooling, wind damage and physical abrasion (Vigh et al., 1981; Mendgen, 1996; Barnes & Cardoso-Vilhena, 1996; Scherbatskoy & Tyree, 1990; Kerstiens, 1996; Eglinton & Hamilton, 1967). Epicuticular waxes may be amorphous, have a flat format covering the entire surface area of the leaf, or have the shape of a crystal or blade. Involved in the formation of these waxes are alkanes, esters, ketones and alcohols (Shepherd et al., 1995).

An important aspect of plant morphology is the density of leaf veins. Angiosperms average 8 mm of vein per mm^2 of leaf area, while non-angiosperms have consistently averaged close to 2 mm mm^{-2} throughout 380 million years of evolution (Boyce et al., 2009). This was an important ecological strategy for the increment of photosynthesis, especially under conditions of higher temperatures and transpiration rates.

The most important anatomical tools to optimise the plant water use is the stomatal complex. Variations in opening, size and position of the stomata in the leaf help plants grow under conditions of water deficit (Larcher, 1995).

3.3.2 Stomata metabolism

Stomatal complexes are present in green stems, flowers and fruits, averaging 30 to 400 per square millimetre. They are formed by an opening, called a *pore*; two *guard cells* that are responsible for opening and closing the pore; and in some situations neighbouring cells called *subsidiary cells*, whose function is to support the guard cells. The stomata liberate into the atmosphere around 95% of the water that exits from the leaves and causes 90% of gas exchange (in this case, also involving carbon dioxide), while only 5% of the water output is through the cuticle. There are two types of stomata morphology: kidney-shaped and grass-like, sized between 5 and 15 μm wide and 20 μm long (Fig. 7).

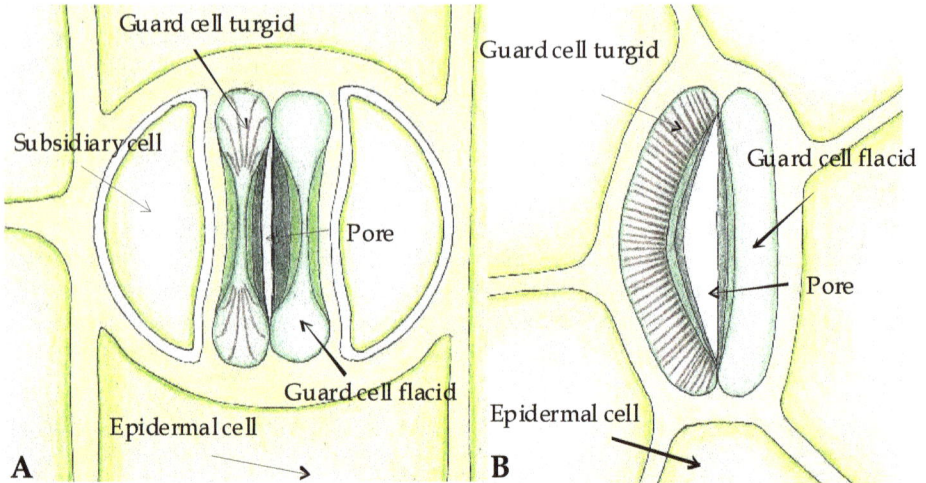

Fig. 7. Stomatal complexes: grass-like (A) and kidney-shaped (B), with the left guard cells turgid (open pore) and the right flaccid (closed pore).

The stomata works as a hydraulic valve regulated by water. When turgid, the stomatal pore opens, and when flaccid due to water loss it closes (Fig. 7). In the early morning, light in the blue band (440-490nm) indicates the arrival of solar radiation and possibility of photosynthetic activity. Thus, the plant opens the stomatal pores to allow influx of carbon dioxide. The first theory about stomatal opening control was called the "starch-sugar hypothesis," and it was widely accepted during the early 1940s. This hypothesis suggested that the hydrolysis of starch in soluble sugars decreased the osmotic potential of cell, promoting water absorption and stomatal opening. However, that theory lost strength as advances were made by studies of potassium movement in the guard cells. Currently, three modes covering the osmoregulation processes of guard cells are accepted: 1) the influx of potassium and chloride through proton pump activation and the synthesis of malate from starch breakdown; 2) sucrose synthesis by starch hydrolysis; and 3) sucrose synthesis by carbon dioxide fixation due to photosynthetic activity.

The blue light signalling process occurs through receptor pigments (phototropins). Thereafter, the opening process occurs by the reduction in the osmotic potential of guard cells by potassium input, consuming ATP by ATPase located in the plasmatic membrane. This ATPase release protons inside the guard cells, causing a variation in pH around 0.5 to 1, which enables membrane hyperpolarisation and, consequently, the opening of channels favours potassium absorption. The resulting osmotic gradient causes water movement towards these cells. Therefore, the guard cells become turgid and the stomatal pore opens, due to the action of cellulose microfibrils (Fig. 7). However, it is important to emphasise that chloride and malate also contribute to the establishment of the osmotic gradient in guard cells exposed to blue light.

When solar radiation begins to decrease this affects the photochemical phase reducing levels of ATP and $NADPH^+$ resulting in losses for the biochemical phase. Internally, the carbon dioxide content begins to rise as they are not being used in the carboxylation process in the Calvin-Benson cycle (C_3). At the same time, calcium ions play an important role when they enter the guard cells, causing solute output and decreasing the osmotic potential of these cells, making the stomata close.

An atmosphere enriched with carbon dioxide can favour the photosynthetic activity of plants with the C_3 mechanism. However, excessive carbon availability will result in stomatal closure in some species, even with a C_3 mechanism (Ainsworth & Rogers, 2007). An example of natural CO_2 enriched conditions, where the plants can be submitted to excessive concentrations of carbon dioxide, occurs near volcanic activity (Miglietta & Raschi, 1993).

3.3 Hormonal and molecular responses in different water conditions

The plants that display a higher production capacity, due to the morphological, physiological and metabolic changes of their organs and cells, tend to present higher demands on available resources and, consequently, possess greater vulnerability to conditions of water restriction. However, there are several strategies for the adaptation to dry environments that can be considered as a tool for progress in overcoming limiting conditions on growth and production. Generally, some plants can accumulate water to delay or escape such stress conditions, while others can deal with the stress through decreased metabolic activity (Bartels, 2005). The effect of water restriction depends on the degree and duration of the stress, the stage of plant development, the genotypic ability of the species and environmental interactions. In recent years, and mainly due to climate changes, several studies have looked to understand the biochemical and molecular basis of water stress (Yokota et al., 2006; Xoconostle-Cázares et al., 2010). Water stress can influence a plant on several levels, with cell expansion and growth being the first processes to respond to water limitation. With the gradual increase of stress, other processes are also affected, such as photosynthesis and allocation of assimilates. At the cellular level, membranes and proteins can be damaged with the increment of reactive oxygen species or peroxidation. These responses are common with other abiotic stresses, such as the effects of salinity and low temperatures, resulting in the synthesis of a similar group of proteins (Artlip & Wisniewski, 2002).

Tolerance to water deficit can be manifested in four ways: 1) the seasonal adjustment of growth to avoid stressful conditions; 2) morphological adaptations, such as an increase in the root:shoot ratio, a reduction in the leaf area and wax accumulation on the leaf surface; 3) physiological adaptations, such as stomatal responses and leaf abscission; and 4) metabolic changes. Among the metabolic changes of adaptation, osmotic adjustment represents the most common change and results from the accumulation of certain metabolites.

The root has been generally accepted as the organ which acts on the perception of water stress, although so far there is no knowledge about how the cells in roots perceive the soil moisture content. However, when the plant water potential is influenced by the water restriction of soil or saline stress, the stomata must respond quickly to avoid water deficit. Signalling between the root and stomata is carried out by abscisic acid (ABA), which has been considered to be a *"plant hormone of stress"* due to its participation in the signalling networks of other factors of stress (Artlip & Wisniewski, 2002). ABA is synthesized from carotenoid by the synthesising enzyme of ABA (zeaxanthin epoxidase, 9-cis-epoxycarotenoid dioxygenase and aldehyde oxidase) which is induced in the root apex or in the parenchyma cells of vascular bundles by water or saline stress. After the synthesis of ABA in the roots, it is transported through the xylem to the leaves alone or conjugated with glucose (the latter being more appropriate for transport over long distances). The proportion of each form of transport (alone or conjugated) is variable between species (Sauter et al., 2002). Once in the leaf, conjugated ABA is hydrolysed into its free form by the apoplastic enzyme β-D-glucosidase, inducing stomatal closure through a signalling system in the guard cells of chloroplasts (Yokota et al., 2006).

In addition to this long distance signalling between the roots and the stomata, recent studies also point out that the leaves must act as sensors of relative humidity in order to avoid desiccation. This is clear from observations of the extremely quick closure of stomata with increments in the vapour pressure gradient between the leaf and the air, even when there is adequate water availability in the soil (Assmann et al., 2000). Recent works expose the possibility that leaf sensors of relative humidity are located in or near their own stomata guard cells (Yokota et al., 2006). In addition, there is genetic evidence that leaf sensitivity to relative humidity (RH) is related to ABA metabolism, exerting a hormonal effect over a short distance. A recent study on genetic selection, based on infrared thermal imaging, identified two genes (*OST1* and *ABA2*) that are directly involved in the signalling route of RH sensing in guard cells (Xie et al., 2006). *OST1* codes a protein kinase that is involved with stomata closure, while *ABA2* codes an enzyme involved in ABA biosynthesis. This reinforces the involvement of ABA as a mediator in a signalling network of guard cells, which can be shared between different stimuli to control stomata closure (Yokota et al., 2006).

ABA has been related to quantitative and qualitative variations in the gene expression and protein synthesis stimulated by water stress (Artlip & Wisniewski, 2002). Meanwhile, it is notable that some of the proteins that are *de novo* synthesized do not appear in the responses to the application of abscisic acid (ABA) and these signals/response routes to the water stress have ABA-dependent and ABA-independent routes (Yokota et al., 2006). In ABA-independent routes, the signal molecule provided by the roots is still unknown. Bioinformatics analysis has promoted advances in the identification of several factors of transcription that are induced by water deficit, classified in six major groups: AP2/ERF (APETALA2/ethylene-response factor); bZIP (Basic leucine-zipper protein); MYB/MYC Zinc-finger protein; CDT-1; NAC and Dreb (Xoconostle-Cázares et al., 2010). In ABA-dependent routes, the promoters of genes containing a *cis*-sequence of six nucleotides are known as the *ABA response element* (ABRE). The genes' expression of ABA-dependence is activated by the AREB/ABF link - a transcription factor of the type bZIP - on the ABRE sequence. Further, in a gene ABA-dependent RD22 expression there is involvement with the transcription factors MYB and MYC, and they are related to the final stages of responses to drought that are ABA dependent (Yokota et al., 2006). The promoters of genes related with responses to drought also have an alternative regulatory sequence of nine nucleotides called a *dehydration response element* (DRE). The DREs are involved in the ABA-independent expressions of genes that are induced under drought conditions. The *trans*-factors to the *cis*-elements are CBF/DREB1 and DREB2, which are expressed transiently after the detection of drought and thereby stimulate the targeted genes involved in drought tolerance (Taiz & Zeiger, 2002; Yokoda et al., 2006).

Genes that are stimulated by drought can be categorised into two groups: 1) coding genes of those proteins responsible for protecting cells and organs against stress; and 2) coding genes of those proteins necessary for signals' translation and regulation of gene expression. The proteins of the first group act directly on membrane functions, the maintenance of water potential, proteins' protection and oxidative stress control. Stands out in this group the family of the *Embryogenesis Abundant protein* (LEA), which is formed by five types of proteins based on the structural domain and which are suspected of acting to protect the cell membrane (Taiz & Zeiger, 2002). Beyond that, and by their hydrophilic properties, they act in water retention and prevent the crystallisation of other proteins and molecules during drying. Within the LEA family itself, the D-11/RAB/Dehydrins group stands out, whose function has been related with the stabilisation of proteins and membranes. This group presents a wide distribution between plant species, and can be considered to be an alternative to the constitutive defence against rapid changes in the *water status* of tissue (Artlip & Wisniewski, 2002). The *aquaporins* family represents another important protein in cell protection against water stress, facilitating water absorption by the plasmatic membrane. Its importance in water relations was recently evidenced by the differential accumulation of aquaporins in relation to the degree of drought tolerance in varieties of beans (Montalvo-Hernández et al., 2008). With regard to membrane protection and the water status of the plant we may also highlight lipid transfer proteins (LTPs) that catalyse the transfer of several classes of phospholipid and glycoproteins for deposits in cell walls or between membrane vesicles (Kader, 1997). The results with LTPs show that these proteins are induced during different conditions of stress and that they can act to increase membrane fluidity, decreasing water loss by increasing tissue impermeability, and as a physical barrier to biotic stress (Treviño & O'Connell, 1998; Maghuly et al., 2009; Gong et al., 2010).

Another protein family has been associated with responses to drought, namely Heat Shock Proteins (HSPs), which are widely distributed in nature. These proteins are known as molecular chaperones, acting in the folding and assembly of functional proteins and in the removal of non-functional proteins, facilitating the recovery of cellular functions after stress. Several HSPs - classified according to their molecular weight - are induced in conditions of water and saline stress, such as HSP70 (the DnaK family), the chaperones GroEL and HSP60, HSP90 and HSP100 and the small HSP (sHSP) (Alamillo et al., 1995; Campalans et al., 2001, Wang et al., 2004). Within these proteins, there is the cyclophilin, which is a chaperone protein with systemic properties and which is highly induced during water stress, conferring multiple tolerances to abiotic stress (Gottschalk et al., 2008; Sekhar et al., 2010). During conditions of stress, the recycling of macromolecules which lose their function to maintain cellular homeostasis is essential. In this process, and under conditions of water stress, an increase of protease activity has been observed (Campalans et al., 1999; Seki et al., 2001), which is important in the destruction of denatured proteins and in the recycling of amino acids necessary to synthesize proteins for water deficit responses. In addition, the ubiquitin and polyubiquitin proteins are also induced with water restriction, both of which act marking proteins for proteolytic degradation (Campalans et al., 1999, Barrera-Figueroa et al., 2007).

3.4 Osmotic regulation

As discussed above, the osmotic regulation process of stomatal opening occurs by the movement of solutes - called *osmolytes* - thereby influencing water movement between the cells. This water movement has several purposes in relation to cell hydration inside plants, such as the stomatal opening and higher water absorption via root system. Several compounds that work in the osmotic regulation of plants are known: carbohydrates (sucrose, sorbitol, mannitol, glycerol, arabinitol, pinitol) (Hare et al., 1998), nitrogen compounds (proteins, betaine, glutamate, aspartate, glycine, choline, putrescine, 4-gamma aminobutiric acid) (Kinnersley & Turano, 2000) and organic acids (malate and oxalate) and so on (Sairam & Tyagi, 2004).

From the standpoint of thermodynamics, there are concepts relating to water potential that are elementary in the explication of water movement. In turgid cells the water potential (Ψw) is composed by pressure ($\Psi \rho$) and osmotic ($\Psi \pi$) potentials. The greatest possibility of interference in the Ψw of plant cells is by varying the amount of internal solutes. Thus, osmotic regulation is a process in which variation of the amount of solutes - and consequently in osmotic potential - interferes in water movement.

The higher the concentration of solutes in a solution, the greater will be the disorder of the system (entropy). Furthermore, this condition leads to more negative osmotic potential and, consequently, to a more negative water potential. The water inside the plant flows from the sites with higher water potentials to those with lower ones. Thus, an increase in solute concentration in the cell favours the entry of water.

The solutes that plants use to decrease the water potential and control the water flow must have a low molecular weight. Among these solutes, the best-studied is potassium (K^+), the lack of which is related to tissue dehydration. However, most references point out that the main role of potassium within the osmoregulation process is the regulation of stomata opening (Läuchli, 1984; Hsiao & Läuchli, 1986). It also acts in water absorption by the roots, in transpiratory control and in the cells' ability to resist low temperatures (Grewal & Singh, 1980). In particular, under conditions of potassium deficit, sugars from starch hydrolysis become more relevant in osmotic regulation than in the process of stomatal opening (Poffenroth et al., 1992).

The proline amino acid also stands out as an osmotic regulator, and is linked to stress both by water deficit and salinity (Molinari et al., 2004; Zhu et al., 2005). Nevertheless, proline does not only exert the function of osmoregulation in plant cells during periods of water deficit, but it can also protect against the activity of free radicals, regulate the pH in the cytoplasm, protect against the denaturation of macromolecules and also act as a source of carbon and nitrogen under conditions of stress (Sharma and Dietz, 2006; Vanrensburg et al., 1993; Sivakumar et al., 2000; Díaz et al., 1999). Another compound with importance in osmotic regulation is glycinebetaine, which, like proline, does not just exert or exercise its function in osmotic potential. These compounds also act in the reduction of reactive oxygen species (ROS) produced under conditions of stress in plants. With the process of tissue dehydration, there is production of ROS, such as the singlet oxygen ($1O_2$), the superoxide radical (O_2-), hydrogen peroxide (H_2O_2) and the hydroxyl radical (OH-) (Liu & Huang, 2000). Reactive oxygen species (ROSs) are produced from oxygen metabolism and play an important role as indicators in the stress process of water deficit. Among these are oxygen ions, peroxides and free radicals, and these compounds will cause oxidative stress in cells and prejudice their operation.

4. Practical aspects of measuring plant water status

4.1 Pression chamber

The most widely used tool for determining water potential is the pressure chamber (Scholander et al., 1965). The measurement is made on the section between a stem or a branch and a vegetative part, which is placed inside a pressure chamber filled with an inert gas (Fig. 8).

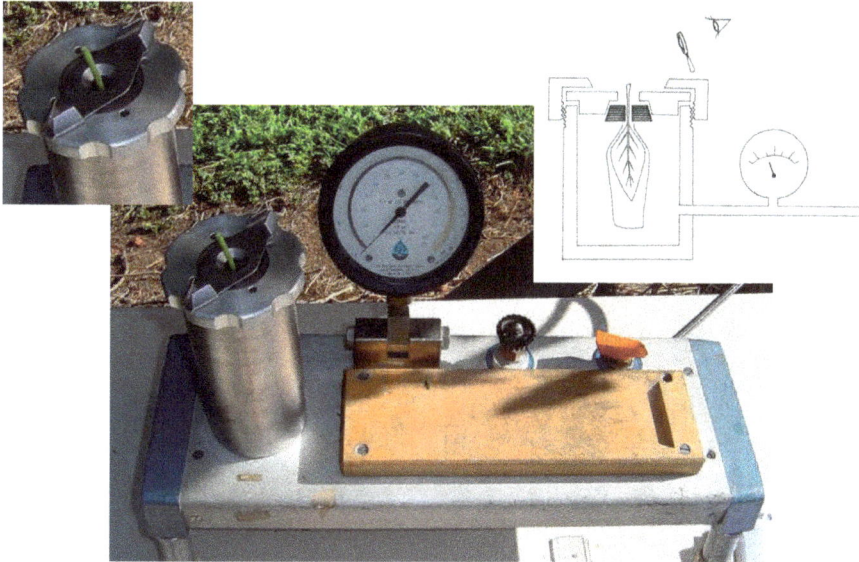

Fig. 8. Pressure chamber used to determine water potential in plants. Photo: Geraldo Chavarria.

After leaf preparation, leaf bagging and chamber sealing, the chamber pressure is increased gradually, resulting in a force inverse to the entry of water, in order to expel water through the observed section of the xylem. Generally, the evaluation of the potential requires the use of a magnifying glass. The pressure exercised to expel water is considered the pressure which is retained in the cells. This technique is used in research for the characterisation of the *water status*, but it may be useful in some crops as a tool for determining the appropriate time for irrigation. Furthermore, some works also point to a high correlation between leaf temperature measured with an infrared thermometer and water potential in the leaf measured with a pressure chamber (Lafitte & Courtois, 2002).

4.2 Porometry

Another technique widely used to measure the water availability in plants is that of stomatal diffusive resistance, using a device called a *porometer*, since the main route of gas exchange between the plant and atmosphere occurs through the stomata. The resistance to water diffusion by the stomata is measured in both sides of the leaf. The equipment may evaluate four different processes: 1) mass flow (air forced through the leaf); 2) vapour diffusion (dry air pumped into the chamber, which is equipped with a sensitive device to detect variations in humidity inside the chamber); 3) maintenance flow (dry air pumped into the chamber and steadily retained, where the flow changes are used to calculate resistance to stomatal diffusion); and 4) state of equilibrium (to monitor the time necessary for the occurrence of equilibrium by applying a dry air flow in the chamber attached to the leaf) (Fig. 9).

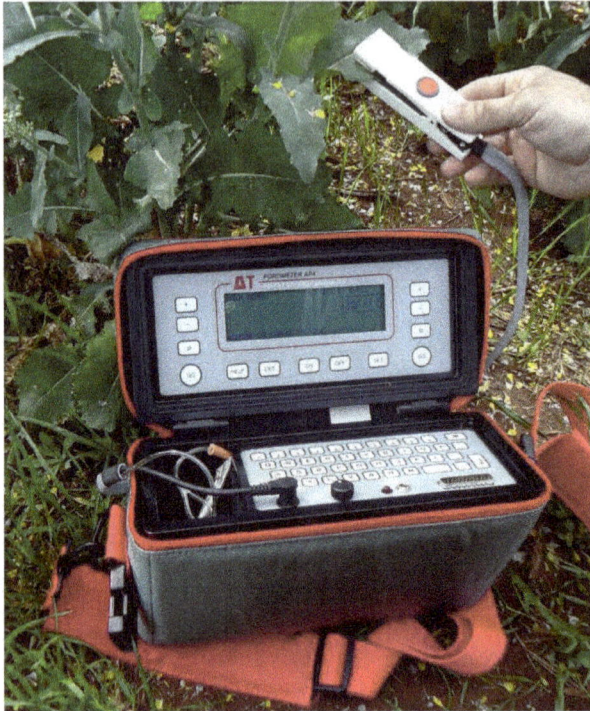

Fig. 9. Stomatal conductance determination using a porometer. Photo: Ana Cláudia Pedersen.

5. Acknowledgment

The authors would like to acknowledge the images of Figures 1, 3 and 4, which were provided by Luciana Mendonça Prado, and 7 and 8 which were provided by Claudinei Crespi. Also thanks to Ana Cláudia Pedersen for the support in the translation and for providing Figures 4 and 9 and to Flavio Bello Fialho for the English review.

6. References

Abdala, G.C.; Caldas, L.S.; Haridasan, M.; Eiten, G. (1998). Above and belowground organic matter and root:shoot ratio in cerrado in central Brazil. *Brazilian Journal of Ecology*, Vol. 2, No. 2, pp. 11-23.

Ainsworth, E. A. & Rogers, A. (2007). The response of photosynthesis and stomatal conductance to rising (CO2): mechanisms and environmental interactions. Plant, Cell and Environment, Vol. 30, N° 1, pp. 258-270.

Alamillo, J.; Almogura, C.; Bartels, D. & Jordano, J. (1995). Constitutive expression of small heat shock proteins in vegetative tissues of the resurrection plant Craterostigma plantagineum. Plant and Molecular Biology, Vol.29, pp. 1093-1099, ISSN 0167-4412

Artlip, T.S. & Wisniewski, M. E. (2002). Induction of Proteins in Response to Biotic and Abiotic Stresses, In: Handbook of Plant and Crop Physiology. M. Pessarakli, (Ed.), 657-679, Marcel Dekker Inc., ISBN 0-8247-0546-7, New York.

Assmann, S.M.; Snyder, J.A. & Lee, Y.R.J. (2000). ABA-deficient (aba1) and ABA-insensitive (abi1-1, abi2-1) mutants of Arabidopsis have a wild-type stomatal response to humidity. Plant Cell and Environment, Vol.23, No. 4, pp. 387-395, ISSN 1365-3040

Barnes, J.D. & Cardoso-Vilhena, J. (1996). Interactions between electromagnetic radiation and cuticle. In: Plant cuticles: an integrated functional approach. G. Kerstiens, (Ed.), 157-174, Bios Scientific Publishers, ISBN 1-85996-130-4, Lancaster

Barrera-Figueroa, B.; Pena-Castro, J.; Acosta-Gallegos, J.A.; Ruiz-Medrano, R. & Xoconostle-Cazares, B. (2007). Isolation of dehydration-responsive genes in a drought tolerant common bean cultivar and expression of a group 3 late embryogenesis abundant mRNA in tolerant and susceptible bean cultivars. Functional Plant Biology, Vol. 34, No. 4, pp. 368-381, ISSN 1445-4408

Bartels, D. & Sunkar, R. (2005). Drought and salt tolerance in plants. Critical Reviews in Plant Sciences, Vol.24, No. 1, pp. 23-58, ISSN 1549-7836

Bartels, D. (2005). Desiccation tolerance studied in the resurrection plant Craterostigma plantagineum. Integrative and Comparative Biology, Vol. 45, No. 5, pp. 696-701, ISSN 1540-7063

Bongers, F.; Engelen, D.; Klinge, H. (1985). Phytomass structure of natural plant communities on spodosols in southern Venezuela: the Bana woodland, Vegetatio, Vol.63, pp. 13-34.

Boyce, C.K.; Brodribb, T.J.; Feild, T.S. & Zwieniecki, M.A. (2009). Angiosperm leaf vein evolution was physiologically and environmentally transformative. Proceedings of the Royal Society B, Vol. 276, pp. 1771–1776, ISSN 1471-2954

Boyer, J.S. (1985). Water transport. Annual Review of Plant Physiology, Vol.36, pp. 473-516, ISSN 0066-4294

Burgess, S. & Bleby, T. (2006). Redistribution of soil water by lateral roots mediated by stem tissues. Journal of Experimental Botany, Vol. 57, No. 12, pp. 3283-3291, ISSN 002-0957

Burghardt, M.; Burghardt, A.; Gall, J.; Rosenberger, C. & Riederer, M. (2008). Ecophysiological adaptations of water relations of Teucrium chamaedrys L. to the hot and dry climate of xeric limestone sites in Franconia (Southern Germany). Flora. Vol. 203, pp. 3–13, ISSN 0367-2530

Campalans, A.; Messeguer, R.; Goday, A. & Pagès, M. (1999). Plant responses to drought, from ABA signal transduction events to the action of the induced proteins. Plant Physiology and Biochemistry, Vol. 37, No. 5, pp. 327–340, ISSN 0981-9428

Campalans, A.; Pagès, M. & Messeguer, R. (2001). Identification of differentially expressed genes by the cDNA-AFLP technique during dehydration of almond (Prunus amygdalus). Tree Physiology, Vol. 21, No. 10, pp. 633-643, ISSN 0829-318X

Chartzoulakis, K.; Patakasb, A.; Kofidisc, G.; Bosabalidisc, A. & Nastoub, A. (2002). Water stress affects leaf anatomy, gas exchange, water relations and growth of two avocado cultivars. *Scientia Horticulturae*, Vol. 95, pp. 39–50, ISSN 0304-4238

Chaumont, F.; Moshelion, M. & Daniels, M.J. (2005). Regulation of plant aquaporin activity. *Biology of the Cell*, Vol. 97, No. 10, pp. 749-764, ISSN 0248-4900

Chavarria, G.; Cardoso, L.S.; Bergamaschi, H.; Santos, H.P. das; Mandelli, F. & Marodin, G.A.B. (2009). Microclimate of vineyards under protected cultivation. *Ciência Rural*, Vol. 39, No. 7, pp. 2029-2034, ISSN 0103-8478

Cochard, H. (2002). A technique for measuring xylem hydraulic conductance under high negative pressures. *Plant, Cell and Environment*, Vol. 25, No. 6, pp. 815–819, ISSN 1365-3040

Devine, M.; Duke, O. S. & Fedtke, C. (1993). *Physiology of Herbicide Action*. Englewood Cliffs, NJ: P.T.R Prentice–Hall.

Díaz, P.; Borsani, O. & Monza, J. (1999). Acumulación de prolina en plantas en respuesta al estrés osmotico. *Agrociencia*, Vol. 3, N° 1, pp. 1-10.

Dichio, B.; Xiloyannis, C.; Sofo A. Montanaro G. (2006). Osmotic regulation in leaves and roots of olive tree (*Olea europaea* L.) during water deficit and rewatering. *Tree Physiology*, 26:179–185.

Eglinton, G. & Hamilton, R.J. (1967). Leaf epicuticular waxes. *Science*, Vol. 156, No. 3780, p. 1322-1335, ISSN 0036-8075

Ehleringer, J.R. (1984). Ecology and ecophysiology of leaf pubescence in North American desert plants. In: *Biology and Chemistry of Plant Trichomes*, E. Rodriguez, P.L. Healy & I. Mehta, (Eds.), 113-132, Plenum Press, ISBN 030-6413-93-0, New York.

Ennajeh, M.; Vadel, A.M, Cochard, H. & Khemir, H. (2010). Comparative impacts of water stress on the leaf anatomy of a drought-resistant and a drought-sensitive olive cultivar. *Journal of Horticultural Science & Biotechnology*, Vol. 85, No. 4, pp. 289–294, ISSN 1462-0316

Enstone, D.E.; Peterson, C.A. & Ma, F. (2003). Root endodermis and exodermis : structure, function, and responses to the environment. *Journal of Plant Growth Regulation*, Vol. 21, No. 4, pp. 335-351, ISSN 0721-7595

Evans, J.R.; Kaldenhoff, R.; Genty, B. & Terashima, I. (2009). Resistances along the CO2 diffusion pathway inside leaves. *Journal of Experimental Botany*. Vol. 60, pp. 2235-2248.

Faludi-Daniel, A. (1981). Stomatal behaviour and cuticular properties of maize leaves of different chilling-resistance during cold treatment. *Physiologia Plantarum*, Vol.24, pp. 287-290, ISSN 0031-9317

Fitter, A. & Hay, R. (2002). *Environmental Physiology of Plants*, Academic Press, ISBN 0-12-257766-3, San Diego, California

Gong, P.; Zhang, J.; Li, H.; Yang, C.; Zhang, C.; Zhang, X.; Khurram, Z.; Zhang, Y.; Wang, T.; Fei, Z. & Ye, Z. (2010). Transcriptional profiles of drought-responsive genes in modulating transcription signal transduction, and biochemical pathways in tomato. *Journal of Experimental Botany*, Vol. 61, No. 13, pp. 3563-3575, ISSN 0022-0957

Gottschalk, M.; Dolgener, E.; Xoconostle-Cazares, B.; Lucas, W.J.; Komor, E. & Schobert, C. (2008). Ricinus communis cyclophilin: functional characterisation of a sieve tube protein involved in protein folding. *Planta*, Vol. 228, No. 4, pp. 687-700, ISSN 0032- 0935

Grewal, J.S. & Singh, S.N. (1980). Effect of potassium nutrition on frost damage and yield of potato plants on alluvial soils of the Punjab (India). *Plant Soil*, Vol. 57, pp. 105–110

Hare, P.D.; Cress, W.A. & Staden, J.V. (1998). Dissecting the roles of osmolyte accumulation during stress. *Plant, Cell and Environment*, Vol. 21, pp. 535–553, ISSN 1365-3040

Hsiao, T.C. & Läuchli, A. (1986). Role of potassium in plant–water relations. In: *Advances in Plant Nutrition*, B. Tinker & A. Läuchli (Eds.), Vol. 2, 281-311, Praeger Scientific, ISBN 0-27592-069-0, New York

IPCC (August 2011). Intergovernmental Panel of Climatic Change WGII, fourth assessment report, 12/08/2011. Available from http://www.ipcc.ch/publications_and_data/ publications_and_data_reports.htm

Kader, J.C. (1997). Lipid-transfer proteins: a puzzling family of plant proteins. *Trends in Plant Science*, Vol. 2, No. 2, pp. 66-70, 1360-1385

Kerstiens, G. (1996). Diffusion of water vapour and gases across cuticles and through stomatal pores presumed closed. In: *Plant cuticles: an integrated functional approach*. G. Kerstiens, (Ed.), 121-134, Bios Scientific Publishers, ISBN 1-85996-130-4, Lancaster

Kinnersley, A.M. & Turano, F.J. (2000). Gamma aminobutyric acid (GABA) and plant responses to stress. *Critical Reviews in Plant Sciences*, Vol. 19, pp. 479–509, ISSN 1549-7836

Kirkham, M.B. (2005). Principles of soil and plant water relations. Elsevier Academic Press, ISBN 0-12-409751-0, San Diego, California.

Kramer, P.J. & Boyer, J.S. (1995). *Water relations of plants and soils*. Elsevier Academic Press, ISBN 0-12-425060-2, San Diego, California

Kutlu N, Terzi R, Tekeli C, Senel G, Battal P & Kadioglu A (2009). Changes in anatomical structure and levels of endogenous phytohormones during leaf rolling in Ctenanthe setosa. *Turkish Journal of Biology*, Vol. 33, pp. 115-122, ISSN 1300-0152

Lafitte, H. R.; Courtois, B. (2002) Interpreting cultivar x environment interactions for yield in upland rice: assigning value to drought-adaptive traits. Crop Science, Vol. 42, No. 1, pp. 1409-1420, ISSN 0011-183X

Larcher W (1995). *Physiological Plant Ecology*. Springer-Verlag, ISBN 978-3540435167, Berlin.

Läuchli, A. (1984). Mechanism of nutrient fluxes at membranes of the root surface and their regulation in the whole plant. In: Roots, Nutrient and Water Influx, and Plant Growth, S.A. Barber; D.R. Bouldin (Eds.), 1-25, ASA Special Publication, ISBN 0891180826, Madison

Levine, L.H.; Richards, J.T. & Wheeler, R.M. (2009). Super-elevated CO_2 interferes with stomatal response to ABA and night closure in soybean (*Glycine max*). Journal of Plant Physiology, Vol.166, No. 9, pp. 903-913, ISSN 0176-1617

Liu X.; Huang B (2000). Heat stress injury in relation to membrane lipid peroxidation in creeping bentgrass. Crop Science, Vol. 40, N°1, pp. 503-510.

Maghuly, F.; Borroto-Fernandez, E.G.; Khan, M. A.; Herndl, A.; Marzban, G. & Laimer, M. (2009). Expression of calmodulin and lipid transfer protein genes in Prunus incisa x serrula under different stress conditions. *Tree Physiology*, Vol. 29, No. 3, pp. 437-444, ISSN 0829-318X

Marschner, H. (1995). *Mineral nutrition of higher plants*. (2Ed). Academic Press, ISBN 0-12-4735428, Michigan matter and root:shoot ratio in cerrado in central Brazil. *Brazilian Journal of Ecology*, Vol. 2, No. 2, pp. 11-23.

Maurel, C.; Verdoucq, L.; Luu, D. & Santoni, V. (2008). Plant aquaporins: membrane channels with multiple integrated functions. *Annual Review of Plant Biology*, Vol. 59, pp. 595-624, ISSN 0066-4294

Mendgen, K. (1996). Fungal attachment and penetration. In: *Plant cuticles: an integrated functional approach.* G. Kerstiens, (Ed.), 175-188, Bios Scientific Publishers, ISBN 1-85996-130-4, Lancaster

Miglietta, F. & Raschi, A. (1993). Studying the effect to elevated CO_2 in the opening a naturally enriched environment in central Italy. *Vegetatio*, Vol. 104, No. 105, pp. 391–400, ISSN 0042-3106

Molina-Montenegro, M.A.; Ávila, P.; Hurtado; Valdivia, A.I. & Gianoli, E. (2006). Leaf trichome density may explain herbivory patterns of Actinote sp. (Lepidoptera: Archaeidae) on Liabum mandonii (Asteraceae) in a montane humid forest (Nor Yungas, Bolivia). *Acta Oecologica*, Vol. 30, pp. 147–150, ISSN 1146-609X

Molinari, H.B.C.,; Marur, C.J.; Bespalhok Filho, J.C.; Kobayashi, A.K.; Pileggi, M.; Leite Júnior, R.P.; Pereira, L.F.P. & Vieira, L.G.E. (2004). Osmoti adjustment in transgenic citrus rootstock Carrizo citrange (Citrus sinensis Osb x Poncirus trifoliate L. Raf) overproducing proline. *Plant Science*, Vol. 167, pp. 1375–81, ISSN 0168-9452

Montalvo-Hernández, L.; Piedra-Ibarra, E.E.; Gómez-Silva, L.; Lira-Carmona, R.; Acosta-Gallegos, J.A.; Vazquez-Medrano, J.; Xoconostle-Cázares, B. & Ruíz-Medrano, R. (2008). Differential accumulation of mRNAs in drought-tolerant and susceptible common bean cultivars in response to water deficit. *The New Phytologist*, Vol. 177, No. 1, pp. 102-113, ISSN 0028-646X

Mullins, G.M.; Bouquet, A.; Williams, L.E. (1992). Biology of the grapevines. Cambridge University Press, New York. 239 p.

Nobel, P.S. (2009). *Physicochemical and environmental plant physiology.* (4th Ed). Elsevier, ISBN 978-0-12-374143-1, London

Ogbonnaya, C.I.; Nwalozie, M.C.; Roy-Macauley, H. & Annerose, D.J.M. (1998). Growth and water relations of Kenaf (Hibiscus cannabinus L.) under water deficit on a sandy soil. *Industrial Crops and Products*, Vol. 8, pp. 65–76, ISSN 0926-6690

Poffenroth, M.; Green, D.B. & Tallman, G. (1992). Sugar concentrations in guard cells of *Vicia faba* illuminated with red or blue light. *Plant Physiology*, Vol. 98, pp. 1460–1471, ISSN 0032-08894

Richter, H. (2001). The cohesion theory debate continues: the pitfalls of cryobiology. *Trends in Plant Science*, Vol. 6, No. 10, pp. 456–457, ISSN 1360-1385

Sairam, R.K. & Tyagi, A. (2004). Physiology and molecular biology of salinity stress tolerance in plants. *Current Science*, Vol. 86, pp. 407–421, ISSN 0011-3891

Sauter, A.; Dietz, K.J. & Hartung, W. (2002). A possible tress physiological role of abscisic acid conjugates in root-to-shoot signalling. *Plant Cell and Environment*, Vol. 25, No. 2, pp. 223-228, ISSN 1365-3040

Scherbatskoy, T. & Tyree, M.T. (1990). Kinetics of exchange of ions between artificial precipitation and maple leaf surfaces. *New Phytologist*, Vol. 114, pp. 703-712, ISSN 0028-646X

Scholander, P.F.; Hammel, H.T.; Hemmingsen, E.A. & Bradstreet, E.D. (1965). Hydrostatic pressure and osmotic potencials in leaves of mangroves and some other plants. *Proceedings of the National Academy Science*, Vol. 51, N° 1, pp.119-125.

Sekhar, K.; Priyanka, B.; Reddy, V.D. & Rao, K.V. (2010). Isolation and characterization of a pigeonpea cyclophilin (CcCYP) gene, and its over-expression in Arabidopsis confers multiple abiotic stress tolerance. *Plant Cell and Environment*, Vol. 33, No. 8, pp. 1324-1338, ISSN 1365-3040

Seki, M.; Narusaka, M.; Abe, H.; Kasuga, M.; Yamaguchi-Shinozaki, K.; Carninci, P.; Hayashizaki, Y. & Shinozaki, K. (2001). Monitoring the Expression Pattern of 1300 Arabidopsis Genes under Drought and Cold Stresses by Using a Full-Length cDNA Microarray. *Plant Cell*, Vol. 13, No. 1, pp.61-72, ISSN 1040-4651

Shao, H.B.; Chu, L.Y.; Jaleel, C.A. & Zhao, C.X. (2008). Water-deficit stress-induced anatomical changes in higher plants. Comptes Rendus Biologies, Vol. 331, No. 3, pp. 215-25, ISSN 1631-0691.

Sharma, S.S. and& K.J. Dietz, 2006. The significance of amino acids and amino acid-derived molecules in plant responses and adaptation to heavy metal stress. *Journal of Experimental Botany*. Vol. 57, pp. 711-726.

Shepherd, T.; Robertson, G.W.; Griffiths, D.W.; Birch, A.N.E. & Duncan, G. (1995). Effects of environment on the composition of epicuticular wax from kale and swede. *Phytochemistry*, Vol. 40, pp. 407-417, ISSN 0031-9422

Sivakumar, P.; Sharmila, P.; Saradhi, P.P. (2000). Proline alleviates salt-stress induced enhancement in Rubisco oxygenase activity. *Biochem Biophys Res Commun*, Vol. 279, N° 1, pp. 512-5.

Steudle, E. (2001). The cohesion-tension mechanism and the acquisition of water by plant roots. *Annual Review of Plant Physiology and Plant Molecular Biology*, Vol. 52, pp. 847–875, ISSN 1040-2519

Syvertsen, J.P.; Lloyd, J.; Mcconchie, C.; Kriedemann, P.E. & Farquhar, G.D. (1995). On the relation between leaf anatomy and CO_2 diffusion through the mesophyll of hypostomatous leaves. *Plant Cell and Environment*, Vol. 18, pp. 149–157, ISSN 1365-3040

Taiz, L. & Zeiger, E. (2002). *Plant physiology*. (3rd Ed.). Sinauer Associates, ISBN 0-87893-823-0, Sunderland.

Treviño, M.B. & O'Connell, M.A. (1998). Three drought-responsive members of the nonspecific lipid-transfer protein gene family in Lycopersicon pennellii show different developmental patterns of expression. *Plant Physiology*, Vol. 116, No. 4, pp. 1461-1468, ISSN 0032-0889

Tyree, M.T. & Sperry, J.S. (1989). Vulnerability of xylem to cavitation and embolism. *Annual Review of Plant Physiology and Plant Molecular Biology*, Vol. 40, pp. 19–38, ISSN 1040-2519

Tyree, M.T. & Zimmermann, M.H. (2002). *Xylem Structure and the Ascent of Sap*. (2Ed). Springer series in wood science, Springer-Verlag, ISBN 3-540-43354-6, Berlin

Tyree, M.T. (2003). Plant hydraulics: the ascent of water. *Nature*, Vol. 423, No. 6943, pp. 923-923, ISSN 0028-0836

Vanrensburg, L.; Kruger, G.H.J.; Kruger, R.H. (1993). Proline accumulation the drought tolerance selection: its relantionship to membrane integrity and chloroplast ultra structure in *Nicotiana tabacum* L. Journal of Plant Physiology, Vol. 141, N°1, pp.188–94.

Vigh, L.; Horváth, I.; Farkas, T.; Mustardy, L.A. & Vogelmann, T.C. (1993). Plant-tissue optics. *Annual Review of Plant Physiology and Plant Molecular Biology*, Vol. 44, pp. 231–251, ISSN 1040-2519

Vogelmann, T.C. (1993). Plant tissue optics. *Annual Review Plant Physiological and Plant Molecular biology*. Vol. 44, pp. 231-251.

Wang, W.X.; Vinocur, B.; Shoseyov, O. & Altman, A. (2004). Role of plant heat-shock proteins and molecular chaperones in the abiotic stress response. *Trends in Plant Science*, Vol. 9, No. 5, pp. 244-252, ISSN 1360-1385

Xie, X.; Wang, Y.; Williamson, L.; Holroyd, G.H.; Tagliavia, C.; Murchie, E.; Teobald, J.; Knight, M.R.; Davies, W.J.; Leyser, H.M. & Hetherington, A.M. (2006). The identification of genes involved in the stomatal response to reduced atmospheric relative humidity. *Current Biology*, Vol. 16, No. 9, pp. 882-887, ISSN 0960-9822

Xoconostle-Cázares, B.; Ramírez-Ortega, F.A.; Flores-Elenes, L. & Ruiz-Medrano, R. (2010). Drought tolerance in crop plants. *American Journal of Plant Physiology*, Vol. 5, No. 5, pp. 241-256, ISSN 0176-1617

Yokota, A.; Takahara, K & Akashi, K. (2006). Water Stress. In: *Physiology and Molecular Biology of Stress Tolerance in Plants*, K.V. M. Rao; A.S. Raghavendra & K.J. Reddy (Eds.), 15-39, Springer, ISBN-10 1-4020-4224-8, Netherlands

Zhu, X.; Gong, H.; Chen, G.; Wang, S. & Zhang, C. (2005). Different solute levels in two spring wheat cultivars induced by progressive field water stress at different developmental stages. *Journal of Arid Environments*, Vol. 62, pp. 1–14, ISSN 0140-1963

Zimmermann, U.; Schneider, H; Wegner, L.H. & Haase, A. (2004). Water ascent in tall trees: does evolution of land plants rely on a highly metastable state? *New Phytologist*, Vol. 162, No. 3, pp. 575–615, ISSN 1469-8137

6

Defence Strategies of Annual Plants Against Drought

Eszter Nemeskéri[1], Krisztina Molnár[2], Róbert Víg[3],
Attila Dobos[3] and János Nagy[2]
*[1]Research Institute and Model Farms, Center of Agricultural Sciences
and Engineering, University of Debrecen
[2]Institute for Land Utilisation, Technology and Regional Development,
Center of Agricultural Sciences and Engineering, University of Debrecen
[3]Hungarian Academy of Sciences - University of Debrecen
Research Group of Cultivation and Regional Development
Hungary*

1. Introduction

The irregular occurrence of drought periods accompanied by high temperature due to climatic changes influences the growth and yields of cultivars grown in agricultural fields. Many papers have reported the morphological and physiological response of plants to drought (Costa-Franca et al., 2000; Lin & Markhart, 1996; Lizana et al., 2006; Martinez et al., 2007; Sariyeva et al., 2009), but the effects of water supply on the physiological processes during different stages of crop development under field conditions have not been adequately studied. At field scale, drought and high (air/soil) temperature stress occur concurrently and not easily separable. Reduced evapotranspiration from a field due to drought is associated with increases in the temperature of the soil surface and that of the plant. In well watered plants, high temperature increases the water vapour pressure within the leaf and hence the rate of transpiration increases with the consequence of lower leaf water potential (Hsiao, 1994). Water stress may arise from either excess or deficit. Some days of waterlogging induce a rapid increase in the stomatal resistance which in turn reduces transpiration and net photosynthesis as a result of stomatal closure. These symptoms are similar to the responses to water deficit stress (Takele & McDavid, 1995). The flooding resulted in stomatal closure due to the leaf dehydration caused by increased resistance to water uptake as a result of the lowered permeability of roots (Amico et al., 2001). However, the more common water-related stress is the water deficit.

Plant responses to water stress differ significantly depending on (i) the intensity and duration of stress, on (ii) the plant species, and (iii) its stage of development. The most common field species can be categorized into three groups based on their sensitivity of drought. The tolerant group includes wheat, barley, sorghum, oat and alfalfa; the group with moderate sensitivity includes species as corn, sunflower, soybean, beans, peas; while potato, tomato and rice are belong to the extremely drought sensitive group (Heszky, 2007). The species ranged as tolerant are able to escape drought while those with moderate sensitivity can prevent the water deficit in their cells and tissues.

Various defence mechanisms have been developed in legumes in order to tolerate drought. The soil water content, high air/soil temperature and humidity adversely affect the shoot growth and yield of plants. Water deficit in the soil (-14.85 kPa) hampers the development and operation of root nodules of legumes. The high temperature (33oC) in the upper soil layer increases the number of nodules but decreases their size and the growth of the plants (Piha & Munns, 1987). Under severe dry conditions (-0.03- -0.5 MPa rooting medium water potential), the nitrogen-binding activity of root nodules decreases (Smith et al., 1988) resulting in the decrease of leaf area and leaf weight of the plants. Root characteristics, especially root length, root length density, and the number of thick roots, are important for plants to develop their aboveground parts by exploiting the soil available water. Expansive growth of leaves is also sensitive to water stress (Hsiao, 1994) and a short period (7 days) mild water stress (soil field capacity SFC 40%) already reduces the rate of leaf area development and leaf mass (Nemeskéri, 2001, Nemeskéri et al., 2010).

The changes in the morphology and anatomy of leaves as induced by drought are similar in many crops due to the same defence mechanisms (e.g. leaf movements). Most responses of plants to climatic factors such as water deficiency and high temperature under the field conditions are often different from that of plants grown in greenhouses (Nemeskéri et al., 2010). The cropping system (outdoor or greenhouse) can affect plant response to drought, this creates further complexity as concern the appraisal of on-set of stress. Therefore, it is important to examine plant traits variation in response to drought in order to create a sort of stress marker which can be used for the irrigation schedules and the modelling of crop transpiration. The determination of drought-stress markers is important in vegetables to maintain their yield production and food quality. In this chapter the defensive strategies of some annual crops (i.e. green bean and green pea) against drought and the relationship markers as measured in various developmental stages and yield level will be presented.

2. Defence strategies against drought

Plants have developed different defence strategies against drought during their evolution. The one possibility for them is to escape the drought when their most sensitive stage of development such as reproductive stage is completed before the drought. However, these crops with short ripening periods have generally low yields. Another one is the drought avoidance that is an ability of plants to maintain relatively high tissue water potential despite a shortage of soil-moisture. Two essential defence mechanisms operate in the plants to avoid the drought; one of them is the maintenance of water circulation in the plants that provide the deep root system and another one is to decrease the loss of water in the plants that can be achieved by restraining of transpiration and morphological changes. During long-term dry periods, the plants try to tolerate the water deficiency with low tissue water potential. The mechanisms of drought tolerance are maintenance of turgor through osmotic adjustment, increased cell elasticity and decreased cell size as well as desiccation tolerance by protoplasmic resistance.

2.1 Maintenance of the water status

The development of the root system is responsible for the water uptake and it contributes to the maintenance the water circulation inside plants. In summer days, the midday leaf water potential usually decreases compare to early morning values due to higher rate of transpiration at noon. In the case of well-watered plants, the water deficiency in the leaves is promptly compensated. In the case of droughted plants, the dehydration of the roots increases their abscisic acid (ABA) synthesis which moves toward the leaves where accumulates leading to stomatal closure (Parry et al., 1992) and consequently the maintenance of water status. Under poor soil water content, the lives microorganisms in the soil are also retarded that causes low activity of mycorrhiza therefore, the nitrogen uptake through the root system decreases possibly causing a reduction of plant growth. In legumes plants, drought causes premature senescence of nodules and disturbance in the mechanisms of oxygen control that are essential for active nitrogen fixation hence the production of reactive oxygen species (ROS) may increase resulting in oxidative damage of nodules (Becana et al., 2000; Hernandez- Jiménez et al., 2002; Matamoros et al., 2003). Drought stress decreased a number of structural and functional traits (e.g. shoot dry weight, root dry weight, nodule dry weight, nitrogen fixation), however the nodular peroxidase (POX) and ascorbate peroxidase (APX) activities are increased significantly in nodules when droughted plants are in symbiosis with chickpea (Esfahani & Mostajeran, 2011). It was found that chickpea plants inoculated with various rhizobial strains the symbiosis showed difference tolerance level under drought condition. All rhizobial strains enhanced the tolerance of symbioses to drought stress, however the local strain contributed to higher increase in the antioxidant enzyme activities than the others (Esfahani & Mostajeran, 2011).

The proportion of root weight can also be changed as a consequence of drought; the root-to-shoot ratio increases under water-stress conditions (Nicholas, 1998). The proportion of soybeans roots was found to increase and that of the stem to decrease in the early reproductive development periods without irrigation. However, drought stress does not affect either the growth of roots or stem in the early stages of vegetative development (Hoogenboom et al., 1986). As opposed to soybean, there was a remarkable decrease in root and stem mass and pod weight of beans under water deficiency, respectively (Nemeskéri, 2001). The one strategy to overcome the limitation of leaf growth occurring under drought could be the rapid root growth, to allow the water uptake from deeper layers of the soil (Reid & Renquist, 1997). The growth rate of wheat and maize roots was found decreasing under moderate and high water deficit stress (Noctor & Foyer, 1998; Shao et al., 2008). It is another possibility is to decrease the endogen ethylene levels in the plants. It was found that accelerated ethylene levels could be responsible for growth inhibition, premature senescence and abscission induced by water deficits (Morgan et al., 1990). 1-aminocyclopropane-1-carboxylate (ACC)-deaminase enzyme impeding the ethylene production can be found in certain microorganisms. The plant growth promoting rhizobacteria containing this deaminase resulted in greater increase in root elongation and in root weight in the inoculated pea plants than uninoculated ones under drought stress conditions (Arshad et al., 2008).

2.2 Avoidance/reduction of the tissue water loss

During short-term drought intensive root elongation to the deeper part of the soil profile and the partial or total stomatal closure provide the avoidance of water loss for the plants. However, during longer dry periods the plants try to prevent the water deficit in the cells of vegetative and generative organs tissues by morphological and physiological changes. Even mild water stress can reduce the rate of leaf area development leading to lower photosynthetic activity and low amount of produced biomass. Therefore the restricted growth of plants results smaller yield.

2.2.1 Stomatal functions

Transpiration provides continuous water and nutrients transportation. Stomata have important role in transpiration because they ensure the prevention of excessive water loss and also help control the leaf temperature. During the past decades, stomatal size and density have been used as an indicator of water loss (Singh & Sethi, 1995; Venora & Calcagno 1991; Wang & Clarke 1993a) but stomatal pore and width were rather considered to determine the capacity of stomata to reduce water loss (Aminian et al., 2011; Mohammady et al., 2005). However, the stomatal conductance showing the speed of water vapour evaporation depends on more plant-specific traits such as stomata density, leaf age and size, sub-stomatal CO_2 concentration, guard cell and cell turgor (Jones, 1992). The water use efficiency (WUE) as a key in determining productivity of a crop species is different in higher plants and relates to the stomata density under water limiting conditions. Higher stomatal frequency is associated with photosynthetic pathways and higher water use efficiency in C4 compared with C3 plants (Hardy et al., 1995).

The stomatal behaviour is directly influenced by signals received from the environment (e.g. light intensity) or mediated by roots (soil water deficit). Dehydrated root triggers the abscisic acid (ABA) synthesis which helps to the discharge potassium ions from bodygard cells that causes stomatal closure and the retention of the water in the leaves (Gomes et al., 2004; Parry et al., 1992). This defence mechanism take place during short-term water deficiency (<7 days) when the dehydration of cells can be achived by the restraining of transpiration. The long-term stomatal closure (10 days) results partial or total cessation of transpiration stream hence plants are not able to uptake water and nutrient from the soil. Consequently, the photosynthetic rate declines reducing the amount of photoassimilates for plant growth and reproduction. At high photosynthetically active radiation (PAR), drought combined with heat stress (40-45oC) result an increase in leaf temperature and temperature oscillations (±3-4oC), attributed to opening and closing of stomata (Reynolds-Henne et al., 2010). In general, highly variable leaf temperatures and stomatal opening within the realtively short intervals demonstrate a high stomatal sensitivity. However, the absolute stomatal responses are species-specific; for example bean stomata are more responsive to heat stress than clover (Reynolds-Henne et al., 2010). The rate of stomatal conductance with decreasing substrate water content decreased more in common bean genotypes than in cowpea ones. It means that beans (*Phaselous vulgaris*) have a rapid and complete stomatal closure causing a decrease in the assimilation rate during drought while cowpeas (*Vigna unguiculata*) keep their stomata partially opened and have a lower decrease in their net photosynthetic rate under the same conditions (Cruz de Carvalho et al., 1998).

2.2.2 Changes in morphology

The maintenance of osmotic adjustment in the cells can be achieved by some attributes changed such as reduction of leaf size, leaf movement or leaf rolling under a short-term water deficiency. The trichome density (leaf hairs) not only provides a mechanical barrier to insects but protects the plant from sunlight injury in the summer, decreases water loss by evaporation and enhances the transpiration-resistance (Du et al., 2009; Pfeiffer et al., 2003). Leaf movements are common defensive responses to drought stress in plants. Paraheliotropic leaf movement in response to stress occurs mainly in legumes and other species such as rice, maize, wheat and sorghum exhibit leaf rolling (Kadioglu et al., 2011; Matthews et al., 1990). Leaves move by means of turgor pressure changes at the pulvinus at the base of each lamina. In bean, paraheliotropism is an important property of the plants to avoid the photoinhibition. The extent of the leaf movements is increased as the water potential drops reducing light interception (Pastenes et al., 2005). Light and heat driving paraheliotropism can also occur in well-watered plants but it has a lesser extent than in water stressed plants. The other type of leaf movement is the leaf rolling operating as a dehydration avoidance mechanism reduces the effective leaf area therefore, the transpiration also reduces. The degree of leaf rolling in maize is linearly correlated with leaf water potential (Fernandez & Castrillo, 1999) while this is correlated with leaf osmotic potential and leaf temperature in rice (Ekanayake et al., 1993). Leaf rolling not only impedes the large water loss but protects the leaves of non-irrigated plants grown in the field from photodamage (Corlett et al., 1994).

The plant species living in arid habitat form epicuticular waxes on the outermost layer of the organs of plants to control the water flow across the cuticle and protect themselves from high radiation levels and prevent damage caused by UV light. It was found that water stress induced the accumulation of waxes on the leaf surface in peas and wax-rich varieties have significantly lower canopy temperature (Sánchez et al., 2001). A greater quantity of waxes also increases the reflection of the photosynthetically active radiation and UV-B light (Grant et al., 1995) which leads to the alleviation of heat stress.

2.2.3 Osmotic adjustment

Osmotic adjustment is an important physiological mechanism by which plants synthesize and accumulate compounds acting as osmolytes in cells in response to water deficits (Seki et al. 2007). When the decrease in water potential in plants is higher than 0.1 MPa then osmotic adjustment is generally occurs (Tari et al., 2003), causing the accumulation of sugars, amino acids, sugar alcohols and quaternary amonium to lower the osmotic potential (Morgan, 1984; Gomes et al., 2010) to increase the osmotic pressure of the cells, drawing water into the cells and tissues hence contributing to the maintenance of turgor. Proline's role as an osmolyte or osmoprotectant in leaves of drought-stressed plants has been debated (Seki et al., 2007; Szabados & Savoure, 2009; Gomes et al. 2010). Indeed, proline has been demonstrated to confer drought stress tolerance to wheat plants by increasing the antioxidant system rather than increasing osmotic adjustment (Vendruscolo et al., 2007; Szabados & Savoure, 2009). However, other species (e.g. pea, castor bean) exposed to water deficit accumulate soluble sugars and proline contributing to the turgor maintenance by osmotic adjustment (Sánchez et al., 1998; Babita et al., 2010). Glycine betaine, which is one of the quarternary ammonium compounds, is considered to be the most effective osmoprotectant in many higher plants, because it protects the cells from environmental stresses by maintaining osmotic balance and stabilizing proteins, enzymes and membranes (Winzor et al., 1992; Yang et al., 1996; Gao et al., 2004). Many compounds were shown to have osmoregulator attributes; for example galactinol, raffinose have important roles in improving stress tolerance in plants (Pennycooke et al., 2003; Nemeskéri et al., 2010), pinitol sugar alcohol accumulates in water-stressed legumes (Guo & Oosterhuis, 1995, 1997; Smith & Phillips, 1982) and in redbud trees (*Cercis canadensis*) to protect the plants from drought (Griffin et al., 2004). Under soil water stress of -0.6 MPa, maintenance of turgor in chick pea was due to a significant decrease in osmotic potential (osmotic adjustment) while in common bean it was due to maintenance of high leaf water potential but not to osmotic adjustment (Amede, 2003). Nevertheless, osmotic adjustment in broad bean in response to water deficit has not been found (Katerji et al., 2002; Khan et al., 2010).

Turgor maintenance can be achieved by osmotic adjustment and changes in the volumetric size of cells. The plants subjected to dehydration may avoid reduced water potential and maintain turgor by the reduction of their turgor-loss volume via shrinkage associated with elastic adjustment of their cell walls (Fan et al., 1994). Cell contraction means a reduction in cell size which is associated with plant resistance to water stress (Lecoeur et al., 1995). Cell size reduction has been reported in cassava plants (*Manihota esculenta*) grown under water stress (Alves & Setter, 2004). Citrus plants are able to develop elastic adjustment in response to water stress by the decrease in the volumetric modulus of elasticity(ϵ) (increase in cell wall elasticity) but do not develop osmotic adjustment as a drought tolerance mechanism under water stress conditions (Savé et al., 1995). Most of the bean varieties that presented high values of volumetric elastic modulus under well-watered conditions (above 3.7 ϵ) reduced significantly this parameter under water stress. Those bean cultivars decrease significantly their ϵ value and therefore increase their cell wall elasticity, while they present a positive drought resistance index and better resistance to the water stress (Martínez et al., 2007).

2.3 Drought tolerance

Drought tolerance is a complex property of the varieties including genetics, physiological or biochemical factors. Nevertheless the response of plant depends on the stage of development and strength and duration of the drought stress. For example, the water supply under flowering and seed development has important effect on the pods and seed yield and also on their quality in grain legumes. Under long-term drought periods (above 14 days), the growth of pants is retarded and they try to overcome the water deficiency by the decrease in water loss and leaf area or by increasing water use efficiency. During flowering, a longer period of water deficiency results in the loss of leaves and flowers that was much more intensive in large-leaved bean varieties than in those with smaller leaves (Nemeskéri et al., 2010).

When the avoidance of drought seemed to be insufficient by the changes of morphology and physiology, the plants undergo different biochemical and molecular genetic changes to maintain osmotic adjustment and the structure of cell membranes in order to avoid cell dehydration. Sugars act as osmotic compounds in protecting plants against drought and they contribute to the stabilization of cell membrane structures. A strong correlation between sugar accumulation and osmotic stress tolerance has been reported (El-Tajeb, 2006; Streeter et al., 2001). However, the accumulation and components of carbohydrates differ according to the individual responses of plant species. Decreases in sugars and oligosaccharides in Arabidopsis have been shown (Anderson & Kohorn, 2001) while in others, such as maize and rice, sugars accumulated during drought (Pelleschi et al., 1997;

Vue et al., 1998). Raffinose and sucrose content of leaves are increased by drought stress in the beans during flowering (Nemeskéri et al., 2010). Oxidative damage of the plant tissue is alleviated by a concerted action of both enzymatic and non-enzymatic antioxidant mechanisms. These antioxidant compounds react with free radicals and neutralize them, thus overcoming the damage caused by stress. Water deficit stress increases the lipid peroxidation in the leaves of young bean plants; however, the activity of catalase and superoxide-dismutase enzymes was high in the drought tolerant bean varieties (Zlatev et al., 2006). Numerous yet undetected water-soluble antioxidant compounds (ACW antioxidant capacity of water-soluble substances) may be responsible for the adaptation of plants to environmental stress factors. The ACW antioxidant contents in the leaves can be influenced by the stomata closure, because this is related to the ascorbic acid redox state of guard cells that control the stomatal movement (Chen & Gallie, 2004).

All responses to drought are controlled by complex mechanisms involved in the changes in gene expression. Micheletto et al., (2007) found that fewer genes (n = 64) were responsive in *Phaseolus vulgaris*, the more sensitive species, compared with *P. acutifolius* (n = 488) and only 25 genes were drought responsive in the roots of both species, at severe drought stress level of -2.5 MPa (midday) leaf water potential. Many genes (328) were induced in roots of *P. acutifolius* at severe drought compared with well-watered plants and 160 of them were repressed while in that of *P. vulgaris* the number of up-regulated gene was 49 and 15 genes were repressed. However, others (Meglič et al., 2008) revealed that nine genes were up-regulated in the leaves of drought stressed bean plants (*P. vulgaris*) in comparison with the well-watered plants and eight genes were repressed. The genes up-regulated belong to various previously reported functional categories characteristic for drought stress such as the late embryogenesis abundant proteins (LEA), synthesis of osmolytes, transcription factors, protein kinases, cellular- and carbohydrate metabolism while five of eight down-regulated genes belong to the functional category related to photosynthesis (Meglič et al., 2008).

2.4 Heat stress and tolerance

Under field conditions, high temperature stress (30-35°C) is frequently associated with reduced water availability (Simoes-Araujo et al., 2003), thus the effects of drought and heat stresses on plant varieties is difficult to evaluate separately in the field. Anatomical changes under high ambient temperatures are generally similar to those under drought stress. At the whole plant level, there is a general tendency of reduced cell size, closure of stomata and reduced water loss, increased stomatal and trichomatous densities, and greater xylem vessels of both root and shoot (Añon et al., 2004). High temperature considerably affects anatomical structures not only at the tissue but also at cellular levels which result in poor plant growth and productivity (Wahid et al., 2007). This causes significant declines in shoot dry mass, relative growth rate and net assimilation rate in maize, pearl millet and sugarcane, though leaf expansion is minimally affected (Ashraf & Hafeez, 2004; Wahid, 2007). However, no doubt the reproductive phases such as gametogenesis and fertilization in various plants are the most sensitive to high temperature because a short period of heat stress can already cause significant increases in floral buds and opened flowers abortion (Foolad, 2005; Guilioni et al., 1997; Young et al., 2004).

The changes in the photosynthetic pigments indicate the thermo-tolerance of some species. An increased chlorophyll *a:b* ratio and a decreased chlorophyll carotenoids ratio were observed in tolerant tomato genotypes under high temperatures (Camejo et al., 2005). At high temperature stress combined with soil water deficiency, chlorophyll *b* content in the leaves of French beans decreased during flowering, nevertheless the chlorophyll a:b ratio did not change in the genotypes in comparison with the control plants (Nemeskéri et al., 2010). However, common bean genotypes were detected to have significant differences in photochemical sensitivity to heat stress and also in the ability to modify their photochemical apparatus (Ribeiro et al., 2008). Plants are capable of adapting to different light environments by changing their photosynthetic pigment compositions. During continuous irradiation, the rapid transformation of chlorophyll *b* to chlorophyll *a* occurred in the leaves of light sensitive varieties and carotene ensures the great protection against photodamage (Procházková & Wilhelmová, 2004). One of the functions of carotenoids including β-carotene and xanthophylls is to act as an accessory pigment, capturing light and transferring energy to chlorophylls to drive photochemistry (Carvalho et al., 2011). However, carotenoids have another important function such as photoprotection of the reaction-centers, pigment-protein antennae, and cells and tissues (Li et al., 2009) and essentially as a non-enzymatic antioxidant (Arruda & Azevedo, 2009; Gratão et al., 2005). Other antioxidant compounds, enzymes and proteins also contribute to the defence of tissue against heat and light stresses. The activity of superoxide-dismutase (SOD) and ascorbate peroxidise enzymes was found to increase different extent with prolonged irradiation of the leaves and stimulated by high temperature, however, the activity of catalase enzyme was more sensitive to high temperature than to high irradiation (Ye et al., 2000).

A rapid heat stress results in the synthesis and accumulation of specific proteins designated as heat shock proteins (HSPs). Heat shock reduces the amount of photosynthetic pigments, soluble proteins, rubisco binding proteins (RBP) in darkness but increases them in light, indicating their roles as chaperones and HSPs (Kepova et al., 2005; Todorov et al., 2003). An increasing number of studies suggest that the protective effects of HSPs can be attributed to the network of the chaperone machinery, in which many chaperones (proteins) act in concert. The HSPs/chaperones also interact with other stress-response mechanisms such as production of osmolytes (Diamant et al., 2001) and antioxidants (Panchuk et al., 2002). The more HSPs activated in the distinct cellular compartments like cytosol, chloroplast, endoplasmic reticulum (ER), mitochondria and membranes the higher heat tolerance plants have. Five mitochondrial low molecular weight HSPs (from 19 to 28 kDa) were expressed in maize seedlings subjected to heat shocks (42ºC), only one (20 kDa) was expressed in wheat and rye, suggesting the reason for higher heat tolerance in maize than in wheat and rye (Korotaeva et al., 2001). The stages of development of the plants in which heat shock genes cannot be induced are generally sensitive to heat stress but those involved by these genes expressed in high level are heat tolerance (Györgyey, 1999). In some cases, the heat shock genes do not provide advantages for the plants. The heat tolerant genes result in shorter internodes of main stem of the plants and reduce of biomass production in cowpea (*Vigna unguiculata* (L.) Walp.), on the other hand, after the first pod setting the ripening is accelerated and many pods are produced with few seeds as a result of these gene actions (Ismail & Hall, 1998).

The changes in anatomical structures at the tissue and cellular levels result in poor plant growth. There are great variations in heat sensitivity within and among plant species; for example, in tomato, though plants are sensitive to high temperatures throughout the plant ontogeny, flowering and fruit set are the most sensitive stages; fruit set is severely affected by high day/night temperature above 35/26°C (Berry & Rafique-Uddin, 1988). A close significant correlation (0,51< r >0,81) was found between the high temperature and damage of the leaves in many of legumes such as soybean, pea, lupine and faba bean which expressed the extent of heat tolerance of these species. In legume plants, the mechanism of damage in cell membrane was the same independently of stress factors (Grzesiak et al., 1996). In general, heat stress singly or in combination with drought, is a common constraint during the anthesis and grain filling stages in pea and many cereal crops of temperate regions (Guilioni et al., 2003), and causes yield loss in common bean, *Phaseolus vulgaris* (Rainey & Griffiths, 2005) and groundnut, *Arachis hypogea* (Vara-Prasad et al., 1999).

3. Defence strategies in beans

In common beans, the mechanisms of drought tolerance principally include the development of an extensive root system (Micklas et al., 2006; Nemeskéri, 2001; Sponchiado et al., 1989) and efficient stomata closure (Barradas et al., 1994; Costa-Franca et al., 2000; Miyashita et al., 2005). It has been shown that the two high yielding *Phaseolus acutifolius* lines may rely on two different strategies. The first is characterized by a thin, deeply penetrating root with large mass and increasing stomatal conductance for one of the lines, while the other line developed a great mass of deeply penetrating roots and reduced the leaf expansion remarkably and stomata conductance (Mohamed et al., 2002). The bean variety that is more sensitive to drought has more intensive and earlier paraheliotropic leaf movement and the reduction in the water content of the leaves is faster under water deficiency than drought tolerant variety (Lizana et al., 2006). In spite of the fact that the many papers analysed the morphological and physiological response of beans to drought, the effects of water supply on the physiological processes during different stage of development of the plants have slightly studied under field conditions. Water use efficiency (WUE) is traditionally defined as the ratio of dry matter accumulation to water consumption over a season. Stomatal frequency positively correlates with the rate of water loss and stomatal conductance and associated with the water use efficiency (Hardy et al., 1995; Wang & Clarke, 1993b). Nevertheless these seemed to be cultivar-specific characteristics.

The pod numbers per plant are controlled mainly by genetic factors but the water use efficiency of the plant and stomatal function influenced the amounts of stock sized pods in French beans. Investigation of five French bean varieties grown in the field conditions revealed that the stomata density is higher on the abaxial surface of the leaves in the green-podded beans and lower on the adaxial surface than that of yellow-podded ones. Nevertheless, in both bean types, the stomata density was significantly larger on the adaxial surface of the leaves under mild water deficiency in comparison with irrigated plants in a dry year (Table 1). The analyse of WUE of green pod yield (WUEy) shows that the green podded varieties use less water for one kilogram of yield than the yellow podded ones. Under water deficiency and serious terminal drought, increase in the stomatal resistance in the leaves of both bean types is related to the decrease in the pods weight. The transpiration is restrained as a result of the increase in stomatal resistance. That results the decline of CO_2 flow into the plants and the reduction of photosynthesis which leads to the decrease in yield.

Trait	Pod colour	Wet year			Dry year		
		I	WD	D	I	WD	D
Yield components							
yield (tons/ ha)	green	7.44 a	5.47 c	7.51 a	12.96 a*	11.13 a*	3.31 bc*
	yellow	7.36 ab	7.77a	8.18 a	7.00 b	4.73 bc*	1.46 c*
pod number/ plant	green	16.87 a	14.63 ab	14.11 ab	14.85 a	13.25 a	10.57 b*
	yellow	13.17 b	13.50 b	13.83 b	9.70 b*	9.92 b*	5.58 c*
total pod weight/ plant (g)	green	55.03 a	42.98 b	42.85 b	44.30 a*	35.15 ab	20.73 c*
	yellow	43.65 b	37.43 bc	31.20 c	35.19 ab*	30.98 b	13.02 c*
stock sized pods/ plant (g)	green	46.68 a	36.54 b	36.31 b	37.60 a*	28.12 b*	16.02 c*
	yellow	31.81 bc	30.10 bc	28.39 c	29.83 ab	25.08 b	9.32 c*
Water relations							
stomata number/mm^2							
abaxial surface	green	380.38 b	395.89 a	379.81 b	379.99 bc	412.56 ab*	402.54 b*
adaxial surface		23.98 f	34.68 f	26.68 f	69.26 c*	93.43 b*	109.14 a*
stomata number/mm^2							
abaxial surface	yellow	282.20 c	259.85 c	237.50 c	350.72 d*	370.46 cd*	429.90 a*
adaxial surface		55.63 e	65.58 d	73.52 d	73.98 c*	88.52 b*	85.15 b
stomatal resistance (s/cm)	green	1.638	1.781	1.759	1.079 d	1.289 cd	3.382 a*
	yellow	no data	no data	no data	1.497 c	1.527 c	3.098 b
SPAD value	green	33.36 a	33.53 a	33.42 a	36.27 c	38.70 b	42.95 a*
	yellow	34.54 a	35.86 a	38.60 a	33.73 d	35.61 c	38.65 b
WUEy (l/ kg/ m^2)	green	329.59 c	503.47 a	368.36 bc	206.14 d*	223.39 d*	624.03 b*
	yellow	286.37 bc	271.37 b	168.69 d	359.22 c*	441.17 c*	947.28 a*

I=irrigation, WD=water deficiency (I/2), D= dry, without irrigation, Data based on five varieties arranged in three replication in the field experiments.
Values in each row and coloumn having different letters are significantly different at the P<0.05 level using Duncan's multiple range test.
* Significant differences between years at the P<0.05 level

Table 1. Effect of water supply on yield components and water relation variables of green beans grown in the field conditions

Ramirez-Valejo & Kelly (1998) also found that stomatal conductance showed a positive association with yield, pod number, seed number and total biomass of common bean genotypes grown under water stress conditions. Reduction of the chlorophyll content in the leaves caused by water and heat stress and high irradiation contributes to the decrease in photosynthesis. The leaf of green bean varieties with yellow-green colour leaf is turned to dark green under prolonged severe drought in the field conditions as a result of the changes in the photosynthetic pigments. Eghball & Maranville, (1991) state that the effectiveness of utilising nitrogen and water are often associated. Chlorophyll itself also contains nitrogen; therefore, the measurement of chlorophyll content shows the utilisation of nitrogen and at the same time indicates the water utilisation. A close correlation (r^2=0.9029) was found between the SPAD values measured by a chlorophyl meter equipment and the chlorophyl content in the leaves so the SPAD values can be used for evaluation for the response of species to the drought and heat stresses in the field (Hawkins et al., 2009; Yadava 1986,). In additions, high positive correlations were proved between the chlorophyll content (SPAD) and the photochemical reaction indexes, and nitrogen demand (Perry & Davenport, 2007), and the measurement of drought stress in wheat (Ommen et al., 1999), the efficiency of transpiration in peanut (Krishnamurthy et al., 2007), the rate of photosynthesis, and the stomatal conductance in rice (Kato et al., 2004) during drought stress. During dry weather, the increment of chlorophyll content expressed by SPAD values indicates the nitrogen supply of the plant (Berzsenyi & Lap, 2003), and the changes in the glucose + fructose level in leaves (Nemeskéri et al., 2009). Under water deficiency, the trend in the increase of the stomatal resistance and SPAD values in the bean leaves is the same therefore, SPAD values can also be used as a water stress marker for the evaluation of drought tolerance of bean genotypes in field experiments.

In different stages of development, the plants respond in various ways to environmental stresses. It is known that bean yields are influenced by the climatic conditions during flowering. Before and during flowering, the stomatal resistance and chlorophyll content (SPAD) in the leaves significantly affect the yield. During the flowering period, a good correlation was found between the stomatal resistance and SPAD values (r^2=0.587) and pod weight per plants (r^2=0.446). According to these correlations, the chlorophyll content in leaves increased up 3.7 s/cm stomatal resistance values then it decreased. The pods weight of plants also decreased remarkably by the increasing stomatal resistance (above 3.7 s/cm). Another marker related to the pods weight per plants is the stomata number on the adaxial surface of leaves in bean. A higher number of stomata can be found on the abaxial surface than the adaxial surface of the bean leaves. However, that on the adaxial surface leaf seems to be more important in the restraining of transpiration. During flowering periods, the beans stomata number/mm^2 have significant influence on the stomatal resistance (r^2=0.399) and the lowest stomatal resistance (2.6-3 s/cm) can be measured by 85-105 stomata/mm^2 on the adaxial surface of the leaves. Larger stomata density on the adaxial surface of the leaves results in larger stomatal resistance producing the decrease in the pods weight of plants. During the pod ripening period, the stomata density on the adaxial surface of the leaves still have an influence on the yield of beans.

There are differences in the responses of the bean types to drought during the flowering period. Both the phytotron and field experiments illustrated that the yellow-podded green bean varieties respond to drought more sensitively than green-podded ones (Nemeskéri et al., 2008, 2010). This was confirmed by the decrease in chlorophyll a and b components and the antioxidant capacity of water soluble (ACW) substances in the leaves of yellow-podded beans compared to the well-watered plants during flowering at severe drought stress in the phytotron, while there was not remarkable change in comparison with green-podded ones (Table 2). Under field conditions during the flowering period, the temperature (as minimum and maximum value) was similar to the treatment of 30/15 ᵒC (day/night) in the phytotron that accompanied by water deficiency and often occurs in temperate zone. The average leaf weight and antioxidant capacity in the leaves were higher under field conditions than in the phytotron at 30/15 ᵒC (day/night). However, the significant differences between the two bean types in photosynthetic pigment content and lipid soluble antioxidant (ACL) level in the leaves were also remained under field conditions (Table 2). The defence against drought had already begun prior to flowering when the correlation between the content of ACL in the leaves and seed yield ($r^2=0.607$) is so significant that it determined 61% of seed yield (Nemeskéri et al., 2010). Although the level of ACL antioxidants in the leaves decreased in nearly all genotypes during flowering, only those genotypes have high tolerance to drought accompanied with high temperature in which, the extent of the decrease is small or they are capable of re-increasing the production of ACL antioxidants during pod ripening (Fig. 1). These genotypes are able to produce seed yield above 1.0 tons /hectare under dry field conditions. The changes in the ACW antioxidants level in the leaves decreased similarly to ACL during flowering (Fig 2) but these antioxidants ensure primarily the protection against water deficiency (Chen & Gallie, 2004, Nemeskéri et al., 2010).

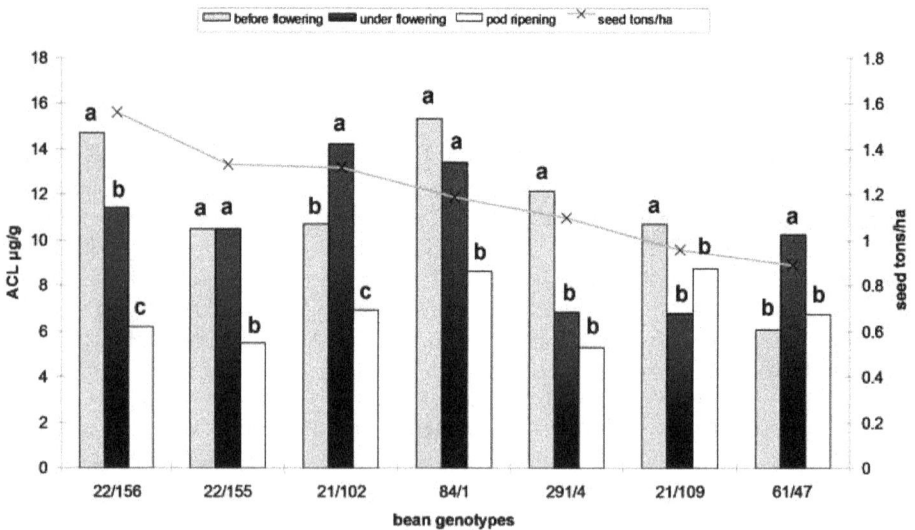

Fig. 1. Change in the level of lipid-soluble antioxidants (ACL) of bean leaves under dry field conditions.

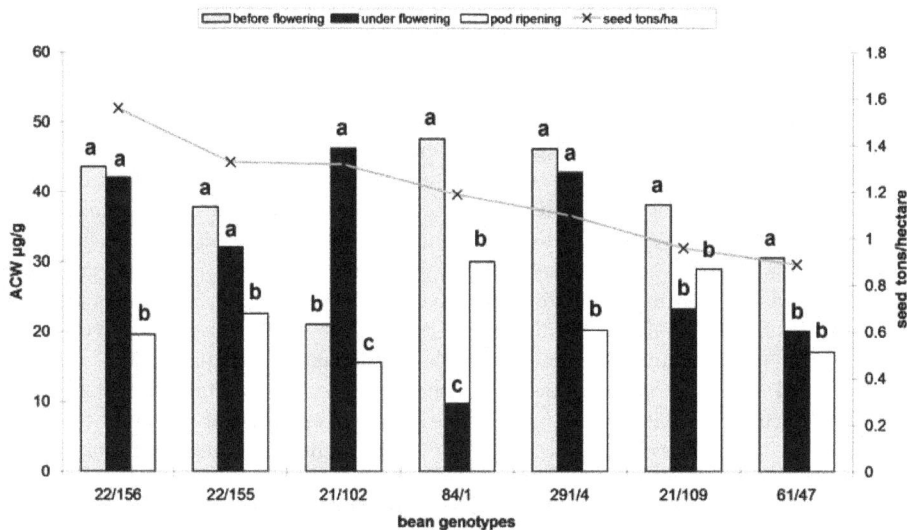

Significant differences between phases of development are indicated by different letters at the P <0.05 level

Fig. 2. Change in the level of water-soluble antioxidants (ACW) of bean leaves under dry field conditions.

Bean groups	A						B	
	Yellow-podded beans			Green-podded beans			Yellow-podded beans	Green-podded beans
Properties	25/15*	30/15	35/25	25/15	30/15	35/25		
Average leaf weight (g)	0.34 b	0.19 c	0.12 d	0.41 a	0.18 c	0.20 c	0.64 a**	0.67 a**
Chlorophyll *a* (mg/g)	30.34 b	31.19 b	25.97 c	37.92 a	39.21 a	37.24 a	34.72 b	51.54 a**
Chlorophyll *b* (mg/g)	4.39 b	4.82 b	3.85 c	5.99 a	6.97 a	5.30 a	4.68 b	7.37 a
Carotene (mg/g)	25.15 c	35.63 b	22.33 c	31.30 b	41.77 a	29.27 bc	37.74 b	55.14 a**
Chlorophyll *a/b*	6.92 a	6.54 a	6.75 a	6.34 ab	5.74 b	7.14 a	7.44 a	7.01 a**
Chlorophyll/ carotene	1.38 a	0.99 c	1.34 a	1.40 a	1.14 b	1.46 a	1.05 a	1.07 a
ACW (µg/mg)	17.07 a	9.81 b	8.11 b	3.51 d	4.70 c	2.64 e	18.23 a**	17.96 a**
ACL (µg/mg)	3.61 c	4.48 b	8.42 a	2.61 d	4.18 b	4.64 b	6.67 b**	8.05 a**

* 25/15 ºC day/night with irrigation=control, 30/15 ºC day/night combined with water deficiency=mild drought stress, 35/25 ºC day/night= severe drought stress ** Significantly different from the mild drought stress (30/15 ºC day/night) at the P <0.05 level

Table 2. Reactions of French beans with different pod types for drought stresses during flowering in the phytotron (A) and field (B) growing conditions.

4. Defence strategies in peas

The pea of the food legumes has been less studied for drought tolerance than soybean and bean. Chickpeas (*Cicer arietinum* L.) and cowpea (*Vigna unguiculata* L. Walp.) cultivated almost completely arid and semi-arid zones of the world are continuously exposed to drought and high temperature during the flowering and maturity stages. Therefore these are well studied for drought tolerance. In cowpea, two types of drought tolerant mechanisms have been described at the seedlings age; one of them the drought tolerant pea lines stopped their growth after the onset of drought stress and turgidity was declined in all tissues of the plants for over two weeks while the other pea line was remained green for a longer time and the growth of the trifoliate leaves continued slowly under drought stress (Agbicodo et al., 2009; Mai-Kodomi et al., 1999). The second reaction involved the combination of more defence mechanisms such as stomatal regulation (partial opening), osmotic control and material mobilization to the younger leaves (Agbicodo et al., 2009). Water stress induces the accumulation of soluble sugars in epicotyls of pea and the increase in free proline content contributing to osmotic adjustment, however, the degree of variation depends on the given cultivar (Sanchéz et al., 2004). Genetic modifications of canopy structure of peas are expected to result in enhanced yield when water is a limiting factor. Genotype *af*, which has the leaflets transformed into tendrils, showed a faster CO_2 exchange rate, lower stomatal resistance, and lower canopy temperature under water stress than the genotypes with normal leaf (Alvino & Leone, 1993). Decrease in the soil water content resulted in the significant decrease of the leaf water potential and relative water content in the leaves in semi-leafless pea (*af*) variety that interrupted the vegetative growth and increased leaf senescence while the pea with normal leaf did not (Baigorri et al., 1999).

The green peas planted in spring utilize the soil moisture content well at the beginning of their growth. However, the next time the intermittent drought results in a decrease in the growth and yields. This may be the reason why few studies have been made to evaluate the factors influencing the water use and drought tolerance of green peas. Early maturing pea varieties exposed to intermittent water stress during the vegetative growth stage grow poorly and when this occurs during early stages of their reproductive phase, it results low yield. In wet years the chlorophyll content in the pea leaves and stomatal resistance are low but these are rising in dry years. The investigation of six green pea varieties grown in the field showed that excess of water supply results in larger increase in stomatal resistance in early ripening pea plants compared to the middle and late ripening ones, but the chlorophyll content in the leaves expressed in SPAD values is low in all groups (Table 3). Under dry weather conditions, the early and late ripening peas respond by high stomata density and stomatal resistance to the moderate water deficiency in comparison with well-watered ones, but the differences in WUE can be detected only in the late ripening group. Water deficiency significantly decreased the pod numbers and pod weights of the plant in the middle and late ripening groups. During both intermittent and prolonged drought periods the semi-leafless pea variety produced fewer pods and seeds with low weight in comparison with well-watered plants due to the high stomata density and stomatal resistance. While drought and heat resistance of chickpea genotypes have been evaluated by early flowering, plant vigor and pod setting under field conditions (Canci & Toker, 2009), these should be done by the stomata function and the changes in the chlorophyll content of the leaves used as a drought stress markers in green pea genotypes. Green peas are

Trait	Maturity	Wet year			Dry year		
		I	WD	D	I	WD	D
Yield components							
yield (tons/ha)	early	8.89 c	10.87 ab	12.38 ab	8.25 a	7.69 a*	6.13 bc*
	middle	10.61 b	12.07 ab	12.76 a	9.86 a	7.03 ab*	6.58 bc*
	late	12.04 ab	12.47 ab	11.35 ab	8.78 a*	6.67 bc*	4.61 c*
pod number/plant	early	3.75 b	3.80 b	3.80 b	3.19 c*	3.17 c*	2.83 c*
	middle	5.63 a	5.68 a	5.42 a	7.20 a*	5.98 b	5.44 b
	late	4.74 ab	5.39 a	4.89 ab	5.70 b*	3.80 c*	3.48 c*
average pod weight(g)	early	19.46ab	19.70 a	18.87 ab	15.29 b*	15.33 b*	12.80 c*
	middle	16.73 ab	15.13 ab	14.84 b	20.62 a*	14.78 bc	14.98 b
	late	16.34 ab	18.25 ab	15.29 ab	17.93 ab	12.30 c*	9.58 c*
seed number/plant	early	18.68 b	19.53 b	19.43 b	17.48 cd	18.40 cd	14.90 d*
	middle	26.86 a	23.62 ab	22.80 ab	40.08 a*	30.00 b*	29.20 b*
	late	22.63 ab	27.42 a	23.39 ab	33.85 b*	23.48 bc*	19.10 c*
100 green seeds weight (g)	early	43.68 a	41.30 a	42.71 a	43.54 a	44.57 a	44.77 a
	middle	25.03 cd	25.14 cd	24.39 d	20.96 d*	21.52 cd	23.52 c
	late	34.53 b	31.72 b	27.68 c	25.32 b*	29.53 b	27.58 b
Water relations							
stomata density (number/mm²)	early	189.20 d	161.40 e	181.40 d	180.60 c	164.50 d	176.00 c
	middle	225.60 b	211.90 c	206.10 c	224.70 b	222.30 b	223.86 b
	late	226.90 b	229.00 b	240.30 a	218.08 b	243.70 a	236.48 a
stomatal resistance (s/cm)	early	1.613 a	1.540 ab	1.625 a	2.757 c*	3.538 ab*	2.653 c*
	middle	1.243 c	1.338 bc	1.316 bc	3.04 bc*	3.538 ab*	3.225 ab*
	late	1.163 c	1.174 c	1.188 c	2.885 bc*	3.538 a*	3.861 a*
SPAD value	early	39.45 c	41.50 b	42.86 a	45.42 c*	45.69 c	46.16 c
	middle	39.10 c	43.16 a	43.67 a	53.71 a*	53.25 a*	53.71 a*
	late	37.65 d	38.69 cd	40.52 b	50.52 b*	50.26 b*	50.24 b*
WUEy (l/ kg/ m²)	early	231.27 a	189.14 a	166.07 b	66.25 cd*	65.58 cd*	75.69 c*
	middle	193.79 ab	172.69 ab	167.72 b	73.73 c*	90.33 bc*	99.67 b*
	late	218.97 a	206.77 a	233.17 a	89.92 bc*	103.85 ab*	137.98 a*

I= irrigation, WD=water deficiency, D=without irrigation, Values in each row and coloumn having different letters are significantly different at the P<0.05 level using Duncan's multiple range test. * Significant differences between years at the P<0.05 level

Table 3. Effect of water supply on yield components and water relation variables of green peas grown in the field conditions

generally grown for canning and freezing processing under irrigated conditions. However, the WUE of the varieties influences the yield quantity and quality, which is related to the drought stress markers in different stages of development. During flowering the chlorophyll content in the leaves (SPAD) and the stomata density can be related to the pods yield of the pea under dry weather conditions. This was confirmed by a close significant correlation found between SPAD and stomata density (r^2=0.674) and SPAD and pods weight (r^2=0.375) of green peas plants grown under non-irrigated conditions, respectively. The wax layer covering the adaxial surface of leaves contributes to the prevention of excessive water loss and because no stomata can be found here, the transpiration goes through the stomata on the abaxial surface of leaves. Under dry conditions (SFC 40%), during pod ripening the stomatal function seems to have a greater influence on the pods weight of the pea than during flowering period. This time the pod yields of the plants decreased significantly when the stomata resistance based on the high significant correlation between the stomata resistance and pods weight (r^2=0.557) was rising above 4 s/cm.The stomatal function influences the water supply of the green pods in the late ripening cultivars during flowering periods, however the chlorophyll content in the leaves affects the green seed weight which is confirmed by a close correlation between stomata resistance and pod weight (r^2=0.994) and SPAD and seed weight (r^2=0.634), respectively. During the pod ripening period, above 4 s/cm stomata resistance, the pod weight and seed weight of plants decreased significantly in the middle ripening group but the decrease was very intensive in late ripening ones under dry field conditions. During the pod ripening period the late ripening green pea varieties are much more sensitive to drought than the middle ripening ones that was confirmed by the relationship between the drought stress markers and yield.

5. Drought stress markers and yield

Breeding for drought tolerant varieties has been accomplished by selection for seed yield under field conditions but such procedure requires full season field data and this is not always an efficient approach. The changes in morphology including the decrease in growth and leaf size, leaf movements etc. are less appropriated to measure the water status of plants. These can be used with low efficiency for evaluation the drought adaptation and irrigation scheduling. The changes in physiological attributes of the plants rather indicate the disturbance of water supply and these can be used to determine the degree of drought tolerance of the varieties even though the reproducible experimental environments can be hardly achieved in the field. The identification of stress markers is difficult because they can vary on the crop species and in many cases the factors operating in the defense of damage caused by environmental stresses are or are not activated under different periods of development. The legumes of the field crops with moderate sensitivity to drought have the most various defence strategies. The deeply penetrating roots with large mass and large density in trichomes on the leaves in soybeans ensure the maintenance of the sub-optimal water status of the plants (Du et al., 2009; Nemeskéri, 2001) also after the onset of a moderate water deficit. However, the root characteristics cannot be considered as drought stress markers because their measurement is rather difficult. The stomata density on the adaxial surface on leaves and the high trichomes density on the leaves decreases the water loss in green beans. Although the stomata cannot be found on the adaxial surface on the leaves in green peas, the wax layer contributes to the defense against high temperature and water deficiency.

The responses of plants to stresses depended on many factors such as the phenological stage and time and strength of stresses. Under mild drought, the higher stomata density on the abaxial surface of the leaves in green beans can already be detected during flowering but in the green pea this occurs only during green pod ripening (Table 4). In terminal drought stomatal action expressed as stomata resistance is intensive in all stages of development in beans but it is very intensive in green peas only during green pod ripening. The water supply, during flowering period, is considered to determine the yield in the legumes. However, the crop species use different defence strategies against drought occurred this time. For example, during flowering the stomatal resistance and the stomata density on the adaxial-side of leaves are related to the quantity of pod yields in bean but at the same time the chlorophyll content in leaves (SPAD) in green peas has a great effect on the pods yield while the stomata resistance has not. The changes in the photosynthetic pigment seemed to be determinative to the green pod yields in both species during pod ripening and these can be used as drought stress markers for the selection of genotypes with drought tolerance. During flowering the defence of plants against damage from climatic stresses occurs in many stages in bean varieties. At first, the carotene and raffinose contents of leaves increase, then with rises in temperature additional protective mechanisms are activated e.g. the production of raffinose, glucose and sucrose are increased and the level of ACL antioxidants in the leaves rises (Nemeskéri et al., 2010). During the seed development in the pods the defence is going on by the increase in the level of antioxidants against severe drought and high temperature. The tolerance to water deficiency is related to the content of ACW antioxidants in the leaves and the tolerance to high temperature by the ACL antioxidants ensures the high seed yields so the changes in the antioxidants levels can be used as stress markers in the breeding for drought resistance.

Species	Treatment	Stomata number/mm² abaxial side			Stomata number/mm² adaxial-side			Stomatal resistance s/cm		
		BF	F	PR	BF	F	PR	BF	F	PR
Green bean	Irrigation	359.58 cd	424.89 b	311.12 e	72.92 c	84.96 b	56.73 d	1.74 cd	1.34 de	1.43 d
	Water deficit	369.40 c	458.31 a	346.86 d	87.66 b	112.80 a	72.63 c	1.79 c	1.40 d	1.18 e
	Dry	410.05 b	455.71 a	382.89 c	105.96 a	103.55 a	81.97 bc	4.11 a	3.39 b	3.40 b
Green pea	Irrigation	205.94 b	165.70 d	262.68 a	-	-	-	1.90 d	0.90 e	4.02 c
	Water deficit	199.58 b	181.57 c	263.74 a	-	-	-	2.43 d	1.41 de	4.59 bc
	Dry	205.07 b	183.11 c	259.98 a	-	-	-	1.92 d	1.29 de	5.21 b

BF= before flowering, F=under flowering, PR=pod ripening

Table 4. Stomata density and stomatal resistance in comparison with green pea and green beans

6. Conclusion

The negative impact of climatic stresses on agricultural productivity can be reduced by a combination of genetic improvement. Genetic improvement of crops for stress tolerance is relatively a new endeavour that would be successful if the relationship between the stress markers and yield were revealed. The plants adapt their morphology and physiology processes to limit damage(s) caused by environmental stresses, however, this requires high energy (depending on the stress level) and leads to the decrease in yield. Crop plants generally use more than one defence mechanism at a time to cope with drought. Defence has physiology and molecular basis depending on stage of development. The defence strategies of field crops with moderate sensitivity to drought have been presented especially in green beans and green pea populations. These species have different light and heat demands and use different defence strategies in their stages of development under field conditions. During flowering periods, the changes in the photosynthetic pigments, stomatal density, stomatal resistance and antioxidants levels can be considered to the drought stress indicators in the case of green bean. At the same time, the chlorophyll contents in leaves and stomatal resistance are related to the yield so these can be regarded as drought stress indicators in green peas. The correlations between the drought stress marker attributes and yields can be used for screening large number of genotypes and selecting them for drought tolerance under field conditions. These stress markers measured in the field indicate the disturbance of water supply in the crops and can also be used either for the irrigation schedules in crop production or the development of a transpiration crop model. These can also be used for the selection of genotypes with drought tolerance in the course of breeding.

7. References

Agbicodo, E.M., Fatokun, C.A., Muranaka, S., Visser, R.G.F., & Linden van der, C.G. (2009). Breeding drought tolerant cowpea: constraints, accomplishment, and future prospects, *Euphytica* Vol. 167 353-370. DOI 10.1007/s10681-009-9893-8

Alves, A.A.C., & Setter, T.L. (2004). Response of cassava leaf area expantion to water deficit: cell proliferation, cell expansion and delayed development, *Ann. Bot.* Vol. 94 605-613.

Alvino, A., & Leone, A. (1993). Response to low soil water potential in pea genotypes (*Pisum sativum* L.) with different leaf morphology, *Scientia Horticulturae* Vol. 53 (No. 1-2): 21-34.

Anderson, C.M., & Kohorn, B.D. (2001). Inactivation of Arabidopsis SIP1 leads to reduced levels of sugars and drought tolerance, *J Plant Physiol.* Vol.158 1215–1219.

Amede, T. (2003). Mechanisms of drought resistance in grain:II Stomatal regulation and root growth, *African J. Science* Vol. 26 (No. 2): 137-144.

Aminian, R., Mohammadi, S., Hoshmand, S., & Khodombashi, M. (2011). Chromosomal analysis of photosynthesis rate and stomatal conductance and their relationships with grain yield in wheat (*Triticum aestivum* L.) under water-stressed and well-watered conditions, *Acta Physiol Plant.* Vol. 33 755-764.

Amico, J.D., Torrecillas, A., Rodríguez, P., Morales, D., & Sánchez-Blanco, M.J. (2001). Differences in the effects of flooding the soil early and late in the photoperiod on the water relations of pot-grown tomato plants, *Plant Science* Vol. 160 481–487.

Añon, S., Fernandez, J.A., Franco, J.A., Torrecillas, A., Alarcón, J.J., & Sánchez-Blanco, M.J. (2004). Effects of water stress and night temperature preconditioning on water relations and morphological and anatomical changes of *Lotus creticus* plants, *Sci. Hortic.* Vol. 101 333–342.

Arshad, M., Shaharoona, B., & Mahmood, T. (2008). Inoculation with *Pseudomonas* spp. containing ACC-deaminase partially eliminates the effects of drought stress on growth, yield, and ripening of pea (*Pisum sativum* L.) *Pedosphere* Vol. 18 (No. 5): 611–620.

Arruda, M.A.Z., & Azevedo, R.A. (2009). Metallomics and chemical speciation: towards a better understanding of metal-induced stress in plants. *Ann Appl Biol.* Vol. 155 301–307.

Ashraf, M., & Hafeez, M. (2004). Thermotolerance of pearl millet and maize at early growth stages: growth and nutrient relations. *Biol. Plant.* Vol. 48 81–86.

Babita, M., Maheswari, M., Rao, L.M., Shanker, A.K., & Rao, D.G. (2010). Osmotic adjustment, drought tolerance and yield in castor (*Ricinus communis* L.) Hybrids, *Environ. Exp. Bot.* Vol. 69 243-249.

Baigorri, H., Antolín, M.C., Sánchez-Díaz, M. (1999). Reproductive response of two morphologically different pea cultivars to drought, *European Journal of Agronomy* Vol. 10 119–128.

Barradas V.L., Jones, H.G., & Clark, J.A. (1994.) Stomatal responses to changing irradiance in *Phaseolus vulgaris* L., *J Exp Bot.* Vol. 45 931-936.

Becana, M., Dalton, D.A., Moran, J.F., Iturbe-Ormaetxe, I., Matamoros, M.A., & Rubio, M.C. (2000). Reactive oxygen species and antioxidant in legume nodules, *Physiol Plant.* Vol. 109 372–381.

Berry, S.Z., & Rafique-Uddin, M. (1988). Effect of high temperature on fruit set in tomato cultivars and selected germplasm, *HortScience* Vol. 23 606–608.

Berzsenyi, Z., & Lap, D.Q. (2003). A N-műtrágyázás hatása a kukorica- (Zea mays L.) hibridek szemtermésére és N-műtrágyareakciójára tartamkísérletben (Effect of N fertilisation on the grain yield and N fertiliser response of maize (Zea mays L.) hybrids in a long-term experiment) *Növénytermelés* Vol. 52 (No. 3-4): 389-407.

Camejo, D., Rodríguez, P., Morales, M.A., Dellámico, J.M., Torrecillas, A., & Alarcón, J.J., (2005). High temperature effects on photosynthetic activity of two tomato cultivars with different heat susceptibility, *J. Plant Physiol.* Vol. 162 281–289.

Canci, H., & Toker, C.C. (2009). Evaluation of annual wild Cicer species for drought and heat resistance under field conditions, *Genet Resour Crop Evol.* Vol. 56. 1–6.

Carvalho, R.F. Takaki, M., & Azevedo, R.A. (2011). Plant pigments: the many faces of light perception. *Acta Physiol Plant.* Vol. 33 241–248.

Chen, Z., & Gallie, D.R. (2004). The ascorbic acid redox state controls guard cell signaling and stomatal movement, *Plant Cell* Vol. 16 (No. 5): 1143–1162.

Corlett, J.E., Jones, H.G., Masssacci, A., & Masojidek, J. (1994). Water deficit, leaf rolling and susceptibility to photoinhibition in field grown sorghum, *Physiol. Pl.* 92. 423-430.

Costa-Franca, M.G., Pham, Thi, A.T., Pimentel, C., Rossiello, R.O.P., Zuily-Fodil, Y., & Laffray, D. (2000). Differences in growth and water relations among *Phaseolus vulgaris* cultivars in response to induced drought stress, *Environ Exp Bot.* Vol. 43 227-237.

Cruz de Carvalho, M.H., Laffray, D., & Louguet, P. (1998). Comparison of the physiological responses of *Phaseolus vulgaris* and *Vigna unguiculata* cultivars when submitted to drought conditions, *Environmental and Experimental Botany* Vol. 40 197–207.

Diamant, S., Eliahu, N., Rosenthal, D., & Goloubinoff, P. (2001). Chemical chaperones regulate molecular chaperones in vitro and in cells under combined salt and heat stresses, *J. Biol. Chem.* Vol. 276 39586–39591.

Du, W-J.,Yu, D-Y., & Fu, S-X. (2009). Analysis of QTLs for the Trichome Density on the Upper and Downer Surface of Leaf Blade in Soybean [*Glycine max* (L.) Merr.] *Agricultural Sciences in China* Vol. 8 (No. 5): 529-537.

Eghball, B., & Maranville, J.W. (1991). Interactive effects of water and nitrogen stresses on nitrogen utilization efficiency, leaf water status and yield of corn genotypes. *Commun. Soil. Sci. Plant Anal.* Vol. 22 1367-1382.

El-Tajeb, N. (2006). Differential response of two Vicia faba cultivars to drought: growth, pigments, lipid peroxidation, organic solutes, catalase and peroxidase activity, *Acta Agronomica Hungarica* Vol. 54 (No. 1): 25–37.

Ekanayake, I.J., De Datta,S.K. & Steponkus, P. L. (1993). Effect of water deficit stress on diffusive resistance, transpiration, and spikelet desiccation of rice (*Oryza sativa* L.), *Ann. Bot.* Vol. 72 73-80.

Esfahani, M.N. & Mostajeran, A. (2011). Rhizobial strain involvement in symbiosis efficiency of chickpea–rhizobia under drought stress: plant growth, nitrogen fixation and antioxidant enzyme activities Rhizobial strain involvement in symbiosis efficiency of chickpea–rhizobia under drought stress: plant growth, nitrogen fixation and antioxidant enzyme activities, *Acta Physiol Plant.* Vol. 33 1075–1083.

Fan, S., Blake, T.J., & Blumwald, E. (1994). The relative contribution of elastic and osmotic adjustments to turgor maintenance in woody species. *Physiol. Plant.* Vol. 40 408-413.

Fernandez, D. & Castrillo, M. (1999). Maize leaf rolling initiation, *Photosynthetica* Vol. 37 493-497.

Foolad, M.R. (2005). Breeding for abiotic stress tolerances in tomato. In: Ashraf, M., & Harris, P.J.C. (eds.), *Abiotic Stresses: Plant Resistance Through Breeding and Molecular Approaches*, The Haworth Press Inc., New York, USA, pp. 613–684.

Gao, X-P., Yan, J-Y., Liu, E-K., Shen, Y-Y., Lu, Y-F., & Zhang, D-P., (2004). Water stress induces in pear leaves the rise of betaine level that is associated with drought tolerance in pear, *J. Hortic. Sci. Biotechnol.* Vol. 79 (No. 1): 114-118.

Gomes, A.M.M., Lagoa, A.M.M.A., Medina, C.L., Machado, E.C., & Machado, M.A. (2004). Interactions between leaf water potential, stomatal conductance and abscisic acid content of orange trees submitted to drought stress, *Braz. J. Plant Physiol.* Vol. 16 (No. 3): 155-161.

Gomes, F.P., Oliva, M.A., Mielke, M.S., Almeida, A.A.F., & Aquino, L.A. (2010). Osmotic adjustment, proline accumulation and cell membrane stability in leaves of Cocos nucifera submitted to drought stress, *Scientia Horticulturae* Vol. 126 379-384.

Grant, R.H., Jenks, M.A., Rich, P.J., Peters, P.J., & Ashworth, E.N. (1995). Scattering of ultraviolet and photosynthetically active radiation by sorghum bicolor: influence of epicuticular wax, *Agric. For. Meteorol.* Vol. 75 263–281.

Gratão, P.L., Polle, A., Lea, P.J., & Azevedo, R.A. (2005). Making the life of heavy metal-stressed plants a little easier, *Funct Plant Biol.* Vol. 32 481-494.

Griffin, J.J., Ranney, T.G., & Pharr, D.M. (2004). Heat and drought influence photosynthesis, water relations and soluble carbohydrates of two ecotypes of redbud (Cercis Canadensis), *J. Amer. Soc. Hort. Sci.* Vol. 129 (No. 4): 497-502.

Grzesiak, S., Filek, W., Skrudlik, G., & Niziol, B. (1996). Screening for drought tolerance: Evaulation of seed germination and seedling growth for drought resistance in Legume plants, *Journal of Agronomy and Crop Science* Vol. 177 245-246.

Guilioni, L., Wery, J., & Tardieu, F., (1997). Heat stress-induced abortion of buds and flowers in pea: is sensitivity linked to organ age or to relations between reproductive organs? *Ann. Bot.* Vol. 80 159-168.

Guilioni, L., Wéry, J., & Lecoeur, J. (2003). High temperature and water deficit may reduce seed number in field pea purely by decreasing plant growth rate, *Funct. Plant Biol.* Vol. 30 1151-1164.

Guo, C. & Oosterhuis, D.M., (1995).Pinitol occurrence in soybean plants as affected by temperature and plant growth regulators, *J. Expt. Bot.* Vol. 46 249-253.

Guo, C. & Oosterhuis, D.M., (1997). Effect of water-deficit stress and genotypes on pinitol occurrence in soybean plants, *Environ. Expt. Bot.* Vol. 37 147-152.

Györgyey, J. (1999). A hőstressz molekuláris alapjai, In: *Molekuláris növénybiológia* eds. Balázs,E., Dudits, D.) pp 303-321. (In Hungarian) Akadémiai Kiadó, ISBN 963 05 7374 4 Budapest, Hungary

Hardy, J.P., Anderson, V.J., & Gardner, J.S. (1995). Stomatal characteristics, conductance ratios, and drought-induced leaf modifications of semiarid grassland species, *Am J Bot.* Vol. 82 1-7.

Hawkins, T.S., Gardiner, E.S., & Comer, G.S. (2009). Modeling the relationship between extractable chlorophyll and SPAD-502 readings for endangered plant species research, J. *Nature Conservation* Vol. 17 123-127.

Hernandez-Jiménez, M.J., Lucas, M.M., & Felipe, M.R. (2002). Antioxidant defence and damage in senescing lupin nodules, *Plant Physiol. Biochem.* Vol. 40 645-657.

Heszky, L. (2007). Szárazság és a növény kapcsolata, *Agrofórum* Vol. 18. (11/M): 37-41. (In Hungarian)

Hoogenboom, G., Huck, M.G., & Peterson, M.C. (1986). Measured and simulated drought stress effects on daily shoot and root growth rate of soybean, *Netherland J. Agric. Sci.* Vol. 34 497-500.

Hsiao, T. (1994). Crop productivity and future world of elevated CO2 and changed climate, In: *Third Congress of the European Society for Agronomy Proceedings* Padova University, (eds. M. Borin and M. Sattin) France, Colmar Cedex. pp 6-17. ISBN 2 9505124 1 0

Ismail, A.M, & Hall, A.E., (1998). Positive and potential negative effects of heat-tolerance genes in cowpea, *Crop Science* Vol. 38 381-390.

Jones, H.G. (1992). Plants and Microclimate, second ed. Cambridge University Press ISBN 0 521415020

Kadioglu, A., Terzia, R., Saruhanb, N., & Saglama, A. (2011). Current advances in the investigation of leaf rolling caused by biotic and abiotic stress factors, *In Press: Plant Sci.* doi:10.1016/j.plantsci.2011.01.013

Kateriji, N., Mastrorilli, M., Hamdy, A., & Hoorn, J.W.van, (2002). Water status and osmotic adjustment of broad bean (*Vicia faba* L.) in response to soil fertility, *Acta Hortic.* Vol. 573 305-310.

Kato, M., Kobayashi, K., Ogiso, E., & Yokoo, M. (2004). Photosynthesis and dry-matter production during ripening stage in a female-sterile line of rice, *Plant Prod. Sci.* Vol. 7 (No. 2): 184-188.

Kepova, K.D., Holzer, R., Stoilova, L.S., & Feller, U. (2005). Heat stress effects on ribulose-1,5-bisphosphate carboxylase/oxygenase, Rubisco bindind protein and Rubisco activase in wheat leaves, *Biol. Plant.* Vol. 49 521-525.

Khan, H.R., Paull, J.G., Siddique, K.H.M., & Stoddard, F.L. (2010). Faba bean breeding for drought-affected environments: A physiological and agronomic perspective, *Field Crops Research*, Vol. 115 279-286.

Korotaeva, N.E., Antipina, A.I., Grabelynch, O.I., Varakina, N.N., Borovskii, G.B., & Voinikov,V.K. (2001). Mitochondrial low-molecular-weight heat shock proteins and tolerance of crop plant's mitochondria to hyperthermia, *Fiziol. Biokhim Kul'turn. Rasten.* Vol. 29 271-276.

Krishnamurthy, L., Vadez, V., Devi, M.J., Serraj, R., Nigam, S.N., Sheshshayee, M.S., Chandra, S., & Aruna, R. (2007). Variation in transpiration efficiency and its related traits in a groundnut (*Arachis hypogaea* L.) mapping population, *Field Crops Res.* Vol. 103 189-197.

Lecoeur, J., Wery, J., Turc, O., & Tardieu, F. (1995). Expansion of pea leaves subjected to short water deficit: cell number and cell size are sensitive to stress at different periods of leaf development, *J. Exp. Bot.* Vol. 46 1093-1101.

Li, Z., Wakao, S., Fischer, B.B., & Niyogi, K.K. (2009). Sensing and responding to excess light, *Ann. Rev. Plant Biol.* Vol. 60 239-260.

Lin, T-Y., & Markhart, III. A.H. (1996). Phaseolus acutifolius A. Gray is more heat tolerant than P. vulgaris l. in the absence of water stress, *Crop Sci.* Vol. 36 110-114.

Lizana, C., Wentworth, M., Martinez, J.P., Villegas, D., Meneses, R., Murchie, E.H., Pastenes, C., Lercari, B., Vernieri, P., Horton, P., & Pinto, M. (2006). Differential adaptation of two varieties of common bean to abiotic stress, I. Effects of drought on yield and photosynthesis, *J Exp Bot.* Vol. 57 (No. 3): 685-697.

Mai-Kodomi, Y., Singh, B.B., Myers, O., Yopp, J.H., Gibson, P.J., & Terao, T. (1999). Two mechanisms of drought tolerance in cowpea, *Indian J Genet.* Vol. 59 309-316.

Martínez, J.P., Silva, H., Iedent, J.F., & Pinto, M. (2007). Effect of drought stress on the osmotic adjustment, cell wall elasticity and cell volume of six cultivars of common beans (*Phaseolus vulgaris* L.), *Europ. J. Agronomy* Vol. 26 30-38.

Matamoros, M.A., Dalton, D.A., Ramos, J., Clemente, M.R., Rubio, M.C., & Becana, M. (2003). Biochemistry and molecular biology of antioxidants in the rhizobia–legume symbiosis, *Plant Physiol.* Vol. 133 449-509.

Matthews, R. B., Azam-Ali, S. N. &. Peacock, J. M. (1990). Response of four sorghum lines to midseason drought: II. leaf characteristics, *Field Crop Res.* Vol. 25 297-308.

Meglič, V., Budic, M., Kavar, T., Maras, M., Sustar-Vozlic, J., & Kidric, M. (2008). Evaluation of *Phaseolus* sp. germplasm response to water deficit. In: Prohens, J., Badenes, M.L. (eds) *Modern variety breeding for present and future needs. Proceedings.* ISBN 978-84-8363-302-1 Editorial Universidad Politéchnica de Valencia, Valencia, Spain, September, 2008

Micheletto, S., Rodriguez-Uribe, L., Hernandez, R., Richins, R.D., Curry, J., & O'Conell, M.A. (2007). Comparative transcript profiling in roots of *Phaseolus acutifolius* and *P. vulgaris* under water deficit stress, *Plant Sci.* Vol. 173 510-520.

Micklas, N.P., Kelly, J.D., Beebe, S.E., & Blair, M.W. (2006). Common bean breeding for resistance against biotic and abiotic stresses: From classical to MAS breeding, *Euphytica* Vol. 147 105-131.

Miyashita, K., Tanakamaru, S., Maitani, T., & Kimura, K. (2005). Recovery responses of photosynthesis, transpiration, and stomatal conductance in kidney bean following drought stress, *Envir Exp Bot.* Vol. 53 205-214.

Mohamed, M.F., Keutgen, N., Tawfik, A.A., & Noga, G. (2002). Dehydration-avoidance responses of tepary bean lines differing in drought resistance, *J Plant Physiol.* Vol. 159 31-38.

Mohammady, S., Moore, K., Ollerenshaw, J., & Shiran, B. (2005). Backcross reciprocal monosomic analysis of leaf relative water content (LRWC), stomatal resistance (SR) and carbon isotope discrimination (D%) in wheat under pre-anthesis water-stress conditions, *Aust. J Agric. Res.* Vol. 10 1059-1068.

Morgan, J.M. (1984). Osmoregulation and water stress in higher plants, *Ann. Rev. Plant Physiol.* Vol. 35 299-319.

Morgan, P.W., He, C.J., De Greef, J.A. & De Proft, M.P. (1990). Does water deficit stress promotes ethylene synthesis by intact plants? *Plant Physiol.* Vol. 94 1616-1624.

Nemeskéri, E. (2001). Water deficiency resistance study on soya and bean cultivars, *Acta Agronomica Hungarica* Vol. 49 (No. 1): 83-93.

Nemeskéri, E., Remenyik, J., & Fári, M. (2008). Studies on the drought and heat stress response of green bean (*Phaseolus vulgaris* L.) varieties under phytotronic conditions, *Acta Agronomica Hungarica* Vol. 56 (No. 3): 321-328.

Nemeskéri, E., Sárdi, É., Kovács-Nagy, E., Stefanovits Bányai, É., Nagy, J., Nyéki, J., & Szabó, T. (2009). Studies on the drought responses of apple trees (*Malus domestica* Borkh.) grafted on different rootstocks, *Int. J. Hortic. Sci. in Hungary*, Vol. 15 (No. 1-2): 29-36.

Nemeskéri, E., Sárdi, É., Remenyik, J., Kőszegi, B., & Nagy, P. (2010). Study of defensive mechanisms against drought of French bean (*Phaseolus vulgaris* L.) varieties, *Acta Physiol. Plant.* Vol. 32 (No. 6): 1125-1134. DOI:10.1007/s11738-010-0504-z

Nicholas, S. (1998). Plant resistance to environmental stress, *Curr. Opin. Biotechnol.* Vol. 9 214-219.

Noctor, G., & Foyer, C.H. (1998). Ascorbate and glutathione: keeping active oxygen under control, *Ann. Rev. Plant Physiol. Plant Mol. Biol.* Vol. 49 249-270.

Ommen, O.E., Donnelly, A., Vanhoutvin, S., Oijen, M., & Manderscheid, R. (1999). Chlorophyll content of spring wheat flag leaves grown under elevated CO_2 concentrations and other environmental stresses within the 'ESPACE-wheat' project, *Eur. J. Agron.* Vol. 10 197-203.

Panchuk, I.I., Volkov, R.A., & Schöffl, F. (2002). Heat stress- and heat shock transcription factor-dependent expression and activity of ascorbate peroxidase in *Arabidopsis*, *Plant Physiol.* Vol. 129 838-853.

Parry, A.D., Griffiths, A., & Horgan, R. (1992). Abscisic acid biosynthesis in roots. II The effects of water stress in wild type and abscisic acid deficient mutant (*notabilis*) plants of *Lycopersicon esculentum* Mill., *Planta* Vol. 187 192-197.

Pastenes, C., Pimentel, P., & Lillo, J. (2005). Leaf movements and photoinhibition in relation to water stress in field-grown beans, *J Exp Bot.* Vol. 56 425-433.

Pelleschi, S., Rocher, J.P., & Prioul, J.L. (1997). Effect of water restriction on carbohydrate metabolism and photosynthesis in mature maize leaves, *Plant Cell Environ* Vol. 20 493–503.

Pennycooke, J.C., Jones, M.L., & Stushnoff, C. (2003). Down-regulating galactosidase enhances freezing tolerance in transgenic petunia 1. *Plant Physiol.*, Vol. 133 901-909.

Perry, E.M., & Davenport, J.R. (2007). Spectral and spatial differences in response of vegetation indices to nitrogen treatments on apple, *Computers and Electronics In Agriculture*, Vol. 59 56-65.

Pfeiffer, T.W., Peyyala, R., Ren, Q., & Ghabrial, A.S. (2003). Increased soybean pubescence density: yield and soybean mosaic virus resistance effects, *Crop Science* Vol. 43 2071-2076.

Piha, M.I., & Munns, D.N. (1987). Sensitivity of the common bean (*Phaseolus vulgaris* L) symbiosis to high soil temperature, *Plant Soil* Vol. 98 183-194.

Procházková, D., & Wilhelmová, N. (2004). Changes in antioxidative protection in bean cotyledons during natural and continuous irradiation-accelerated senescence, *Biol Plant.* Vol. 48 (No. 1): 33–39.

Ramirez-Vallejo, P. & Kelly, J.D. (1998). Traits related to drought resistance in common bean, *Euphytica* Vol. 99 127–136.

Rainey, K., & Griffiths, P. (2005). Evaluation of *Phaseolus acutifolius* A. Gray plant introductions under high temperatures in a controlled environment, *Genet. Resour. Crop Evol.* Vol. 52 117–120.

Reid, J. B. & Renquist, A. R. (1997). Enhanced root production as a feed-forward response to soil water deficit in field-grown tomatoes. *Aus. J. Plant Physiol.* Vol. 24 685–692.

Reynolds-Henne, Ch.E., Langenegger, A., Mani, J., Schenk, N., Zumsteg, A., & Feller, U. (2010). Interactions between temperature, drought and stomatal opening in legumes, *Environmental and Experimental Botany* Vol. 68 37–43.

Ribeiro, R. V., Santos, M. G., Machado, E. C. & Oliveira, R. F. (2008). Photochemical Heat-Shock Response in Common Bean Leaves as Affected by Previous Water Deficit, *Russian Journal of Plant Physiology* Vol. 55 (No. 3): 350–358.

Sánchez, F.J., Manzanares, M., Andrés, E.F., Tenorio, J.L., & Ayerbe, L. (1998). Turgor maintenance, osmotic adjustment and soluble sugar and proline accumulation in 49 pea cultivars in response to water stress, *Field Crops Res.* Vol. 59 225-235.

Sánchez, F.J., Manzanares, M., Andrés, E.F., Tenorio, J.L., & Ayerbe, L. (2001). Residual transpiration rate, epicuticular wax load and leaf colour of pea plants in drought conditions. Influence on harvest index and canopy temperature, *Europ. J. Agronomy* Vol. 15 57–70.

Sánchez, F.J. de Andrés, E.F., Tenorio, J.L., & Ayerbe, L. (2004). Growth of epicotyls, turgor maintenance and osmotic adjustment in pea plants (*Pisum sativum* L.) subjected to water stress, *Field Crops Research* Vol. 86 1–90.

Sariyeva, G., Kenjebaeva, S. & Lichtenthaler, H.K. (2009). Photosynthesis performance of non-rolling and flag leaf rolling wheat genotypes during temperature stress, *Acta Botanica Hungarica*, Vol. 51 (No. 1-2): 185-194.

Savé, R., Biel, C., Domingo, R. Ruiz-Sánchez, M.C., & Torrecillas, A. (1995). Some physiological and morphological characteristics of citrus plants for drought resistance, *Plant Science* Vol. 110 167-172.

Seki, M., Umezewa, T., Urano, K., & Shinozaki, K. (2007). Regulatory metabolic networks in drought stress responses, *Curr. Opin. Plant Biol.* Vol. 10 296-302.

Shao, Hong-Bo., Chu, Li-Ye., Jaleel, C.A., Zhao, & Chang-Xing. (2008). Water-deficit stress-induced anatomical changes in higher plants, *C. R. Biologies* Vol. 331 215-225.

Singh, S., & Sethi, G.S. (1995). Stomatal size, frequency and distribution in Triticum aestivum, Secale cereale and their amphiploids, *Cereal Res. Commun.* Vol. 23 103-108.

Simoes-Araujo, J.L., Rumjanek, N.G., & Margis-Pinheiro, M. (2003). Small heat shock proteins genes are differentially expressed in distinct varieties of common bean, *Braz. J. Plant Physiol.* Vol. 15 33-41.

Smith, A.E. & Phillips, D.V., (1982). Influence of sequential prolonged periods of dark and light on pinitol concentration in clover and soybean tissue, *Physiol. Plant.* Vol. 54 31-33.

Smith, D.L., Dijak, M., & Hume, D.J. (1988). The effect of water deficit on N2 (C2H2) fixation by white bean and soybean, *Canadian J Plant Sci.* (Ottawa), Vol. 68 (No. 4): 957-967.

Sponchiado, B.N., White, J.W., Castillo, J.A., & Jones, P.G. (1989). Root growth of four common bean cultivars in relation to drought tolerance in environments with contrasting soil types, *Exp Agric.* Vol. 25 249-257.

Streeter, J.G., Lohnes, D.G., & Fioritto, R.J. (2001). Pattern of pinitol accumulation in soybean plants and relationships to drought tolerance, *Plant Cell Environ.* Vol. 24 429-438.

Szabados, L., & Savoure, A. (2009). Proline: a multifunctional amino acid. *Trends Plant Sci.* Vol. 2 89-97.

Takele, A. & McDavid, C.R. (1995). The response of pigeonpea cultivars to short durations of waterlogging, *African Crops Science Journal* Vol. 3 (No. 1): 51-58.

Tari, I., Csiszár, J., Gallé, Á. Bajkán, Sz., Szepesi, Á., & Vashegyi, Á. (2003). Élettani megközelítések gazdasági növényeink szárazságtűrésének genetikai transzformációval történő javítására, *Bot. Közlem.* Vol. 90 (No. 1-2): 139-158. (In Hungarian)

Todorov, D.T., Karanov, E.N., Smith, A.R., & Hall, M.A. (2003). Chlorophyllase activity and chlorophyll content in wild type and *eti 5* mutant of *Arabidopsis thaliana* subjected to low and high temperatures, *Biol. Plant.* Vol. 46 633-636.

Vara-Prasad, P.V., Craufurd, P.Q., & Summerfield, R.J. (1999). Fruit number in relation to pollen production and viability in groundnut exposed to short episodes of heat stress, *Ann. Bot.* Vol. 84 381-386.

Vendruscolo, E.C.G., Schuster, I., Pileggi, M., Scapim, C.A., Molinari, H.B.C., Marur, C.J., & Vieira, L.G.E. (2007). Stress-induced synthesis of proline confers tolerance to water deficit in transgenic wheat, *J. Plant Physiol.* Vol. 164 1367-1376.

Venora, G., & Calcagno, F. (1991). Study of stomatal parameters for selection of drought resistant varieties in Triticum Durum DESF, *Euphytica* Vol. 57 275-283.

Vue, J.C.V., Baker, J.T., Pennanen, A.H., Allen, L.H., Bowes, G., & Boote, K.J. (1998). Elevated CO2 and water deficit effects on photosynthesis ribulose bisphosphate carboxylase-oxygenase, and carbohydrate metabolism in rice, *Physiol. Plant.* Vol. 103 327-339.

Wahid, A., Gelani, S., Ashraf, M., & Foolad, M.R. (2007). Heat tolerance in plants: An overview, *Environmental and Experimental Botany* Vol. 61 199-223.

Wahid, A. (2007). Physiological implications of metabolites biosynthesis in net assimilation and heat stress tolerance of sugarcane sprouts, *J. Plant Res.* Vol. 120 219-228.

Wang, H., & Clarke, J.M. (1993a). Relationship of excised-leaf water-loss and stomatal frequency in wheat, *Can J Plant Sci.* Vol. 73 93–99.

Wang, H., & Clarke, J.M. (1993b). Genotypic, intra plant and environmental variating in stomatal frequency and size in wheat, *Can. J Plant Sci.* Vol. 73 671–678.

Winzor, C.L., Winzor, D.J., Paleg, L.G., Jones, G.P., & Naidu, B.P. (1992). Rationalization of the effects of compatible solutes on protein stability in terms of thermodynamic nonideality, *Arch. Biochem. Biophys.* Vol. 296 102-107.

Yadava, U.L. (1986): A rapid and nondestructive method to determine chlorophyll in intact leaves, *HortScience* Vol. 21 (No. 6): 1449-1450.

Yang, G., Rhodes, D., & Joly, R.J., (1996). Effects of high temperature on membrane stability and chlorophyll fluorescence in glycinebetaine-deficient and glycine-betaine-containing maize lines, *Aust. J. Plant Physiol.* Vol. 23 437-443.

Ye, L., Gao, Hiu-Vuan, & Qui, Zou (2000). Responses of the antioxidant systems and xanthophyll cycle in *Phaseolus vulgaris* L. to the combined stress of high irradiance and high temperature, *Photosynthetica* Vol. 38 (No. 2): 205-210.

Young, L.W.,Wilen, R.W., & Bonham-Smith, P.C. (2004). High temperature stress of *Brassica napus* during flowering reduces micro- and megagametophyte fertility, induces fruit abortion, and disrupts seed production, *J. Exp. Bot.* Vol. 55 485–495.

Zlatev, Z., Lidon, F., Ramalho, J., & Yordanov, I. (2006). Comparison of resistance to drought of tree bean cultivars, *Biol. Plant.* Vol. 50 (No. 3): 389–394.

Section 3

Mineral Nutrition and Root Absorption Processes

7

Soil Fungi-Plant Interaction

Giuseppe Tataranni, Bartolomeo Dichio and Cristos Xiloyannis
The University of Basilicata
Italy

1. Introduction

One of the main European agricultural problems is the decline in soil fertility due to the reduction of the natural soil harmony. When natural soil composition is altered, few cultivated plants replace spontaneous populations of numerous species. Man becomes the only regulator of a new fragile equilibrium between the simplified biocenosis elements. Agronomic techniques (fertilisation, irrigation, soil tillage, etc.) become instruments to achieve such an improbable equilibrium. Frequently, these techniques are just the ones responsible for environmental pollution, disequilibrium among mineral elements and the general decrease in soil fertility. They are based on simplification of the relationships between the plant and other components of the natural habitat. This simplification should make agricultural systems easier to be controlled, but, indeed, it creates conditions of extreme weakness for plant life. It is clear that life on emerged lands has been possible thanks to complex relationships and, for plants in particular, to microorganism symbiosis. On the other hand, decreasing relationships between cultivated plants and other components of the natural habitat give rise to environmental degradation and pollution (due to the need of using high amounts of chemical inputs).

Soil fungi – plant roots have evolved forms of symbiosis, namely mutualistic associations. The partners in these relations are members of the *Zygomycota*, *Ascomycota* or *Basidiomycota* divisions and most vascular plants (Harley and Smith, 1983; Kendrick, 1985). In these symbioses, the host plant receives mineral nutrients, while the fungus obtains photosynthetically derived carbon compounds (Harley, 1989; Harley and Smith, 1983). At least seven different fungi – plants associations have been recognized, with distinct morphology patterns, involving different groups of organisms (Harley, 1989; Harley and Smith, 1983). The most common ones are: i) vesicular arbuscular mycorrhizas (VAM), in which *Zygomycetes* fungi produce arbuscules, hyphae and vesicles among root cortex cells, between cell wall and plasmatic membrane; ii) ectomycorrhizas (ECM), where *Basidiomycetes* and other fungi form a mantle around roots and a so-called Hartig net among root cells; iii) orchid mycorrhizas, where fungi produce coils of hyphae within roots (or stems) of orchidaceous plants; (iv) ericoid mycorrhizas, developing hyphal coils in outer cells of Ericales hair roots (Brundrett, 1991). Factors that can influence the establishment and persistence of mycorrhizal associations are various, besides symbiont compatibility: external factors - edaphic or microclimatic conditions, presence of further soil organisms, nutrient competition; and internal factors - organism phenology. Infective propagules must be present when root growth activity occurs, since roots have a limited period of susceptibility

(Brundrett and Kendrick, 1990) and rapid colonisation of the root system is required for an effective association (Bowen, 1987).

This chapter focuses on some effects of that symbiosis on plant physiology. The general positive influence of symbiosis is remarkable, especially in terms of nutritional function. Moreover, the microorganisms of the rhyzosphere play an important role in root hydraulics and plant water status: directly - through uptake and transport, exceeding or regulating part of the cellular barriers; or/and indirectly - the hyphae increase the water potentially available by exploring a bigger soil volume and interstices between little agglomerates. Possible mechanisms by which symbiosis improves plant performance are discussed. Experimentations have been carried out to estimate the progressive colonisation and persistence of the symbiotic microorganisms on roots in agricultural conditions. Studying artificial symbiosis, induced during the fruit tree production process, could concur to enhance the plant tolerance to abiotic and biotic stresses. In this regard, *Glomus intraradices* and *Trichoderma harzianum* fungi were found to be effective as native inoculants in many Mediterranean areas. Changes in morphological and physiological plant characteristics could increase the general quality and viability of planting material in sustainable management.

1.1 The genus *Trichoderma*

The genus *Trichoderma* characterises filamentous fungi that can be isolated from many soil types. As part of a healthy soil environment, *Trichoderma* species have been found worldwide. They are able to colonise plant roots and debris, including decomposition of organic substrates (saprobes) (Kendrick, 1985). The fungi of this genus are genetically quite diverse, with a number of different capabilities (Harman et al., 2004). The biocontrol mechanisms exercised by *Trichoderma* could be attributed to competition for nutrients, release of extracellular hydrolytic enzymes, and secondary metabolites toxic to plant pathogens at very low concentrations (Mathivanan et al., 2008). *Trichoderma* also induces defence responses in host plants (e.g., 'systemic acquired resistance' - SAR) (Mathivanan et al., 2008). In particular, *T. harzianum* produces a variety of antibiotic antifungal peptides that interact with cell membranes of plant fungal pathogens, so inhibiting their growth (Harman et al., 2004). Furthermore, *T. harzianum* has been shown to inhibit wood rots and other fungal plant pathogens by up to 60% through production of volatile antibiotics (Morrell, 1990).

Selected strains of *T. harzianum* showed to suppress plant diseases, but also not to be adaptable to various plant species and pathogens. To overcome these limitations, researchers at Cornell University produced a hybrid strain with enhanced vigour and larger adaptability, called strain T-22 (T22). It is the active ingredient of commercial biocontrol products. T22 acts as a deterrent, protecting the root system from the attack of pathogen fungi like *Fusarium, Pythium, Rhizoctonia* and *Thielaviopsis*.

The main applications of T22 are as preventive and defence of the root against pathogen agents. The biofungicide protects the plant by establishing itself in the rhizosphere and occupying that soil ecological niche. Since it is a living organism, T22, as saprophyte, can grow along the entire length of the root system, establishing a physical barrier against pathogen attack. As long as root systems remain actively growing, *T. harzianum* T22 continues to act by feeding plant wastes released. In this way, it also takes nutrients away from the pathogens. Plant tissues, which already are under attack before the application of T22 spores, will not experience the full benefits of using the biocontrol agent. T22 can damage root-rotting pathogen fungi through the releases of digestive enzyme chitinases, which dissolve the chitin in cell walls. Once damaged, the pathogen itself becomes prey of other soil organisms. *Trichoderma* spp. can also attack and parasitize other fungi directly. Plant root system is even more efficient, by T22 enhanced root growth and development. A larger root system is likely able to explore soil, and to use water and nutrients efficiently. In fact, in addition to the disease control, T22 increases the overall health of a plant. Its use allows reducing the nursery bench time, mineral fertilisation and fungicide applications, and getting larger leaves and stems as well as faster rooting in outdoor conditions (Harman et al., 2004; Mathivanan et al., 2008; Morrell, 1990; Sofo et al., 2010, 2011).

1.2 The genus *Glomus*

Glomus is the largest genus of arbuscular mycorrhizal fungi (AMF). All species establish symbiotic relationships with plant roots. The establishment of a functional symbiosis involves a sequence of recognition events, leading to the morphological and physiological integration of the two organisms (Giovannetti & Sbrana, 1998). The life cycle of an AMF begins with spore germination, and follows with a pre-symbiotic mycelia growth phase, hyphal branching, appresorium formation, root colonisation, and finally arbuscule development (Giovannetti et al., 1994). Arbuscules, the interface for nutrient exchange, are formed by repeated dichotomous fine branching. They grow inside individual cells of the root cortex, but remain outside their cytoplasm, due to plasma membrane invaginations. Hyphae with arbuscules contain numerous nuclei and structures, polyvesicular bodies and electron dense granules inside small vacuoles (Fig. 1, 2), the site of intense alkaline phosphatase and ATPase activities (Gianinazzi et al., 1979). Mycorrhizal fungi form, also in soil, a hyphal network that can obtain and transport nutrients, propagate the association and interconnect plants (Newman, 1988). The production of plant-external hyphae varies with species and isolates of fungi, can be influenced by soil properties and is an important determinant of mutualistic effectiveness (Gueye et al., 1987).

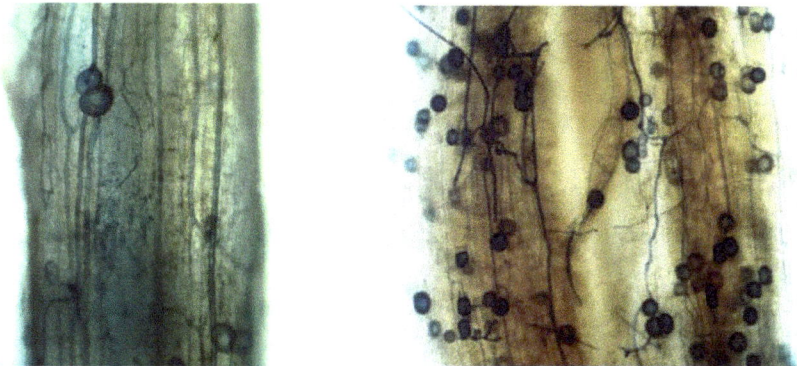

Fig. 1. Roots of *Prunus* rootstocks with hyphae and spores of *Glomus intraradices* (right; 100x magnification). In particular, (left) the multiple walls of spores (200x magnification).

Researchers have highlighted a particular infective capacity of the species *Glomus intraradices* (Estaùn et al., 2003). *G. intraradices* is autochthon to the Mediterranean basin; hence it is well adapted to the temperate climate.

Beneficial effects, in particular of some *Glomeromicota* spp. fungi, have been reported on growth, tissue hydration and leaf physiology (Allen, 2007).

The activity of VAM fungi in soils is usually quantified by measuring structures formation within roots with a microscope using a clearing step and staining procedures (Phillips & Hayman, 1970; Trouvelot et al., 1986).

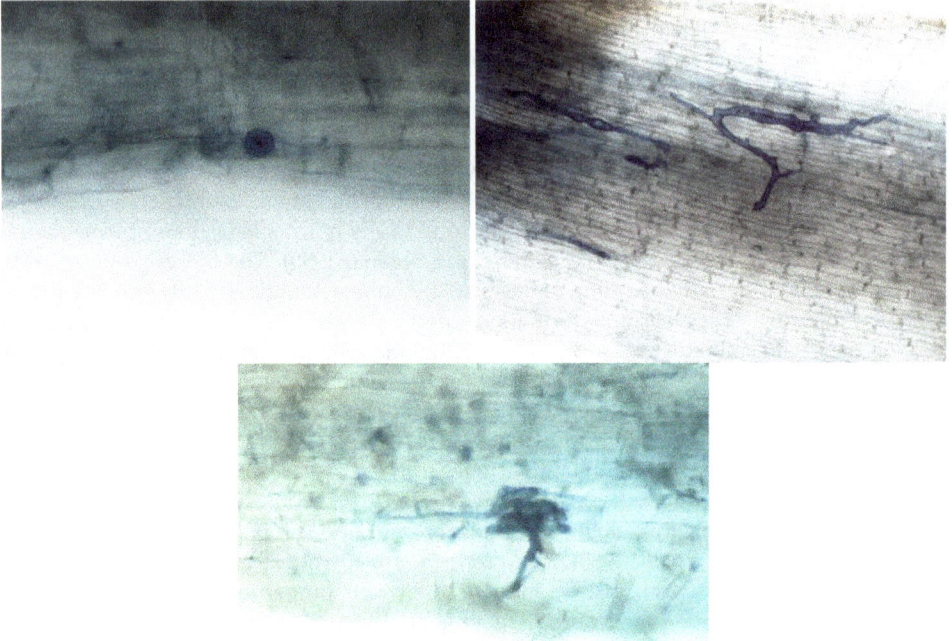

Fig. 2. Roots of *Prunus* rootstocks with *Glomus intraradices* spore (left), appresorium (right) and arbuscule (below; 400x, 200x and 400x magnification, respectively).

2. The influence of mychorrizas on plants

2.1 Use of *Trichoderma harzianum* strain T-22 in *Prunus* spp. micropropagation processes

T. harzianum is utilized as inoculant for crop production purposes, and for the improvement of plant nursery processes (Mathivanan et al., 2008). In fact, it is necessary to increase the quality of nursery plant material, in terms of higher and faster root and shoot development. To this end, the growth substrate (peat) can be inoculated with *Trichoderma* few days before rootstocks transplanting at the concentration of about 1 kg m^{-3} of substrate (see Section 3.1).

The inoculation of *T. harzianum*, during the rooting phase when plants are cultured in vitro under sterile conditions, could be another way to minimize losses in the acclimatisation phase. This method could avoid the competition of *T. harzianum* with other microorganisms usually found in soil, hence allowing a better interaction with plant roots and a better induction of plant growth.

The first steps in understanding the interactions between a plant and *T. harzianum* are to define and optimize the most appropriate method and time of inoculation, in order to verify the effects of this fungal symbiont on the micropropagated plant material, in a way that fits existing production processes. Inoculation of T22 was applied on *in vitro*-cultured shoots of GiSeLa6® (*Prunus cerasus x Prunus canescens*) and GF677 (*Prunus amygdalus x Prunus persica*), important commercial rootstocks utilized for stone fruits (Sofo et al., 2010). The results showed that early inoculation with T22 (at the time of setting up the in vitro cultures) damaged the plants severely (Sofo et al., 2010). Early inoculation was not successful, as the fungus did not establish a symbiotic relationship with the plants but damaged them by acting first as a competitor for nutrients in the agar medium, then as a saprophyte. On the contrary, 7 days after shoot transfer in the root-inducing medium, plants survived and increased shoot growth and root development (about 160% of non-inoculated controls). A co-metabolic system with mutual benefits for plant and fungus was likely established (Sofo et al., 2010). In a different, aseptic hydroponic system, *Trichoderma* treated cucumber plants showed similar effects (Yedidia et al., 1999). Specific *Trichoderma* spp. strains have shown to be capable of plant growth enhancement, and of the bio-control of a range of wood-rot fungi when grown on a low-nutrient medium. *T. virens* or *T. atroviride* increased biomass production and stimulated production of lateral roots in *Arabidopsis* seedlings (Contreras-Cornejo et al., 2009). In our case, a simple MS medium was used, which is quite representative of fresh softwood (Harman et al., 2004). The colonisation frequency (number of colonised root fragments/total root fragments*100) of the *in vitro* system GiSeLa6®-T22 was 20%, applying the fungus 7 days after shoot transfer in the root-inducing medium. Although the measured fungal colonisation frequency was not high, the morphological effects of fungus colonisation on GiSeLa6® plants were evident, and inoculation mass and hyphae of T22 along roots of GiSeLa6® rootstock were clearly observed (Sofo et al., 2010). *T. harzianum* is capable of invading roots, but is typically restricted to the outer layers of the cortex (Yedidia et al., 1999).

T. harzianum cultured *in vitro* under controlled and stable conditions, without any nutrient limitations or other competing microorganisms, can display maximum growth and interact optimally with plants. Even if a further step in the micropropagation process, with a later treatment, might not be profitable.

2.2 Biochemical and morphological changes in *Prunus* rootstock inoculated with *Trichoderma harzianum* strain T-22

Trichoderma harzianum has been successfully used for the biological control of many plant pathogens through chemiotropic mycoparasitic interactions with some target fungal organisms (Thangavelu et al., 2004; Sahebani and Hadavi, 2008). Enzymes, which are active during mycoparasitism, degrade the cell wall (Belén Suárez et al., 2005; Yang et al., 2009). *T. harzianum* strain T-22 (T22) has the ability to directly enhance root growth and plant development also in the absence of pathogens (Harman, 2000; Sofo et al., 2010). It has also been suggested that this could be due to the production of some unidentified growth-regulating factors by the fungus or to the induction of the production of these factors in plants (Windham et al., 1986). All these findings indicate the versatility by which *T.*

harzianum can directly manifest biological control activity. In spite of their theoretical and practical importance, the mechanisms responsible for the growth response due to the direct action of *T. harzianum* on agronomic plants have not been investigated extensively.

Both endo- and ectomycorrhiza can enhance adventitious root formation. Ectomycorrhizas are able to produce auxins, gibberellins and other phytohormones (Hartmann et al., 2002).

It is well known that high values of the IAA/CKs ratio in plants promote root formation, whereas low contents induce shoot-bud formation. Thus, the growth-promoting action of T22 could be due to changes in phytohormone levels and balance in the plant. Another hypothesized positive direct effect of *T. harzianum* on plants is the solubilisation of some insoluble or sparingly soluble minerals by acidification of the medium, which could provide a better nutrient availability and uptake for the plants (Altomare et al., 1999; Kücük et al., 2008; Singh et al., 2010).

This research investigated the biochemical basis of the direct plant-growth-promoting activity of T22 on the genotype GiSeLa6® (*P. cerasus x P. canescens*), one of the most important commercial rootstocks used for sweet and sour cherry varieties.

The application of *T. harzianum* strain T-22 (T22) during the rooting phase of GiSeLa6® rootstocks resulted in greater mean root and shoot length if compared to un-inoculated plants (Sofo et al., 2010).

It is probable that the up-regulation of key genes for hormone biosynthesis or the down-regulation of the genes involved in hormone catabolism was induced by the T22 secretion of elicitors diffused into the medium or directly transferred from the fungal hyphae to the root cells, as suggested by Harman et al. (2004). There is evidence that the change in phytohormone levels is one of the direct mechanisms by which *T. harzianum* promoted root and shoot growth.

Both auxins and cytokinins are involved in shoot and root growth, and morphology. Indole-3-acetic acid (IAA) is the most widely naturally-occurring auxin in vascular plants, and it is involved in lateral and adventitious roots initiation and emergence, as well as in shoot development by changes in cell division, expansion and differentiation (Hedden and Thomas, 2006). The results showed that after T22-inoculation, IAA in leaves and roots significantly increased by 150 and 120%, respectively, whereas DHZR decreased by 80%. Increases in t-ZR were found only in leaves (90%). The auxin/cytokinins ratios changed from 29 to 47 in leaves, and from 15 to 21 in roots of un-inoculated and T22-inoculated plants, respectively. Root activity determined a decline of medium acidity, and this effect was more marked in T22-inoculated plants (up to pH 4) (Sofo et al., 2011). Among cytokinins, trans-zeatin (t-ZR) and dihydrozeatin (DHZR) - two of the most active in plants - control cell division and are involved in reducing apical dominance, inhibiting xylem formation and root growth, promoting leaf expansion and chloroplast development, and delaying senescence (Srivastava, 2002). As root induction and growth are stimulated by auxins and inhibited by cytokinins, the observed increase in IAA and decreases in t-ZR and DHZR could explain the higher root growth observed in T22-treated plants.

These results could also explain the higher shoot elongation observed (previous section) and the results of Sofo et al. (2010) who found increases in the number of leaves and in stem diameter of T22-treated GiSeLa6® rootstocks. Generally, abscisic acid (ABA) acts as a general inhibitor of growth and metabolism, and negatively affects the synthesis of proteins and nucleic acids, even though these effects vary with tissue and developmental stage (Kobashi et al., 2001; Srivastava, 2002). Notwithstanding the significant differences in ABA levels between leaves and roots, T22 did not induce a higher ABA accumulation in both the tissues and thus did not determine growth inhibition (Sofo et al., 2010, 2011).

Altomare et al. (1999) emphasised that the plant-growth-promoting capacity of *T. harzianum* was associated with in vitro solubilisation of certain insoluble minerals accomplished via production of chelating metabolites and fungus redox activity. In our experiment, we clearly demonstrated that a strong acidification in the medium inoculated with T22 occurred. This acidification could determine the solubilisation of some salts and their higher availability for plants (Kücük et al., 2008; Singh et al., 2010). It is noteworthy that pH values markedly decreased also in the presence of T22 alone, but it appears that the synergistic action by plant and T22 caused a greater acidification of the medium (Sofo et al., 2011). This might depend on the fact that T22 enhances the acidifying capacity of the root due to the proton extrusion of the root cells through the plasmalemma (Kücük et al., 2008). The pH decrease could allow the solubilisation of MnO_2, Fe_2O_3, metallic zinc, and calcium phosphate, and the reduction of Fe(III) and Cu(II), with evident benefits for plant nutrient uptake (Kücük et al., 2008; Singh et al., 2010). The minimum pH values found by Altomare et al. (1999) in sucrose-yeast extract liquid cultures plus *T. harzianum* but without plants, were approximately 5.0, then similar to our values. In our experiment, we observed a further pH decrease of approximately 1 unit of pH during T22-plants interaction (Sofo et al., 2011). We hypothesize that in natural soils, which usually have a pronounced buffering capacity, the acidification due to T22 could be less pronounced, as recently suggested by Singh et al. (2010).

Microscopic analyses were carried out to compare root systems of T22-inoculated and un-inoculated plants. Plant overall morphology of T22-inoculated and un-inoculated plants grown in cherry medium differed significantly (Sofo et al., 2011). These analyses revealed changes in root cell wall suberification of the exoderm and endoderm, with an increase in suberized cellular layers from 1 to 2-3, and an enhancement of cell wall epifluorescence (Sofo et al., 2011). Root tissues contain abundant alkaloids: berberine, chelerythrine, sanguinarine and chelidonine (along with other isoquinoline alkaloids), and some of them act as fluorochromes for suberin and lignin, providing numerous potential natural dye sources for fluorescence microscopic techniques (O'Brien and McCully, 1981; Brundrett et al., 1988). In our case, the observed cell wall epifluorescence (Fig. 3) indicated that T22 seems to induce the ex novo synthesis of phenolic compounds in the plants, likely by the secretion of elicitors and the following induction of defence responses, as suggested by Mathivanan et al. (2008). The enzymes of the phenylpropanoid pathway, involved in lignin biosynthesis (Gianinazzi-Pearson et al., 1994), are some of the compounds induced and involved in plant defence. The increased lignification of root endodermal cells induced by mycorrhiza was already found by Dehne (1982), but it was never demonstrated for T22. We suggest that the accumulation of protective molecules, such as lignin and suberin, in plants inoculated with T22 could accelerate their hardening. The accumulation of structural substances may be of key importance in the resistance process, by increasing the mechanical strength of the host cell walls (Dalisay and Kuc, 1995).

During the acclimatisation phase of nursery processes, all these observed biochemical and morphological changes induced by T22 could be an advantage. Plants could acclimatise better to new and hostile environments, so increasing plant survival in the absence of pesticide applications.

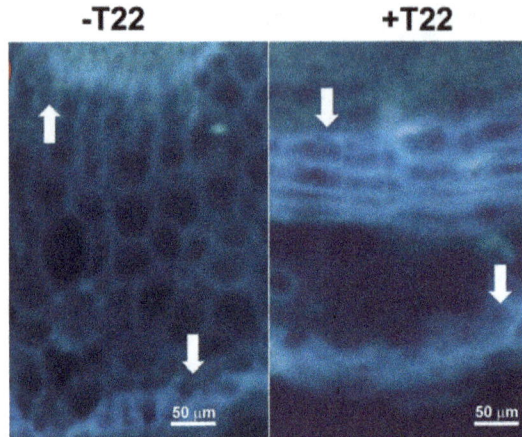

Fig. 3. Root cross-sections (diameter ≤ 1 mm; 6 mm from the root tip) of micropropagated plants inoculated with T22 (right) and un-inoculated plants (left) observed at 100x magnification with a mercury lamp. The arrows indicate the endodermic (above) and exodermic (below) layers (Sofo et al., 2011).

2.3 The influence of mychorrizas on vegetative growth

The artificial mychorrization, in the nursery phase, becomes useful particularly when the populations of native fungi are not present or have been reduced by intensive agricultural practices. Beneficial effects, in terms of positive plant growth responses and enhanced nutrient uptake, have been obtained with many agricultural crops.

The presence of arbuscular endo-mycorrhizal fungi was already evident in roots of olive tree plants, in many regions of the globe (Roldàn-Fajardo and Barea, 1986). The dependence of this species on these symbiotic microorganisms, relatively to mycotrophic function, is remarkable (Hayman et al., 1976; Roland-Fajardo and Barea, 1986; Briccoli Bati et. al., 1992). The positive influence of mycorrhizas was also demonstrated both on rooted olive cuttings (Di Marco et al., 2002) and during their growth in nursery (Briccoli Bati and Godino, 2002; Briccoli Bati et al., 2003). The study of the mycorrhizal symbiosis influence induced in olive tree, could concur to optimize the development of the plant and to improve its tolerance to some a-biotic and biotic stresses.

Trials were carried out on the effects of mychorrizic symbiosis induced in five cultivars of *Olea europaea*: Carolea, Coratina, Maiatica of Ferrandina, Leccino and Tondina. During the experimentation, self-rooted cuttings were inoculated, at two transplants in pot, with: *Glomus intraradices*, *Glomus* spp., *Glomus* spp. plus *Trichoderma* sp. plus bacteria (*Pseudomonas* spp., *Bacillus* spp.). Using the ordinary management techniques, the plants grew under optimal water status and limited fertilization (no phosphorus).

Morph-anatomically, by observations and microscopic analyses, the roots of the mycorrhized olive plants exhibited just few peculiar characteristics: the distribution in the substrate was always homogenous, the tip diameters were bigger, the adhesion of soil particles in the rhyzosphere was determined by a widespread presence of mycelium. Other authors do not give special particulars on that, but only report some differences related to root length and depth in other tree species (Andersen et al., 1988). Observations were made to compare root systems of different inoculated and un-inoculated plants just before sampling. Morphology and behaviour were significantly different. Control plants: the roots were branched, of 0.5-1.0 mm in mean diameter; their main colour was pale yellow-brown; they came out of the pot bottom; soil particles, especially of peat, tightly adhered to them. *Glomus intraradices* treatment: the roots, homogeneously of 0.2-0.5 mm in mean diameter, also in this case were branched, presented tip thickenings; root growths, even if coming out of the pot bottom, were more reduced in number than the control; their colour was the same, pale yellow-brown; mycelium seemed to wrap up and hold soil ryzosphere particles, which consequently didn't adhere directly to the epidermal layer. *Glomus* spp. treatment: the roots, homogeneously of 0.2-0.5 mm in mean diameter, well distributed in the pot volume and branched, presented tip thickenings; root growths, even if coming out of the pot bottom, were more reduced in number than the control; their colour was the same, pale yellow-brown; the soil ryzosphere particles didn't adhere directly to the epidermal layer. *Glomus* spp. + *Trichoderma* sp. + bacteria treatment: the roots were branched and well distributed in the whole soil volume; they came out of the pot bottom; the root number with a diameter smaller than 0.5 mm was greater; their colour did not change; the soil ryzosphere particles didn't adhere directly to the epidermal layer.

The percentage of mychorrizal colonisation reached values up to 100%, while the intensity of mychorrization was variable, from 20 to 50%. Destructive and not-destructive biometric measurements estimated the variations, if any, in terms of plant growing rate and stored dry matter. Altogether, in the first eighteen months, all the inoculated treatments showed greater increments, up to 20%, than the controls. *Glomus intraradices* gave the best result, 20% more (Fig. 4). However, a strong inoculum-cultivar interaction emerges; Maiatica cultivar, in fact, reached higher stem diametrical increments, 20% more, with the *Glomus* spp.; Carolea cultivar, 10% more, and Coratina, 30% more, with the *Glomus* spp. plus *Trichoderma* sp. plus bacteria; Leccino and Tondina confirmed the efficiency of *Glomus intraradices*, 30% and 25% more, respectively (Tataranni et al., 2010; Fig. 5). AM fungi prove to be better competitors than others when several isolates are inoculated together in pot culture experiments (Lopez-Aguillon & Mosse, 1987; Wilson, 1984). In these studies, the success is generally due to the ability to colonise roots most rapidly (Wilson, 1984). The outcome of competition should be expected to depend on the placement and amount of the inoculum, hyphal growth rates in soil and root interactions (Abbott & Robson, 1991; Hepper et al., 1988).

The quality of the plant material used to establish an orchard is of utmost importance for its success. The use of symbiotic and biocontrol microorganisms can help to improve rootstock production and quality (Whipps and Gerhardson, 2007; Kapoor et al., 2008).

Myrobolan 29C plants grown in pots, treated with *Glomus intraradices* and *Trichoderma harzianum*, showed better physiologic and development performance as compared to the control (Dichio et al., 2011). Mycorrhization with *Glomus* and soil colonisation with *Trichoderma* enhanced plant growth (total biomass) by about 24% and 38%, respectively. The inoculation of micropropagated plantlets with selected fungi has been tested to improve the quality of many cash crops (Augé, 2001). Calvente et al. (2004) supported above all the effectiveness of native *G. intraradices* fungi as inoculants in many Mediterranean areas. Sofo et al. (2010) also demonstrated that the application of *T. harzianum* T22 increased the numbers of leaves, roots, and stem diameters.

Fig. 4. *Olea europaea* cv. Coratina control plants (left) and inoculated with *Glomus intraradices* (right) after about eighteen months (Tataranni et al., 2010).

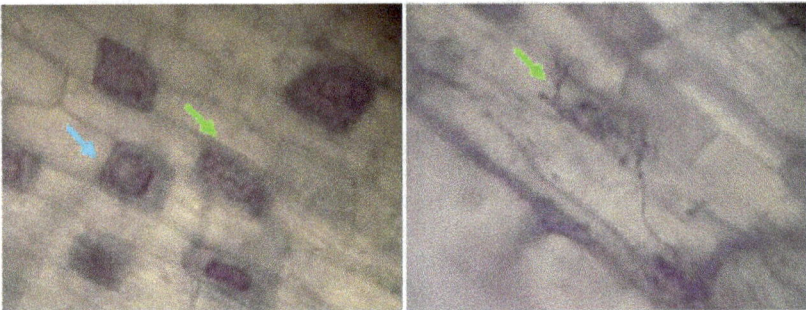

Fig. 5. *Glomus intraradices* colonisation. Hyphae and fungus structures in roots of *Olea europaea* cv. Coratina stained with Trypan blue; green arrows: arbuscules; blue arrow: spore in formation (400x and 600x magnification; Tataranni et al., 2010).

2.4 Effects of fungicides application on mycorrhization rate and persistence parameters of *Glomus intraradices*

Agricultural practices, such as pesticides, fungicide use and fertilizer applications, can eliminate or severely reduce the incidence of mycorrhizal activity. Mycorrhizal fungi can be quite sensitive to some fungicides, but not to at all of them. Some fungicides can actually stimulate mycorrhizal fungi. Two mechanisms were proposed to account for beneficial effects of fungicides on mycorrhizas: i) root colonization may be stimulated if the fungicide alters the host plant physiology by increasing root exudates (Jabaji-Hare & Kendrick, 1985; Schwab et al., 1982); ii) the fungicide reduces populations of organisms antagonistic to AMF (Hetrick & Wilson, 1991). However, the indiscriminate use of pesticides and fungicides usually leads to a reduction in spore numbers (Manjunath & Bagyaraj, 1984) and diversity (Schreiner & Bethlenfalvay, 1996). In addition, high levels of these substances also reduce

AMF colonisation of roots, resulting in reduced AMF activity (Newsham et al., 1995; Udaiyan et al., 1995). Phosphorus influxes into mycorrhizal roots are quickly reduced after the application of systemic fungicides (Hale & Sanders, 1982), such as benomyl [methyl 1-(btylcarbamoyl)-2-benzimidazole carbamate]. According to the literature, alterations induced by fungicides on the health and physiology of plants have significant effects on symbionts (Garcia-Romera & Ocampo, 1988; Schwab et al., 1982). The response of mycorrhizal fungi to chemical treatments can be influenced by the different species of host plants, the variety of compounds used, their modes of action, different application methods and rates, the growth phases of mycorrhizal fungi, environmental conditions (Giovannetti et al., 2006; Jalali & Domsch, 1975). Fungicides, applied as soil drench, reduced the proportion of active external hyphae, expressed as alkaline phosphatase (ALP) active fractions (Kjøller & Rosendahl, 2000). Internal fungal structures seem to be less vulnerable to toxicants, when these are applied to the soil. The loss of activity in external structures might inhibit the transfer of nutrients and water towards the plant, but the survival of fungal structures in the root cortex is important for the re-establishment of symbiosis.

Effects of fungicide applications applied in combination during acclimatisation - a crucial phase of micropropagated rootstock production - were evaluated on mycorrhization rate and persistence parameters (Phillips & Hayman, 1970; Trouvelot et al., 1986) of G. intraradices, in potted GF677 rootstocks (Fig. 6). Plants in the acclimatisation phase, despite fungicide applications, are particularly sensitive to diseases favoured by high relative humidity conditions. G. intraradices was able to better colonise the roots of non-fungicide plants, during the first months from inoculation (April – July). In July, the frequency of mycorrhization system revealed to be by 10% higher in non-treated plants. The intensity of root colonisation ranged from 80 to 85% without fungicide applications; it remained at 70% when applying fungicides. Fungicides also affected arbuscular formation; in fact, arbuscular abundance parameters revealed a constant 10% reduction as compared to non-treated plants. External hyphae of AM fungi should be more sensitive to fungicides, above all during presymbiotic growth and the establishment of symbiosis (Kjøller & Rosendahl, 2000). Compounds with a systemic mode of action were also reported to inhibit AM activity (Hale & Sanders, 1982; Newsham & Watkinson, 1995). Boscalid is a systemic fungicide belonging to the anilide class, effective against botrytis rot and crown rot by inhibiting the electron transfer of the succinate dehydrogenase complex in the mitochondrial inner membrane. Successively, in December, differences on mycorrhization parameters between fungicide- and non-treated plants became not significant. At that point, the colonisation intensity of plants with fungicides was found to be even higher by 30%. Arbuscular abundance parameters were equivalent. In non-host condition, the impact of Iprodione, a contact fungicide that inhibits DNA and RNA synthesis, was reduced on AM fungi (Giovannetti et al., 2006). According to some authors, Metalaxyl even increased radical colonisation and the length of external hyphae (Giovannetti et al., 2006; Groth & Martinson, 1983). So, the better root colonisation of treated rootstocks observed in December could be determined by indirect effects due to the control that fungicides have on pathogens and antagonists of AM fungi (Giovannetti et al., 2006; Hetrick & Wilson, 1991).

The combination of the used fungicides caused a decrease in hyphal colonisation by Glomus intraradices during the first months after the treatments. AMF show different sensitivity to fungicides at the various phases of the biological cycle. In particular, the presymbiotic growth and the establishment of symbiosis seem to be very delicate phases, since fungi essentially develop outside the host coming into direct contact with residual products. The beneficial effects of fungicides on mycorrhizas, as verified later on, can probably be explained by the reduction in competition of antagonistic organisms and pathogens. For nursery plant production, although inoculations with AMF and fungicide treatments may protect against pathogens at the same time, in certain phases, it is necessary to consider all involved interactions carefully and prefer more environmental friendly methods.

Fig. 6. Fungicide phytotoxic effects in the acclimatisation phase of rootstock production process, after 20 days from application (left). Rootstock treated with biological agents (right).

2.5 Effects of mycorrhizas on plant water relation

Fungi are able to alter the water relations of the plant (Safir & Nelsen, 1985), but not all works compared symbiotic and control plants of similar size and/or nutritional status. So, mycorrhizal effects indirectly related to nutrition could not always be distinguished (Augé, 2001). However, non-nutritional symbiotic effects have been reported. As for the mechanism of influence, studies suggest hormonal involvement (Allen et al., 1982; Section 2.2); stimulation of gas exchange through increased sink strength (Kaschuk et al., 2009), reduction of resistances to water flow in plant (Allen, 2007; Uehlein et al., 2007). Substantial evidence has been obtained in *Bouteloua gracilis*, *Helianthus annuus*, *Citrus jambhiri* and *Prunus*, species in which mycorrhization increased the transpiration rate and, in some cases, the leaf or root conductance (Levy & Krikun, 1980; Allen, 1981, 1982; Koide, 1985; Tataranni et al., 2011).

2.5.1 Potentials, stomatal conductance and photosynthesis

Plant water status is typically quantified by measuring potentials (Ψ). Leaf Ψ of non-stressed plants is usually not affected by symbiosis (Levy & Krikun, 1980; Allen, 1982; Davies et al., 1993; Augé, 2001), but results are controversial in non water-limiting conditions. When host total leaf Ψ is similar to non-mycorrhized controls, because of frequently different photosynthetic rates observed, leaves of symbiotic plants might develop dissimilar osmotic potentials (Augé, 2001).

Symbiosis, instead, postponed declines in leaf Ψ during drought stress (Davies et al., 1993; Subramanian et al., 1997). Leaf Ψ was also reported to recover control levels more quickly in mycorrhized than control plants after drought (Subramanian et al., 1997).

Drought can cause an active accumulation of solutes in plant cells (osmotic adjustment; Dichio et al., 2006). This plant response to stress conditions was shown to be more marked during mycorrhization (Allen & Boosalis, 1983; Davies et al., 1993), resulting in higher leaf turgor (Davies et al., 1993).

When root systems is limited to small soil volumes (pots), leaf or shoot Ψ did not differ or declined more quickly in mycorrhized plants (Safir & Nelsen, 1985; Goicoechea et al., 1997).

Stomatal conductance and leaf potentials (Ψ) are related. Leaf hydration should be, in fact, naturally associated with altered stomatal behaviour. Despite information on potentials, symbiotic compared to control plants often display different transpiration rates and stomatal conductance to water vapour (Augé, 2001). Published data show different fungus–host species sensitivity, but mycorrhizas have, in general, induced increases in transpiration and stomatal conductance (Augé, 2001). Soil water was consequently observed to be depleted more in those plants. Symbiotic plants of similar leaf areas, transpiring at higher rates, led to more rapid soil water losses (Augé, 2001).

Stomatal parameters were altered by symbiosis also without changes in leaf potentials (Allen and Boosalis, 1983; Allen, 2006). In this case, mycorrhization probably influenced intrinsic hydraulic or biochemical properties of plant by signals directly (Augé, 2001). During drought, Duan et al. (1996) reported concentrations of ABA in xylem sap of mycorrhized plants lower than controls; Goicoechea et al. (1997) detected the same trend in leaves and roots.

In several experiments, differences in stomatal conductance between symbiotic and control plants were observed only under drought (Henderson & Davies, 1990; Davies et al., 1993), as it is the case for potentials. Stomata showed different thresholds at which they began to close or closed fully (Allen and Boosalis, 1983). Moreover, stomatal conductance in mycorrhized plants remained unaffected by declines in available soil water longer than control plants (Duan et al., 1996).

Photosynthesis is stimulated by symbiosis. Symbiotic plants often show higher photosynthetic rates, which is consistent with effects on stomatal conductance. Sink stimulation has been identified from various researches as one possible explanation for the differences in photosynthesis observed between nutrient-fertilised plants and symbiotic plants (Kaschuk et al., 2009). The inoculation of legumes with rhizobia and/or AM fungi resulted in sink mediated stimulation of photosynthesis, improving the photosynthetic nutrient use efficiency and the proportion of seed yield in relation to the total plant biomass. The carbon invested in the symbiosis was insufficiently compensated by enhanced nutrient acquisition, evidencing a nutrient-independent effect, in which the C costs are compensated directly by increased photosynthetic rates. Symbiosis could act by removing the limitation of rubisco activity and electron transport rates (increase in leaf N and P mass fraction and/or removal of the triose-P) (Kaschuk et al., 2009). Enzymatic activities and total protein concentrations were found to be typically higher in symbiotic compared to control plants during drought (Goicoechea et al., 1997; Subramanian & Charest, 1997).

Both non-hydraulic and hydraulic symbiosis - induced effects probably influence leaf water relations and gas exchange in host physiology at the same time, especially during drought (Davies et al., 1993; Saliendra et al., 1995). Stomata responses (variation in leaf Ψ, turgor or water content) seem to involve ABA pathway (Saliendra et al., 1995), xylem - apoplast pH (Green et al., 1998; Hartung et al., 1998) and/or auxin/cytokinin balance (Sofo et al., 2010, 2011). These are among the factors that can further regulate hydraulic conductivity and, in particular, water channels activities (Tournaire-Roux et al., 2003).

2.5.2 Hydraulic conductivity

Sap flow depends on several plant and environmental factors (e.g. soil water availability, evaporative demand, etc.), and their interactions. Among plant factors, hydraulic capacity of the conductive tissues is involved in plant water relations since it contributes to leaf water supply. At least one-half of the resistance to hydraulic flow through plant vascular systems occurs in the root system (radial and axial pathways). These resistances are regulated by the physics of conduits (e.g. size and density of vessels), by some physiological traits, such as aquaporin-induced changes of membrane permeability, and depend on the time scale considered (Vandeleur et al., 2005). Directly, through absorption and transport, exceeding part of the cellular barriers, or indirectly, exploring a bigger soil volume and interstices between little agglomerates, the hyphae increase the water flow potentially available to plants. The experimental evidences are not always in agreement but, as Safir & Nelsen (1985) suggest, the positive effects due to the fungus could become important in limiting conditions of water stress. The best uptake of an essential element - phosphorus - seems to be the base mechanism proposed to explain these phenomena. Phosphorus could increase the permeability of root cells to water; at the same time, the development of hyphae, besides the already mentioned advantages, could reduce the inner root resistances that the water flow normally meets, above all crossing the walls (Safir & Nelsen, 1985). In optimal nutritional conditions (high phosphorus), no significant differences in root conductivity between mycorrhized plants and controls have been found; in *Fraxinus pennsylvanica*, differences of root conductivity were just found in relation to the accumulated phosphorus concentrations measured in the various plant organs, and in relation to dry matter partitioning (Andersen et al., 1988).

We tested the hypothesis that mycorrhizas can increase the hydraulic conductance (i.e. reduce resistance) of inoculated *Prunus* plants in pots without phosphorus applications. Myrobolan 29C micropropagated plants were treated with commercial *Glomus intraradices*, during the acclimatisation stage in the nursery, and grafted (Portici, apricot cv.) by chip budding technique. *G. intraradices* effectively and persistently infected the root in our system (Tataranni et al., 2011). Hydraulic conductance was measured in rootstock and rootstock + grafting point by a Hydraulic Conductance Flow Meter. A Hydraulic Conductance Flow Meter (HCFM; Dynamax, Inc. Houston, TX - U.S.A.) measured water flow (F) forcing distilled and degassed water into the excised "rootstock + grafting point" or root system (opposite direction to the transpiration stream) and changing the applied pressure (P) simultaneously. The slope of the linear regression (over the range of 0.1 - 0.5 MPa pressures applied) between water flux (F) and pressure (P) represented the hydraulic conductance (K). Values of K for each individual (rootstock + grafting point) were determined after cutting the scion 3-4 cm above the grafting point and carefully mounting the compression coupling head on the cut surface. After K was recorded, also rootstocks were cut, 1 cm below the grafting point, and rootstock conductance (K_R) was determined. Both K and K_R were standardised per unit of dry weight and referred to as conductivity (kg s^{-1} MPa^{-1} g^{-1}), thus avoiding the effects directly due to the size (Augé, 2001). The rootstock conductivity of the inoculated plants was double compared to the control, reaching the value of 1.4×10^{-7} kg s^{-1} MPa^{-1} g^{-1}. The conductivity of the rootstock + grafting point combination of the inoculated plants was 5-times greater than the controls (Fig. 7, 8, 9; Tataranni et al., 2011). In order to understand possible mechanisms by which mycorrhizas enhance plant hydraulics, it is important to consider the resistances that water might meet in a symbiotic system. It is well known that hyphal tips penetrate the wall, reducing apoplastic resistances to water flow, but not the plasmalemma of cortical root cells. External hyphae extend into the soil, explore it, and constitute the largest biomass fraction of the fungus. Fungal hyphae, moreover, can wrap roots (Allen, 2006), rapidly transporting water and nutrients (Duddridge et al., 1980). Water can be taken up by a hyphal tip in the soil and transferred either through the cytoplasm or through the fungus inner wall layers to a cortical cell, without encountering further resistances. Membrane structures, indeed, are the site of extensive exchanges of mineral nutrients, carbohydrates and water. Marked changes in symbiotic membrane specialisation were already detected and related to aquaporins (Allen, 2007; Uehlein et al., 2007). Immunolocalisation showed the accumulation of aquaporin proteins in mycorrhized root fragments, in the vascular part and especially in the root cortex. A subsequent clear increase in water permeability was recorded, too (Uehlein et al., 2007). Symbiosis could so modify the cell-to-cell contribution to water transport and therefore influence root hydraulic conductivity.

Positive interactions between the root system and the *Glomus* fungi increased hydraulic conductivity, hence improved leaf specific conductivity and tree water use efficiency may be expected.

Fig. 7. Micrografted apricot plant on biotizated rootstock (Portici/Myrobolan 29C). The compression head of the HCFM was installed on the scion (K determination) and on the rootstock (K$_R$ determination; Tataranni et al., 2011).

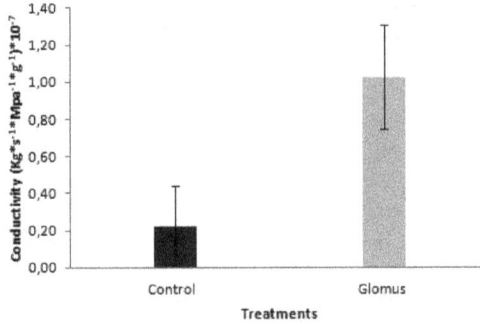

Fig. 8. Conductivity of rootstock + grafting point in control and *Glomus intraradices* inoculated plants (Portici/Myrobolan 29C). Mean ± SE, n=10 (Tataranni et al., 2011).

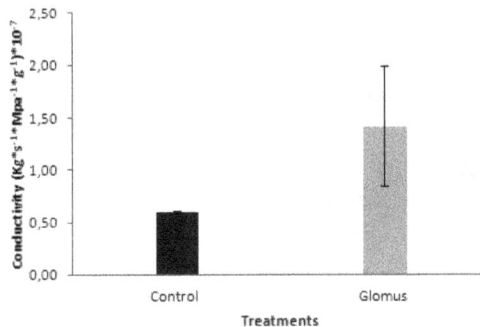

Fig. 9. Conductivity of Myrobolan 29C rootstock in control and *Glomus intraradices* inoculated plants. Mean ± SE, n=10 (Tataranni et al., 2011).

The effects of beneficial fungi, *Glomus intraradices* and *Trichoderma harzianum* T22, on plant water uptake and transport were also compared (Tataranni et al., 2011b).

Persistent *G. intraradices* - Myrobolan 29C root interaction and *T. harzianum* soil colonisation were established after six months from inoculation in acclimatisation. All the parameters assessed to evaluate colonisation were higher in the treated plants than in the controls (Phillips & Hayman, 1970; Trouvelot et al., 1986). For *Glomus* treated plants, the frequency of root system mycorrhization (F%) was 100% in the treated plants, while it was about 40% for the control. The colonisation intensity (M%) achieved 40% in the treated rootstocks, near 0% for the control. The arbuscular abundance in root segments (a%) was about double in the treated plants, reaching the value of 13%. For *Trichoderma* treated plants, CFUs measured from soil samples were 2.6 x 10^4 g^{-1}, significantly different compared to the controls, 0.4 x 10^4 g^{-1}. Control contamination, which was however uninfluential to our trials, can be due to the presence of natural microorganisms in the soil mix used (Briccoli et al., 2009). The root conductivity of the inoculated plants was double compared to the controls (Fig. 10). There were not significant differences between *Glomus* and *Trichoderma* treatments (Tataranni et al., 2011b).

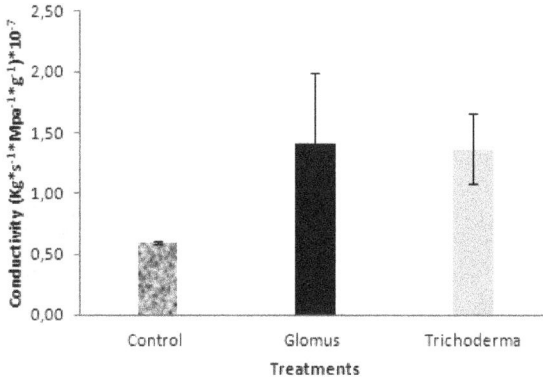

Fig. 10. Hydraulic conductivity of micropropagated Myrobolan 29C in control and inoculated plants. Comparison between *Glomus* and *Trichoderma treatments*. Mean ± SE, n=10 (Tataranni et al., 2011b).

3. Fungus application and persistence

3.1 Inoculation materials and methods

Criteria for optimizing biotisation methods in plant production processes must combine qualitative, environmental and economic benefits. The source of bioagents (commercial or fresh culture) and the possible combination with other microorganisms represent a key topic for setting up successful production procedures.

The main methods of inoculation, both for *Trichoderma* and *Glomus*, have been compared in experimentations during the acclimatisation phase of micropropagated rootstocks. The microorganism source can be dissolved into a water solution and plants dipped in it just before planting (dipping method). Alternatively, the source can be mixed with the peat substrate four days before planting (mixing method). Other studies evaluated *Trichoderma* application in micropropagation during the rooting phase (see Section 2.1).

The inoculation by dipping, on the basis of the experiments carried out, was more efficient and effective against pathogens (see Section 3.3; Bardas et al., 2011b).

Concerning the possibility to propagate and use fresh *Trichoderma*, T22 fresh culture revealed the most promising results in root colonisation, compared with commercial treatments at the same dose, either mixed in turf (fresh T22 colonisation reached 100%, 15 days after transplantation, commercial product was 50%) or inoculated by dipping (fresh T22 colonisation reached 100%, 15 days after transplantation, commercial 40%). Fresh T22, when mixed in turf, revealed the most promising results in substrate colonisation: colonisation reached 100%, 60 days after transplanting, 65% by dipping method. The specific enhanced colonisation rates of fresh T22 cultures also influenced several plant growth characteristics (see Section 3.3; Bardas et al., 2011b). That was probably due to an early activation of the "fresh" microorganisms and subsequent plant protection mechanisms.

3.2 Persistence of biocontrol agents

The presence of *Glomus intraradices* in roots and *Trichoderma harzianum* in pot soil was confirmed to be significant after six months from the inoculum carried out during the acclimatisation phase of some commercially used rootstock (GF677, Myrobolan 29C, GiSeLa®6, OHF 19-89; Fig. 11 and 12).

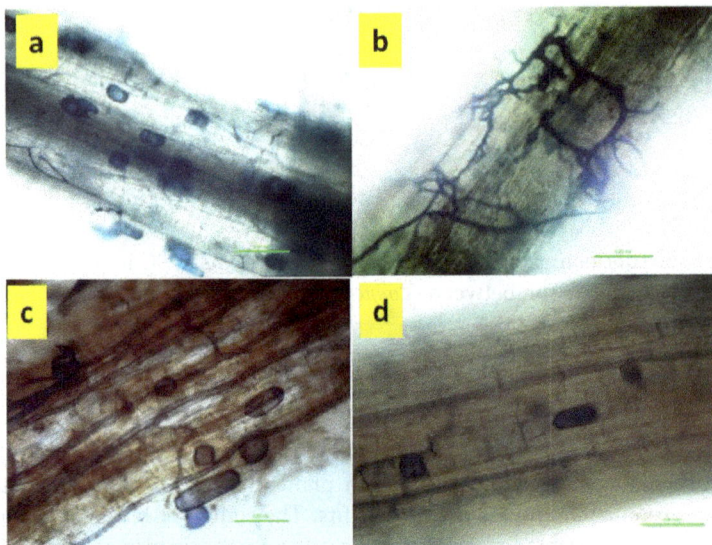

Fig. 11. *Glomus intraradices* colonisation (6 months from inoculation). Roots of GF677 (a), Myrobolan29C (b), GiSeLa®6 (c) and OHF19-89 (d), with variable fungus structures, observed under transmitted light microscope (200x magnification; scale bars = 0.01 cm).

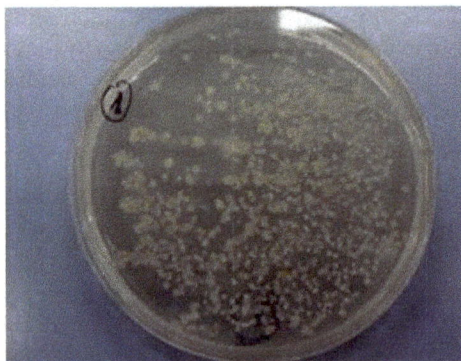

Fig. 12. CFUs of *Trichoderma harzianum* on semi-selective culture media.

The following parameters are calculated to evaluate *G. intraradices* colonisation (Trouvelot et al., 1986): F% is the frequency of root system mycorrhization; M% is the intensity of root colonisation; a% is the arbuscular abundance in root fragments analysed; A% is the root arbuscular abundance.

All treated rootstocks showed 100% mycorrhization frequency (F%), while controls about 50%. Controls were probably infected by natural agents already present in the soil or in the irrigation water. Colonisation intensity (M%), the most important parameter to evaluate the colonisation, ranged between 40 and 85%, reaching the highest values in GF677, compared to controls with only 20%; the differences observed were statistically significant. The arbuscular presence (a% and A%) was about 50% in *Glomus* treated GF677, 10% in Myrobolan 29C and 20% in OHF19-89. Arbuscular structures were limited in number for GiSeLa®6 and control plants.

T. harzianum CFU mean values ranged between 1.45 in GF677 rootstocks and 3.0×10^4 g^{-1} for GiSeLa®6, while controls contamination rate was low, between 0.2 and 0.55×10^4 g^{-1}.

The results of two-year experimentation on the persistence of microorganisms in potted plants showed a general colonisation trend of *Glomus intraradices* and *Trichoderma harzianum* with a peak in June-July months, followed by a decrease in December. These microorganisms, however, survive during the winter period and reactivate in spring (Fig. 13 and 14). Seasonal variations in the periodicity of root growth and mycorrhizal activity occur in ecosystems (Brundrett and Kendrick, 1988). Literature reports that there are no significant seasonal variations in the degree of mycorrhizal colonisation of many herbaceous plants, because only a fraction of their roots are replaced each year (Brundrett and Kendrick, 1988, 1990). Other species of deciduous plants have annual roots with well-defined periods of growth and senescence, resulting in abrupt transitions in mycorrhizal colonisation levels. Moderate seasonal variations in the extent of mycorrhizas for roots of European deciduous forest species (Mayr and Godoy, 1989) and salt marsh plants (Van Duin et al., 1989) were also associated with new root production during the growing season.

Moreover, mycorrhizal strategies of plants may be correlated with the environmental conditions prevailing when plants produce new roots, as was observed in a temperate deciduous community (Brundrett, 1991). In this community, most species with root growing in summer were mycorrhizal, while those with active roots in spring or fall had little or no mycorrhizas.

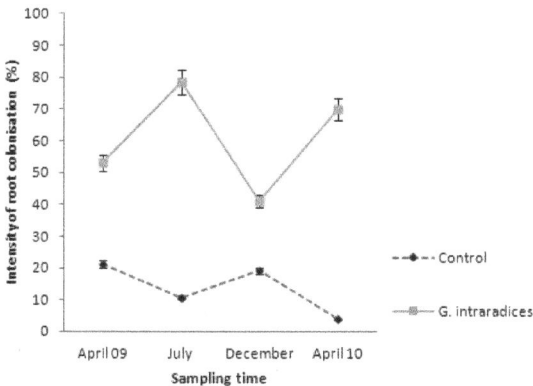

Fig. 13. *Glomus intraradices* trend of persistence (intensity of root colonisation, M%) during a year period in treated and control GF677 rootstocks.

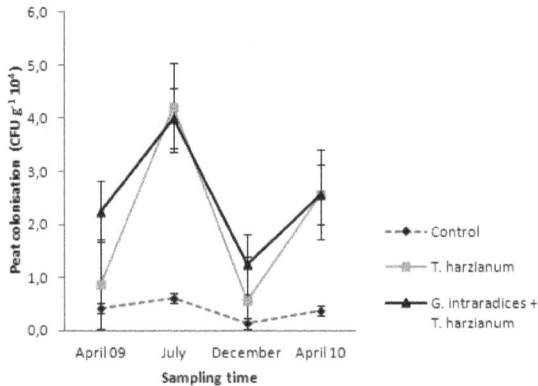

Fig. 14. *Trichoderma harzianum* trend of persistence (colony forming units, CFU) during a year period in treated and control Myrobolan 29C rootstocks.

3.3 Interaction of biocontrol agents with soil pathogens in nursery

Biological control has emerged as an important method in the management of plant pathogens. *Pythium ultimum, Fusarium oxysporum* and *Rhizoctonia solani* diseases are limiting factors during the acclimatisation phase of rootstock production, when further problems can be due to the absence of beneficial microorganisms on plants obtained by micropropagation and in greenhouse conditions. Mycorrhiza and *Trichoderma* species have long been recognized for their ability to reduce pathogen growth, survival or infections by different mechanisms (Harman et al., 2004). Chitinolytic enzymes play an important role in fungal – fungal interactions and especially in mycoparasitism (Duo-Chuan, 2006), hydrolyzing chitin, which is an essential component of fungal cell wall (Bowman et al., 2006). So, specific treatments can be a potential means in integrated rootstock production systems.

Micropropagated plants of Myrobolan 29C treated at transplantation with *Trichoderma harzianum* and *Glomus intraradices* alone or in combination, were inoculated with *Fusarium oxysporum, Rhizoctonia solani* and *Pythium ultimum* (Dichio et al., 2011). Disease incidence was measured 15 and 30 days after transplantation. However, the biocontrol agents induced a greater plant growth (see Section 2.3) compared to controls, with *T. harzianum* T-22 treatment showing the best promoting effect. *T. harzianum* T-22 decreased values of *P. ultimum* and *R. solani* diseased plants. All biocontrol treatments showed lower *F. oxysporum* disease incidence values compared with the untreated plants (Dichio et al., 2011).

Various *Trichoderma* species (*T. harzianum* T22 and *T. asperellum* B1) were also tested with GF677 rootstocks, *in vitro* and *in planta*, against single and combined inoculations of the pronounced plant pathogens (Bardas et al., 2011a). *In vitro* experiments revealed differentiations regarding the mode of action of the tested biocontrol agents. *In planta* experiments showed that both *Trichoderma* spp. are excellent root and peat colonisers, resulting in promotion of specific plant growth characteristics and suppression of *P. ultimum* and *R. solani* mediated disease development. *T. harzianum* T22 and *T. asperellum* B1 combined treatment was statistically different compared to single biocontrol treatments, showing higher colonisation rates, greater plant growth and enhanced pathogen control (Bardas et al., 2011a).

By employing especially *Trichoderma* as a biocontrol agent, it is possible to successfully control *P. ultimum* and *R. solani* without use of fungicides. This leads not only to lower production costs and better control of damping off diseases, but also to environmental benefits, such as minimal direct or indirect soil and ground water contamination, and improved personnel safety practically eliminating the restriction time for personnel re-entry after pesticide application. In addition, we observed that avoiding the use of fungicides, the quality and uniformity of the plants were improved and the duration of the acclimatisation phase became 4-5 days shorter. The integration of novel sustainable technologies in the production scheme of commercial rootstock production leads to significant economic, operational and environmental benefits, which could be extended to other crops.

4. Conclusion

The most appropriate inoculation time and method were defined to confirm the effect of mycorrhization on micropropagated rootstocks. At the time of planting from the jars to the soil, micropropagated rootstocks were inoculated with *T. harzianum* and *G. intraradices* by plant dipping, in inoculum – water solution, or by mixing the inoculum with the peat substrate before planting (four days before transplanting). There were low levels of *T. harzianum* detected shortly after application with dipping, compared to application by mixing with peat. T22 fresh culture revealed the most promising results in root colonisation, compared with commercial treatments at the same dose.

The use of biological agents (*T. harzianum* T22 and *G. intraradices*) in nursery production process by micropropagation, significantly decreased the percentage of dead plants during the acclimatisation phase, also compared to fungicide treated plants. It was evident that the growth rate of the inoculated plants, without fungicide applications, was not different from those treated with fungicides. Moreover, the treatment with fungicides reduced root colonisation and arbuscular formation. Therefore, the plant production process can be innovated by eliminating the use of fungicides during the acclimatisation phase.

The highlighted result was the control effect obtained for the biological agents against fungal pathogens. Both *Glomus* and *T. harzianum* (in combination and alone), but mostly *T. harzianum*, promoted plant growth and conferred protection from soil-borne pathogens (*Fusarium oxysporum*, *Rhizoctonia solani* and *Pythium ultimum*).

Based on the results obtained, the higher shoot and root growth induced for plants inoculated with T22 could be due to a different phyto-hormonal balance or to acidification and redox fungal activity of the growth medium. The levels of IAA increased significantly after inoculation with T22 in both leaves and roots. The auxin/cytokinin ratios increased strongly in leaves and roots of plants treated with T22. Production of hormone-like metabolites and/or induction of the production of these compounds in the plants have been proposed as one of the direct mechanisms by which symbiosis promotes plant growth. Experiments demonstrated that strong acidification occurred in the medium inoculated with T22.

In pots, the presence of *G. intraradices* in roots and *T. harzianum* in soil was confirmed to be significant in June, after about one year from inoculation. The results of two-year experimentation on the persistence of microorganisms showed a general colonisation trend of *G. intraradices* and *T. harzianum* with a peak in June-July followed by a decrease in December. These microorganisms, however, survive during the winter period and reactivate in spring. In nursery fields, *G. intraradices* colonisation was very high in all treatments.

Mycorrhization of rootstocks showed effects on cultivar quality and general vigour after shoot development. The results of the plant growth characteristics showed positive effects of the inoculum on all parameters measured (plant height, number of leaves, shoot dry weight, root dry weight and the root-shoot ratio). *Glomus* and *Trichoderma* biotised plants reached a total mean dry weight greater than the controls, by about 20% and 30%, respectively, at the end of the production cycle in pots. The field management using biocontrol agents was shown to induce the best growth rates compared to the normal growers' management practices. In the nursery, in general, at the early stages of growth, mycorrhized plants showed high level of fresh and dry matter accumulation. There were no significant differences between the mineral element concentrations in the dry matter of the various treatments compared to the controls; but, considering the mean total amount per plant, quantities of macro-elements were greater in treatments with *Glomus* or *Trichoderma* than in the controls. This was due to the greater growth rate of the inoculated plants.

Beneficial fungi can enhance efficiency of plant roots to absorb water, macro- and microelements from the soil or container media. Plants treated with *G. intraradices* and *T. harzianum* showed an increase in water conductance. External inputs can thus be reduced. Plant tolerance to pathogens increases. The plant could be more efficient in surviving drought conditions. Other benefits include enhanced seedling growth, increased adventitious root formation of cuttings and enhanced transplant establishment. Mycorrhiza or biocontrol microrganisms enhance, in general, plant health and vigor. Advantages in utilizing mycorrhizal fungi or other beneficial microorganisms, during propagation and production, include marketing higher value plants.

In a vision of sustainable environment and new market demands, it becomes possible to improve plant quality and anthropic input efficiency. Symbiosis represents an excellent example of ecological adaptation. Plants and microorganisms interact in the same environment in order to obtain evolutionary advantages.

5. Acknowledgments

The research leading to these results has received funding from the European Community's Seventh Framework Programme (FP7-SME-2007-1) under grant agreement n. 222048 - Sitinplant Project, and the Project "Processi innovativi di gestione dei sistemi frutticoli ed olivicoli finalizzati al miglioramento della qualità e della sostenibilità ambientale" – SAFO Project.

6. References

Abbott, L.K. & Robson, A.D. (1991). Field management of VA mycorrhizal fungi. In: *The Rhizosphere and Plant Growth* (Ed, by D.L. Keister and P.B. Cregan) pp. 355–362

Allen, M.F. (1982). Influence of vesicular-arbuscular mycorrhizae on water movement through *Bouteloua gracilis* (H. B. K.) Lag ex Steud. *New Phytologist*, Vol. 91, pp. 191-196

Allen, M.F. (2006). Water dynamics of mycorrhizas in arid soils. In *G.M. Gadd (ed.) Fungi in biogeochemical cycles.* Cambridge Univ. Press (New York), pp. 74–97

Allen, M.F. (2007). Mycorrhizal fungi: highways for water and nutrients in arid soils. *Vadose Zone Journal,* Vol. 6, No. 2, pp. 291–297

Allen, M.F. & Boosalis, M.G. (1983). Effects of two species of VA mycorrhizal fungi on drought tolerance of winter wheat. *New Phytol.*, Vol. 93, pp. 67–76

Allen, M.F., Moore, J.T.S. & Christensen, M. (1982). Phytohormone changes in *Bouteloua gracilis* infected by vesicular-arbuscular mycorrhizal fungi. II. Altered levels of gibberellin-like substances and abscisic acid in the host plant. *Can. J. Bot.*, Vol. 60, pp. 468–471

Allen, M.F., Smith, W.K., Moore, T.S. & Christensen, M. (1981). Comparative water relations & photosynthesis of mycorrhizal & non-mycorrhizal *Bouteloua gracilis* H.B.K. Lag ex Steud. *New Phytologist*, Vol. 88, pp. 683-693

Andersen, C.P., Markhart, A.H., Dixon, R.K. & Sucoff, E.I. (1988). Root hydraulic conductivity of vesicular-arbuscular mycorrhizal green ash seedlings. *New Phytol.*, Vol. 109, pp. 465-471

Augé, R.M. (2001). Water relations, drought and vesicular-arbuscular mycorrhizal symbiosis. *Mycorrhiza*, Vol. 11, pp. 3–42

Bardas, G.A., Ballas, E., Mavrodimos, K. and Katis, N. (2011a). Effect of several biocontrol agents on plant growth characteristics and disease severity of *Pythium ultimum* and *Rhizoctonia solani* disease complex during acclimatization phase of GF-677 rootstocks. *The 15th Hellenic Phytopathological Congress*, in press

Bardas, G.A., Ballas, E., Mavrodimos, K. & Xiloyiannis, C. (2011b). Effect of *Trichoderma harzianum* strain T22 on *Pythium ultimum* disease severity during acclimatization phase of GF-677 rootstocks. *The 15th Hellenic Phytopathological Congress*, in press

Belén Suárez, M., Sanz, L., Chamorro, M.I., Rey, M., González, F.J., Llobell, A. & Monte, E. (2005). Proteomic analysis of secreted proteins from *Trichoderma harzianum*: Identification of a fungal cell wall-induced aspartic protease. *Fungal Genet. Biol.*, Vol. 42, pp. 924-934

Blake, P.S., Browning, G., Benjamin, L.J. & Mander, L.N. (2000). Gibberellins in seedlings and flowering trees of Prunus avium L. *Phytochemistry*, Vol. 53, pp. 519-528

Bowen, G.D. (1987). The biology and physiology of infection and its development. In: *Ecophysiology of VA Mycorrhizal Plants* (Ed. by G. R. Safir). pp. 27-57. CRC Press. Boca Raton. Florida

Bowman, M.B. & Free, J.S. (2006). The structure and synthesis of the fungal cell wall. *BioEssays*, Vol. 28, pp. 799-808

Briccoli Bati, C., Rinaldi, R. & Sirianni, T. (1992). Prime osservazioni sulla presenza di micorrize di tipo VAM in oliveti dell'Italia meridionale. *Atti Giornate scientifiche S.O.I.*, Ravello 8-10 Aprile, pp. 46-47

Briccoli Bati, C. & Godino, G. (2002). Influenza delle micorrize sull'accrescimento in vivaio di piante di olivo. *Italus Hortus*, Vol. 9, No.3, pp. 20-21

Briccoli Bati, C., Godino, G. & Belfiore, T. (2003). Ruolo della simbiosi micorrizica nella produzione vivaistica di piante di olivo. *Italus Hortus*, Vol. 10, No.4, pp. 160-164

Briccoli Bati, C., Santilli, E., Varlaro, M.E. & Alessandrino, M. (2009). Effects of a commercial arbuscular mycorrhizal fungi inoculum on vegetative growth of three young olive cultivars. *XXXIII CIOSTA - CIGR V Conference 2009* (Reggio Calabria, Italy), pp.2015-2019

Brundrett, M.C. (1991). Mycorrhizas in natural ecosystems. In *Advances in ecological research*. Vol. 21, pp. 171-313, Begon, M, Fitter, A. H. & Macfadyen, A. (Eds.), *Academic Press Limited*, ISBN 0-12-013921-9

Brundrett, M.C., Enstone, D.E. & Peterson, C.A. (1988). A berberine-aniline blue fluorescent staining procedure for suberin, lignin, and callose in plant tissue. *Protoplasma*, Vol. 146, pp. 133-142

Brundrett, M.C. & Kendrick, B. (1988). The mycorrhizal status, root anatomy, and phenology of plants in a sugar maple forest. *Can. J. Bot.*, Vol. 66, pp. 1153-1173

Brundrett, M.C. & Kendrick, B. (1990). The roots and mycorrhizas of herbaceous woodland plants I. Quantitative aspects of morphology. *New Phytol.*, Vol. 14, pp. 457-468

Calvente, R., Cano, C., Ferrol, N., Azcón-Aguilar, C. & Barea, J.M. (2004). Analysing natural diversity of arbuscular mycorrhizal fungi in olive tree (*Olea europaea* L.) plantations and assessment of the effectiveness of native fungal isolates as inoculants for commercial cultivars of olive plantlets. *Applied Soil Ecology*, Vol. 26, pp. 11–19

Contreras-Cornejo, H.A., Macías-Rodríguez, L., Cortéspenagos, C. & López-Bucio, J. (2009). *Trichoderma virens*, a plant beneficial fungus, enhances biomass production and promotes lateral root growth through an auxin-dependent mechanism in *Arabidopsis*. *Plant Physiology*, Vol. 149, pp. 1579–1592

Dalisay, R.F. & Kuc, J.A. (1995). Persistence of induced resistance and enhanced peroxidase and chitinase activities in cucumber plants. *Physiol. Mol. Plant Pathol.*, Vol. 47, pp. 315-327

Davies, F.T., Potter, J.R., Linderman, R.G. (1993). Drought resistance of mycorrhizal pepper plants independent of leaf P-concentration – response in gas exchange and water relations. *Physiol Plant*, Vol. 87, pp. 45–53

Dichio, B., Tataranni, G., Bardas, G.A., Mavrodimos, C., Katis, N., Ipsilantis, I. & Xylogiannis, V. (2011). Production of micropropagated and mycorrhisated Myrobolan 29C rootstock. *Acta Hort*, in press

Dichio, B., Xiloyannis, C., Sofo, A. & Montanaro, G. (2006). Osmotic adjustment in leaves and roots of olive tree (*Olea europaea* L.) during drought stress and rewatering. *Tree Physiol*, Vol. 26, pp. 179-185

Dehne, H.W. (1982). Interaction between vesicular-arbuscular mycorrhizal fungi and plant pathogens. *Phytopathology*, Vol. 72, pp. 1115-1119

Di Marco, L., Policarpo, M., Corso, A. & Torta, L. (2002). Indagini preliminari sull'inoculazione artificiale di talee di olivo con funghi VAM. *Convegno Intern. Olivicoltura. Atti VI Giornate scientifiche SOI*. Spoleto 22-25 Aprile, pp. 117-122

Duan, X., Neuman, D.S., Reiber, J.M., Green, C.D., Saxton, A.M., Augé, R.M. (1996). Mycorrhizal influence on hydraulic and hormonal factors implicated in the control of stomatal conductance during drought. *J Exp Bot*, Vol. 47, pp. 1541–1550

Duddridge, J.A., Malibari, A. & Read., D.J. (1980). Structure and function of mycorrhizal rhizomorphs with special reference to their role in water transport. *Nature*, Vol. 287, pp. 834–836

Duo – Chuan, L. (2006). Review of fungal chitinases. *Mycopathologia*, Vol. 161, pp. 345–360

Estaún, V., Camprubí, A., Calvet, C. & Pinochet, J. (2003). Nursery and field response of olive trees inoculated with two arbuscular mycorrhizal fungi, *Glomus intraradices* and *Glomus mosseae*. *J. Amer. Soc. Hort. Science*, Vol. 28, No. 5, pp. 767-775

Garcia-Romera, I. & Ocampo, J.A. (1988). Effect of the herbicide MCPA on VA mycorrhizal infection and growth of *Pisum sativum*. *Z Planzenenrnäl Bodenk*, Vol. 151, pp. 225-228, ISSN 0044-3263

Gianinazzi-Pearson, V., Gollotte, A., Dumas-Gaudot, E., Franken, P. & Gianinazzi, S. (1994). Gene expression and molecular modifications associated with plant responses to infection by arbuscular mycorrhizal fungi. pp. 179-186. In: M. Daniels J.A. Downic and A.E. Osbourn (eds.), *Advances in Molecular Genetics of Plant-microbe Interactions.* Kluwer, Dordrecht, Netherlands

Giovannetti, M., Turrini, A., Strani, P., Sbrana, C., Avio, L. & Pietrangeli, B. (2006). Mycorrhizal fungi in ecotoxicological studies: soil impact of fungicides, insecticides and herbicides. *Prevention Today*, Vol. 2, pp. 47–61, ISSN 1120-2971

Goicoechea, N., Antolin, M.C. & Sánchez-Díaz, M. (1997). Gas exchange is related to the hormone balance in mycorrhizal or nitrogen-fixing alfalfa subjected to drought. *Physiol Plant*, Vol. 100, pp. 989–997

Green, C.D., Stodola, A. & Augé, R.M. (1998). Transpiration of detached leaves from mycorrhizal and nonmycorrhizal cowpea and rose plants given varying abscisic acid, pH, calcium and phosphorus. *Mycorrhiza*, Vol. 8, pp. 93–99

Groth, D.E. & Martinson, C.A. (1983). Increased endomycorrhizal infection of maize and soybeans after soil treatment and Metalaxyl. *Plant Dis.*, Vol. 67, pp. 1377-1378, ISSN 0191-2917

Gueye, M., Diem, H.G. & Dommergues, Y.R. (1987). Variation in N2 fixation, N and P contents of mycorrhizal *Vigna unquiculata* in relation to the progressive development of extra-radical hyphae of *Glomus mosseae*. *Mircen J.*, Vol. 3, pp. 75-86

Hayman, D.S., Barea, J.M. & Azcon-Aguilar, R. (1976). Vesicular-arbuscular mycorrhiza in Southern Spain: its distribution in crops growing in soil of different fertility. *Phytopath. medit.*, Vol. 15, pp. 1-6

Hale, K.A. & Sanders, F.E. (1982). Effects of Benomyl on Vesicular-Arbuscular Mycorrhizal infection of Red Clover (*Trifolium pratense* L.) and consequences for phosphorus inflow. *Journal of Plant Nutrition*, Vol. 5 No. 12, pp. 1355-1367, ISSN 0190-4167

Harley, J.L. (1989). The significance of mycorrhiza. *Mycol. Res.*, Vol. 92, pp. 129-139

Harley, J.L. & Smith, S.E. (1983). *Mycorrhizal Symbiosis*. Academic Press, Toronto, ISBN

Harman, G.E. (2000). Myths and dogmas of biocontrol. Changes in perceptions derived from research on *Trichoderma harzianum* T-22. *Plant Dis.*, Vol. 84, pp. 377-393

Harman, G.E., Lorito, M. & Lynch, J.M. (2004). Uses of *Trichoderma* spp. to alleviate or remediate soil and water pollution. p.313-330. In: A.I. Laskin, J.W. Bennett and G.M. Gadd (eds.), *Advances in Applied Microbiology*. Vol. 56. Elsevier Academic Press, San Diego, CA, USA

Hartmann, H.T., Kester, D.E., Davies, F.T.Jr. & Geneve R.L. (2002). *Plant Propagation - Principles and Practices*. 7th edition. Prentice Hall. Englewood Cliffs, New Jersey

Hartung, W., Wilkinson, S. & Davies, W.J. (1998). Factors that regulate abscisic acid concentrations at the primary site of action at the guard cell. *J Exp Bot*, Vol. 49, pp. 361–367

Hedden, P. & Thomas, S.G. (2006). *Plant Hormone Signaling*. Blackwell Publishing, Oxford, UK

Henderson, J.C. & Davies, F.T. (1990). Drought acclimation and the morphology of mycorrhizal *Rosa hybrida* L. cv Ferdy is independent of leaf elemental content. *New Phytol*, Vol. 115, pp. 503–510

Hepper, C.M., Azon-Aghuilar, C. Rosenclahl, S. & Sen, R. (1988). Competition between three species of *Glomus* used as spatially introduced and indigenous mycorrhizal inocula for leek (*Allium porrum* L.). *New Phytol.*, Vol. 110, pp. 207-215

Hetrick, B.A.D. & Wilson, G.W.T. (1991). Effects of mycorrhizal fungus species and metalaxyl application on microbial suppression of mycorrhizal symbiosis. *Mycologia*, Vol. 83, pp. 97-102, ISSN 0027-5514

Jabaji-Hare S.H. & Kendrick W.B. (1985). Effects of fosetyl-AI on root exudation and on composition of extracts of mycorrhizal and nonmycorrhizal leek roots. *Can. J. Plant Pathol.*, Vol. 7, pp. 118-126, ISSN 0706-0661

Jalali, B.L. & Domsch, K.H. (1975). Effect of systemic fungi-toxicants on the development of endotrophic mycorrhiza. In *Sanders et al. Endomycorrhizas*, New York Academic press, pp. 619-626, ISBN 0-12-618350-3

Kapoor, R., Sharma, D., & Bhatnagar, A.K. (2008). Arbuscular mycorrhizae in micropropagation systems and their potential applications. *Sci. Horticult.*, Vol. 116, pp. 227-239

Kaschuk, G., Kuyper, T.W., Leffelaar, P.A., Hungria, M. & Giller, K.E. (2009). Are the rates of photosynthesis stimulated by the carbon sink strength of rhizobial and arbuscular mycorrhizal symbioses? *Soil Biology and Biochemistry*, Vol. 41, pp. 1233-1244, ISSN 0038-0717, 10.1016/j.soilbio.2009.03.005

Kendrick, B. (1985). The Fifth Kingdom. Mycologue Publications. Waterloo, Ontario

Kjøller, R. & Rosendahl, S. (2000). Effects of fungicides on arbuscular mycorrhizal fungi: differential responses in alkaline phosphatase activity of external and internal hyphae. *Biol. Fertil. Soils*, Vol. 31, pp. 361–365, ISSN 0178-2762

Kluwer, Dordrecht. Altomare, C., Norvell, W.A., Bjorkman, T. & Harman, G.E. (1999). Solubilization of phosphates and micronutrients by the plant-growth-promoting and biocontrol fungus Trichoderma harzianum Rifai 1295-22. *Appl. Environ. Microb.*, Vol. 65, pp. 2926-2933

Kobashi, K., Sugaya, S., Gemma, H. and Iwahori, S. (2001). Effect of abscisic acid (ABA) on sugar accumulation in the flesh tissue of peach fruit at the start of the maturation stage. *Plant Growth Regul.*, Vol. 35, pp. 215-223

Koide, R. (1985). The effect of VA mycorrhizal infection and phosphorus status on sunflower hydraulic and stomatal properties. *Journal of Experimental Botany*, Vol. 36, No. 168, pp. 1087-1098

Kücük, C., Kivanc, M., Kinaci, E. & Kinaci, G. (2008). Determination of the growth and solubilization capabilities of *Trichoderma harzianum* T1. *Biologia - Section Botany*, Vol. 63, No. 2, pp. 167-170

Levy, Y. & Krikun, J. (1980). Effect of vesicular-arbuscular mycorrhiza on *Citrus jambhiri* water relations. *New Phytologist*, Vol. 85, pp. 25-31

Lopez-Aguillon, R. & Mosse, B. (1987). Experiments on competitiveness of three endomycorrhizal fungi. *Plant Soil*, Vol. 97, pp. 155-170, ISSN 0032-079X

Mayr, R. & Godoy, R. (1989). Seasonal paiterns in vesicular-arbuscular mycorrhiza in melic beech forest. *Agric. Ecos. Envir.*, Vol. 29, pp. 281-288

Manjunath, A. & Bagyaraj D.J. (1984). Effect of fungicides on mycorrhizal colonization and growth of onion. *Plant Soil*, Vol. 80, pp. 147-150, ISSN 0032-079X

Mathivanan, N., Prabavathy, V.R. & Vijayanandraj, V.R. (2008). The effect of fungal secondary metabolites on bacterial and fungal pathogens. p. 129-140. In: P. Karlovsky (ed.), *Secondary Metabolites in Soil Ecology. Soil Biology*, Vol. 14, Springer-Verlag, Berlin-Heidelberg 2009, ISBN 978-3-540-89620-3

Morrell, J.J. (1990). Effects of volatile chemicals on the ability of microfungi to arrest basidiomycetous decay. *Material und Organismen*, Vol. 25, pp. 267–274, ISSN 0025-5270

Newman, E.I. (1988). Mycorrhizal links between plants: their functioning and ecological significance. *Adv. Ecol. Res.*, Vol. 18, pp. 243-270

Newsham, K.K., Watkinson, A.R., West, H.M. & Fitter, A.H. (1995). Symbiotic Fungi Determine Plant Community Structure: Changes in a Lichen- Rich Community Induced by Fungicide Application. *Funct. Ecol.*, Vol. 9, pp. 442-447, ISSN 0269-8463

O'Brien, T.P. & McCully, M.E. (1981). The Study of Plant Structure Principles and Selected Methods. *Termarcarphi Pry*, Melbourne, Australia

Phillips, J.M. & Hayman, D.S. (1970). Improved procedures for clearing roots and staining parasitic and vesicular-arbuscular mycorrhizal fungi for rapid assessment of infection. Trans. Brit. Mycol. Soc., Vol. 55, pp. 158–161, ISSN 0007-1536

Roland-Fajardo, B.E. & Barea, J.M. (1986). Mycorrhizal dependency in the olive tree (*Olea europaea* L.). in *"Mycorrhizae: Physiology and Genetic"*. eds. Gianninazzi-Pearson V., Gianninazzi S., INRA, Paris, pp. 323-326

Saliendra, N.Z., Sperry, J.S. & Comstock, J.P. (1995). Influence of leaf water status on stomatal response to humidity, hydraulic conductance, and soil drought in *Betula occidentalis*. *Planta*, Vol. 196, pp. 357–366

Safir, G.R., Boyer, J.S. & Gerdemann, J.W. (1972). Nutrient status and mycorrhizal enhancement of water transport in soybean. *Plant Physiol.*, Vol. 49, pp. 700–703

Safir, G.R. & Nelsen, C.E. (1985). VA mycorrhizas: Plant and fungal water relations: In: Proceedings of the Sixth North American Conference on Mycorrhizae (Ed. by R. Molina), pp. 161-164. Forest Research Laboratory, Corvallis, Oregon, USA

Sahebani, N. & Hadavi, N. (2008). Biological control of the root-knot nematode *Meloidogyne javanica* by *Trichoderma harzianum*. Soil Biol. Biochem., Vol. 40, pp. 2016-2020

Schreiner, R.P. & Bethlenfalvay, G.J. (1996). Mycorrhizae, biocides, and biocontrol. Effects of three fungicides on developmental stages of three AM fungi. *Biol. Fertil. Soils*, Vol. 23, pp. 189-195, ISSN 0178-2762

Schwab, S.M., Johnson, E.L.V. & Menge J.A. (1982). Influence of simazine on formation of vesicular-arbuscular mycorrhizae in *Chenopodium quinona* wild. *Plant Soil*, Vol. 64, pp. 283-287, ISSN 0032-079X

Singh, V., Singh, P.N., Yadav, R.L., Awasthi, S.K., Joshi, B.B., Singh, R.K., Lal, R.J. & Duttamajumder, S.K. (2010). Increasing the efficacy of Trichoderma harzianum for nutrient uptake and control of red rot in sugarcane. *J. Hortic. Forest.*, Vol. 2, pp. 66-71

Sofo, A., Scopa, A., Manfra, M., De Nisco, M., Tenore, G., Troisi, J., Di Fiori, R. and Novellino, E. (2010). *Trichoderma harzianum* strain T-22 induces changes in phytohormone levels in cherry rootstocks (*Prunus cerasus x P. canescens*). *Plant Growth Regul.* In press. doi: 10.1007/s10725-011-9610-1

Sofo, A., Milella, L. and Tataranni, G. 2010. Effects of *Trichoderma harzianum* strain T-22 on the growth of two *Prunus* rootstocks during the rooting phase. *J. Hortic. Sci. Biotech.*, Vol. 85, pp. 497-502

Sofo A., Tataranni G., Scopa A., Dichio B. & Xiloyannis C. (2011). Direct effects of *Trichoderma harzianum* strain T-22 on micropropagated GiSeLa6® (*Prunus* spp.) rootstocks. *Environmental and Experimental Botany*, DOI: 10.1016/ j.envexpbot.2011.10.006

Spanu, P. & Bonafante-Fasolo, P. (1988). Cell wall bound peroxidase activity in roots of mycorrhizal *Allium porrum*. *New Phytol.*, Vol. 109, pp. 119-124

Srivastava, L.M. (2002). Plant, Growth and Development - *Hormones and Environment*. Elsevier Academic Press, San Diego, CA, USA

Subramanian, K.S. & Charest, C. (1997). Nutritional, growth, and reproductive responses of maize (*Zea mays* L.) to arbuscular mycorrhizal inoculation during and after drought stress at tasselling. *Mycorrhiza*, Vol. 7, pp. 25–32

Tataranni, G., Montanaro, G., Dichio, B. & Xiloyannis, C. (2011). Hydraulic conductivity in mycorrhisated *Prunus* plants. *Acta Hort*, in press

Tataranni, G., Montanaro, G., Dichio, B. & Xiloyannis, C. (2011b). Effects of mycorrhizas on hydraulic conductivity in micrografted Myrobolan 29C rootstocks. *Acta Hort*, in press

Tataranni, G., Santilli, E., Briccoli Bati, C. & Dichio, B. (2010) Influenza della simbiosi micorrizica sulla risposta vegetativa di cinque cultivar di *Olea europaea* L. *Italus Hortus*, Vol. 17, pp. 60-63

Thangavelu, R., Palaniswami, A. & Velazhahan, R. (2004). Mass production of *Trichoderma harzianum* for managing fusarium wilt of banana. *Agr. Ecosyst. Environ.*, Vol. 103, pp. 259-263

Tournaire-Roux, C., Sutka, M., Javot, H., Gout, E., Gerbeau, P., Luu, D., Bligny, R.& Maurel, C. (2003). Cytosolic pH regulates root water transport during anoxic stress through gating of aquaporins. *Nature*, Vol. 425, pp. 393-397

Trouvelot, A., Kough, J.L. & Gianinazzi-Pearson, V. (1986). Mesure du taux de mycorhization VA d'un systéme radiculare. Recherche de methods d'estimation ayant une signification fonctionelle. In: Mycorrhizae: physiology and genetics-Les mycorhizes: physiologie et génétique. *Proceedings of the 1st ESM/1er SEM*, Dijon, 1-5 July 1985-INRA, Paris, pp. 217-221, ISBN 2853407748

Udaiyan, K., Manian, S., Muthukumar, T. & Greep, S. (1995). Biostatic effect of fumigation and pesticide drenches on an endomycorrhizal-*Rhizobium*-legume tripartite association under field conditions. *Biol. Fertil. Soils*, Vol. 20, pp. 275-283, ISSN 0178-2762

Uehlein, N., Fileschi, K., Eckert, M., Bienert, G.P., Bertl, A. & Kaldenhoff, R. (2007). Arbuscular mycorrhizal symbiosis and plant aquaporin expression. *Phytochemistry*, Vol. 68, pp. 122-129

Vandeleur, R., Niemetz, C., Tilbrook, J. & Tyerman, S.D. (2005). Roles of aquaporins in root responses to irrigation. *Plant Soil*, Vol. 274, pp. 141-161

Van Duin, W.E. Rozema, J. & Ernst, W.H.O. (1989). Seasonal and spatial variation in the occurrence of vesicular-arbuscular (VA) mycorrhiza in salt marsh plants. *Agric. Ecos. Environ.*, Vol. 29, pp 107-110

Whipps, J.M., & Gerhardson, B. (2007). Biological pesticides for control of seed- and soil-borne plant pathogens. p. 479-501. In: *J.D. van Elsas, J.K., J.K. Janson and J.T. Trevors (eds), Modern Soil Microbiology*, CRC Press, Boca Raton

Wilson, J.M. (1984). Comparative development of infection by three vesicular-arbuscular mycorrhizal fungi. *New Phytol.*, Vol. 97, pp. 413-426

Windham, M.T., Elad, Y. & Baker, R. (1986). A mechanism for increased plant growth induced by *Trichoderma* spp. *Phytopathology*, Vol. 76, pp. 518-521

Yang, H-H., Yang, S.L., Peng, K-C., Lo, C.T. & Liu, S-Y. (2009). Induced proteome of *Trichoderma harzianum* by *Botrytis cinerea*. *Mycol. Res.* Vol. 113, pp. 924-932

Yedidia, I., Benhamou, N. & Chet, I. (1999). Induction of defense responses in cucumber plants (*Cucumis sativus* L.) by the biocontrol agent *Trichoderma harzianum*. *Applied and Environmental Microbiology*, Vol. 65, pp. 1061-1070

8

Plant-Soil-Microorganism Interactions on Nitrogen Cycle: *Azospirillum* Inoculation

Elda B. R. Perotti and Alejandro Pidello
Rosario National University (UNR) - Research Council of UNR
Argentine

1. Introduction

When soil is analyzed as a system, it shows great complexity. It is constituted by aggregates of mineral (sand, clay, lime) and organic (live and decomposing organisms) elements, which interact in the soil atmosphere and the water.

The most dynamic chemical fraction constituents of soil, like the compounds from organic matter degradation, the ions in the soil solution (including protons) and the gas concentration of pore atmosphere, define the main physico-chemical variables of soil.

As soil microorganisms are active transformation agents of both the mineral and the organic components of the soil, they play an essential role in plant nutrient cycles, i.e., soil fertility. The study of microbial functions, considered as the activity expression of microorganisms under the environment's conditions, constitutes an important issue in soil and plant microorganism interaction. By means of the expression of their specific functions, microbial communities may act on the environment, for example, by producing or consuming organic or inorganic compounds, or by releasing enzymes into the environment. On the other hand, the environment acts on function expression by means of biotic, abiotic and anthropic conditioners. In this context, soil inoculation with microorganisms is a biotechnological practice that may be performed within the framework of productive soil handling (Lynch, 1987). The goals to achieve include contributing to plant nutrition and soil fertility, controlling phytopathogenic microorganisms or biodegrading compounds -organic or inorganic- which contaminate the soil (Meiri & Altman, 1998).

The recognition of the benefits gained by *Leguminous* plant inoculation with *Rhizobium,* and widespread communication have led to a search for interaction *Gramineous* plants-bacteria, which may be similarly promising. The discovery of *Spirillum* (currently *Azospirillum*), a bacterium that can fix N_2 in the roots of a tropical forage gramineous plant with lush growth (Döbereiner & Day, 1975), led to the hypothesis which states that growth in those plants was directly related to the association *Gramineous* plant-N_2 fixation bacterium (Klucas, 1991).

With full conviction about the importance of *Azospirillum* as a biotechnological tool, studies were conducted in several countries (Bashan & de-Bashan, 2010; Jain et al., 2010). As a result of these studies, information was rapidly spread about inoculation tests performed globally in several gramineous plants in soils, with different species of the genus *Azospirillum*

(Harmann & Baldani, 2006). These studies were especially carried out at an agronomic scale (Dhale et al., 2011; Okon et al., 1998). The results obtained were dissimilar, and this was attributed to different *Azospirillum* strains behavioural patterns in the various soils used (Bashan & Dubrovsky, 1996; Bashan, 1999; Hungria et al., 2010).

There was also a random effect in the inoculation with *Azospirillum* on plant production in the Argiudoll soil studied. In inoculation tests performed in field plots which are representative of the Argentine Humid Pampa, a significant increase in biomass (18-22%) was observed only in some grasses crop cuts. Wheat yield, expressed by ear weight and number, and grain weight were not affected by inoculation with these bacteria. On the other hand, when forage sorghum was inoculated under laboratory growth conditions, it was observed that the inoculated strain infected the roots and that, modifications in soil inorganic nitrogen availability were produced (Perotti, 2001).

Even if this vast experimental global study showed the relativity of the initial idea (i.e., that the effect of *Azospirillum* was directly related to its ability to fix N_2), it also evidenced the multiple capabilities these bacteria have. As well as having the potential to fix N_2, they can produce siderophores, bacteriocins, and plant growth hormones (Bashan & de-Bashan, 2010; Jain et al., 2010). They can also increase ion absorption (e.g., K^+ and NO_3^-, avoid plant hydric *stress,* and modify soil redox potential (Bagheri, 2011; Bashan, 1999; Hungria et al., 2010; Pidello et al., 1993). *Azospirillum* also showed that it can assimilate NO_3^- in aerobiosis, reduce NO_3^- to NO_2^- in anaerobiosis, and produce polyphenol oxidase (Bashan, 1999; Hartmann & Baldani, 2006). These multiple capabilities led to its classification as a bacterium that promotes plant growth (PGPR, Plant Growth Promoting Rhizobacterium). The previously described potential capabilities of *Azospirillum* spp. showed these bacteria as an optimal tool to modify the functional population structure in a complex system such as the agricultural soil.

In this work, the greatest importance was given to the study of inoculation effect on: (i) plant; (ii) soil inorganic nitrogen; (iii) soil urease activity; (iv) N_2 fixation activity in soil, and (v) indigenous rhizospheric diazotrophic microorganisms. The study was based on the hypothesis that the expression of functional capabilities in microorganisms introduced into the soil depends on the accomplishment of two basic conditions: (i) the effective settlement of the bacterium in the rhizosphere (i.e., depending on its ability to survive when introduced into the soil) and (ii) that this fact affects the rizosphere environment (Perotti, 2001; Pidello, 2011). In general terms, and from the soil biotechnology perspective, it is important to know whether the bacterium introduced into the soil is able to functionally "settle" within the system (Bashan & Vazquez, 2000).

This study is intended to help understand the relationship between plant and microorganisms in soil. This contribution was meant to go beyond a descriptive approach and to advance on explanatory aspects, so as to establish some causality relationships in the bilateral relationship between the biotic and abiotic components which would best define the system and inform about the feasibility of manipulating it with a biotechnological objective. Even though these relationships focused on the non-symbiotic diazotrophic functional group, a methodological base was provided to conduct the study of other heterotrophic microbial functional groups which play a significant role in plant nutrition. The ability to survive and modify the agricultural system researched of *Azospirillum,* bacteria chosen as a model of a functional diazotrophic population, was analyzed based on the current knowledge available.

We will first provide a brief description of the nitrogen cycle in the soil-plant-microorganism system, and later a critical analysis of the results obtained from short-term to mid-term tests with plants in molisol soil pots (Typic Argiudoll from the Argentine Humid Pampa), in a laboratory controlled setting and with soil microcosms.

2. The nitrogen cycle

Plant growth and development is related to soil fertility, i.e., nutrient accessibility (chemical elements which, in the form of small ions or molecules are taken by roots to contribute to plant nutrition), and water availability (the water present in the soil profile, that fill soil micro-pores, and is available for plant roots). Nitrogen is one of the most important chemical elements in the soil because it appears in large amounts in various compounds (aminoacids, nucleic acids). The largest nitrogen reservoir in our planet is found in the lithosphere, the mother rock, representing 97.8% Earth's total nitrogen, which is not available to organisms. The atmosphere's diatomic nitrogen (N_2), only representing 2% total N, despite existing in a high proportion in air (79%), is used as a nitrogen source by several microorganisms, and is considered the main source of nitrogen in the biosphere (Frioni, 1990).

In general terms, it may be stated that the various nitrogenated compounds may undergo the following transformations:

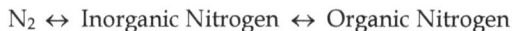

$$N_2 \leftrightarrow \text{Inorganic Nitrogen} \leftrightarrow \text{Organic Nitrogen}$$

These transformations, which are the result of very slow processes (25, 100 or 200 years may elapse before an atom produces every interconversion) (Jenkinson, 1990), guarantee the presence and diversity of nitrogenated compounds in the biosphere. As mentioned before, biological N_2 fixation is the main process allowing the incorporation of atmospheric N_2 into the nitrogen cycle, and is carried out by microorganisms called "diazotrophs", which have a nitrogenase enzyme complex, and whose most important habitat is the soil.

2.1 The nitrogen cycle in soil fertility

Nitrogen, together with carbon, hydrogen and oxygen are the fundamental components of organisms in the biosphere. The inorganic forms, ions NH_4^+ and NO_3^-, or the diatomic form (N_2) are incorporated into the biosphere by plants and microorganisms, which receive the ions from the soil or air, respectively. Moreover, the degradation of plant and animal remains, and microbial cells (soil organic matter), provides inorganic nitrogen and small nitrogenated molecules (amines and amides) which can also be taken by microorganisms and plants (Heller et al., 1998).

The transformations undergone by the various nitrogen species in nature are either biotic (mainly microorganisms and plants) or abiotic (physico-chemical) processes, which is called the "Nitrogen Cycle". This cycle was conceived in 1913 by Lohnis, who identified the various compounds and the biochemical processes involved in its operation. Abiotic factors conditioning these processes (e.g., ion adsorption and lixiviation) were included later into the nitrogen cycle concept (Paul & Clark, 1989).

From the microbial ecology perspective, understanding the significance of the various stages in this cycle, in each agro-ecosystem, and establishing the causality relationships linking them are the main objectives in the study of this cycle. The availability of this information is necessary to "manipulate" this cycle by developing management practices that enhance its efficiency, and guarantee its operation within a "sustainable" approach, without affecting other element cycles that characterize the agricultural system's operation.

2.2 The cycle processes

As shown in Table 1, nitrogen has several oxidation states (between +5 and -3) and, based on this fact, compounds in nature are extremely diverse. This diversity and the type of chemical bonds occurring among atoms (simple, double or triple covalence) produce significant differences in relation to the stability in compounds. The most stable forms found in soil are NO_3^- and N_2, the former as a free ion in the soil solution, and the latter as part of the soil atmosphere. Ion NO_3^- stablility occurs because a nitrogen atom has the highest oxidation degree, whereas N_2 stability is related to the high bond dissociation energy that the molecule has (941 kJ mol^{-1}).

(+5) nitric acid (HNO$_3$)	(+2) nitric oxide (NO)	(-1) diimide (N$_2$H$_2$)
(+4) nitrogen dioxide (NO$_2$)	(+1) nitrous oxide (N$_2$O)	(-2) hydrazine (N$_2$H$_4$)
(+3) nitrous acid (HNO$_2$)	(+1) nitrous oxide (N$_2$O)	(-3) ammonia (NH$_3$)

Table 1. Nitrogenated compounds based on nitrogen oxidation degree

Transformation of N_2 into NO_3^- is thermodynamically possible, but no microorganisms are known which, in the presence of O_2, have developed a metabolic strategy to do it (McBride, 1994). For N_2 to become NH_4^+, there must be a highly reductive medium, and great energy to break the triple bond in the molecule ($N\equiv N$), which is a condition that only occurs inside diazotrophic microorganisms. These microorganisms are able to create an optimal environment through the nitrogenase complex.

According to the type of transformation that occurs in relation to element oxidation degree, the nitrogen cycle processes in the soil may be classified into: (i) oxidative pathways, where energy is released; and (ii) reductive pathways, where energy is consumed.

2.2.1 Oxidative pathways

In oxidative pathways, nitrogen electrons in an oxidation state of (-3) are attracted to oxygen, and nitrogen gains an oxidation state of (+3) or (+5). In this process, which is called nitrification, energy is generated and used by microorganisms to perform anabolic processes and thus sustain their biomass growth. Even though these microorganisms, called "chemolithotroph" (Heller et al., 1998), are not numerous, they are important from the agricultural soil microbial ecology viewpoint, because to gain metabolic energy they are independent of soil organic matter availability.

This process includes two steps. In the first step, soil nitrificating microorganisms take NH_4^+ and, in the presence of O_2, oxidate it into NO_2^-, performing the process called nitritation (Equation 1). The oxidation of NO_2^- into NO_3^- constitutes the second step called nitratation

(Equation 2). The negative values of standard free energy variation ($\Delta G^{o'}$) indicate that this chemical pathway can be used by microorganisms for growth.

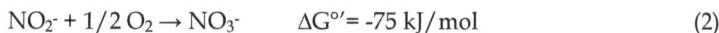

$$NH_4^+ + 3/2\, O_2 \rightarrow NO_2^- + 2\, H^+ + H_2O \qquad \Delta G^{o'} = -352 \text{ kJ/mol} \qquad (1)$$

$$NO_2^- + 1/2\, O_2 \rightarrow NO_3^- \qquad \Delta G^{o'} = -75 \text{ kJ/mol} \qquad (2)$$

2.2.2 Reductive pathways

In the reductive pathways, nitrogen has a lower oxidation degree with energy consumption and reductive power. These requirements may be given by the catabolism of carbon compounds ("heterotrophic" microorganisms), the transformation of light energy ("phototroph" microorganisms), or the oxidation of elements in their elementary state ("chemolithotroph" microorganisms), as it is the case of the denitrificating bacterium *Thiobacillus denitrificans*, which obtains energy from sulfur compound oxidation, or *Micrococcus denitrificans*, which obtains energy from hydrogen oxidation (Glinski & Stepniewski, 1983). In agricultural soil, chemoheterotrophic microorganisms from the nitrogen cycle are proportionally greater in number. In the processes of heterotrophic microbial catabolism, energy is obtained from electron transfer of carbon-reduced-to-O_2 compounds (aerobic respiration) to an acceptor of alternative electrons such as NO_3^- (anaerobic respiration) or to organic compounds (fermentation) (Leclerc et al., 1995).

The reductive pathways of the nitrogen cycle are the most important pathways from the microbial ecology perspective due to the number of processes in which they are involved (biological N_2 fixation, ammonification, NO_3^- and NH_4^+ assimilation, denitrification), and to their close dependence on the transformations of soil organic matter.

2.2.2.1 Biological N_2 fixation

In this process, the atmosphere's N_2 is reduced to NH_4^+ by the nitrogenase enzyme complex, as described in the following equation:

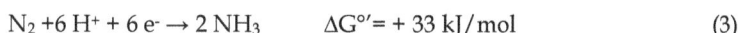

$$N_2 + 6\, H^+ + 6\, e^- \rightarrow 2\, NH_3 \qquad \Delta G^{o'} = +33 \text{ kJ/mol} \qquad (3)$$

The energy required in this process is provided by the adenosine triphosphate (ATP) molecule, and each transferred electron requires a minimum of 2 ATP (a total of 12 ATP per fixed N_2 molecule). As a coupled reduction (2 H^+/H_2) occurs, the global theoretical reaction of this process is:

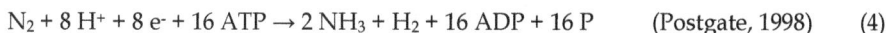

$$N_2 + 8\, H^+ + 8\, e^- + 16\, ATP \rightarrow 2\, NH_3 + H_2 + 16\, ADP + 16\, P \qquad \text{(Postgate, 1998)} \qquad (4)$$

Based on the previous discussion, energy dependence of N_2 fixation is high, and it is ultimately furnished by organic matter degradation or by the photosynthetic process.

2.2.2.2 Inorganic nitrogen assimilation

As it was previously described, microorganisms and plants add nitrogen into their cells, mainly as ions NO_3^- or NH_4^+.

Ammonium that may be obtained from the soil or by means of N_2 fixation, is transformed into organic nitrogen by the following mechanisms: (i) from α-ketoglutarate together with glutamate dehydrogenase (GDH) to render glutamate, as shown by the following reaction:

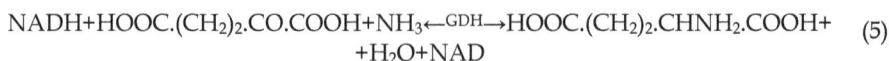

$$NADH+HOOC.(CH_2)_2.CO.COOH+NH_3 \xleftarrow{GDH} HOOC.(CH_2)_2.CHNH_2.COOH+ \\ +H_2O+NAD \tag{5}$$

(ii) from a glutamate molecule together with the glutamate synthase (GS) enzyme to render glutamine. As shown in Equation 6, the reaction requires energy as ATP. Glutamine is later transformed with a reductive power (NADH) and α-ketoglutarate with the paticiparion of glutamine-α-ketoglutarate aminotransferase (GOGAT) in glutamate (Equation 7). This pathway (called GS/GOGAT) is frequently used in the metabolism of soil microorganisms.

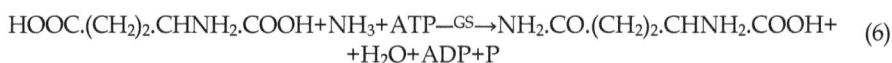

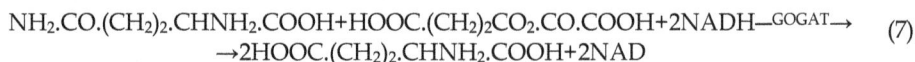

$$HOOC.(CH_2)_2.CHNH_2.COOH+NH_3+ATP \xrightarrow{GS} NH_2.CO.(CH_2)_2.CHNH_2.COOH+ \\ +H_2O+ADP+P \tag{6}$$

$$NH_2.CO.(CH_2)_2.CHNH_2.COOH+HOOC.(CH_2)_2CO_2.CO.COOH+2NADH \xrightarrow{GOGAT} \\ \rightarrow 2HOOC.(CH_2)_2.CHNH_2.COOH+2NAD \tag{7}$$

When NH_4^+ comes from the soil solution (where it is naturally found in low concentration) the low affinity for the sustrate having the GDH enzyme ($K_{m\ (GDH)}$ = 10^{-3} M) and the high affinity of GS ($K_{m(GS)}$ = 10^{-5} M) lead to the fact that the main NH_4^+ addition mechanism in soil microorganisms is the pathway GS/GOGAT (Postgate, 1998), even though there are other nitrogen uptake pathways (e.g., small organic molecules).

2.2.2.3 Ammonification

NH_4^+ release in soil from organic nitrogen is originated in the decomposition of animal, plant or microbial cells. It is produced by glutarate deamination (Equation 5) or by the action of hydrolytic enzymes on amino acid molecules (deaminase enzymes) (Equation 8):

$$R-CHNH_2.COOH+H_2O \rightarrow R-CH_2OH+CO_2+NH_3 \tag{8}$$

or amides (deamidases) as in the case of urea hydrolysis by the action of the urease enzyme:

$$O=C(NH_2)_2 \rightarrow 2NH_3+CO_2+ H_2O \tag{9}$$

2.2.2.4 Denitrification

This process, considered as a disassimilative pathway of the N cycle (Postgate, 1998), comprises the enzymatic reduction of NO_3^- into NO_2^-, then to N_2O and N_2. In this process, NO_3^-, instead of O_2, acts as a final electron acceptor. Electrons may come from organic matter oxidation, photosynthesis or mineral compounds (Paul & Clark, 1989). The whole process is described by the following equations (10.a) or (10.b) with or without nitrite oxidoreductase involvement. The enzymes involved in each stage are: (1) nitrate reductase; (2) nitrite reductase; (3) nitrite oxidoreductase; and (4) nitrose oxidoreductasa:

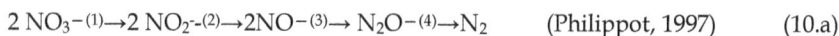

$$2\ NO_3^{-\ (1)} \rightarrow 2\ NO_2^{-\ -(2)} \rightarrow 2NO^{-\ (3)} \rightarrow N_2O^{-\ (4)} \rightarrow N_2 \qquad \text{(Philippot, 1997)} \qquad \text{(10.a)}$$

$$2\ NO_3^{-\ (1)} \rightarrow 2\ NO_2^{-\ -(2)} \rightarrow N_2O^{-\ (4)} \rightarrow N_2 \qquad \text{(Philippot, 1997)} \qquad \text{(10.b)}$$

3. Results

3.1 Inoculated strain survival

The bacterium used in this study was *A. brasilense* strain 7001 (streptomycin-resistant derivate of *A. brasilense* sp 7-ATCC 29145), obtained by C. Elmerich, Pasteur Institute, France). Cultures were performed in Nfb (nitrogen free broth) medium (Döbereiner, 1980). The survival and growth of the inoculated bacteria were evaluated with serial dilutions, and plate techniques on Nfb agar medium amended with streptomycin (1 mg mL^{-1}), and congo red (Rodriguez Cáceres, 1981). Survival of bacteria inoculum was studied in pots with and without plants, and in sterilized soil by gamma radiation (25 KGy γ-irradiation from a ^{60}Co source, Atomic Energy Comission, Ezeiza, Argentina).

3.1.1 Survival in rhizospheric soil

The survival of inoculated strain in fescue plant rhizospheres was different in each experiment but, in general, during the first 2 weeks, the initial cell number decreased in one or two orders. From that period until the end of the experiment (Fig. 1), cell number increased, fluctuating between 10^5-10^7 CFU per dry soil gram (Colony-Forming Unit, CFU). This result of inoculum survival, which may be regarded as high in degree, was not significantly affected by plant density (Perotti et al., 1987). A similar result was obtained when the strain was inoculated in soil supplemented with glucose. Even though this carbohydrate is not used by *A. brasilense* (Hartmann & Zimmer, 1994), it may be used by a large number of soil heterotrophic bacteria, suggesting that the competitive pressure with which the inoculum was confronted in this situation might be significantly increased.

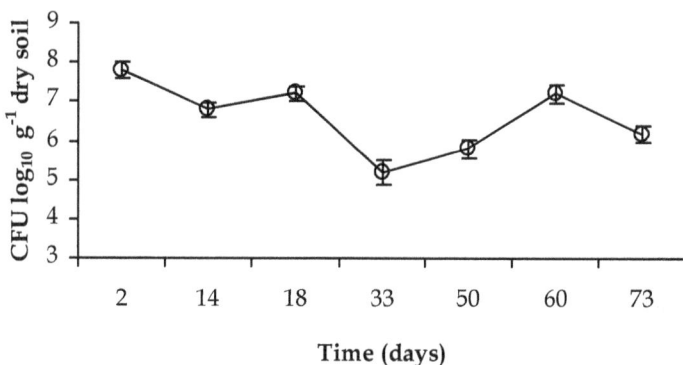

Fig. 1. Evolution of *A. brasilense* survival in fescue rizhospheric soil.

The inoculum survival was not different in the supplemented soil compared to the soil without supplement (Perotti & Pidello, 1999a).

These results suggest that rhizospheric carbon excretions would not significantly affect the survival capability of the inoculated strain.

3.1.2 Survival in non-rhizospheric soil

The survival of the studied *Azospirilum* strain in non-rhizospheric soil was tested in Typic Argiudoll soil samples with naturalized grassland vegetation taken in three different months during the same year: March and May (autumn), and September (spring); the soil characteristics are shown in Table 2. The soil samples selected from those three months differed in relation to global microbial activity, which was defined by soil dehydrogenase activity (DA) (Casida et al., 1964), and by the oxidizable carbon (CO)/NH_4^+-N ratio. Redox potential intensity (Eh) corresponds to aerobic soils (McBride, 1994) with a lower value in September soil, and pH is within the neutrality range.

Sample month	CO (%)	NH_4^+-N ($\mu g\ g^{-1}$ dry soil)	CO/NH_4^+-N ratio	Dehydrogenase activity (μg triphenyl-formazan $g^{-1}h$)	Eh (mV)	pH
March	2.34	4.8	0.585	120	407	7.1
May	1.83	5.3	0.366	450	437	7.6
September	1.27	26.5	0.048	60	305	7.2

Table 2. Main features of Typic Argiudoll soil obtained in three different months representing the seasons autumn (March and May) and spring (September) during a year.

Survival of the *Azospirillum* strain introduced in the soils was assessed in inoculated soil incubations, under controlled laboratory conditions (20 °C, twice the equivalent humidity, in a dark place). As time elapsed, the inoculated strain population decreased in the three samples of soil (Fig. 2). There was an approximate two order-reduction at 41 days after incubation. The second day after incubation, there was a population of approximately 10^7 cells per soil gram. Overall, the population evolution decreased to an approximate value of 10^5 on day 41. In September soil, three weeks after the inoculation, the inoculated strain evolution was more significant than the one from other soil samples. The different soil behaviour observed in September soil might be justified by means of the value shown by the CO/NH_4^+-N ratio, which was significantly lower than the other cases. This result suggests that in heterotrophic populations the survival potencial would be related to sustrate availability. As *Azospirillum* was introduced in soil samples that had different amounts of oxidable carbon (Table 2), the inoculum probably competed with an indigenous heterotrophic population that was temporarily stimulated.

A positive relation between the inoculated strain population and dehydrogenase activity in each of the sample dates during the incubation period was also observed. This relationship suggests that the active inoculum evidently affected an enzymatic activity that shows the total system's microbial activity, i.e., the inoculum may have affected the structure of a functional population which can oxidate organic carbon, and by this pathway have an impact on the significant decrease in soil reduction potential (Eh), as shown in Table 2.

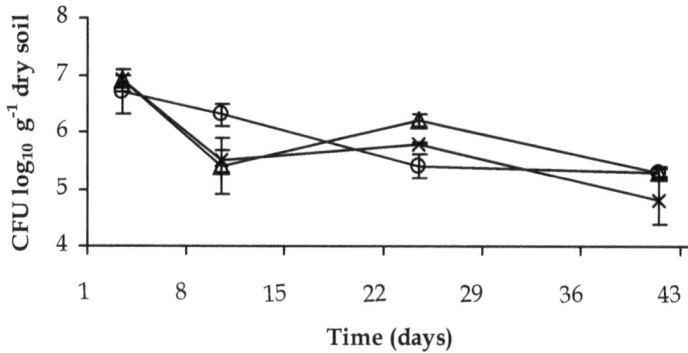

Fig. 2. Evolution of *A. brasilense* survival in soil samples taken in the months March (o), May (x) and September (Δ).

3.1.3 Survival in soil sterilized by gamma radiation

Inoculation experiments in soil sterilized by gamma radiation were conducted under controlled laboratory conditions (20 °C, twice the equivalent humidity, in a dark place). The *Azospirillum* population introduced in sterilized soil constantly increased during the incubation period. The population of the inoculated strain was approximately 10^6 CFU per gram of dry soil the second day after inoculation, and approximately 10^8 CFU per gram of dry soil 16 days after inoculation (Perotti & Pidello, 1999b). This result might be explained if the fact that the inoculated strain does not have indigenous microbial competitor in the irradiated soil is accepted. As reported by other authors (Cawse, 1975; Lensi et al., 1991), in the soil sterilized by gamma radiation there is a significant increase in soluble organic carbon and in ammonium ion content. In this situation, the availability of these easily assimilative sustrates and the absence of microbial competitors may have produced the significant growth of the strain introduced.

3.2 Effects of *Azospirillum* inoculation on plant systems

As fescue is an important *Gramineous* plant in the grasslands of the Argentine Pampa region, it was used as a plant model to study the effects of *Azospirillum* inoculation. In the next experiments, shoot/root ratio, nitrate content in shoot and nitrate reductase activity in leaves were examined in presence or absence of denitrifying *Azospirillum* strains under different plant growth conditions. Seeds of *Festuca arundinacea* L. CV El Palenque (INTA, Argentina) were desinfected with hydrogen peroxide for 30 min. and washed three times with distilled water. They were later sowed in plastic pots with a mixture of soil and sand. The soil (Typic Argiudoll, 0.16-0.27 %N; 2.19% C and pH (H_2O) 6.1) was sieved with a 150 mess sieve. The pots were placed into a growth cabinet at 28-22 °C (day/night), 70% relative humidity, with a photoperiod of 16 hs. The light levels that the plantules received were 197 or 274 $\mu Em^{-2} s^{-1}$ photosynthetically active radiation, within a wavelength range of 400-700 nm (PAR), which were provided by cool-white fluorescent tubes and incandescent bulbs. In the first series of experiments ("high plant density"), twenty germinated seeds per pot were sowed in 150 g of soil-sand mixture (1:5). Plants in Experiment 1 received 197 μE m^{-2} s^{-1} and in Experiment 2 received 274 μE m^{-2} s^{-1}. In the second series of experiments ("low plant density"), four germinated seeds per pot were sowed in 150 g soil-sand (1:5), in Experiments 3 and 4, and fifteen seeds per pot in 1200 g soil-sand (1:3) in Experiment 5. In Experiment 3 the illumination regime was 197 μE m^{-2} s^{-1}, and 274 μE m^{-2} s^{-1} in Experiments 4 and 5.

In every experiment the soil was inoculated simultaneously with 10^8 cell/pot using two strains: *A. brasilense* sp 7001 and *A. lipoferum* sp G (A. Pidello and J. Balandreau. Abstr. Proc. Steenboock-Kettering Int. Simp. N_2 Fixation, Madison, 1978). The control pots received the same concentration of inoculums, but these were autoclaved. To guarantee the physiological state of the plants, the soil was irrigated by capillary absorption with a nutrient N-free solution. Only in Expt. 5 (the longest experiment in duration) 50 mg L^{-1} NH_4^+-N and 33 mg L^{-1} NO_3-N were added in the nutrient solution at the 5th and 6th weeks. In Expt. 2 and 4, plants were harvested after 27 days, and in Expts. 1, 3 and 5 after 22, 23 and 74 days, respectively.

Shoot and washed root dry weights were obtained by drying the plants at 70 °C for 48 h in a forced air oven stove. "In vivo" nitrate reductase activity in leaves was determined by a method proposed by Neyra & Hageman (1974). Nitrate-N in leaves was determined by a colorimetric method (Cataldo et al., 1975). Inorganic-N in soil (Expt. 5) was determined after extraction with 2N KCl (10/100; P/V) during 1 h by a distillation method (Bremmer & Keeney, 1965).

3.2.1 Effect on shoot/root ratio and shoot nitrate content at high and low plant density

Table 3, shows that plant density produced significant modifications ($p<0.05$) in the shoot/root ratio in both inoculated plants and control plants, regardless of the PAR values used in the experiments. According to current literature data, these values coincide with CO_2 interchange values of 38% and 41% maximum value for *F. arundinacea* under experimental conditions, similar to those used in these experiments (Wilhelm & Nelson, 1978), which suggests that the plants were not in the appropriate condition to express their maximum potential to synthesize biomass. This leads to the assumption that the values observed may be conditioned by this fact. However, the ratios presented in Table 3 were the quotients of ratio of the shoot/root ratio in inoculated treatments vs controls (Bashan & Dubrovsky, 1996), which reveal that the variable showed a clearly different behaviour in both cases (in high density treatments the value was 1.125±0.035 and in low density it was 0.823±0.052, with significant differences between both values at $p< 0.05$). Consequently, the results obtained suggest that there are two opposite effects on the shoot/root ratio. When plant density was higher, the bacteria introduced increased the ratio indicating that they may, directly or indirectly, increase efficiency in root absorption (Bashan & Dubrovsky, 1996). When plant density was low, the bacterial inoculum lowered the ratio, suggesting that there would be an increase in the functional aspects of the root, increasing root mass, which might lead to an indirect increase in aerial mass.

The NO_3^--N content in shoot was higher in inoculated plants, and this increase was not modified by PAR difference or plant number (Table 3). This increase effect in nitrogen uptake coincides with the results reported by other authors (Roy Mihir Lal & Srivastava Ramesh, 2010; Saubidet et al., 2002), and was explained in association with root mass stimulation (Fages, 1994).

Inoculated plant / Non-inoculated plant ratio

PAR ($\mu E\ m^{-2}\ sec^{-1}$)	Expt.	High plant density Shoot/Root Shoot NO_3^-		Expt.	Low plant density Shoot/Root Shoot NO_3^-	
197	[1]	1.09 a[1]	1.26 b	[3]	0.91 c	1.13 d
274	[2]	1.16 a	1.16 b	[4]	0.83 c	1.22 d
				[5]	0.73 c	1.35 e
mean (SEM)		1.12(0.03)[*]	1.21(0.05)ns		0.82 (0.05)	1.23 (0.06)

[1] The same letters in the same column indicate that the mean values are not significantly different between them (p=0.05).
[*] Significant differences between high and low plant density at p=0.05 level; ns: Non-significant differences between high and low plant density.

Table 3. Effect of *Azospirillum* inoculum on shoot/root ratio in inoculated plants vs non-inoculated plants, and NO_3^--N shoot content in inoculated plants vs non-inoculated plants at two values or photo-synthetically active radiation, wave length range 400-700 nm (PAR).

3.2.2 Effect on nitrate reductase activity in leaves

Nitrate reductase activity in leaves, measured during a 10 week-period after inoculation (Expt. 5) is shown in Fig. 3. The results suggest that the differences observed in NO_3^--N content in shoots were not related to nitrate reductase activity (NRA). In inoculated treatments, only after the tenth week, NO_3^--N content in shoots significantly increased above the level observed during the experiment in both treatments (Fig. 3A). NRA constantly increased in inoculated and control plants after five weeks. This fact suggests that NO_3^--N in leaves was not a limiting factor in NRA. The NO_3^--N content measured possibly showed a total pool in leaves but not the nitrate flux which is related to NRA. In this experiment with nitrogen supply, inoculated rhizosphere showed there was a similar level of soil inorganic-N at different times (Fig. 3B and Fig. 3C).

3.2.3 Effect on rhizospheric diazotrophic microorganisms

An indigenous diazotrophic population was selected in order to study the impact of inoculation with *A. brasilense* on preexistent microorganisms in soil. The study was conducted in the rhizospheric soil of fescue inoculated with *Azospirillum*. A comparison was made between population evolution of microorganisms and their potential for growth in a nitrogen-free medium (Nfb). Colony morphology in Nfb agar plates, as well as microbial cell reactions to Gram stain and oxidase and catalase tests were used to individualize indigenous rhizospheric diazotrophic populations. Growth continued during the 22-day growth period in fescue plants. This fact should be considered in order to analyse the significance of the result from the population ecology perspective. Apart from the individualized colonies mentioned above, a high number of very small colonies (oligonitrophilic microorganisms) were found, which was included in the total number of possible diazotrophic microorganisms (this data was not shown).

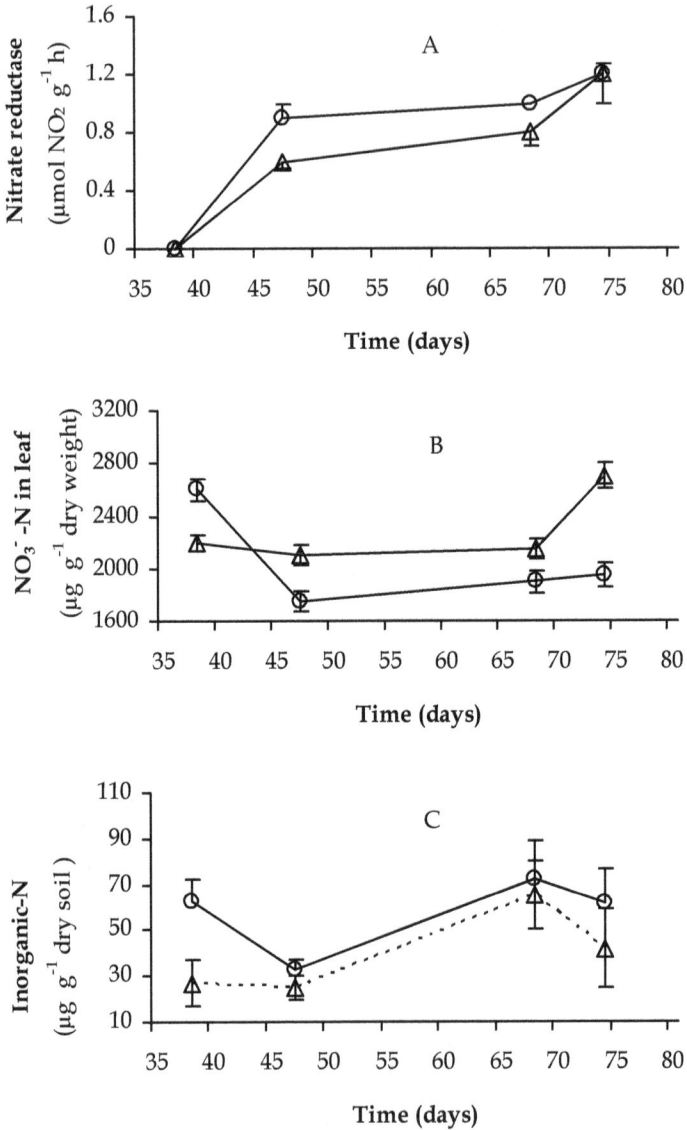

Fig. 3. Nitrate reductase activity (A), NO_3^- -N content of F. arundinacea leaf (B), and soil inorganic nitrogen (NH_4^+-N, NO_2^-.NO_3^--N) (C) in the inoculated (Δ) and the control treatment (o).

The results indicated that immediately after inoculation the number of indigenous diazotrophs showed the maximum values observed, approximately 10^8 CFU g^{-1} dry soil decreasing one order in the first week, with stability around this value until day 22. It was also observed that 10 days after inoculation, the presence of the inoculated strain produced a significant decrease in the population of two colonies identified as Gram negative. This was a reduction of approximately two orders. The difference described vanished during the course of the study showing that there were other types of interaction, e.g., an inhibiting effect of *A. brasilense* selectively affecting some microorganisms but not the whole indigenous diazotrophic population. This effect appeared to decrease towards the end of the study, when both systems (inoculated and non-inoculated) reached a similar distribution in density in the studied populations.

3.3 Effect on bio-edaphic parameters

The bio-edaphic parameters assessed were inorganic nitrogen, dehydrogenase, nitrogenase and urease activities, and soil reduction potential (Eh). Evolution of these parameters in root-free soil microcosms inoculated with *A. brasilense* 7001 and in non-inoculated controls was assessed. Short-term to mid-term experiments were conducted, under controlled laboratory conditions.

3.3.1 Effect on inorganic nitrogen content

In the study soil, there were significant fluctuations in inorganic nitrogen during the whole year (Perotti, 2001), which shows the dynamic nature of soil microbial activity in this ecosystem. Nitrite, nitrate and ammonium ion evolution, in soil taken in March, May (autumn) and September (spring) (Table 2), was studied for 10 weeks after inoculation with *A. brasilense*.

The evolution of ammonium in inoculated soil was different with regard to the evolution in control soil (Fig. 4A). On day 3, there was a significant difference between the inoculated and the uninoculated systems. A rapid immobilization phase was observed, which might be associated to metabolic activity of indigenous microorganisms in the controls, and indigenous and introduced microorganisms in the inoculated soil. After this initial period, inoculated soils showed a constant reduction in ammonium content during a 40-day period. This situation was not observed in controls where, from day 10, there was a new mineralization phase continuing until the end of the period assessed, which showed higher ammonium concentration values than the values recorded in the inoculated soil.

These results show that *Azospirillum* modified the mineral nitrogen dynamics, leading to stabilization in the ammonium immobilization phase, independently of the existing initial differences in available oxidable carbon levels in the three soil samples (Table 2). On the other hand, inoculated bacteria produced a clearer nitrite and nitrate accumulation rate than controls (Fig. 4B). Nitrite and nitrate concentrations in inoculated soils show a linear relation with time ($r=0.83$; $p<0.01$). As it was previously reported, a greater ammonium accumulation suggests that in these soils, ammonium concentration was not a limiting factor for nitrification (Dommergues & Mangenot, 1974). This result should be highlighted, as it shows that *Azospirillum* has the potential to help functional autotrophic groups which are important in soil fertility (Perotti & Pidello, 1991).

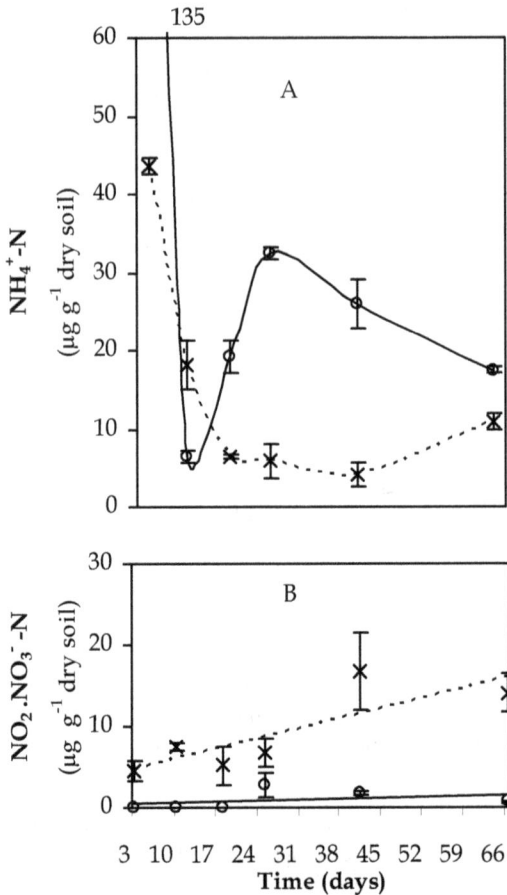

Fig. 4. Effect of *A. brasilense* inoculation on inorganic nitrogen in Argiudoll. NH_4^+-N (A) and NO_2^-.NO_3^--N (B). Inoculated soil (x) and control soil (o). Mean from March, May (autumn) and September (spring) soil samples.

3.3.2 Effect on dehydrogenase activity

Inoculation produced an initial depression in this enzymatic activity in relation to controls, manifested during the first 10 days (Fig. 5). Lower activity shows there was a reduction in total microbial activity (Casida et al., 1964).

On the other hand, the relationship between dehydrogenase activity and ammonium content in the control soil was positive ($r=0.74$; $p<0.05$). This would indicate that the lowest concentrations observed in ammonium concentration (e.g., on day 10) appeared when there was greater microbial activity. This relationship was not observed in the inoculated soil, where ammonium appeared in lower concentrations in relation to the concentrations observed in controls. This fact would explain why in the inoculated soil there are no phases that show a clear predominance of mineralization or ammonium immobilization in soil, with their corresponding increase or disappearance of ammonium concentration (Fig. 4A). Under these conditions, the lack of relationship between the changes in ammonium concentration and total microbial activity is logical.

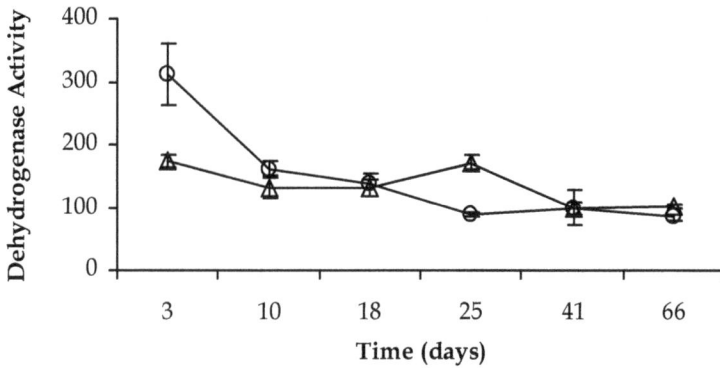

Fig. 5. Dehydrogenase activity evolution (μg triphenyl formazan chloride g^{-1} soil 24 h) in incubations of Typic Argiudoll inoculated with *A. brasilense*. Control soil (o) inoculated soil (Δ). Bars represent the standard error.

3.3.3 Effect on nitrogenase activity

Non-symbiotic biological N_2 fixation (Equations 3 and 4) in the studied soil is naturally expressed at low levels (Perotti, 2001). This does not necessarily mean that this activity is low; it probably responds to a transitory sustrate flux of readily assimilated compounds, originating in soil radicular exhudates or organic matter transformation (Pidello & Perotti, 1997). This flux would produce sustrate competence favouring the expression of heterotrophic microbial functions in the form of "pulses", with spatial heterogeneous distribution when located outside the strictly rhizospheric area ("hot spot") (Drewitt & Warland, 2007).

Potential nitrogenase activity (the soil contained in microcosms was supplemented with 1mg glucose per g), measured 24 h after inoculation with *Azospirillum*, was higher than the activity occurring in the control soil (28 nmol C_2H_4 g^{-1} dry soil h^{-1} vs 23 nmol C_2H_4 g^{-1} dry soil h^{-1}, respectively, $p<0.05$). Twelve days after inoculation, the inoculated soil maintained the trend observed, with higher values than those found in control soils, although the differences were not statistically significant, which shows there was a potentially functional indigenous population.

3.3.4 Effect on urease activity

Urease is one of the enzymes which are responsible for organic nitrogen mineralization (Equation 9), and its activity is responsible in part for the presence of ammonium in soil. It is associated to the presence of microbial or plant cells, and it is also adhered as a free enzyme to soil coloides (Nannipieri, 1993).

We studied the effect of *Azospirillum* strain introduction in soil on urease activity as a model which would show the action of this microorganism on active enzyme in soil, whose expression does not exclusively depend on the specific functional group number. The experimental model used was inoculation in soil sterilized by gamma radiation and in non-sterilized soil. In both treatments, 15 days after the start of the experiment, there was a significant increase in urease activity in inoculated systems (Perotti & Pidello, 1999b), and in general, the study suggests that the bacteria potential is strongly conditioned by factors such as NH_4^+-N content and the indigenous microbial population.

3.3.5 Effect on redox and acid-base intensity

Oxidation-reduction conditions in any microbial ecosystem are considered a "master" variable (McBride, 1994), which helps anticipate innumerable potential interrelationships between abiotic and biotic components. Within certain limits, reduction potential (Eh) - which is an intensity component defining the redox state– is a useful tool to delimit the expression range of various functional microbial groups (Pidello, 2011).

In microcosms of soil inoculated with *A. brasilense*, the values of apparent reduction potential (Eh´) remained lower than in uninoculated control soils, even though these values were higher than 400 mV in both systems, which shows that in the studied systems aerobiosis conditions predominated (Pidello et al., 1993). The values of apparent reduction potential corrected for pH=7 (Eh´7) in inoculated systems were also lower, which proves that the reduction observed is due to differences in electroactive compounds associated with the presence of *Azospirillum* and not to a proton concentration effect (Glinski & Stepniewski, 1983).

4. Conclusions

Plant inoculation with beneficial bacteria is a biotechnological practice involving critical aspects in the relationships between plant-bacteria and soil. It involves aspects related to both, plant physiology and bacteria as well as the bilateral effects produced in the soil environment. A fundamental aspect for the end result in soil-plant microorganism inoculation depends on two conditions: the first is that the effective settlement of bacteria in the rhizosphere is produced, and the second is that the bacteria significantly affect the environment. So, from the soil biotechnology perspective it is important to know whether the bacteria introduced into the soil will be able to act over the system.

In naturally fertile soils, as in the case of the Typic Argiudoll assessed, there are mineralization-immobilization step alternations in the nitrogen cycle that may temporarily increase nitrogen concentration levels. An increase in inorganic nitrogen content in soil may compromise the expression of microbial activities related to the nitrogen cycle, as well as other physiological aspects of the bacteria introduced, which are linked with their survival and potential to act on the system.

The results obtained regarding *A. brasilense* survival in the studied soil indicate that during the longest assessment periods (approximately 40 days) there was a reduction of approximately two orders in the number of bacteria in the soil compared with the concentration in the inoculant used. This may be considered a highly positive behaviour which does not appear to be significantly affected by plant density. During the 24 hours following the inoculant introduction into an environment favouring heterotrophic population development in general, there was no change in inoculant concentration evolution. When the *Azospirillum* strain was introduced into gamma-sterilized soil, survival was higher, and the inoculated population significantly increased towards the end of the assessment period. These results showed that the indigenous functional population affects inoculated strain survival. The results also suggest that there would be an active interaction between inoculum-native microorganisms. This would be evident in two more significant phenomena: on the one hand, the indigenous microflore affected *Azospirillum* strain survival and, on the other, the *Azospirillum* strain affected the indigenous population of diazotrophic microorganisms and their funcional expression nitrogenase activity, and also the total number and functional level of total diazotrophic microflore as suggested by the differences observed in dehydrogenase activity.

In relation to the impact of inoculation on the other hierarchical variables in this study, the following conclusions may be drawn: (i) the tests conducted to assess the effect on plants showed that *Azospirllum* influenced on the partition of plant dry matter, and that this effect of *Azospirillum* on modifications in shoot/root rate was not related to the intensity of photosynthetically active radiation (PAR); and (ii) the amount of NO_3-N was higher in inoculated plants, a fact observed in plants that grew in soils with and without nitrogen supplement, which would be related to an increase observed in inoculated non-rhizospheric soil NO_3^- availability. On the other hand, an increase in root development observed in inoculated plants through an increase in the ion absorption surface might explain this effect. These facts clearly show that in the system studied there are significant changes in the structure of the functional population. This suggests that these variables, undoubtedly important for the system's eco-physiology, would operate by means of changes produced by the inoculum on the structure of the indigenous microbial population.

5. Acknowledgements

This work was partially supported by the Consejo de Investigaciones de la Universidad Nacional de Rosario (CIUNR) and the Secretaría de Ciencia y Técnica de la República Argentina (SECyT).

6. References

Bagheri, A.R. (2011). Effect of salinity and inoculation with *Azospirillum* on carbohydrate production, nitrogen status and yield of barley. *African Journal of Biotechnology*, 10 (45): 9089-9096, ISSN 1684-5315.

Bashan, Y. & Dubrovsky J.C. (1996). *Azospirillum spp.* participation in dry matter partitioning in grasses at the whole plant level. *Biol. Fertil. Soils*, 23, 435-440, ISSN: 0178-2762.

Bashan, Y. (1999). Interactions of *Azospirillum spp.* in soils: a review. *Biol. Fertil. Soils*, 29: 246-256, ISSN: 0178-2762.

Bashan, Y. & Vazquez, P. (2000). Effect of calcium carbonate, sand, and organic matter levels on mortality of five species of *Azospirillum* in natural and artificial bulk soils. *Biol. Fertil. Soils*, 30 (5-6):450-459, ISSN: 0178-2762.

Bashan, Y. & de-Bashan, L.E. (2010). How the plant growth-promoting bacterium *Azospirillum* promotes plant growth-a critical assesment. *Advances in Agronomy*, 108:77-136, ISSN: 0065-2113.

Bremmer, J.M. & Keeney, D.R. (1965). Steam distilation methods for determination of ammoniun, nitrate and nitrite. *An. Chem. Acta*, 32:485-495. ISSN: 0003-2670.

Casida, L.E.; Klein, D.A. & Sanntoro, T. (1964). Soil dehydrogenase activity. *Soil Sci.*, 98(6):371-376. ISSN: 0038-075X.

Cataldo, D.A.; Haroo, R. H.; Schrader, L. E. & Youngs, V. L. (1975). Rapid colorimetric determination of nitrate in plant tissue by nitration of salicylic acid. *Commun. Soil Sci. and Plant analysis*, 6:71-80.ISSN: 0010-3624.

Cawse, P.A. (1975). Microbiology and biochemistry of irradiated soils. In: *Soil Biochemisry*, vol 3, Paul, E.A.; Douglas, M.c.; Laren, A.. (Eds.) Marcel Dekker, New York, pp. 213-267.

Dhale, D. A.; Chatte, S. N. & Jadhav, V. T. (2011). Response of bioinoculants on growth, yield and fiber quality of cotton under irrigation. *Agric. Biol. J. N. Am.*, 2(2):376-386, ISSN: 2151-7517.

Döbereiner, J. & Day, J.M. (1975). Associate symbiosis in tropical grasses, characterization of micro-organisms and dinitrogen fixing sites. *Proceedings of International Symposium on N2 fixation*, Newton, W.E.; Hyman, C.J. (Eds.), Washington State Univ. Press. Pullman, pp. 518-539.

Döbereiner, J. (1980). Forage grass and grain crops. Bergensen F.J. (Ed.). In: *Methods for evaluating biological nitrogen fixation*. John Wiley & Sons, Chinchester, England, pp. 535-555.

Dommergues, I. & Mangenot, F. (1974*). Ecologie Microbiènne du sol*, Masson. et Cie., París, 795 pp.

Drewitt, G. & Warland, J.S. (2007). Continous measurements of belowground nitous oxide concentrations. *Soil Sci. Soc. Am. J.*, 71: 1-7, ISSN: 0361-5995.

Fages, J. (1994). *Azospirillum* inoculants and field experiments. In: *Azospirillum Associations*. Okon, Y. (Ed): CRC Press, USA, pp. 88-105.

Frioni, L. (1990). *Ecología microbiana del suelo*. Ed. Universidad de la República, Montevideo, Uruguay, 519 pp.

Glinski, J. & Stepniewski, W. (1983). *Soil Aeration and Its Role for Plants*. CRS Press, Inc. Boca, 229 pp.

Hartmann, A. & Zimmer, W. (1994). Physiology of *Azospirillum*. In: *Azospirillum/Plant Associations*. Okon, Y. (Ed.) CRC Press Boca Raton, pp.5-39.

Hartmann, A. & Baldani J. I. (2006). The genus *Azospirillum*. In: *The Prokaryotes*, Springer, New York, 5: 115-140.

Heller, R.; Esnault, R. & Lance, C. (1998). *Physilogie végétale.1. Nutrition* 6e édition. Dunod, París, 323 pp.

Hungria, M.; Campo, R.J.; Souza, E.M. & Pedrosa, F.O. (2010). Inoculation with selected strains of *Azospirillum brasilense* and *A. lipoferum* improves yields of maize and wheat in Brazil. *Plant & Soil*, 331 (1-2): 413-424, ISSN: 0032-079X.

Jain, V.; Khetrapal, S.; Aravind, S. & Saikia, S.P. (2010). Enhanced levels of soil nitrogen and endogenous phytohormones in maize (*Zea mays* L.) inoculated with *Azospirillum brasilense. Indian Journal of Plant Physiology*, 15 (2) 0019-5502, ISSN: 0019-5502.

Jenkinson, D.S. (1990). An introduction to the global nitrogen cycle. *Soil use manag.*, 6 (2): 56-61, ISSN: 022-0032.

Klucas, R. V. (1991). Associative Nitrogen Fixation in plants. In: *Biology and Biochemistry of Nitrogen Fixation*. Dilworth, M. J. & Glenn, A. R. (Eds.) pp. 187-198.

Leclerc, H.; Gaillard, J.L. & Simonet, M. (1995). *Microbiologie générale. La bactérie et le monde bactérien*. Doin éditeurs, París, 535 pp.

Lensi, R.; Lescure, C.; Steinberg, C.; Savoie, J. M. & Faurie, G. (1991). Dynamics of residual enzyme activities, denitrification potential, and physico-chemical propierties in a γ-sterilized soil. *Soil Biol. Biochem.* 23:367-373, ISSN: 0038-0717.

Lynch, J.M. (1987). Soil Biology: Accomplishments and Potential. *Soil Sci. Soc. Am. J.*, 51:1409-1412.

McBride, M.B. (1994). *Environmental Chemistry of Soils*. Oxford University Press, New York, 406 pp.

Meiri, H. & Altman, A. (1998). Agriculture and Agricultural Biotechnology: Development Trends Toward the 21st Century. In: *Agricultural Biotechnology*, A. Altam (Ed.), Marcel Dekker, Inc., pp. 1-17. ISBN: 0-8247-9439-7, New York.

Nannipieri, P. (1993). *Ciclo della sostanza organica nel suolo: aspetti agronomici chimici ecologici e selvicolturali*. Patron Editore, Bologna, 334 pp.

Neyra, C.C. & Hageman, R.H. (1974). Dependence of nitrite reduction on electron transport in chloroplast. *Plant Physiol.* 54:480-483, ISSN: 0032-0889.

Okon, Y.; Bloemberg, G. V. & Lugtemberg, B. J. J. (1998) Biotechnology of Biofertilization. In: *Agricultural Biotechnology*, A. Altam (Ed.), Marcel Dekker, Inc., .pp. 327-348. ISBN: 0-8247-9439-7, New York.

Paul, E.A. & Clark, F.E. (1989). *Soil Microbiology and Biochemistry*. HB Jovanovich (Ed.). Academic Press INC., San Diego, 275 pp.

Perotti, E.B.R.; Cortés, V.; Chapo, G.; Menédez, L. & Pidello, A. (1987). Effect of *Azospirillum* strains in *Festuca* rhizosphere on plant and soil nitrogen content. *Rev.Lat-amer. Microbiol.*, 29: 210-215, ISSN: 0187-4640.

Perotti, E.B.R. & Pidello, A. (1991). Evolución del Nitrógeno mineral del suelo en presencia de *Azospirillum. Ciencia del Suelo*, 8 (1): 41-46, ISSN: 1850-2067.

Perotti, E.B.R. & Pidello, A. (1999a). Efecto y supervivencia de *Azospririllum brasilense* a las 24 horas de su inoculación en un suelo suplementado con glucosa. In: *Biología del Suelo*. Stegmayer, A.R.; Pernasetti, D.S.; Gomez Bello, C. (Eds.) Catamarca, Argentina, pp.181-184.

Perotti, E.B.R. & Pidello, A. (1999b). Effect of *Azospirillum brasilense* inoculation on urease activity in soil and gamma-sterilized soil. *Rev. Arg. Microb.*, 31: 36-41. ISSN: 0325-7541.

Perotti, E.B.R. (2001). Estudio y manipulación de las funciones microbianas del suelo pampeano relacionadas con la fertilidad. PhD. Thesis. Facultad de Ciencias Bioquímicas y Farmacéuticas. Universidad Nacional de Rosario. 151 pp.

Philippot, L. (1997). Rôle d´une function microbienne dans l´ecologie des micro-organismes du sol: valeur selective des gènes de la dénitrification. Thèse Diplôme de Doctorat. Úniversité Claude-Bernard-Lyon I, 132 pp.

Pidello, A.; Menéndez, L. & Lensi, R. (1993). *Azospirillum* affects Eh and potential denitrification in a soil. *Plant & Soil*, 157:31-34, ISSN: 0032-079X.

Pidello, A. & Perotti, E.B.R. (1997). Influence de la plante sur l´état rédox du sol. *Proceeding of Colleque Rhizosphere Aix´97*- Société Francaise de Microbiologie. Aix en Provence (Francia), november, 26-27 p. 17.

Pidello, A. (2011). *Ecología Microbiana- Química redox*. Corpus, Rosario, Argentina, 144 pp.

Postgate, J.R. (1998). *Nitrogen Fixation*. Third Edition. Cambridge University Press, Cambridge, 112 pp.

Rodriguez Cáceres, E.A. (1981). Improved medium for isolation of *Azospirillum spp. Appl. Environ. Microbio.*, 44:990-991.

Roy Mihir Lal & Srivastava Ramesh C. (2010). Influence of *Azospirillum brasilense* on biochemical characters of rice seedlings. *Indian J. Agric. Res.*, 44(3):183-188, ISSN: 0367-8245.

Saubidet, M.L.; Fatta, N. & Barneix, A.J. (2002). The effect of inoculation with *Azospirillum brasilense* on growth and nitrogen utilization by wheat plants. *Plant & Soil*, 245 (2): 215-222, ISSN: 0032-079X.

Wilhelm, W.W. & Nelson, C.J. (1978). Irradiance of Tall Fescue Genotypes with contrasting Levels of photosynthesis and Yield. Publications from USDA-ARS/UNL Faculty, pp.1-5.

9

Fruit Transpiration: Mechanisms and Significance for Fruit Nutrition and Growth

Giuseppe Montanaro*, Bartolomeo Dichio and Cristos Xiloyannis
Department of Crop Systems, Forestry and Environmental Sciences
University of Basilicata
Italy

1. Introduction

Water and minerals transport in plant occur through the transpiration stream as triggered mainly by the environmental conditions. The transpiration of leaves has been studied in great detail over many years and the roles played by the various leaf structures (cuticle, stomata, lenticels etc) are well understood. Similarly well understood are the influences on foliar transpiration of meteorological variables such as temperature, radiation, vapour pressure deficit and wind, also, how these variables can be differently important to foliar transpiration depending on whether one is considering an isolated leaf or an entire canopy (Jarvis, 1985). In contrast with this, the transpiration of fruits has not been studied to nearly the same extent, neither with regard to skin structures and their associated functional properties nor with regard to the transpiration response of fruit to the various meteorological variables.

Most fruit-crop species (including apple, apricot, avocado, kiwifruit, capsicum and tomato), suffer from pre- and post-harvest physiological disorders. Higher incidences of these physiological disorders have many times been reported as being associated with lower concentrations of calcium (Ca) in the fruits (Faust et al. 1968, Tzoutzoukou and Bouranis 1997, Ferguson et al. 2003, Thorpe et al. 2003). Transpiration is the main driving force for the xylem stream (White and Broadley 2003) in which Ca seems to move relatively freely while this ion is also well known to be substantially immobile in the phloem (Buckowak and Wittwer 1957). Along with the observation that fruit are largely phloem fed, calcium's well know xylem mobility and phloem immobility explain in part why fruit are generally low-Ca organs and also why higher fruit transpiration rates are sometimes associated with increased fruit Ca levels (Cline and Hanson 1992, Tromp and Van Vuure 1993, Montanaro et al 2006 and 2010). It is therefore reasonable to hypothesise that for any particular fruit, the seasonal integral of fruit transpiration rate will predict fruit Ca content at harvest and thus (at least potentially) the incidence of Ca-related physiological disorders.

This chapter focuses the mechanisms behind fruit transpiration in some fruit tree species, its seasonal trend and discusses the significance of fruit water loss on mineral composition particularly on Ca accumulation.

---

* Corresponding Author

2. Understanding and measuring fruit transpiration

Based on the Fick's- or Ohm's-law analogy (Fiscus et al., 1982), specific fruit transpiration rate (T_F) (the water vapour efflux per unit area of skin surface) depends on the product of the conductance properties (C) of the fruit skin and the driving force (V):

$$T_F = C \times V \tag{1}$$

Fruit conductance depends on skin properties determining its permeance, hence it is species specific. In addition it varies along with the season. Smith et al (1985) showed that the skin conductance in kiwifruit berry decreases rapidly during the early two weeks after anthesis (Fig. 1), thereafter the conductance decreases again reaching the minimum at about 8 weeks after anthesis where remains relatively constant until harvest.

Back to the Eq. 1, the term V is the difference in water vapour pressure between the airspaces inside the fruit and the air immediately outside it. The driving force depends on environmental conditions such as temperature, radiation, relative humidity, etc. Under most commercial training systems in most of fruits, their intercellular air (which is very close to saturation with relative humidity values close to ~99.4% (Nobel, 2005)) will have a water vapour pressure very close to the theoretical value for water at the temperature of this intercellular air. This means that transpiration will be driven substantially by a vapour pressure gradient whose value is numerically close to the water vapour pressure deficit (VPD) (Fig. 2) of the bulk atmosphere and this in turn is a reasonably exact function just of the surrounding air's temperature and of its relative humidity (Goudriaan and van Laar 1994). Based on Eq. 1, fruits under higher relative humidity are predicted to have a lower transpiration. Results presented by Li et al. (2002), even in a preliminary form, are in line with this prediction. In that study, it has been reported that an increase of relative humidity from 40 to 60% induced a reduction of fruit water loss by approximately 30-50% in two peach varieties under laboratory condition.

Measurement of fruit water loss could be performed in both attached and detached fruit.

Regular leaf gas exchange instrumentation equipped with appropriate chamber helps to measure fruit transpiration in attached fruit, particularly for those species (e.g. kiwifruit) with relatively long peduncle (Photo 1). Since such a instrument requires knowledge of the surface area being analysed, it is required to preliminarily determine an appropriate correlation with fruit area and some geometrical traits of fruit which could be easily measured in the field (e.g. length, diameter). For example, for estimates of fruit surface area in kiwifruit berry a fruit sample (×70) were carefully peeled during the growing season and their skin area measured using a portable area meter (Model Li-3000, LI-COR). For the same fruit, estimated fruit surface area (Es) was calculated as $ES = L*W*3.14$, where L was the fruit length and W the maximum fruit diameter. The fitted linear regression (y = 1.0078 * ES + 0.798; R^2 = 0.97) between estimated and measured fruit surface area was then used in the field for measurements of fruit transpiration (Montanaro et al., 2006).

Fruit transpiration could be inferred in the field also through the weigh loss of detached fruit (Lang 1990). In that case, the fruit water reservoir is able to sustain the transpiration for several hours (~10) (Li et al., 2002) despite the lack of the xylem/phloem supply by the parent plant. This valuable method is also used to assess in/out fluxes of fruit. Briefly, it is based on fruit diameter (and volume) changes occurring in fruits subjected to a sequence of

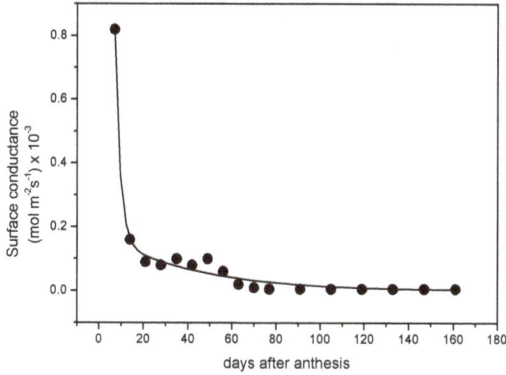

Fig. 1. Surface conductance to water vapour measured in detached kiwifruit. (Redrawn from Smith et al., 1995).

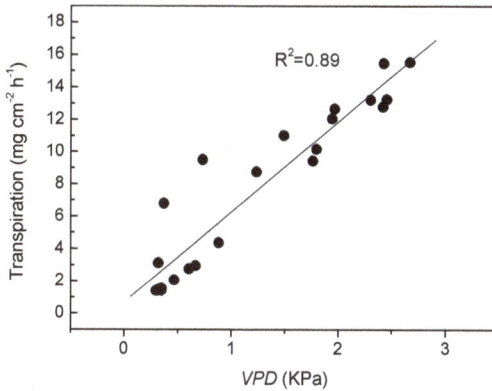

Fig. 2. Relationship between vapour pressure deficit (*VPD*) and fruit transpiration measured in detached kiwifruit berry in 3-week-old fruits. (Redrawn from Montanaro et al., in press).

Photo 1. Measurement of fruit transpiration in attached kiwifruit (left) and detached apricot (right) using a portable gas-exchange equipment (ADC-LCA4, ADC BioScientific Ltd, Hoddesdon, England).

manipulation as concern their vascular connections (Lang 1990). The diameter variations of fruit should be accurately determined using specific highly accurate gauge (Lang 1990). That method requires the measurement of diameter of "intact" (with normal vascular connections), "girdled" (with the phloem connection severed) and "detached" fruit (with all vascular connections severed). The method is based on a subtractive analysis to quantify separately the phloem, xylem and the transpiration contribution, under the assumption that volume growth integrates a fruit in/out fluxes.

3. Daily and seasonal changes in fruit water loss

The absolute value of fruit transpiration is inevitably related to the whole fruit surface area. Regular growing fruits show a rapid expansion of their surface area early after fruit set due to cell division (Fig. 3). Later in the season fruit surface further increases but at a lesser extent.

Specific fruit transpiration varies considerably over a 24-hour period taking relatively high values in the middle part of the day but reducing to almost nothing at night (Montanaro et al., in press). Figure 4 shows the daily variation of fruit water loss in two species (peach and kiwifruit) which is presumably due to diurnal variation in key variables of the fruit's aerial microenvironment (e.g. temperature, relative humidity). A noteworthy fact is that despite fruit transpiration was assessed with different method (weigh loss for peach, gas exchange for kiwifruit) it clearly shows a similar behaviour and a consistent absolute value.

Seasonal fruit transpiration usually shows a maximum rate just after fruit set, thereafter it declines quite promptly reaching a minimum at the half of the whole development stage (Fig. 5). In apricot fruit (cv Tyrithos) that maximum has been recorded to be close to 0.55 mmol m^{-2} s^{-1} in very young fruits (i.e. 6 days after fruit-set) (Fig. 5). During the following 4 weeks, transpiration rapidly decreased to approx. 0.35 mmol m^{-2} s^{-1} accounting for about 80% of the whole transpiration decline. Thereafter, transpiration again decreased but more slowly reaching 0.30 mmol m^{-2} s^{-1}.

In kiwifruit, the rate of fruit water loss is similar even the absolute value significantly higher than stone fruits (apricot/peach). Fruit transpiration per unit of fruit area had a maximum value of 2.3 mmol m^{-2} s^{-1} in the first days after fruit set (Fig. 5). During the following 2 weeks transpiration rapidly decreased by 65% compared to the initial values. In the subsequent 30 days it reached very low values, approximately 10% of the initial rate. The substantial reduction in transpiration of fruit by day 50-60 after fruit set occurs at the same time as the permanent disfunction of the fruit vascular system as reported by Dichio et al. (2003).

There are evidences that the minimum rate of transpiration reached by fruit is species specific. The minimum transpiration rate recorded at the end of the grow period in *Prunus armeniaca*, was relatively high at 55% of the early value according to data on peach (Li et al., 2002). The transpiration profile is also in agreement with findings in grapevine and kiwifruit even though the extent of decline differed (Boselli et al., 1998; Montanaro et al., 2006). In these species, the lowest transpiration rate was about 10% (kiwifruit) and 20% (grapevine) of the initial values. In kiwifruit the large decline of fruit transpiration has been associated with the evolution of a suberised outer cell layers, and death of the outer cells associated with wax biosynthesis (Celano et al., 2009). It is likely that such changes are not so marked in the case of apricot, leading the lowest transpiration rate to remain within about ½ of the initial values.

3.1 Changes in skin properties

Seasonal changes in fruit-skin conductance are probably responsible for the decline in transpiration during the season showed in Figure 5. Xiloyannis et al. (2001) the decrease in fruit transpiration in kiwifruit has been associated with the collapse of the surface tissues of the fruit and the development of a suberized periderm. That is, transversal freeze fractures of frozen-hydrated kiwifruit epidermis proved that the external layer of epidermal cells collapsed during fruit development (Photo 2), producing a bearing-like protection for the fully hydrated parenchymatic cells positioned below them. These epidermis structural changes are associated with the heavy reduction in transpiration rate recorded during fruit growth (Xiloyannis et al., 2001).

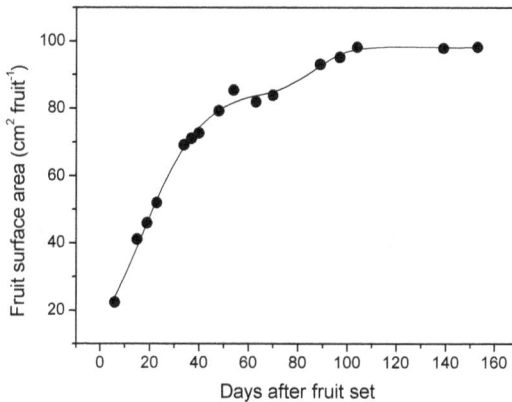

Fig. 3. Seasonal trend of fruit surface area observed in kiwifruit (cv Hayward).

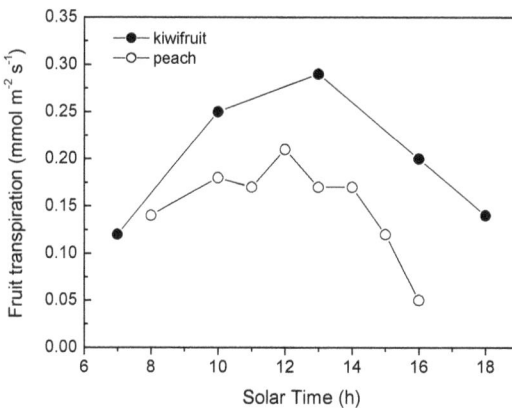

Fig. 4. Daily oscillation of fruit water loss measured 1-week after fruit-set in attached kiwifruit (cv Hayward) and detached peach fruit (cv Dixired). (Redrawn from Li et al., 2002 and Montanaro et al., 2006).

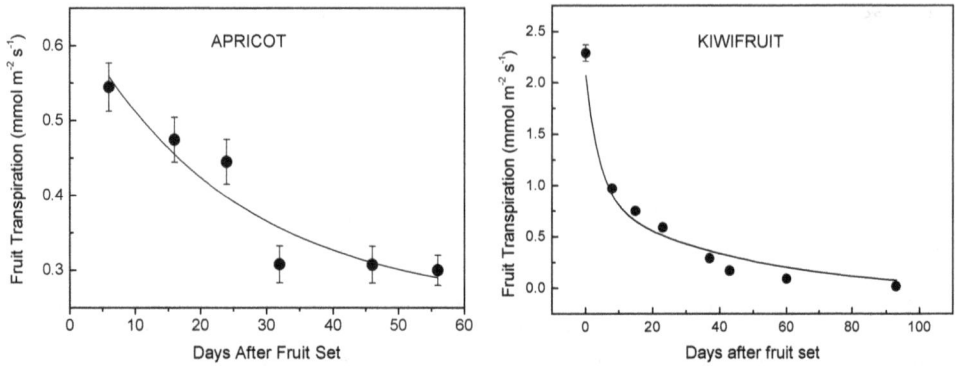

Fig. 5. Seasonal trend of specific fruit transpiration rate measured midday in apricot (left) and kiwifruit (right) as measured in detached and attached fruit, respectively. (Redrawn from Montanaro et al., 2006 and 2010).

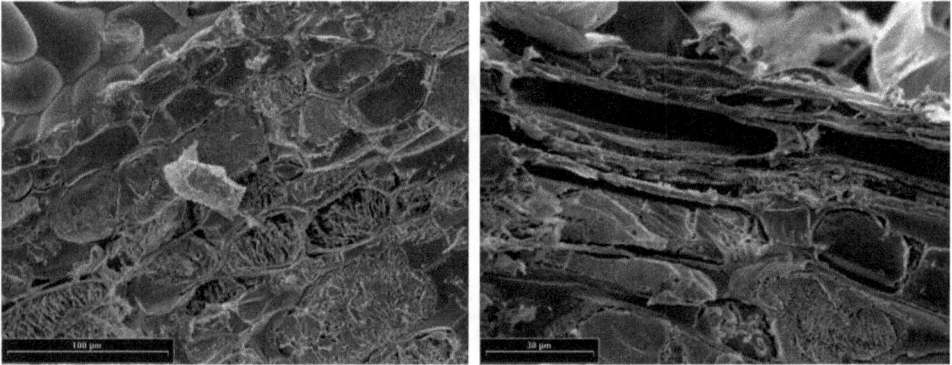

Photo 2. Low Temperature SEM images. Transversal freeze fractures of frozen-hydrated kiwyfruit epidermis sampled at the 4th (left) and at the 17th (right) week from fruit set. Note the collapse of the external layers of epidermal cells during the development of the fruit, which produces a bearing-like protection for the parenchimatic cells positioned below them. (From Xiloyannis et al., 2001).

3.2 Which role for hairs?

In some species, fruits have the skin with dermal structures (hairs) which inevitably are involved in skin conductance properties and in turn affect fruit transpiration. This is the case with some kiwifruit species (e.g. *Actinidia deliciosa*).

From an anatomical point of view the dermal hairs of kiwifruit have been reasonably well characterised (White, 1986) but, as yet no information exists as to their possible influence on the transpiration physiology of the fruit. It is therefore worthwhile at this stage to introduce to the discussion some relevant aspects of their development and physiology.

As previously reported in Xiloyannis et al., (2008), the juvenile kiwifruit already has a full set of hairs (trichomes) at bloom and no new hairs develop from this time on. The hairs are also fully formed and are typically 2.5 to 3.0 mm long and their bases are about 0.1 mm in diameter giving them an aspect ratio (length/basal diameter) of about 25. In the very young fruit, the hairs are packed together so closely that their bases are almost touching. Subsequent surface expansion of the fruit moves the hairs apart so that they become increasingly isolated from one another during the growing season.

Hairs are of quite simple structure (Fig. 6) and contain no vascular tissue. This requires that any evaporative water loss occurring from the more apical cells must be made good by diffusion of water up from the skin through the more basal cells. This diffusion pathway (it is likely that it will be predominantly through the cell wall matrix, not through the protoplasts) is probably sufficient to maintain hair vitality early on when the hairs' close proximity to one another means that they will offer a degree of mutual protection from water loss. However, as they move apart with fruit growth, this protection will become less and less, and so the acropetal diffusion pathway will probably become limiting at some stage. This understanding fits with our observation that cell death (drying) is first evident in the most apical hair cell and this then progresses downwards (basipetally) through the hair eventually reaching the most basal ones last (Xiloyannis et al., 2008).

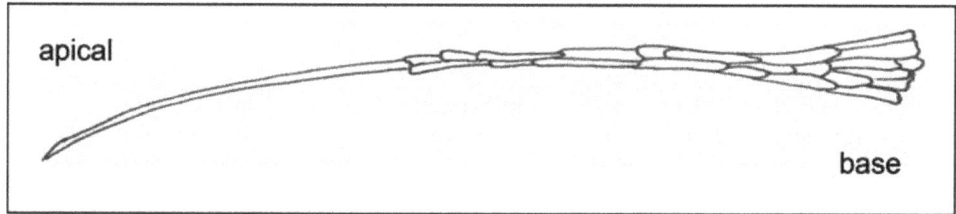

Fig. 6. Drawing of a dermal hair from cv. 'Hayward' kiwifruit. Cell death occurs basipetally, from left to right in this drawing (Redrawn from Xiloyannis et al., 2008).

A simple calculation based upon hair dimensions (the area of a hair $\cong \pi \times$ base radius $\times$ hair length) and hair density (hairs mm^{-2} of fruit surface) (data not shown) proves that, in a very young kiwifruit, the still-living hairs are sufficiently numerous as to increase the 'living' surface area of the fruit by a factor of around 40-times, compared with a hairless fruit of the same dimensions. This massive amplification of fruit surface area is almost certainly responsible for the very high surface conductance of a young kiwifruit. The amplification factor steadily decreases as the fruit surface area increases with growth (data not shown), but the hair dimensions and hair numbers per fruit remain the same (i.e. the hair density decreases). The proportion of living hairs (hairs with some live hair cells) falls from close to 100% at the time of fruit set to about 50% by 30 days after full bloom and by 60 days after full bloom all the cells of all hairs are dead (Fig. 7). We interpret the rather striking skin surface conductance results of Smith et al., (1995) and our own fruit transpiration results, not in terms of the development of a more complex dermal layer (although this too could be involved), but predominantly in terms of the basipetal progression of cell death in the dermal hairs. During this short period of early fruit growth, as hair cell death progresses, the 'hair effect' will change from one in which they *increase* water loss from the fruit skin to one in which they *decrease* it. The fully dead (dry) hairs will serve to reduce fruit water loss by entrapping a thick (c. 3 mm) boundary layer of still, moist air around the fruit. This effect is much the same as that of the layer of fur that reduces the rate of heat loss from the skin of mammals. Moreover, this understanding is consistent with the behaviour, elsewhere reported (Dichio et al., 2007) that, under the influence of an increasing forced air stream, the kiwifruit's boundary layer resistance breaks down progressively and fruit transpiration increases very significantly. This in turn impacts the import of certain minerals (see below).

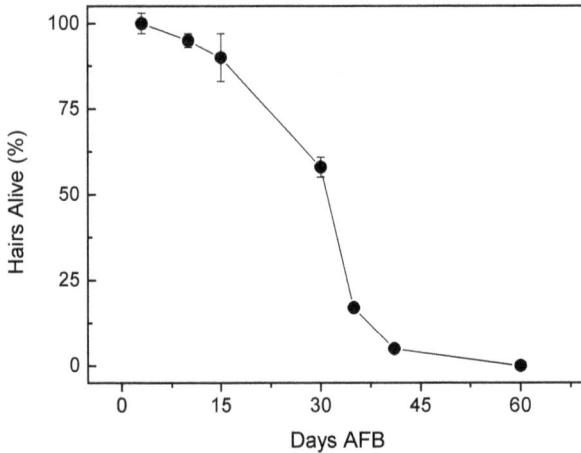

Fig. 7. Hair viability (% of hairs in the sample containing some live cells). Fruit were grown under normal commercial conditions in New Zealand and were not subjected to any special treatment. Each point is the average (± SE) of 100 observations. (Redrawn from Dichio et al., 2003)

3.3 The within-fruit resistances

Late season occurrence of anatomical changes of fruit epidermis produces a reduction of surface conductance (Smith et al., 1995; Celano et al., 2009) contributing to the transpiration decline previously observed during fruit development (Montanaro et al., 2006), this might strongly reduce the driving force of xylem inflow. Reduction of incoming xylem stream would be also the result of various co-occurring phenomena such as disruptions (or occlusions) localized either in the (tomato) pedicel (Van Ieperen et al., 2003) or in the (apple, kiwifruit) fruit tissue (Dichio et al., 2003; Dražeta et al., 2004). In fruit of Vitis sp the reduction of hydrostatic pressure gradients and the high phloem flows at the end of the season have been indicted to repress xylem inflow (Bondada et al., 2005; Keller et al., 2006; Choat et al., 2009). In some crops (e.g. apple, tomato and grape, kiwifruit) it has been argued that high hydraulic resistance inside fruits may be involved in phloem unloading processes and to protect the fruit from excessive backflow (Keller et al., 2006 and references therein; Morandi et al., 2010a).

Figure 8 shows the seasonal changes in the hydraulic conductance of fruit (berry+stalk) of kiwifruit. Within the first 35 days after full bloom, hydraulic conductance increased rapidly. This is presumably due to the differentiation of new xylem vessels as a result of a post bloom rise in the activity of the intrafascicular vascular cambium, probably stimulated by seed set (Dražeta et al., 2004). Conductance rose to about 0.22 cm^3 MPa^{-1} s^{-1} × 10^{-3}. For about 20 days, conductance remained relatively stable but then decreased progressively after day

60 after full bloom, according to Dichio et al. (2003). In the late season, hydraulic conductance (measured using a pressure bomb) was very low (0.05 -0.1 cm^3 MPa^{-1} s^{-1} × 10^{-3}) (Fig. 8). This result is in conflict with the data emerging from our dye studies (Dichio et al., 2003), which show a much earlier cessation of flow. This discrepancy will be addressed more thoroughly in a separate study and may have to do with the rather high axial pressure gradient applied to the xylem vessels resulting from the application of a bomb pressure of 1.3 MPa (this high bomb pressure is a requirement of our method in order to obtain measurable volumes of exuded sap). The problem is that the pressure is applied over a relatively short distance of about 70 mm between the fruit flesh and the proximal end of the fruitstalk – this is an unnaturally high gradient for the xylem that would normally experience axial gradients some 3-orders of magnitude less. An alternative explanation of the discrepancy (Bondada et al., 2005) could be a reduction in the available pressure gradients (the driving force of xylem sap flow), but this explanation requires confirmation through further experimentation and analysis. The picture emerging emphasises that physical alterations along the fruit xylem pathway do account for some of the reduction in xylem water inflow to the fruit in the late season and consequently the lowered rates of Ca import.

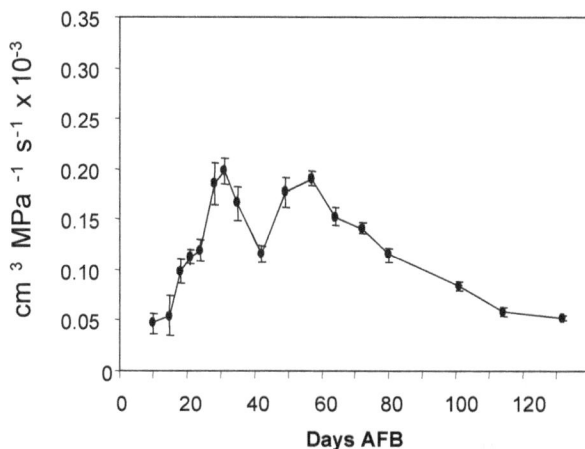

Fig. 8. Hydraulic conductance of kiwifruit fruit (berry + peduncle) measured using a pressure bomb method. Each point is the average of 12 fruit. (Redrawn from Xiloyannis et al., 2008).

3.4 Calculating diurnal fruit transpiration

The diurnal fruit transpiration could be calculated through the integration of the fitting curve of the daily measurements. By plotting the single diurnal values, the seasonal trend of the total fruit transpiration could be estimated. For example, in kiwifruit it has been observed a considerable increase in daily fruit transpiration during the early first 25 days after fruit set when a value of approx. 2 g of water per fruit per day was reached (Fig. 9). Subsequently, fruit transpiration decreased. In fact, at 43 day after fruit set, the reduction in fruit transpiration per day with respect to the peak value observed was 65%, later in the season (93 day) fruit transpiration was almost zero (Fig. 9). The increase in total fruit transpiration in the early part of fruit growth apparently contrasts with the reduction of the rate of fruit water loss (Fig. 5) usually observed in that early stage. The rapid increase of fruit area occurring early of fruit development (Fig. 3) helps to explain such a raise of total amount of water lost by fruit. Measurements of transpiration flux (rate and seasonal trend) of kiwifruit are comparable with findings in stone fruit (detached peach, Li and co-workers 2002) measured in field by weight loss method.

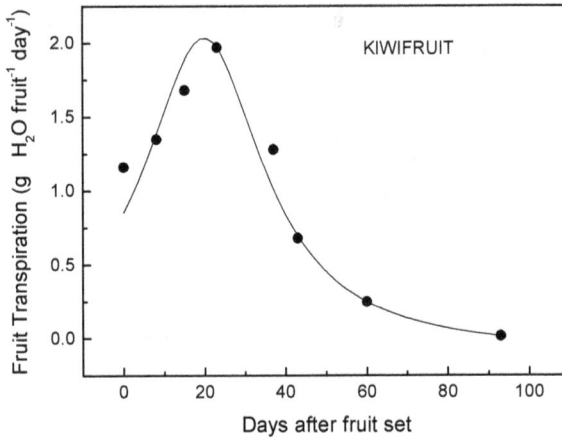

Fig. 9. Seasonal fruit transpiration calculated in kiwifruit (cv Hayward). Each point is the integral of 5 measurements recorded from 7 pm to 6 am.

4. Significance of transpiration on dry matter and minerals accumulation

Fruit transpiration represents the key driver for accumulation of xylem-born minerals, while the growth of fruit depends on phloem stream. However, recently a role of fruit water loss for the phloem unloading into fruit has been proposed. Water losses induce a decrease in fruit water potential (more negative) via an increase in the osmotic concentration and a decrease in the turgor pressure (Morandi et al., 2010b). In this physiological model, as more water is lost by a fruit, more water can be drawn from the phloem and the xylem streams into the fruit itself. However, other regulatory factors, like environmental conditions, related to time of the day, and stem/fruit water potential gradients, may affect fruit inflows. During the central part of the day, the xylem import to the fruit is low, likely due to the high amount of water directed to transpiring leaves, which reach lower water potentials (Morandi et al., 2010b). Xylem inflows cannot balance the high transpiration water losses, with the consequence that fruit shrink (Morandi et al., 2007), increase their concentration and decrease their turgor pressure. This may facilitate translocation and bulk flow phloem unloading into fruit tissues relating fruit transpiration to phloem import, hence transpiration may be viewed as important in determining fruit daily imports of water and dry matter (Morandi et al., 2010b).

Based on this idea, reduction of fruit transpiration would reduce accumulation of dry matter. The Figure 10 shows the effect of reduced transpiration (obtained through bagging which saturates the environment surrounding the fruit leading the relative humidity close to 100%) in apricot and peach fruits.

Optimal mineral nutrition represents a prerequisite for achievement and preservation of high quality plant product. Particular attention should be paid to calcium nutrition due to its involvement in determining tissue mechanical strength and tolerance to biotic and abiotic stresses (Hirschi, 2004). Calcium is a phloem immobile element (Bukovac and Wittwer, 1957), this implies that the amount of Ca reaching a fruit (or a leaf) is almost entirely dependent on supply trough the xylem. Xylem transportation of Ca depends on plant and environmental variables (and their interaction) which often are unfavourable and reduce Ca accumulation in fruit (White and Broadley, 2003).

In kiwifruit, it has been proved the positive effect of windspeed on Ca accumulation (Dichio et al., 2007. That is, for low average windspeeds (0–1.5 m/s), Ca remained almost constant, the level ranging from 18 to 24 mg/fruit. As windspeeds increased beyond this range, the level of mineral rose significantly. Applying a simple cubic model to the data ($y = a + bx^3$) the r^2 value obtained for this rise were 0.72 (Fig. 11). For the highest windspeeds of 3.3 m/s (about 12 km/h) the levels of Ca was almost double that for still air (see Fig. 11). In contrast with Ca, the accumulation of fruit K (mass per fruit) showed no evidence at all of any rise with windspeed. Instead K levels were about 300 mg/fruit (ranging from 212 to 380 mg/fruit) and this level was maintained regardless of windspeed. Applying the same cubic model to the K results gave an r^2 value of only 0.13.

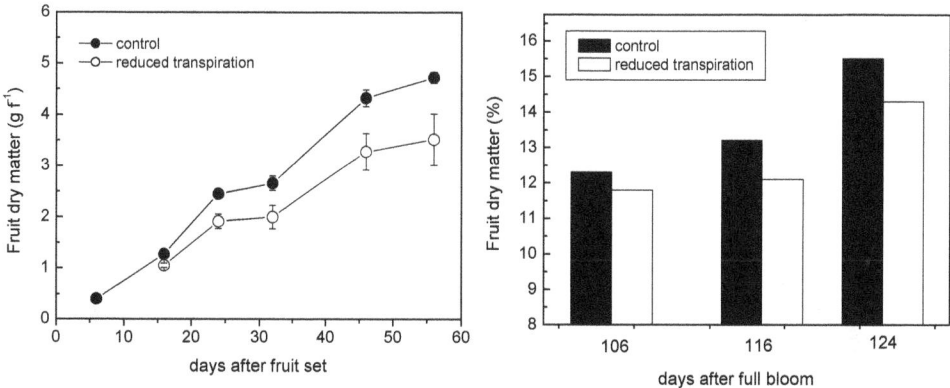

Fig. 10. Dry matter accumulation in fruit of apricot (left) and peach (right) measured in transpiring fruit (control) and under reduced transpiration. (Redrawn from Montanaro et al., 2010 and Morandi et al., 2010b).

Plant water status and transpiration play a key role in supplying the various plant tissue with Ca, particularly those with a low transpiration rate such as fruit (Bangerth, 1979). Mechanisms by which Ca is accumulated into fruit have been searched in a number of fleshy and stone fruit (e.g. apple, tomato, kiwifruit, apricot) (Ferguson and Watkins, 1989; Montanaro et al., 2006 and 2010; Liebisch et al., 2009) with fewer attempts to separate transpirational mechanisms apart from others (e.g. hormonal, metabolic). Recently, a study was undertaken to (i) investigate the relationship between accumulation of Ca and transpiration in developing apricot fruit, and (ii) determine the prominence of transpirational flux upon the whole transpiration-independent transportation processes on Ca nutrition (Montanaro et al., 2010). The hypothesis behind this work was that if the fruit transpiration is the determining factor of the accumulation of Ca (phloem-immobile element) then the import of Ca would be suppressed by restriction of fruit water loss, while the import of phloem-mobile nutrients (i.e. K and Mg) would not be. To test this hypothesis, the seasonal changes of fruit transpiration and Ca, K and Mg concentrations and accumulation were assessed in fruits left to naturally transpire (control) or under restricted transpiration. Restriction of transpiration was obtained through bag application (Photo 3).

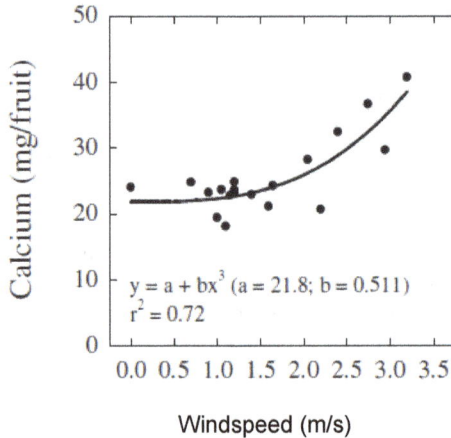

Fig. 11. Whole fruit mineral content for calcium as a function of the mean windspeed to which individual fruit were exposed throughout the growing season. Each point represents a single fruit. (Redrawn from Dichio et al., 2007).

Photo 3. A view of zip-lock bags just mounted on fruit within the canopy of apricot trees (left). After a couple of hours in the microenvironment surrounding the fruit the water vapour condensed (right), hence the relative humidity assumed at saturation point.

Bagging treatments have been widely used to search for fruit response to certain growth condition in terms of skin colour, mineral composition, size, maturity, etc. (Hofman et al., 1997; Amarante et al., 2002). In this study, bags were installed to keep the relative humidity of the air surrounding fruit close to saturation point in order to test whether the limitation of water loss by fruit suppresses the import of some nutrients.

In non-transpiring fruits Ca concentration was significantly affected by restriction of transpiration (it was approximately 30% of that of control fruit), while phloem-mobile elements (K, Mg) were not (Fig. 12).

The evidence that restriction of transpiration did not exert any significant effect on K and Mg concentrations, suggests that the lowest amount of Mg and K accumulated in non-transpiring (bagged) fruit is attributable to the 25% reduction of DM import rather than to the reduced concentration (Fig. 12). At fruit scale, preservation of saturating air humidity significantly lowered the import of Ca due to reduced concentration and reduced DM import according to findings in apple (Tromp and Van Vuure, 1993).

In this study, although plants grew on a Ca-rich soil, the non-transpiring fruits were Ca-deficient compare to control fruits (Fig. 12) and showed the typical visual symptoms of chlorosis. Development of punctual signs of Ca-related disorders have for a long time been associated with mechanisms involved in the movement of Ca within plant parts (Simon, 1977) can conceivably to be explained by differences in transpiration.

The evidence that even though transpiration was negligible, some calcium entered the fruit supports the idea that transpiration is not the only factor governing the movement of Ca through its strong function as driving force of the xylem stream (Bangerth, 1979). Partitioning of Ca amidst plant organs is imputable also to metabolic demand and chemical aspects of the conductive tissues (e.g. adsorption and desorption processes occurring at exchange sites along the walls of the xylem pathway) (McLaughlin and Wimmer, 1999). Calcium delivery to fruit has also been associated to hormonal activities. For example, a mutual relationship between polar basipetal auxin transport and acropetal Ca transport has been reported for tomato, apple and avocado (Stahly and Benson, 1970; Bangerth, 1976; Banuelos et al. 1987; Cutting and Bower, 1989). However, there is still limited information to discriminate (and quantify) the effect of these transpiration-independent stimuli in Ca nutrition. Interpretation of the ratio between concentrations detected in non-transpiring and control fruit would be valuable for assessing of the prominence of transpiration on Ca nutrition.

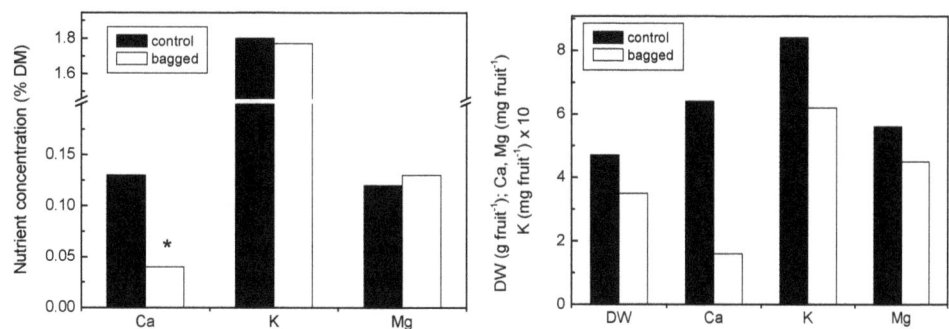

Fig. 12. Concentrations of calcium (Ca), potassium (K) and magnesium (Mg) (left) and amounts per fruit of dry matter (DW), Ca, K and Mg measured at harvest in naturally transpiring apricot fruit (control) and non-transpiring (bagged) fruit. (Redrawn from Montanaro et al., 2010). (For the left figure, * indicates a significant difference at $p < 0.05$, Student's t-test between treatments).

In the case of Ca, that ratio was 0.45 about 50 days after bags were positioned on a whole-bagged period basis. Hence, the fruit transpiration accounted for 55% of the Ca concentration gained by fruit. By contrast, that ratio for K and Mg was not severely affected by the restriction of transpiration, so that the ratio was on average 0.97 and 0.91 for K and Mg respectively, indicating a weak effect of restricted transpiration. For K and Mg the transpiration-independent factors operated as well, however their high phloem mobility does not allow for their unmasking.

Calcium movement in plants has been extensively studied in relation to factors such as transpiration flux, hormonal activities, nutrient demand and chemical properties of the conductive system (Bangerth, 1979; Banuelos et al., 1987; Cutting and Bower, 1989; McLaughlin and Wimmer, 1999), however information is still limited to discriminate the relative prominence of each individual factor. Our data were insufficient to indicate the transpiration-independent mechanism(s) by which 45% of the Ca concentration accrued under high humidity condition, but the hormones were possibly involved. For example, in the case of tomato fruit, the polar basipetal auxin transport has been indicted to promote about 15% of Ca concentration in fruit kept at saturating RH (Banuelos et al., 1987). In addition, in kiwifruit a correlation between Ca accumulation and auxin has been reported (Fig. 13).

Calcium accumulation in fruit could be affected also by other factors such as the position of fruiting sites within the canopy. For example, apple fruit in the upper parts of the canopy tend to have lower Ca concentrations than those in the lower parts, the age (and others features) of leaves of fruiting branches may also affect Ca accumulation in fruit (Volz et al., 1996). It as been observed in apple fruit that reducing the leaf:fruit ratios at specific fruiting sites resulted in lower Ca concentrations, with the activity (presumably transpiration) of spur leaves being especially important (Volz et al., 1996; Lang and Volz, 1998).

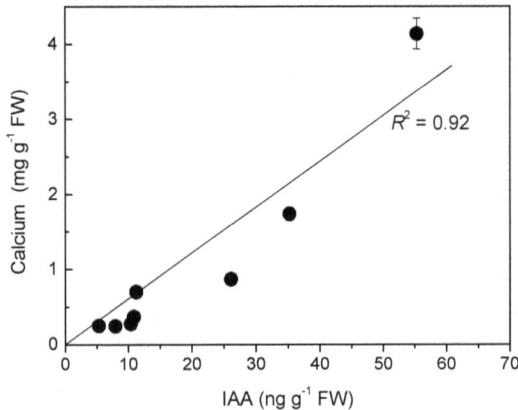

Fig. 13. Correlation between calcium and auxin (IAA) detected in approx. 8-week-old kiwifruit berry (Redrawn from Sorce et al., 2011).

In kiwifruit, relationships exist between fruiting position, leaf:fruit ratios and fruit mineral concentrations. Thorp et al. (2003) demonstrated that a relatively small shift in leaf:fruit ratio was sufficient to affect fruit Ca concentrations, but only in regions of the vine with relatively low fruit mineral levels and low leaf areas. In the same study, it has been highlighted the role of cane size on that relationship. Large diameter canes have axillary shoots with relatively high leaf areas compared with shoots on small diameter canes, especially at basal positions on the cane (Thorp et al., 2003).

Mechanisms behind the effect of leaves on Ca accumulation in fruit need to be clarify. It should be hypothesised that Ca accumulation in a plant organ is proportional to the sap flow unloaded by the organ and to the relative Ca concentration in sap:

$$Ca = [Ca] \times SAP_{Flow}$$

Hence, transpiration of leaves may helps to drive more sap which carries nutrients. To explain the effect of fruit position and shoot type on Ca accumulation, it could be evoked that the relative rate of Ca use along the transport pathway may affect [Ca]. Hence, fruits located at the tip of a shoot tend to receive a Ca-poor sap compare to fruits at the base.

5. Acknowledgments

This work was funded by Ministero dell'Istruzione, dell'Università e della Ricerca - Prin2009 grant.

6. References

Amarante, C., Banks, N. H., Max, S., 2002. Effect of preharvest bagging on fruit quality and postharvest physiology of pears (Pyrus communis). N. Z. Crop Hort. Sci. 30, 99-107.

Bangerth, F. 1976. A role for auxin and auxin transport inhibitors on the Ca content of artificially induced parthenocarpic fruits. Physiol. Plant. 37, 191-194.

Bangerth, F. 1979. Calcium-related physiological disorders of plants. Annu. Rev. Phytopathol. 17, 97-122.

Banuelos, G. S., Bangerth, F., Marschner, H. 1987. Relationship between polar basipetal auxin transport and acropetal Ca transport into tomato fruits. Physiol. Plant. 71, 321-327.

Bondada, B.R., Matthews, M.A. and Shackel, K.A., 2005. Functional xylem in the post-veraison grape berry. J. Experim. Bot. 56:2949–2957.

Boselli, M., Di Vaio, C., Pica, B. 1998. Effect of soil moisture and transpiration on mineral content in leaves and berries of Cabernet Souvignon grapevine. J. Plant Nutr. 21(6), 1163-1178.

Bukovac, M.J., Wittwer, S.H. 1957. Absorption and mobility of foliar applied nutrients. Plant Physiol. 32, 428-435.

Celano, G., Minnocci, A., Sebastiani L., D'Auria, M., Xiloyannis, C. (2009): Changes in the structure of the skin of kiwifruit in relation to water loss. J. Hort. Sci. Biotechnol. 84, 41-46.

Choat, B., Gambetta, G.A., Shackel, K.A., Matthews, M.A., 2009. Vascular function in grape berries across development and its relevance to apparent hydraulic isolation. Plant Physiol., 151, 1677–1687.

Cline, J.A., Hanson, E.J., 1992. Relative humidity around apple fruit influences its accumulation of calcium. J. Amer Soc Hort Sci 117(4), 542-546.

Cutting, J. G. M., Bower, J. P. 1989. The relationship between basipetal auxin transport an calcium allocation in vegetative and reproductive flushes in avocado. Sci. Hort. 41, 27-34.

Dichio B., Remoroni D., Lang A., 2007. Calcium accumulation in fruit of kiwifruit grown under different windspeed conditions. Acta Hort., 753:509-514.

Dichio, B., Remorini, D., Lang, A. 2003. Developmental changes in xylem functionality in kiwifruit fruit: implications for fruit calcium accumulation, Acta Hort. 610:191-195.

Dražeta, L., Lang, A., Hall, A.J., Volz, R.K., Jameson, P.E., 2004. Causes and effects of changes in xylem functionality in apple fruit. Ann. Bot. 93, 275-282.

Faust, M., Shear, C.B., Smith, C.B. 1968. Investigation of corking disorders of apples II. Chemical composition of affected tissues. Proc Amer Soc Hort Sci 92:82-88.

Ferguson, I.B., Thorp, T.G., Barnett, A.M., Boyd, L.M., Triggs, C.M. 2003. Inorganic nutrient concentrations and physiological pitting in 'Hayward' kiwifruit. J Hort Sci Biotech 78:497-504.

Fiscus, E.L., Klute, A., Kaufmann, M.R. 1983. An interpretation of some whole plant water transport phenomena. Plant Physiol. 71: 810-817.

Goudriaan, J., van Laar, H. H. 1994. Modelling Potential Crop Growth Processes. Kluwer Academic Publ., Dordrecht – Boston – London, p. 238.

Hirschi, K. D. 2004. The calcium conundrum. Both versatile nutrient and specific signal. Plant Physiol. 136, 2438–2442.

Hofman, P. J., Smith, L. G., Joyce, D. C., Johnson, G. I., Meiburg, G. F. 1997. Bagging of mango (Mangifera indica cv. "Keitt") fruit influences fruit quality and mineral composition. Postharvest Biol. Technol. 12, 83-91.

Jarvis, PPG., 1985. Coupling of transpiration to the atmosphere in horticultural crops: The Omega factor. Acta Hort. 171, 187-206

Keller, M., Smith, J.P., Bondada, B.R., 2006. Ripening grape berries remain hydraulically connected to the shoot. J. Experim. Bot. 57(11), 2577–2587.

Lang, A., Volz, R.K., 1998. Spur leaves increase calcium in young apples by promoting xylem inflow and outflow. J. Amer. Soc. Hort. Sci. 123, 956-960.

Lang, A. 1990. Xylem, phloem and transpiration flows in developing apple fruits. J. Exp. Bot. 41, 645-651.

Li, S. H., Génard, M., Bussi, C., Lescourret, F., Laurent, R., Basset, J., Habib, R. 2002. Preliminary studies on transpiration of peaches and nectarines. Eur. J. Hort. Sci. 67, 39-43.

Liebisch, F., Max, J. F. J., Heine, G., Horst W. G. 2009. Blossom-end rot and fruit cracking of tomato grown in net-covered greenhouses in Central Thailand can partly be corrected by calcium and boron sprays. J. Plant Nutr. Soil Sci. 172, 140 – 150.

McLaughlin, S. B., Wimmer, R. 1999. Ca physiology and terrestrial ecosystem processes. New Phytol. 142, 373-417.

Montanaro, G., Dichio, B., Xiloyannis, C., 2010. Significance of fruit transpiration on calcium nutrition in developing apricot fruit. Journal of Plant Nutrition Soil Science, 173: 618–622. doi: 10.1002/jpln.200900376

Montanaro, G., Dichio, B., Xiloyannis, C., Celano, G. 2006. Light influences transpiration and calcium accumulation in fruit of kiwifruit plants (Actinidia deliciosa var. deliciosa). Plant Sci. 170, 520-527.

Montanaro, G., Dichio, B., Xiloyannis, C., Lang, A., in press. Preliminary evaluation of the transpiration response of young actinidia fruit to the weather. Proceedings of the VII International Symposium on Kiwifruit, ISHS. 12-17 September 2010, Faenza – Italy.

Morandi, B, Rieger, M.W., Corelli Grappadelli L., 2007. Vascular flows and transpiration affect peach (Prunus persica Batsch) fruit daily growth. J Exp Bot 58: 3941-7.

Morandi, B., , Manfrini, L., Losciale, P., Zibordi, M., Corelli-Grappadelli, L., 2010a. Changes in vascular and transpiration flows affect the seasonal and daily growth of kiwifruit (Actinidia deliciosa) berry. Ann. Bot., 105, 913–923.

Morandi, B., Manfrini, L., Losciale, P., Zibordi, M., Corelli-Grappadelli L. 2010b. The positive effect of skin transpiration in peach fruit growth. J. Plant Physiol., 167:1033-1037.

Nobel, P.S. 2005. Physiochemical and environmental plant physiology - 3rd ed. Burlington: Elsevier Academic Press.

Simon, E. W. 1977. The symptoms of calcium deficiency in plants. New Phytol. 80, 1-15.

Smith, K.U., Clages, T.G.A., Green, E.F., Walton, Changes in abscissic acid concentration, surface conductance, and water content of developing kiwifruit, Sci. Hortic. 61 (1995) 13-27.

Sorce, C., Lombardi, L., Remorini, D., Montanaro, G. 2011. Occurrence of natural auxin and accumulation of calcium during early fruit development in kiwifruit. Australian Journal of Crop Science, 5(7): 895-898.

Stahly, E. A., Benson, N. R. 1970. Calcium levels of "Golden Delicious" apples sprayed with 2,3,5-Triidrobenzoic acid. J. Am. Hort. Sci. 95, 726-727.

Thorp, T.G., Ferguson, I.B., Boyd, L.M., Barnett, A.M., 2003. Fruiting position, mineral concentration and incidence of physiological pitting in "Hayward" kiwifruit. J. Hort. Sci. Biotechnol., 78, 505-511.

Tromp, J., Van Vuure, J. 1993. Accumulation of calcium, potassium and magnesium in apple fruits under various conditions of humidity. Physiol. Plant. 89, 149-156.

Tzoutzoukou, C. G., Bouranis, D. L. 1997. Effect of preharvest application of calcium on the postharvest physiology of apricot. J. Plant Nutr. 20, 295-309.

Volz, R. K., Tustin, D. S. and Ferguson, I. B. 1996. Mineral accumulation in apple fruit as affected by spur leaves. Scientia Horticulturae, 65, 151-161.

White, J. 1986. Ontogeny and morphology of ovarian and fruit hairs in kiwifruit. New Zeal J Bot 24:403-414.

White, P., Broadley, M. R. 2003. Calcium in plants. Ann. Bot. 92, 487-511.

Xiloyannis C., Dichio B., Montanaro G., Lang A., Celano G., Mazzeo M., 2008. Fruit morphological and physiological traits influence calcium transport and accumulation in kiwifruit. Acta Hort., 767:369-378

Xiloyannis, C., Celano, G., Montanaro, G. Dichio, B., Sebastiani, L., Minnocci, A. 2001. Water relations, calcium and potassium concentration in fruits and leaves during annual growth in mature kiwifruit plants. Acta Hort. 564: 129-134.

10

Selenium Metabolism in Plants:
Molecular Approaches

Özgür Çakır[1], Neslihan Turgut-Kara[1]* and Şule Arı[1,2]
[1]Department of Molecular Biology and Genetics, Faculty of Science
Istanbul University, Istanbul
[2]Research and Application Center for Biotechnology and
Genetic Engineering, Istanbul
Turkey

1. Introduction

Selenium (Se) is placed in Group VIA of the Periodic Table. Its chemistry is similar to sulfur (S). Practically all small organic selenium compounds are isologues of corresponding sulfur compounds. With a few exceptions, they are also isologues of sulfur amino acids or derivatives thereof. Selenium plays an indispensable role for humans, animals and microorganisms. It is beneficial for the metabolism at lower concentrations, whereas at higher concentrations it becomes toxic. In other words, the range between deficiency and toxicity is very narrow. Short-term consumption of high levels of Se by human and animals may cause nausea, vomiting, and diarrhea, whereas chronic consumption of high concentrations of Se compounds can result in a disease called selenosis (Goldhaber, 2003). Excess selenium in the environment can be the result of either natural geological processes or industrialization.

Selenium acts as a cancer preventative agent when given in pharmacological amounts. Numerous studies have demonstrated the efficacy of methylselenocysteine (MeSeCys) in preventing mammary cancer in mammalian model systems, and importantly, MeSeCys has been shown to be twice as active as Se-methionine (the primary component of Se-yeast supplements) in preventing the development of mammary tumors (Ip & Ganther, 1992; Lu et al., 1996; Ip et al., 2000; Finley and Davis 2001; Medina et al., 2001; McKenzie et al., 2009). This non-protein seleno amino acid is produced in certain plants including members of the *Astragalus, Allium* and *Brassica* genera (Cai et al., 1995; Clark et al., 1996). While the specific mechanism for the anticancer activity of Se has not been fully elucidated, researchers have speculated that the Se could be effect the cell cycle then induce apoptosis in cancer cell lines (Foster et al., 1986; Cai et al., 1995; Andreadou et al., 1996; Ganther & Lawrence, 1997; Combs & Gray, 1998; Ip 1998; Sinha et al., 1999; Lu & Jiang, 2001; Kim et al., 2001; Wang et al., 2002; Ip et al., 2000). There is also evidence that Se may inhibit tumor angiogenesis (Lu & Jiang 2001). The molecular mechanism of cancer prevention by selenium using the genomics approach was studied on the target organs breast, prostate, colon and lung. The results of the microarray analysis indicated that selenium, independent of its form and the target organ, alters several genes in a manner that can account for cancer prevention. Selenium can

---

* Corresponding Author

up regulate genes related to phase II detoxification enzymes, certain selenium-binding proteins and apoptotic genes, while down regulating those related to phase I activating enzymes, stress responsive genes, cytoskeletal and cell adhesion functions and cell proliferation (El-Bayoumy & Sinha, 2005; Goulet et al., 2007). Also, Goulet and her colleagues were demonstrated that an increase in the occupancy of phospho-histone H3 at selected promoters, which suggest that SeMet can influence gene expression by chromatin remodeling in a manner of epigenetic (Goulet et al., 2007).

Plant roots can take up Se from soil as selenate, selenite, or organoselenium compounds. The biosynthesis of most selenium compounds in nature follows the pathways leading to isologous sulfur compounds in plants (Table 1) as well as yeast, bacteria or animals. Roots

Compounds	Species
Selenocysteine	*Vigna radiata*
Selenocystathionine	*Astragalus praleongus*
	Astragalus pectinatus
	Neptunia amplexicaulis
	Morinda reticulate
	Brassica oleracea capitata
	Stanleya pinnata
	Lecythis ollaria
Se-Methylselenocysteine	*Astragalus crotalariae*
	Astragalus bisulcatus
	Astragalus praleongus
	Brassica oleracea capitata
	Brassica oleracea botrytis
	Allium sativum
	Allium cepa
	Allium tricoccum
	Melilotus indica
	Oonopsis condensata
	Phaseolus lunatus
γ-Glutamyl-*Se*-methylselenocysteine	*Astragalus bisulcatus*
	Allium sativum
	Allium cepa
	Phaseolus lunatus
Selenomethionine	*Brassica juncea*
	Brassica oleracea capitata
	Allium tricoccum
	Melilotus india
Se-Methylselenocysteine *Se*-oxide	*Brassica oleracea capitata*
Selenobiotin	*Phycomyces blakesleeanus*
γ-Glutamylselenocystathionine	*Astragalus pectinatus*
γ-Glutamylselenomethionine	*Allium sativum*
3-Butenyl isoselenocyanate	*Stanleya pinnata*
Selenosinigrins	*Armoracia lapathifolia*
	Stanleya pinnata
Selenosugars	*Astragalus racemosus*

Table 1. Low molecular weight selenium-containing compounds in plants (adapted from Birringer et al., 2002)

take up selenate faster than selenite at the same concentration but acquire organoselenium compounds, such as selenocysteine (SeCys) and selenomethionine (SeMet), most avidly (White et al., 2007). Thereafter it is metabolized (via sulfur assimilation pathway) in that selenocysteine, SeMet and other Se analogues of various S metabolites (Ellis and Salt, 2003). The nonspecific incorporation of seleno amino acids into proteins is thought to contribute to Se toxicity (Brown & Shrift, 1981). Plants differ in their ability to metabolize and tolerate Se, and divided into three groups according to Se accumulation capacity: primary accumulators (hyperaccumulators), secondary accumulators, and non-accumulators. One proposed mechanism of Se tolerance in accumulator plants is the specific conversion of potentially toxic seleno amino acids into nonprotein derivatives such as MeSeCys. Some *Allium* and *Brassica* species, when grown in Se enriched medium, can accumulate 0.1–2.8 μmol g^{-1} dry weight MeSeCys or its functional equivalent γ-glutamylmethylselenocysteine (γ-glutamyl-Se-MeSeCys). However, certain specialized Se accumulating plants, such as *Astragalus bisulcatus*, accumulate up to 68 μmol g^{-1} dry weight Se (6000 mg kg^{-1} dry weight), of which 90–95% is MeSeCys in young leaves. The seeds of these plants also accumulate Se as γ-glutamyl-Se-MeSeCys (Pickering et al., 2003). During to incorporation of the active seleno amino acid SeCys into essential selenoproteins some of the key enzymes play important roles as a regulatory manner. Mutation or overexpression analysis were showed that ATP sulphurylase, selenocystein methyltransferase (SMT), APS reductase, serine acetyltransferase, selenocysteine lyase, selenocysteine transferase, cystathionine-γ-synthase, and chloroplast selenocysteine lyase are important enzymes on the way Se tolerance and accumulation (Figure 1).

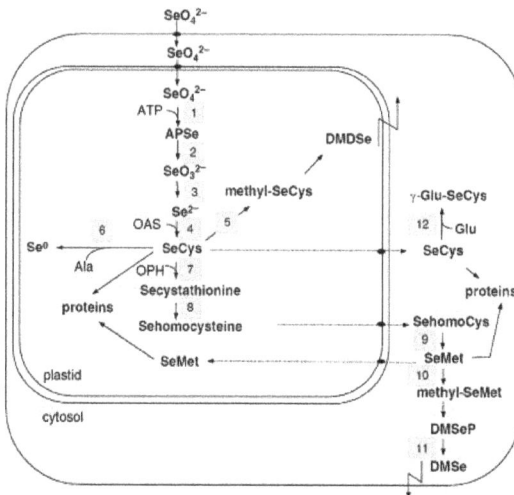

Fig. 1. Schematic overview of Se metabolism in plants. APSe adenosine phospho selenate, OAS O-acetylserine, OPH O-phosphohomoserine, SeCys selenocysteine, SeMet selenomethionine, DMSeP dimethylselenoproprionate, DMSe dimethylselenide, DMDSe dimethyldiselenide. Numbers denote known enzymes. (1) ATP sulfurylase, (2) adenosine phosphosulfate reductase, (3) sulfite reductase (or glutathione), (4) OAS thiol lyase, (5) SeCys methyltransferase, (6) SeCys lyase, (7) cysthathionine-γ-synthase, (8) cysthathionine-β-lyase, (9) methionine synthase, (10) methionine methyltransferase, (11) DMSP lyase, (12) γ-glutamylcysteine synthetase (from Pilon-Smits & Quinn 2010)

SMT is the most important enzyme in Se hyperaccumulating plants. SMT catalyses the methylation of SeCys to MeSeCys, and the gene firstly isolated from hyperaccumulator *A. bisulcatus* (Neuhierl & Bock, 1996; Neuhierl et al., 1999), then isolated some other accumulator and nonaccumulator plant species (Table 2). SMT is constitutively expressed in roots and leaves of *A. bisulcatus*, and does not induced by Se (Pickering et al., 2003). Heterologous expression of AbSMT in transgenic *Arabidopsis thaliana* results in the production of MeSeCys and its derivative γ-glutamyl-Se-MeSeCys, compounds not normally produced in *A. thaliana* (Ellis et al., 2004). Accumulation of MeSeCys was similarly observed in transgenic *Brassica juncea* expressing AbSMT (LeDuc et al., 2004). According to these results, only Se-hyperaccumulating species of *Astragalus* are capable of synthesizing MeSeCys compared with their non-accumulating relatives, and that SMT activity is closely linked with the capacity to hyperaccumulate Se (Sors et al., 2005), it can be hypothesized that Se non-accumulating species do not contain a functional SMT enzyme.

Plants that SMT gene isolated and characterized	Accumulation capacity	Accession number	Reference
Astragalus bisulcatus	Hyperaccumulator	AJ131433.1	Neuhierl et al., 1999
Astragalus chrysochlorus	Secondary accumulator	GQ844862.2	Çakır & Arı (Unpublished data, 2012)
Camellia sinensis	Secondary accumulator	DQ480337.1	Zhu et al., 2008
Brassica oleracea var. italica	Secondary accumulator	AY817737.1	Lyi et al., 2005
Plants that SMT gene isolated			
Astragalus racemosus	Accumulator	GQ398501.1	
Astragalus pectinatus	Accumulator	GQ398502.1	
Astragalus ceramicus	Nonaccumulator	GQ398503.1	Sors et al., 2009
Astragalus drummondii SMT-like	Nonaccumulator	GQ398504.1	
Astragalus leptocarpus	Nonaccumulator	GQ398505.1	

Table 2. Plants that SMT gene isolated and characterized so far

In general, accumulator species likely do not have any Se-specific pathways but take up and metabolize Se and S indiscriminately, also current knowledge demonstrates that Se essentiality in higher plants are still not definitive. The potential health benefits of some Se compounds when combined with the increased application of phytoremediation techniques in contaminated soil were augmented the study of Se biochemistry in plants. Thus, Se metabolism in plants has been reviewed by a number of authors (Terry et al., 2000; Briggers et al., 2002; Germ et al., 2007a; White et al., 2007; Pilon & Quinn, 2010; De Filippis 2010). The present Chapter will focus on basic aspects of molecular selenium metabolism in plants and future perspectives of phytoremediation techniques.

2. Selenium metabolism in plants

2.1 Uptake and transport

Selenium could be occurs in following oxidation states: –2 (selenide), 0 (elemental Se), +4 (selenite), and +6 (selenate). Selenate is accumulated in plant cells via the process of active transport (Brown & Shrift, 1982). Unlike selenate, there is no evidence that the uptake of selenite is mediated by membrane transporters. Alternatively, plants can take up organic forms of Se such as selenomethionine (SeMet) actively, but not effectively (Abrams et al., 1990) (Figure 2). Selenate directly competes with sulfate for uptake by plants. It has been proposed that both anions are taken up via a sulfate transporter in the root plasma membrane. Selenate uptake in other organisms, including *Escherichia coli* (Lindblow-Kull et al., 1985) and yeast (Cherest et al., 1997), is also mediated by a sulfate transporter.

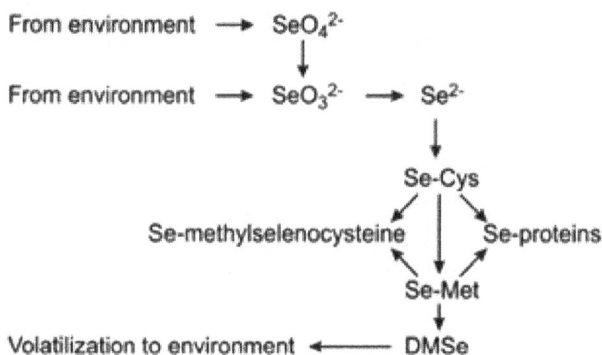

Fig. 2. Schematic representation of the main steps of Se metabolism in plants (from Germ et al., 2007b)

The expression of the sulfate transporter genes is regulated by the S status of the plant, as well as by the regulators, glutathione (GSH) and O-acetylserine. While high levels of sulfate and GSH decrease transcription, high levels of O-acetylserine increase transcription of the high-affinity transporter genes as well as sulfate uptake. Thus, increasing O-acetylserine levels can potentially increase selenate uptake (Davidian et al., 2000). Terry and colleagues reported that application of O-acetylserine increased selenate accumulation in Indian mustard almost two-fold compared to untreated plants, and they speculate that O-acetylserine, a precursor of cysteine (Cys) and a product of the nitrate assimilation pathway, might pivotal importance as a coregulator of the S and nitrogen metabolic pathways. Overexpression studies on the sulfate transporter genes increased selenate accumulation up to two-fold in transgenic plants compared to wild type. These information show that sulfate transporter is involved in selenate uptake (Terry et al., 2000).

Translocation of the Selenium in the plant parts depend on the form of how it is supplied. Zayed et al (1998) showed that the shoot/root ratio of the Se concentrations ranged from 1.4 to 17.2 when selenate was supplied but was only 0.6 to 1 for plants supplied with SeMet and less than 0.5 for plants supplied with selenite. Time-dependent kinetics of Se uptake by Indian mustard showed that only 10% of the selenite taken up was transported from root to shoot, whereas selenate was rapidly transported into shoots (De Souza et al., 1998). Thus, plants transport and accumulate substantial amounts of selenate in leaves but much less selenite or SeMet. Selenite is rapidly converted to organic forms of Se such as SeMet which are retained in the roots (Zayed et al., 1998), this helps to explain why selenite is poorly translocated to shoots. In addition, partitioning of Se in various plant parts is species specific, also depends on the stage of development, and on physiological condition of the plant. In the accumulators, Se is gathered in young leaves during the early vegetative and reproductive stage of growth, and 3,5 fold high levels of Se are found in seeds while the Se content in leaves is drastically reduced. Non accumulating cereal crop plants, often show about the same Se content in grain and in roots, but smaller amounts in the stems and leaves. Distribution of Se in plants also depends on the form and concentration of Se supplied to the roots and on the nature and concentration of other substances, especially sulfates, accompanying the Se (Zayed et al., 1998). Plants can also absorb volatile Se from the atmosphere via the leaf surface. The Se absorbed by the leaves is accumulated in roots as inorganic selenite, selenoglutathione (SeGSH), SeMet, and protein-bound SeMet..

2.2 Accumulation in plants

Hyperaccumulation is the ability of certain plants to accumulate extraordinarily high concentrations of metals and trace elements, even when grown in soil with low concentrations (Baker & Brooks, 1989). This ability for certain elements gives some selective advantage to the hyperaccumulators plants. A selective benefit of hyperaccumulation is predominant occurrence of this kind of plant species on soils that are enriched in the elements. Hypothesis for the ecological significance of hyperaccumulation include drought tolerance, allelopathy, and chemical defense against herbivores and/or pathogens (Boyd & Martens, 1993; Jhee et al., 1999, Galeas et al., 2007). Some plant species are known to hyperaccumulate more than one metals or trace elements. At least 400 plant species in 45 plant families are hyperaccumulators, and these have been found in many different geographic locations (Reeves & Baker, 2000). Hyperaccumulation of Se has been observed in the plant families *Asteraceae, Brassicaceae, Chenopodiaceae, Lecythidaceae, Fabaceae, Rubiaceae* and *Scrophulariaceae*, are only found on seleniferous soils (Beath et al., 1934; Cannon, 1960; Reeves & Baker, 2000). The Se accumulator species can tolerate Se in the field up to 10.000 mg kg⁻¹ DW (Pilon-Smits & Quinn, 2010).

Plants differ in their ability to accumulate Se when they grow on seleniferous soils, and divided into three groups according to Se accumulation capacity: primary accumulators (hyperaccumulators), secondary accumulators, and non-accumulators. Accumulator plants can accumulate from hundreds to several thousand milligrams of Se kg⁻¹ dry weight in their tissues, without any negative effects. That ability is mainly due to the reduction of the intracellular Se concentration of Se–Cys and Se–Met which are normally incorporated into proteins (Pilon-Smits & Quinn, 2010).

Special attention should be paid to *A. bisulcatus*, since this is the best-characterized Se accumulator. This species grows on naturally Se-rich soils in the southwestern part of the USA. Typical for these plants is their strong Se (sweet) odor. In its natural habitat, that species can take up to 0.65% Se dry weight in their shoots (Dumont et al., 2006). When the plants are grown on a selenate rich soil, the older leaves contain mainly inorganic Se (91%), whereas in the young leaves 90–95% of the Se is organic. The roots show the lowest Se level when compared with others tissues. Although, at root level, the Se is mainly organic (92%). There is a presumption that the MeSeCys in the young leaves is metabolized and that the Se is reoxidized to form selenate as the leaves become older. Another explanation would be that the MeSeCys is exported from the young shoots as it ages and accumulates in the even younger shoots. An alternative explanation is the metabolization of MeSeCys to DMSe, which would explain the malodorous nature of the plant used for protection against insect attack. In these plants, the main compound found is MeSeCys, one of the common species found in Se accumulators (Dumont et al., 2006; Pickering et al., 2003).

On the nonaccumulators side, most forage and crop plants, as well as grasses, contain less than 25 mg Se kg^{-1} dry weight and do not accumulate Se much (Brown & Shrift, 1982). Although Se accumulators grow on seleniferous soils, not all plant species on seleniferous soils are Se accumulators: some plants accumulate only a few milligrams of Se kg^{-1} dry weight. For example, the genus *Astragalus* contains both Se accumulating species and nonaccumulating species, and they can grow next to each other on the same soil (Duckart et al., 1992).

Primary accumulators have discrimination coefficients (DC$_i$= [Se/S]$_{plant}$/[Se/S]$_{solution}$) of more than one in solution culture, and have concentrations of Se in the range of thousands of mg per kg dry weight (Ellis & Salt, 2003). Primary accumulators include various *Astragalus* species, which are members of the Fabaceae, as well as *Stanleya pinnata*, a member of the *Brassicaceae* (Feist & Parker, 2001). Secondary accumulators take up Se in proportion to the amount of Se available in the soil, they have a DC$_i$ of less than one, and tissue concentrations of Se in the hundreds of mg kg^{-1} (Bell et al., 1992). Members of this group include species of *Astragalus, Aster, Atriplex* and *Melilotus*, as well as *Brassica juncea* (Indian mustard) (Banuelos & Meek, 1990; Guo & Wu, 1998; Ellis & Salt, 2003). Recently, Ari and her colleagues has identified an *Astragalus* species, *A. chrysochlorus*, as a new secondary Se-accumulator plant with a typical Se concentration of more than several hundred milligrams of Se kg^{-1} dry weight in tissues (DC$_i$=0.95) when grown on tissue culture media containing sodium selenate (Ari et al., 2010).

Selenium hyperaccumulations may increase the surrounding soil Se concentrations (phytoenrichment). The enhanced soil Se contents around hyperaccumulators can impair the growth of Se-sensitive plant species, pointing to a possible role of Se hyperaccumulation in elemental allelopathy (El Mehdawi et al., 2011). Selenium also may increase the tolerance of the plants to drought-induced oxidative damage and high temparature stress by enhancing their antioxidant defense and methylglyoxal detoxification system (Hasanuzzaman & Fujita, 2011; Djanaguiraman et al., 2010).

2.3 Incorporation of Se into protein

In most of the selenoproteins discovered so far, selenium is present as a selenocysteine residue that is integrated into the main chain of amino acids, as was first demonstrated for formate dehydrogenase and glutathione peroxidase (Birrigger et al., 2002). Whenever investigated, the selenocysteine residue was shown to be of pivotal importance for the catalytic efficiency of such proteins. The incorporation of selenocysteine into these selenoproteins is directed by a specific tRNA that recognizes a UGA codon. Normally, the UGA codon acts to terminate translation. In combination with a selenocysteine insertion sequence (SECIS), however, the

UGA codon is recognized by the selenocysteine tRNA, which directs the insertion of selenocysteine (Gladyshev & Kryukov, 1999; Low and Berry, 1996; Ellis & Salt, 2003).

Organisms that require Se for normal cellular function contain essential selenoproteins, such as formate dehydrogenase, glutathione peroxidase, and selenophosphate synthase. Reports have suggested the presence of selenoproteins in plants, but there is no direct evidence for the specific incorporation of selenocysteine in vascular plants. However, plants are thought to assimilate SeCys, where SeCys is metabolized to SeMet, which are both nonspecifically incorporated into proteins. SeCys is formed by the action of Cys synthase, which couples selenide with O-acetylserine (Ng & Anderson, 1978) (Figure 3). GS-Se⁻ may be the physiological substrate of Cys synthase rather than free Se^{2-} (Tsang & Schiff, 1978). Kinetic studies of *in vitro* enzymes were showed that cystathionine- synthase exhibited a preference for SeCys: It had a higher affinity for SeCys (K_m=70 µM) than for Cys (K_m=240 µM) (Dawson & Anderson, 1988). Cystathionine-lyase did not differentiate between the Se and S forms of cystathionine, since the enzyme had a similar affinity for cystathionine (K_m=0.31 mM) and selenocystathionine (K_m=0.35 mM) (McCluskey et al., 1986). The most likely enzyme for the synthesis of SeMet from SeHomoCys is the cytosolic enzyme, Met synthase (Figure 3). Selenium is readily incorporated into proteins in nonaccumulator plants treated with Se (Brown & Shrift, 1982). The incorporation into proteins occurs through the nonspecific substitution of SeCys and SeMet in place of Cys and Met, respectively (Figure 3). Studies showed that both Met and SeMet are substrates for the methionyl t-RNA synthetase (Terry et al., 2000).

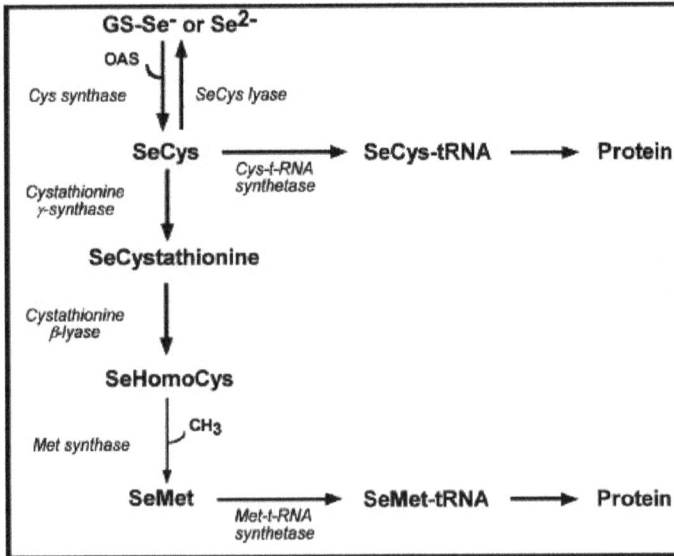

Fig. 3. Schematic pathway of the incorporation of selenide into SeCys, SeMet, and proteins. SeCys lyase is a Se-specific enzyme, whereas all the other enzymes shown recognize both S and Se. The only exception is Met synthase, which is involved in Met synthesis, and is very likely to be involved in SeMet synthesis although there is no evidence supporting this (from Terry et al., 2000).

3. Health benefits of Selenium

Severe selenium deficiency of human and animals have been observed in isolated selenium-poor areas. Although Se deficiency is rare in the US, it does occur in several parts of the world, such as China, where concentrations of Se in the soil are low. Consumption of food containing less than 0.1 mg Se kg^{-1} results in deficiency. Regular consumption of food containing more than 1 mg Se kg^{-1} results in only toxicity, but 1000 mg Se kg^{-1} DW can lead to acute Se poisoning and death for humans and animals (Wilber, 1980).

In the 1960s, selenium was proposed to be an essential trace element as a consequence of human and animal studies (Birringer et al., 2002). Since that time, scientists have showed growing interest on Se studies. The toxic effects of excess Se have been known for some time but, in the past decade, it has become more evident that Se has many potential health benefits beyond meeting basic nutritional requirements. In the seventies, Chinese scientists reported that severe selenium deficiency causes diseases in humans: Keshan disease, which is a fatal cardiomyopathy, and Kashin–Beck disease, a disabling chondronecrosis (Birringer et al., 2002). In addition, Se deficiency can lead to heart disease, hypothyroidism and a weakened immune system (Combs, 1980). Concerns about the health hazards from overexposure now tend to become overwhelmed by a bewildering discussion of the benefits. Adequate alimentary selenium supply is claimed to delay the onset of ageing, cardiovascular diseases and cancer, to enable an optimum immune response, to guarantee an appropriate function of the endocrine system, and to be indispensable for male reproduction. For example, in a long-term double-blind studies, supplemental Se was associated with significant reductions in lung, colorectal and prostate cancers (Ip & Ganther, 1992). In 1996, Clark and co-workers reported that supplementation of people with selenized yeast is capable of reducing the overall cancer morbidity by nearly 50%. The possible anti-cancer effect of Se might be summarize according to critical reviews on this area that appear to be widely accepted (Birringer et al., 2002): optimize somehow glutathione peroxidases (GPx) activities; provides optimum selenoprotein expression; the nature of the Se compound has critical importance; although, synthetic selenium compounds do not support selenoprotein synthesis, but also found to be anticarcinogenic. "Need to be proved" advances have led to several mechanisms being proposed for the anticancer activity of Se: antioxidant protection (via selenoproteins); altered carcinogen metabolism; enhanced immune surveillance; regulation of cell proliferation and tumor cell invasion and inhibition of neoangiogenesis (Zeng & Combs, 2008).

4. Molecular approaches to alter Se metabolism in plants

Several different transgenics have been obtained so far. They were showed enhanced Se tolerance, accumulation, and assimilation from inorganic to organic Se, and volatilization. Selenium accumulation was up to nine-fold higher than wild type and volatilization up to three-fold faster, under laboratory conditions. These findings may be useful for cleaning up of excess levels of Se in the environment and also as fortified foods to prevent Se deficiency related diseases. For example, accumulators of MeSeCys would be especially useful for the anticarcinogenic purpose (Unni et al. 2005). In a first step to assess the transgenics' potential for phytoremediation or as Se-fortified food, they were tested for their capacity to accumulate Se from naturally seleniferous soil and from Se-contaminated sediment.

The genetic engineering strategcy for biofortification and phytoremediation are both the same: higher Se levels in harvestable plant parts are purpose. A significant difference between genetic engineering for biofortification and phytoremediation objectives that Se in tissues of renewable plants should not get through toxic concentrations. In biofortification case, some Se compounds have more powerful anti-carcinogenic properties than the others. For example, MeSeCys is the best form of Se to use in biofortified foods, and according to that reason overexpression of SMT may be the best purpose of biofortification. It may also be possible to overexpress some targeted gene(s) in specific plant tissues, such as in the grain, or to overexpress these targeted genes so that anticarcinogenic Se compounds can be readily extracted for production.

4.1 Phytoremediation and biofortification

Selenium accumulator plants can convert inorganic selenate and selenite to SeCys and other organic selenocompounds, including volatile forms. Se hyperaccumulators may have special metabolic pathways for methylation of SeCys and the conversion of methyl-SeCys to volatile DMDSe. Transgenic approaches have been used to further enhance plant Se accumulation, tolerance, and volatilization (Table 3).

Selenate is translocated without chemical modification through the xylem to the leaves after its root absorption via the sulfate transporter (De Souza et al. 1998, Zayed et al. 1998). Afterwards, selenate is metabolized by the enzymes responsible of sulfate assimilation when it enters chloroplasts. ATP sulfurylase catalyzes the key step in the reduction of selenate by activating it to adenosine phosphoselenate (APSe), an activated form of selenate. *In vitro* ATP sulfurylase has been shown to activate selenate, as well as sulfate (Burnell 1981; Dilworth & Bandurski, 1977; Shaw& Anderson, 1972). A gene construct containing the *A. thaliana aps1* gene (Leustek et al. 1997), with its own chloroplast transit sequence, fused to the *Cauliflower* Mosaic Virus 35S promoter cloned into indian mustard plants to overexpress ATP sulfurylase. Molecular studies provided *in vivo* evidence that ATP sulfurylase is responsible for selenate reduction, and that this enzyme is rate limiting for selenate reduction and Se accumulation (Pilon-Smits et al. 1999). X-ray absorption spectroscopy (XAS) analysis of wild-type Indian mustard plants supplied with selenate showed that selenate was accumulated in both roots and shoots, but when selenite was supplied, an organo-Se compound (similar to SeMet) accumulated (De Souza et al. 1998). It is concluded that the reduction of selenate was rate limiting to selenate assimilation. This rate-limiting step was overcome in transgenic plants overexpressing ATP sulfurylase because these plants accumulated a SeMet-like compound when supplied with selenate (Pilon-Smits et al. 1999). In another study, transgenic *A. thaliana* overexpressing both ATP sulfurylase and APR (APS reductase) had a significant enhancement of selenate reduction as a proportion of total Se, whereas SAT (serine acetyl transferase) overexpression resulted in only a slight increase in selenate reduction to organic forms. In general, total Se accumulation in shoots was lower in the transgenic plants overexpressing ATPS, PaAPR (*P. aeruginosa* APR), and SAT. Root growth was adversely affected by selenate treatment in both ATPS and SAT overexpressors and less so in the PaAPR transgenic plants. It is concluded that ATPS and APR are major contributors of selenate reduction in planta. However, Se hyperaccumulation in *Astragalus* is not driven by an overall increase in the capacity of these enzymes, but rather by either an increased Se flux through the S assimilatory pathway, generated by the biosynthesis of the sink metabolites MeCys or MeSeCys (Sors et al., 2005).

The dominant adenosine 5'-phosphosulfate reductase (APR2) in *A. thaliana* converts activated sulfate to sulfite, a key reaction in the sulfate reduction pathway. apr2-1 transgenic plants had decreased selenate tolerance and photosynthetic efficiency. Sulfur metabolism was perturbed in apr2-1 plants grown on selenate, as observed by an increase in total sulfur and sulfate, and a 2-fold decrease in glutathione concentration. Knockout of APR2 also increased the accumulation of total selenium and selenate. However, the accumulation of selenite and selenium incorporation in protein was decreased in apr2-1 mutants. Decreased incorporation of selenium in protein is typically associated with increased selenium tolerance in plants. However, because the apr2-1 mutant exhibited decreased tolerance to selenate, Grant et al. (2011) proposed that selenium toxicity can also be caused by selenate's disruption of glutathione biosynthesis leading to enhanced levels of damaging reactive oxygen species.

As described above, selenium can be assimilated and volatilized via the sulfate assimilation pathway. Cystathionine-γ-synthase (CgS) is the enzyme which catalyzes the synthesis of Se-cystathionine from Se-cysteine, the first step in the conversion of Se-cysteine to volatile dimethylselenide. Overexpression of CgS in *B. juncea*, the first enzyme in the conversion of SeCys to SeMet, resulted in two to threefold higher volatilization rates compared to untransformed control plants (Van Huysen et al., 2003). The CgS transgenics accumulated 40% less Se in their tissues than wild type probably as a result of their enhanced volatilization. Probably due to their lower tissue Se levels the CgS transgenics were also more Se tolerant than wildtype plants. Van Huysen et al. (2003) studied APS and CgS transgenics to evaluate for their capacity to accumulate Se from soil that is naturally rich in Se. In that study, wild-type Indian mustard and the Se hyperaccumulator *S. pinnata* were used for comparison. After growing 10 weeks on Se soil, similar to those of *S. pinnata*, the APS transgenics contained 2,5-fold higher shoot Se levels than wild type Indian mustard. The CgS transgenics contained 40% lower shoot Se levels than wild type. These findings were very significant that they are the first report on the performance of transgenic plants on Se in soil and they showed the potential of genetic engineering for phytoremediation.

Selenocysteine lyase (SL) catalyzes the removal of selenium from L-selenocysteine to yield L-alanine. This enzyme is proposed to have a role in the recycling of the micronutrient selenium from degraded selenoproteins which contain selenocysteine residue. Selenocysteine lyase has a strict substrate specificity for L-selenocysteine and no activity for L-cysteine. However, it is unknown how the enzyme distinguishes between selenocysteine and cysteine. To manipulate plant Se metabolism, another genetic engineering approach is the prevention of the toxic process of its nonspecific incorporation into proteins. A mouse SL was expressed in *A. thaliana* and *B. juncea* (Pilon et al. 2003; Garifullina et al. 2003). Selenocysteine lyase enzyme specifically breaks down SeCys into alanine and elemental Se. The SL transgenics showed reduced Se incorporation into proteins. Se tolerance increased when mouse SL was expressed in the cytosol of *A. thaliana*, but decreased when it was expressed in the chloroplast (Pilon et al. 2003). All the transgenic SL plants showed enhanced Se accumulation, up to twofold compared to wildtype plants. Similar results were obtained when an *A. thaliana* homologue of the mouse SL (called CpNifS) was discovered and overexpressed: the CpNifS transgenics showed less Se incorporation in proteins, twofold enhanced Se accumulation, as well as enhanced Se tolerance (Van Hoewyk et al., 2005). This enzyme has been cloned from *A. thaliana* and expression of this gene in *B. juncea* originally appeared to reduce selenate toxicity, and Banuelos et al. (2007) attributed this to a reduction in incorporation of Se into proteins. The gene used in this study may be similar to the *AtCpNifS* chloroplast gene used by Van Hoewyk et al. (2005).

In *Arabidopsis* genome, there are three highly conserved homologues of the mammalian 56-kD selenium-binding protein (SBP). A transgenic approach is used to study the function of SBP in this model plant by constitutively overexpressing and down-regulating the endogenous *Atsbp1* gene. It was employed both a conventional antisense method and gene silencing by intron-containing hairpin RNAs. *Atsbp1*-overexpressing and silenced plants were phenotypically normal, under standard growth conditions, when compared with wild type plants. Transgenic plants exhibited different growth responses to exogenously supplied selenite, which correlated with the expression levels of *Atsbp1*. Plants with increased *Atsbp1* transcript levels showed enhanced tolerance to selenite, while plants with reduced levels were more sensitive. Results indicate that *Atsbp1* appears to be involved in processes controlling tolerance of *Arabidopsis* to selenium toxicity (Agalou et al., 2005). A more distant related family of genes that well studied in *A. thaliana*, induce higher levels of binding polypeptides and proteins. It was recently found by Dutilleul et al. (2008) that expression of specific binding proteins for Se also delivered tolerance to cadmium (Cd), most likely also by binding this heavy metal (Dutilleul et al. 2008).

The *Sultr 123* gene family orchestrate sulphate transporters, and by co-operation may also regulate Se transportation. Lydiate et al. (2007) used 'knock-down' technology in *A. thaliana*, determined that *Sultr 123* genes reduced high affinity sulphate transporters transportation of Se and stated that reduced, but had little effect on selenite transportation (Table 3). The *Sultr* gene family are similar to the *SHST* family of sulphate transporter genes.

Se upregulates transcripts that regulate the synthesis and signaling of ethylene and jasmonic acid. *Arabidopsis* mutants which are defective in ethylene or jasmonate response pathways exhibited reduced tolerance to Se, therefore, it suggests an important role for these hormones in Se tolerance. Selenate upregulated a variety of transcripts that were also induced in stress conditions. Selenate seemed to repress plant development, as suggested by the downregulation of genes involved in cell wall synthesis and auxin-regulated proteins. By discovering the Se-responsive genes plants could be created that can better tolerate and accumulate Se, which may enhance the effectiveness of Se phytoremediation or serve as Se-fortified food (Van Hoewyk et al. 2008).

MeSeCys is produced from selenocysteine and S-methylmethionine by SMT enzme. Neuhierl et al. (1999) were cloned successfully the gene encoding SMT from *A. bisulcatus* (AbSMT). This enzyme belongs to a class of methyltransferases involved in metabolism of S-methylmethionine. It shares significant sequence homology with homocysteine S-methyltransferases (HMT). Despite the fact that both SMT and HMT enzymes catalyze methyl transfer using S-methylmethionine as the methyl donor, they exhibit significant Se-containing (for SMT) and S-containing (for HMT) substrate choice as a methyl acceptor *in vitro* (Neuhierl and Bock, 1996; Ranocha et al., 2000). SMT was found to be constitutively

Transgene	Gene origin (plant species)	Transgenic plant species	Effects on Se tolerance and accumulation	Reference
APS2 isoform of ATP sulphurylase	*A. thaliana*	*N. tabacum*	No significant effects on Se accumulation and Se tolerance	Hatzfeld et al. (1998)
APS1 isoform of ATP sulphurylase	*A. thaliana*	*B. juncea*	Increase in Se accumulation and an increase in Se tolerance	Pilon-Smits et al. (1999)
CgS (crystathionine - γ-synthase)	*A. thaliana*	*B. juncea*	Lower Se levels in shoots and increased Se tolerance	Van Huysen et al. (2003)
SMT (selenocysteine methyltransferase)	*A. bisulcatus*	*A. thaliana*	Increase in foliar Se levels and increase in tolerance to selenite, but not selenate	Ellis et al. (2004)
SMT (selenocysteine methyltransferase)	*A. bisulcatus*	*B. juncea*	Increase in total Se levels and increase in tolerance to selenite, but not selenate	LeDuc et al. (2004)
APS isoform of ATP sulphurylase	*A. thaliana*	*B. juncea*	Increase in Se accumulation and an increase in Se tolerance	Van Huysen et al. (2004)
CgS (Crystathionine - γ-synthase)	*A. thaliana*	*B. juncea*	Lower Se levels in shoots and increased Se tolerance	Van Huysen et al. (2004)
APS1 isoform of ATP sulphurylase	*A. thaliana*	*A. thaliana*	Decreased Se accumulation and Se tolerance	Sors et al. (2005)
PaAPR (APS reductase)	*A. thaliana*	*A. thaliana*	Decrease in foliar Se and increase selenate tolerance	Sors et al. (2005)
SATm (Mitochondria serine acetytransferase)	*T. goesingense*	*A. thaliana*	No significant effects on Se accumulation and tolerance	Sors et al. (2005)
Selenium binding polypeptides/proteins (SBP)	*A. thaliana*	*A. thaliana*	Resistance to Se achieved due to overexpression of Se binding proteins	Agalou et al. (2005)
AtCpNifS chloroplast protein like SeCys lyase	*A. thaliana*	*A. thaliana*	Enhanced selenate tolerance by reducing Se incorporation into protein	Van Hoewyk et al. (2005)
ATP sulfurylase *SMT* (selenocysteine methyltransferase)	*A. thaliana* *Astragalus bisulcatus*	*B. juncea*	Substantial improvement in Se accumulation from selenate (4 to 9 times increase)	Le Duc et al. (2006)

Transgene	Gene orIgIn (plant specIes)	TransgenIc plant specIes	Effects on Se tolerance and accumulatIon	Reference
Selenocysteine lyase (SeCyslyase)	A. thaliana	B. juncea	Higher selenate tolerance probably by reducing Se incorporation into protein	Banuelos et al. (2007)
SMT (selenocysteine methyltransferase)	A. thaliana	B. juncea	Increase in total Se levels and increase in tolerance to selenite, but not selenate	Banuelos et al. (2007)
SULTR 1,2,3 Sulphate proton transporters	A. thaliana	A. thaliana (knock-down gene technology)	Selenate accumulation reduced by HAST transport, little effect on selenite	Lydiate et al. (2007)
AtCpNifS chloroplast protein like SeCys lyase	A. thaliana	A. thaliana	Confirm higher selenate tolerance by reducing Se incorporation into protein	Van Hoewyk et al. (2008)
SBP 1,2,3 Se binding protein gene family	A. thaliana	A. thaliana	Elevated tolerance to heavy metal cadmium (Cd) by Se protein also binding Cd	Dutilleul et al. (2008)
ATPS1 SMT (selenocysteine methyltransferase)	A. thaliana A. bisulcatus	Nicotiana tabacum L. cv. Samsun	SMT can be utilised to increase the metabolism of Se into MeSe-Cys, the effects of ATPS activity vary depending on the species involved	McKenzie et al. (2009)
Adenosine 5'-phosphosulfate reductase	A. thaliana	A. thaliana (knock-down gene technology)	decreased selenate tolerance and photosynthetic efficiency	Grant et al. (2011)

Table 3. Molecular genetic studies on selenium tolerance and accumulation, including the origin of the genes (modified from De Filippis, 2010).

expressed in roots and leaves of A. bisulcatus, and appear to be not affected by Se induction (Pickering et al., 2003). Heterologous expression of AbSMT in transgenic A. thaliana results in the synthesis of MeSeCys and its derivative γ-glutamyl-Se-MeSeCys, these are the compounds not natively produced in A. thaliana (Ellis et al., 2004). In transgenic Brassica juncea expressing AbSMT accumulated MeSeCys similarly (Le Duc et al., 2004). The SMT transgenics showed increased Se accumulation, in the form of methyl-SeCys, as well as increased Se tolerance. Se volatilization rates also enhanced with the expression of SMT, with more volatile Se synthesized in the form of DMDSe. In SMT expressing trangenics, Se tolerance, accumulation, and volatilization drawed the attention when the plants were supplied with selenite as opposed to selenate. In this manner, the conversion of selenate to selenite were thought to be a rate-limiting step for the production of SeCys. APS and SMT transgenics were hybridized to create double-transgenic plants that overexpress both APS and SMT (APSxSMT plants) to deal with this rate-limitation. The APS x SMT double transgenics accumulated up to nine times higher Se levels than wild type (LeDuc et al. 2006). The predominant form of the Se compounds in the double transgenics was methyl-SeCys. The APSxSMT double transgenics accumulated up to eightfold more methyl- SeCys than wild type and almost two fold more than the only SMT transgenics. Se tolerance was similar in the single and double transgenics. On the other hand, McKenzie et al (2009) concluded that while the SMT gene from Se hyperaccumulators can probably be utilised universally to increase the metabolism of Se into MeSeCys, the effects of enhancing ATP sulfurylase activity could vary depending on the species involved.

The APS enzyme seems to be rate-limiting for the assimilation of selenate to organic Se compounds, and CgS enzyme is also rate-limiting for DMSe volatilization. Increased APS expression also appears to induce selenate uptake and Se and S accumulation, probably depending on upregulation of sulfate transporter expression. The results from the SL and CpNifS transgenics indicate that SeCys breakdown can decrease nonspecific incorporation of Se into proteins. This situation enhances Se tolerance because elemental Se does not involve with cellular processes. As mentioned above, in plants CpNifS functions in Se tolerance in nature is unknown; it's most serious function is in synthesis of iron–sulfur clusters (Van Hoewyk et al. 2007). The results from the SMT transgenics show that SMT is a key enzyme for Se hyperaccumulation, offering increased Se tolerance and accumulation when expressed in nonaccumulators. Nevertheless, for Se assimilation and detoxification, APS also needs to be overexpressed with SMT. APS x SMT double transgenics link the ability to reduce selenate to selenite and SeCys with the competence to methylate SeCys and thus to detoxify the internal Se. These studies suggest that through genetic manipulation of high biomass, fast-growing plants, Se phytoremediation and biofortification can be improved into a viable option, while producing crops with better nutritional quality.

4.2 Problems and future aspects

The possible transfer of undesirable traits to elite plants and crop cultivars for agriculture is an obvious concern over phytoremediation techniques, especially in using genetically modified plants (Hanson et al. 1997; Terry et al. 2000). The use of phyto-crops for food or animal consumption may be affected by hyperaccumulation and high levels of some elements, for example Se, into plants' part. However technology exists to identify the fate of most of these toxic compounds, and their toxicity as demonstrated by the development of chemo preventitive enriched Se accumulating (fortified) edible crop plants (e.g. potato, radish and other vegetables) in Australia, UK, USA and other parts of the world (Broadley et al. 2004; Lefsrud et al. 2006; Pedrero et al. 2006; Haug et al. 2007; Zhao et al. 2007).

To clean-up Se from constructed wetlands and their waters is the major environmental problem. An affective solution seems to be to use of 'artificially constructed wetlands'. For wetland efficiency for removal of Se the most suitable plant species should be planted and some species like cattail grass (high biomass) and widgeon grass (high amounts hyperaccummulated) removed the most Se in trials (Banuelos 2006; Nyberg 1991). The world Se resources need to be managed so that this non-renewable vulnerable resource is not squandered. Selenium uptake, mobilisation and assimilation are quite well understood and are similar to sulphur, however there are some steps not well understood, especially enzymatic and non-enzymatic steps about to the reduction of intermediates to selenide.

New genes and proteins will be discovered to improve Se tolerance, accumulation, and volatilization with the arrival of the genomic era. Also, comparative studies of Se hyperaccumulators and related nonaccumulators or of Se-tolerant ecotypes and non-tolerant ecotypes of the same species may reveal new genes that upregulate Se uptake, accumulation, and volatilization. Such new genes may not be involved in the commonly studied sulfur metabolism. For example, a Se-binding protein (SBP) homolog, when overexpressed in *A. thaliana*, increased tolerance to Se as well as cadmium (Agalou et al. 2005). SBP's function is unknown so far, but it has been hypothesized to be similar to glutathione. Moreover, recent genetic and genomic studies (Zhang et al. 2006; Tamaoki et al. 2008; Van Hoewyk et al 2008) have identified new quantitative trait loci (QTL) and genes involved in Se tolerance. The plant hormones jasmonic acid (JA) and ethylene are emerging as important players in plant responses to Se tolerance, possibly via their influence on S and Se assimilation. Further studies may reveal key genes that induce the responses that together provide Se tolerance and accumulation in model plants and hyperaccumulators. These key genes could be the candidates for overexpression, producing the complete Se hyperaccumulation in plants. It is desirable to study the potential ecological implications of growing Se accumulating or volatilizing plants before existing and future transgenics are used at a large scale in the field for phytoremediation or as fortified foods. Additional considerations for the use of transgenics for phytoremediation are the same as those involved with growing transgenics for other purposes and should also be evaluated and weighed against the risks of alternative remediation methods.

Many molecular studies have been reported the overexpression of genes encoding proteins involved in Se uptake, transport and assimilation. In this way further strategies for genetic engineering of Se accumulation, transformation and toxicity will become evident, and the use of transgenic plants for use in a variety of ways could be evaluated. Phytoremediation offers a cost effective and environmentally friendly alternative or complementary technology to conventional bioremediation techniques. However the biological processes of phytoremediation are still largely unknown in many cases, and plant-microbe interactions, mechanisms of degradation and transformation, volatilisation, chelation, binding and detoxification need more detailed investigations. In this point of view there is value in enhancement of traits in plants useful in phytoremediation such as high biomass and growth potential in seleniferous soils, which might otherwise be considered agriculturally non-productive land. Se-hyperaccumulating plants (wether naturally occurring or transgenic plants) have possibilities in that they combine pollutant decontamination with production of a product with beneficial properties to humans and animals.

5. Conclusion

Building on the genomic and biochemical studies described above, follow-up research may reveal key genes that trigger the cascade of responses that provide Se tolerance and accumulation in model plants and hyperaccumulators. Also, genes may be found that encode specific transporters of selenocompounds into and within hyperaccumulators. Such key genes will be the ultimate candidates for overexpression studies, with the potential of transferring the complete Se hyperaccumulator profile into high-biomass species. Recent research has elucidated many important ecological interactions involving Se in plants. In this chapter, it has been focussed some important areas for future research. Particularly, more research is desirable on the role of soil microbes in plant Se uptake and volatilization, and the movement of Se through the food chain via Se hyperaccumulators or Se-fortified crop plants. The role of Se in below-ground ecological interactions with microbes and other organisms is also a fairly unexplored area. In addition to effects of Se on root–microbe interactions, Se may protect plants from root feeding herbivores, and selenocompounds released from hyperaccumulator roots may be toxic to surrounding vegetation. Similarly, the effects of Se on pollination ecology will be an interesting field of further study. Better knowledge of the processes involved in plant metabolism, the limiting factors involved, the contributions of ecological partners and the effects of Se on ecological partners are all useful for minimizing potential harmful effects of Se while benefiting from the positive effects of plant Se on animal and human health.

The capacity of plants to accumulate and volatilize Se will be very useful for the phytoremediation of Se-contaminated soils and waters (Banuelos and Meek 1990). When plant Se accumulation is well managed, this offers an efficient and cost-effective way to remove Se from the environment. Since plants are an effective source of dietary Se, Se-enriched plant material from phytoremediation or other sources can be considered fortified food. After being grown on Se-contaminated soil or being irrigated with Se-contaminated water, the Se-laden plant material may be used as a feed supplement for livestock, or as a biofuel. If successful, the potential of this strategy may be further enhanced by the use of selected transgenic lines. Of course, any use of Se-accumulating wildtype or transgenic plants will need to be accompanied by careful risk assessment, to avoid escape of transgenes and any adverse ecological effects of the accumulated Se.

6. References

Abrams, M. M., Shennan, C., Zazoski, J. & Burau, R. G. (1990). Selenomethionine uptake by wheat seedlings, *Agron J*, Vol, 82, pp. 1127–1130, ISSN 0002-1962

Agalou, A., Roussis, A.& Spaink, H.P. (2005) The *Arabidopsis* selenium-binding protein confers resistance to toxic levels of selenium. *Functional Plant Biology*, Vol. 31, pp.881–890, ISSN 1445-4408

Andreadou, I., Menge, W. M. P. B., Commandeur, J. N. M., Worthington, E. A. & Vermeulen, N. P. E. (1996). Synthesis of novel Se-substituted selenocysteine derivatives as potential kidney selective prodrugs of biologically active selenol compounds: evaluation of kinetics of b-elimination reactions in rat renal cytosol, *J Med Chem*, Vol, 39, pp, 2040–2046, ISSN 0022-2623

Ari, S., Cakir, O. & Turgut-Kara, N. (2010). Selenium tolerance in *Astragalus chrysochlorus*: identification of a cDNA fragment encoding a putative Selenocysteine methyltransferase, *Acta Physiol Plant*, Vol 32, pp, 1085–1092, ISSN 0137-5881

Baker, A.J.M. & Brooks, R.R. (1989). Terrestrial higher plants which hyperaccumulate metal elements: A review of their distribution, ecology, and phytochemistry, *Biorecovery*, Vol, 1, pp, 81-126, ISSN 269-7572

Banuelos, G., LeDuc, D.L., Pilon-Smits, E.A.H., Tagmount, A.& Terry, N. (2007). Transgenic Indian mustard overexpressing selenoctsteine lyase or selenocysteine methyltransferase exibit enhanced potential for selenium phytoremediation under field conditions. *Environmental Science & Technology*, Vol.41, pp.599–605, ISSN 0013-936X

Banuelos, G. S. & Meek, D. W. (1990). Accumulation of selenium in plants grown on selenium-treated soil, *J Environ Qual*, Vol, 19, pp, 772-777

Banuelos, G.S. (2006). Phyto-products may be essential for sustainability and implementation of phytoremediation. *Environmental Pollution*, Vol. 144, pp. 19–23, ISSN 0269-7491

Beath, O. A., Draize, J. H., Eppson, H. F., Gilbert, C. S. & McCreary, O. C. (1934). Certain poisonous plants of wyoming activated by selenium and their association with respect to soil types, *J Am Pharm Assoc*, Vol, 23,pp, 94–97, ISSN 1086-5802

Bell, P.F., Parker, D.R. & Page, A.L. (1992). Contrasting selenate sulfate interactions in selenium-accumulating and nonaccumulating plant species. *Soil Sci Soc A. J*, Vol. 56, pp,1818–24, ISSN 0361-5995

Birringer, M., Pilawa, S.& Flohe, I. (2002). Trends in selenium biochemistry. *Natural Product Reports*, Vol.19, pp. 693–718, ISSN 0265-0568

Boyd, R. S. & Martens, S. N. (1993). The raison d'etre for metal hyperaccumulation by plants, In: *The vegetation of ultramafic (serpentine) soils*. A. J. M. Baker, J. Proctor & R. D., pp. 279–289, Reeves, Intercept, Andover, UK

Broadley, M.R., Bowen, H.C., Cotterill, H.L., Hammond, J.P., Meacham, M.C., Mead, A.& White, P.J. (2004) Phylogenetic variation in the shoot mineral concentration of angiosperms. Journal *of Experimental Botany*, Vol. 55, pp. 321–336, ISSN 0022-0957

Brown, T. A. & Shrift, A. (1981). Exclusion of selenium from proteins in selenium-tolerant *Astragalus* species, *Plant Physiol*, Vol, 67, pp. 1951-1953, ISSN: 0032-0889

Brown, T. A. & Shrift, A. (1982). Selenium: toxicity and tolerance in higher plants, *Biol Rev*, Vol, 57, pp, 59–84, ISSN 1464-7931

Burnell, J.N. (1981). Selenium metabolism in *Neptunia amplexicaulis*. *Plant Physiology*, Vol. 67, pp.316–324, ISSN 0032-0889

Cai, X. J., Block, E., Uden, P. C., Zhang, X., Quimby, B. D. & Sullivan, J. J. (1995). *Allium* chemistry: Identification of selenoamino acids in ordinary and selenium-enriched garlic, onion, and broccoli using gas chromatography with atomic emission detection, *J Agric Food Chem*, Vol, 43, pp, 1754–1757, ISSN 0021-8561

Cannon, H. L. (1960). Botanical prospecting for ore deposits, *Science*, Vol, 132, pp, 591–598, ISSN 0036-8075

Cherest, H., Davidian, J. C., Thomas, D., Benes, V., Ansorge, W., Surdin-Kerjan, Y. (1997). Molecular characterization of two high affinity sulfate transporters in *Saccharomyces cerevisiae*. *Genetics*, Vol, 145, pp, 627–635, ISSN 0016-6731

Clark, L.C., Combs, G.F. Jr, Turnbull, B.W., Slate, E.H., Chalker, D.K., Chow, J., Davis, L.S., Glover, R.A., Graham, G.F., Gross, E.G., Krongrad, A., Lesher, J.L. Jr, Park, H.K., Sanders, B.B. Jr, Smith, C.L. & Taylor, J.R. (1996). Effects of selenium supplementation for cancer prevention in patients with carcinoma of the skin a randomized controlled trial – a randomized controlled trial. *Journal of the American Medical Association*, Vol. 276, pp. 1957–1963, ISSN 0098-7484

Combs, G. F. (1980). The search for the nutritional role of selenium: a success story in poultry nutrition, *Feed Management*, Vol, 31, pp, 38-39, ISSN 0014-956X

Combs, G.F. Jr& Gay, W.P. (1998). Chemopreventive agents: selenium. Pharmacology & Therapeutics, Vol.79, pp.179-192, ISSN 0163-7258

Davidian, J.-C., Hatzfield, Y., Cathala, N., Tagmount, A. & Vidmar, J. J. (2000). Sulfate uptake and transport in plants, In: *Sulfur Nutrition and Sulfur Assimilation in Higher Plants: Molecular, Biochemical and Physiological Aspects*, C. Brunold, H. Rennenberg, L. J. De Kok, I. Stuhlen, J.-C. Davidian, pp. 1-19, Bern: Paul Haupt.

Dawson, J. C. & Anderson, J. W. (1988). Incorporation of cysteine and selenocysteine into cystathionine and selenocystathionine by crude extracts of spinach. *Phytochemistry*, Vol, 27, pp, 3453–3460, ISSN 0031-9422

De Fillips, L. F. (2010). Biochemical and molecular aspects in phytoremediation of selenium, In: *Plant Adaptation and Phytoremediation*, M. Ashraf, M. Ozturk, M. S. A. Ahmad, pp. 193-226, Springer, ISBN 9048193699

De Souza, M. P., Pilon-Smits, E. A. H., Lytle, C. M., Hwang, S., Tai, J., Honma, T. S. U., Yeh, L. & Terry, N. (1998). Ratelimiting steps in selenium assimilation and volatilization by Indian mustard. *Plant Physiol*, Vol, 117, pp, 1487–1494, ISSN 0032-0889

Dilworth, G.L.& Bandurski, R.S. (1977). Activation of selenate by adenosine 50- triphosphate sulfurylase from *Saccharmoyces cereviseae*. *Biochemical Journal*, Vol.163, pp.521–529, ISSN 0264-6021

Djanaguiraman, M., Prasad, P. V. V. & Seppanen, M. (2010). Selenium protects *Sorghum* leaves from oxidative damage under high temperature stress by enhancing antioxidant defense system, *Plant Physiol Biohem*, Vol 48, pp, 999-1007, ISSN 0981-9428

Duckart, E. C., Waldron, L. J. & Donner, H. E. (1992). Selenium uptake and volatilization from plants growing in soil, *Soil Sci*, Vol, 153, pp, 94–99, ISSN 0038-075X

Dumont, E., Vanhaecke, F. & Cornelis, R. (2006). Selenium speciation from food source to metabolites: a critical review, *Anal Bioanal Chem*, Vol, 385, pp, 1304–1323, ISSN 1618-2642

Dutilleul, C., Jourdain, A., Bourguignon, J.& Hugouvieux, V. (2008) The *Arabidopsis* putative selenium binding protein family: Expression study and characterisation of SBP1 as a potential new player in cadmium detoxification processes. *Plant Physiology*, Vol. 147, pp.239–251, ISSN 0032-0889

El-Bayoumy, K. & Sinha, R. (2005). Molecular chemoprevention by selenium: a genomic approach, Mutat Res, Vol 591, pp. 224-236, ISSN: 0027-5107

El Mehdawi, A. F., Quinn, C. F. & Pilon-Smits E. A. H. (2011). Effects of selenium hyperaccumulation on plant–plant interactions: evidence for elemental allelopathy?, New Phytologist, Vol, 191, pp, *120–131, ISSN* 1469-8137

Ellis, D.R. & Salt, D. E. (2003). Plants, selenium and human health, *Curr Opin Plant Biol*, Vol, 6, pp. 273–279, ISSN: 1369-5266

Ellis, D.R., Sors, T.G., Brunk, D.G., Albrecht, C., Orser, C., Lahner, B., Wood, K.V., Harris, H.H., Pickering, I.J.& Salt, D.E. (2004). Production of Se-methylselenocysteine in transgenic plants expressing selenocysteine methyltransferase. *BMC Plant Biology*, *Vol. 4*, pp.1–12, ISSN 1471-2229

Feist, L. J. & Parker, D.R. (2001). Ecotypic variation in selenium accumulation among populations of *Stanleya pinnata*, *New Phytol*, Vol, 149, pp, 61-69, ISSN 1469-8137

Finley, J. W. & Davis, C. D. (2001). Selenium (Se) from high-selenium *broccoli* is utilized differently than selenite, selenate, and selenomethionine, but is more effective in inhibiting colon carcinogenesis, *BioFactors*, Vol, 14, pp, 191-196, ISSN 0951-6433

Foster, S. J., Kraus, R. J. & Ganther, H. E. (1986). The metabolism of selenomethionine, Se-methylselenocysteine, their selenonium derivatives, and trimethylselenonium in the rat, *Arch Biochem Biophys*, Vol, 251, pp, 77–86, ISSN 0003-9861

Galeas, M. L., Zhang, L. H., Freeman, J. L., Wegner & Pilon-Smits, E. A. H. (2007). Seasonal fluctuations of selenium and sulfur accumulation in selenium hyperaccumulators and related nonaccumulator, *New Phytol*, Vol, 173, pp,517-525, ISSN 1469-8137

Ganther, H. E. & Lawrence, J. R. (1997). Chemical transformations of selenium in living organisms. Improved forms of selenium for cancer prevention. *Tetrahedron*, Vol, 53, pp, 12299–112310, ISSN 0040-4020

Garifullina, G.F., Owen, J.D., Lindblom, S-D., Tufan, H., Pilon, M. & Pilon-Smits, E.A.H. (2003). Expression of a mouse selenocysteine lyase in *Brassica juncea* chloroplasts affects selenium tolerance and accumulation. *Physiologia Plantarum*, Vol.118, pp.538–544, ISSN 0031-9317

Germ, M., Stibilj, V., Osvald, J. & Kreft, I. (2007a). Effect of selenium foliar application on chicory (*Cichorium intybus* L.), *J Agricult Food Chem*, Vol, 55,pp, 795-798, ISSN 0021-8561

Germ, M., Stibilj, V. & Kreft, I. (2007b). Metabolic Importance of Selenium for Plants, *The European Journal of Plant Science and Biotechnology*, Vol, 1, pp, 91-97, ISSN 1752-3842

Gladyshev, V. N. & Kryukov, G. V. (1999). Evolution of selenocysteinecontaining proteins: significance of identification and functional characterization of selenoproteins, *Biofactors*, Vol 14, pp, 87-92, ISSN 0951-6433

Goldhaber, S.B. (2003). Trace element risk assessment: essentiality vs. toxicity. *Regul Toxicol Pharmacol*, Vol. 38, pp. 232–242, ISSN: 0273-2300

Goulet, A. C., Watts, G., Lord, J. L., Nelson, M. A. (2007). Profiling of Selenomethionine Responsive Genes in Colon Cancer by Microarray Analysis, *Canc Biol Ther*, Vol, 6, pp, 1-10 , ISSN 1538-4047

Grant, K., Carey, N.M., Mendoza, M., Schulze J., Pilon, M., Pilon-Smits, E. A. H. & van Hoewyk D. (2011). Adenosine 5'-phosphosulfate reductase (APR2) mutation in *Arabidopsis* implicates glutathione deficiency in selenate toxicity. *Biochemical Journal*, Vol. 438, pp. 325–335, ISSN 0264-6021

Guo, X. & Wu, L. (1998). Distribution of free seleno-amino acids in plant tissue of *Melilotus indica* L. grown in selenium laden soils, *Ecotoxicol Environ Safety*, Vol, 39, pp, 207-214, ISSN 0147-6513

Hanson, A.D., Trossat, C., Nolte, K.D. & Gage, D.A. (1997). 3-dimethylsulphonioproprionate biosynthesis in higher plants. In: *Sulphur nutrition and assimilation in higher plants: Regulatory agricultural and environmental aspects*, Cram, W.J., De Kok, L.J., Stulen, I., Brunold, C., Rennenberg, H., pp.147-154, Backhuys Publishers, USA

Hasanuzzaman, M. & Fujita, M. (2011). Selenium pretreatment upregulates the antioxidant defense and methylglyoxal detoxification system and confers enhanced tolerance to drought stress in rapeseed seedlings. Biol Trace Elem Res, DOI: 10.1007/s12011-011-8998-9, ISSN 1559-0720

Hatzfeld, Y., Cathala, N., Grignon, C.& Davidian, J.C. (1998). Effect of ATP sulphurylase overexpression in bright yellow 2 tobacco cells. *Plant Physiology*, Vol. 116, pp.1307–1313, ISSN 0032-0889

Haug, A., Graham, R.D., Christophersen, O.A.& Lyons, G.H. (2007). How to use the world's scarce selenium resources efficiently to increase the selenium concentration in food. *Microbial Ecology in Health and Disease*, Vol. 19, pp.209–228, ISSN 0891-060X

Ip, C. & Ganther, H. E. (1992). Relationship between the chemical form of selenium and anticarcinogenic activity, In: *Cancer Chemoprevention*, I. Wattenberg, M. Lipkin, C. W. Boon, G. J. Kellott & R. Boca, pp. 479-488, CRC Press

Ip, C. (1998). Lessons from basic research in selenium and cancer prevention, *J Nutr*, Vol, 28, pp, 1845–1854, ISSN 0022-3166

Ip, C., Birringer, M., Block, E., Kotrebai, M., Tyson, J., Uden, P. C. & Lisk, D. (2000). Chemical speciation influences comparative activity of selenium-enriched garlic and yeast in mammary cancer prevention, *J Agric Food Chem*, Vol, 48, pp, 2062–2070, ISSN 0021-8561

Jhee, E. M., Dandridge, K. L., Christy, Jr., A. M. & Pollard, A. J. (1999). Selective herbivory on low-zinc phenotypes of the hyperaccumulator *Thlaspi caerulescens* (*Brassicaceae*), *Chemoecology*, Vol, 9, pp, 93 – 95, ISSN 0937-7409

Kim, T., Jung, U., Cho, D. Y. & Chung, A. S. (2001). Se-methylselenocysteine induces apoptosis through caspase activation in HL-60 cells, *Carcinogenesis*, Vol, 22, pp, 559–565, ISSN 0143-3334

LeDuc, D.L., AbdelSamie, M., Montes-Bayo'n, M., Wu, C.P., Reisinger, S.J. & Terry, N. (2006). Overexpressing both ATP sulfurylase and selenocysteine methyltransferase enhances selenium phytoremediation traits in Indian mustard. *Environmental Pollution*, Vol. 144, pp. 70–76, ISSN 0269-7491

LeDuc, D.L., Tarun, A.S., Montes-Bayon, M., Meija, J., Malit, M.F., Wu, C.P., Abdel-Samie, M., Chiang, C.Y., Tagmount, A., De Souza, M., Neuhierl, B., Bock, A., Caruso, J.& Terry, N. (2004). Overexpression of selenocysteine methyltransferase in *Arabidopsis* and Indian mustard increases selenium tolerance and accumulation. *Plant Physiology*, Vol. 135, pp.377–383, ISSN 0032-0889

Leustek, T., Smith, M., Murillo, M., Singh, D.P., Smith, A.G., Woodcock, S.C., Awan, S.J. & Warren, M.J. (1997). Siroheme biosynthesis in higher plants. Analysis of an *S*-adenosyl-L-methionine-dependent uroporphyrinogen III methyltransferase from *Arabidopsis thaliana*. *J Biol Chem*, Vol. 272, pp, 2744-2752, ISSN 0021-9258

Lefsrud, M.G., Kopsell, D.A., Kopsell, D.E. & Randle, W.M. (2006). Kale carotenoids are unaffected, whereas biomass production, elemental concentration, and selenium accumulation respond to, changes in selenium fertility. *Journal of Agricultural and Food Chemistry*, Vol.54, pp.1764–1771, ISSN 0021-8561

Lindblow-Kull, C., Kull, F. J. & Shrift, A.(1985). Single transporter for sulfate, selenate, and selenite in *Escherichia coli* K12, *J Bacteriol*, Vol, 163, pp. 1267–1269, ISSN 0021-9193

Low, S. C. & Berry, M. J. (1996). Knowing when not to stop: selenocysteine incorporation in eukaryotes, *Trends Biochem Sci*, Vol, 21, pp, 203-208, ISSN 0968-0004

Lu, J. & Jiang, C. (2001). Antiangiogenic activity of selenium in cancer chemoprevention: metabolite-specific effects, *Nutr Cancer*, Vol, 40, pp, 64–73, ISSN 0163-5581

Lu, J., Pei, H., Ip, C., Lisk, D. J., Ganther, H. & Thompson, H. J. (1996). Effect of an aqueous extract of selenium-enriched garlic on in vitro markers and *in vivo* efficacy in cancer prevention, *Carcinogenesis,* Vol, 17, pp, 1903–1907, ISSN 0143-3334

Lydiate, D., Higgins, E., Robinson, S., Korbas, M., Yang, S.I.& Pickering, I. (2007). Selenium acquisition by *Arabidopsis* plants. *Canadian Light Source,* Vol. 27, pp.106–107

Lyi, S.M., Heller, L.I., Rutzke, M., Welch, R.M., Kochian, L.V. & Li, L. (2005). Molecular and biochemical characterization of the selenocysteine Se-methyltransferase gene and Se-methylselenocysteine synthesis in broccoli. *Plant Physiol,* Vol. 138, pp,409–420, ISSN 0032-0889

McCluskey, T. J., Scarf, A. R. & Anderson, J. W. (1986). Enzyme-catalyzed elimination of selenocystathionine and selenocystine and their sulfur isologues by plant extracts. *Phytochemistry,* Vol, 25, pp, 2063– 2068, ISSN 0031-9422

McKenzie, M.J., Hunter, D.A., Pathirana, R., Watson, L.M., Joyce, N.I., Matich, A.J., Rowan, D.D.& Brummell, D.A. (2009). Accumulation of an organic anticancer selenium compound in a transgenic Solanaceous species shows wider applicability of the selenocysteine methyltransferase transgene from selenium hyperaccumulators. *Transgenic Research,* Vol. 18, (3), pp.407–424, ISSN 0962-8819

Medina, D., Thompson, H., Ganther, H. & Ip, C. (2001). Se-Methylselenocysteine: a new compound for chemoprevention of breast cancer. *Nutrition and Cancer,* Vol, 40, pp, 12 – 17, ISSN 0163-5581

Neuhierl, B.& Bock, A. (1996). On the mechanism of selenium tolerance in selenium accumulating plants. Purification and characterization of a specific selenocysteine methyltransferase from cultured cells of *Astragalus bisulcatus. European Journal of Biochemistry,* Vol. 239, pp. 235–238, ISSN 0014-2956

Neuhierl, B., Thanbichler, M., Lottspeich, F.& Boeck, A. (1999). A family of S-methylmethionine dependent thiol/selenol methyltransferases: Role in selenium tolerance and evolutionary relation. *Journal of Biological Chemistry,* Vol. 274, pp.5407–5414, ISSN 0021-9258

Ng, B. H. & Anderson, J. W. (1978). Synthesis of selenocysteine by cysteine synthases from selenium accumulator and non-accumulator plants, *Phytochemistry,* Vol, 17, pp, 2069–2074, ISSN 0031-9422

Nyberg, S. (1991). Multiple use of plants: Studies on selenium incorporation in some agricultural species for the production of organic selenium compounds. *Plant Foods for Human Nutrition,* Vol. 41, pp.69–88, ISSN 0921-9668

Pedrero, Z., Yolanda, M. & Carmen, C. (2006). Selenium species bioaccessibility in enriched raddish (*Raphanus sativa*). A potential dietary source of selenium. *Journal of Agricultural and Food Chemistry* Vol.54, pp.2412–2417, ISSN 0021-8561

Pickering, I.J., Wright, C., Bubner, B., Ellis, D., Persans, M.W., Yu, E.Y., George, G.N., Prince, R.C.& Salt, D.E. (2003). Chemical form and distribution of selenium and sulfur in the selenium hyperaccumulator *Astragalus bisulcatus. Plant Physiology,* Vol. 131, pp.1-8, ISSN 0032-0889

Pilon, M., Owen, J.D., Garifullina, G.F., Kurihara, T., Mihara, H., Esaki, N. & Pilon-Smits, E.A.H. (2003). Enhanced selenium tolerance and accumulation in transgenic Arabidopsis thaliana expressing a mouse selenocysteine lyase. *Plant Physiology,* Vol. 131, pp. 1250–1257, ISSN 0032-0889

Pilon-Smits, E. H. A. & Quinn, C. F. (2010). Selenium Metabolism in Plants. In: *Cell Biology of Metals and Nutrients, Plant Cell Monographs 17*, R. Hell & R. R. Mendel, Springer-Verlag, Berlin Heidelberg

Pilon-Smits, E.A.H., Hwang, S., Lytle, C.M., Zhu, Y., Tai, J.C,, Bravo, R.C., Chen, Y., Leustek, T. & Terry N. (1999). Overexpression of ATP sulfurylase in Indian mustard leads to increased selenate uptake, reduction, and tolerance. *Plant Physiology*, Vol.119, pp.123–132, ISSN 0032-0889

Ranocha, P., Bourgis, F., Ziemak, M.J., Rhodes, D., Gage, D.A. & Hanson, A.D. (2000). Characterization and functional expression of cDNAs encoding methionine-sensitive and –insensitive homocysteine S- methyltransferases from Arabidopsis. *Journal of Biological Chemistry*, Vol. 275, 15962-15968, ISSN 0021-9258

Reeves, R.D. & Baker, A.J.M. (2000). Metal- Accumulating Plants, In: *Phytoremediation of toxic metals: using plants to clean-up the environment*, I. Raskın & , B.D. Ensley, pp. 193-230, New York, John Wiley and Sons

Shaw, W.H.& Anderson, J.W. (1972). Purification, properties and substrate specificity of adenosine triphosphate sulphurylase from spinach leaf tissue. *Biochemical Journal*, Vol. 127, pp.237–247, ISSN 0264-6021

Shrift,A. & Ulrich, J. M. (1976). Transport of selenate and selenite into *Astragalus* roots, *Plant Physiol*, Vol, 44, pp. 893–896, ISSN 0032-0889

Sinha, R., Kiley, S. C., Lu, J. X., Thompson, H. J., Moraes, R., Jaken, S. & Medina, D. (1999). Effects of methylselenocysteine on PKC activity, cdk2 phosphorylation and gadd gene expression in synchronized mouse mammary epithelial tumor cells, *Cancer Lett* , Vol, 146, pp, 135–145, ISSN 0304-3835

Sors, T.G., Ellis, D.R. & D.E. Salt (2005). Selenium uptake, translocation, assimilation and metabolic fate in plants. *Photosynthesis Research*, Vol. 86, pp.373–389, ISSN 0166-8595

Sors, T.G., Martin, C.P., Salt, D.E. (2009). Characterization of selenocysteine methyltransferases from *Astragalus* species with contrasting selenium accumulation capacity. *Plant J* 59(1):110–122 ISSN 0960-7412

Tamaoki, M., Freeman, J.L. & Pilon-Smits, E.A.H. (2008). Cooperative ethylene and jasmonic acid signaling regulates selenite resistance in Arabidopsis thaliana. *Plant Physiol*, Vol.146, pp,1219–1230, ISSN 0032- 0889

Terry, N., Zayed, A. M., Souza, M. P. & Tarun, A. S. (2000). Selenium in higher plants, *Annu Rev Plant Physiol Plant Mol Biol*, Vol, 51, pp, 401-432, ISSN 1040-2519

Tsang, M. L. S. & Schiff, J. A. (1978). Studies of sulfate utilization by algae. 18. Identification of glutathione as a physiological carrier in assimilatory sulfate reduction by *Chlorella*, *Plant Sci Lett*, Vol, 11, pp, 177–183, ISSN 0304-4211

Unni, E., Koul, D., Alfred Yung, W-K. & Sinha R (2005). Semethylselenocysteine inhibits phosphatidylinositol 3-kinase activity of mouse mammary epithelial tumor cells in vitro. *Breast Cancer Research*, Vol.7, pp.699-707, ISSN 1465-5411

Van Hoewyk, D., Abdel-Ghany, S.E., Cohu, C., Herbert, S., Kugrens, P., Pilon, M.& Pilon-Smits, E.A.H. (2007). The Arabidopsis cysteine desulfurase CpNifS is essential for maturation of iron-sulfur cluster proteins, photosynthesis, and chloroplast development. *Proceedings of the National Academy of Sciences of the United States of America*, Vol. 104, pp.5686–5691, ISSN 0027-8424

Van Hoewyk, D., Garifullina, G.F., Ackley, A.R., Abdel-Ghany, S.E., Marcus, M.A., Fakra, S., Ishiyama, K. Inoue, E., Pilon, M.& Takahashi, H. (2005). Overexpression of

AtCpNifS enhances selenium tolerance and accumulation in *Arabidopsis*. *Plant Physiology*, Vol. 139, pp.1518–1528, ISSN 0032-0889

Van Hoewyk, D., Takahashi, H., Hess, A., Tamaoki, M.& Pilon-Smits, E.A.H. (2008). Transcriptome and biochemical analyses give insights into selenium-stress responses and selenium tolerance mechanisms in *Arabidopsis*. *Physiologia Plantarum*, Vol. 132, pp.236–253, ISSN 0031-9317

Van Huysen, T., Abdel-Ghany, S., Hale, K.L., LeDuc, D., Terry, N. & Pilon-Smits, E.A.H. (2003). Overexpression of cystathionine-gamma-synthase enhances selenium volatilisation in *Brassica juncea*. *Planta*, Vol. 218, pp.71–78, ISSN 0032-0935

Van Huysen T, Terry N, Pilon-Smits EAH (2004). Exploring the Selenium phytoremediationpotential of transgenic Brassica juncea overexpressing ATP sulfurylase or cystathionine g-synthase. *Int J Phytoremed* 6:111–118

Wang, Z., Jiang, C. & Lu, J. (2002). Induction of caspase-mediated apoptosis and cell-cycle G1 arrest by selenium metabolite methylselenol, *Mol Carcinog*, Vol, 34, pp, 113–120, ISSN 1098-2744

White, P. J., Bowen, H. C., Marshall, B. & Broadley, M. R. (2007). Extraordinarily high leaf selenium to sulfur ratios define 'Se-accumulator' plants, *Ann Bot*, Vol, 100, pp, 111–118, ISSN 0305-7364

Wilber, C. G. (1980). Toxicology of selenium: a review. *Clin Toxicol*, Vol, 17, pp, 171–230, ISSN 1556-3650

Zayed, A., Lytle, C.M. & Terry N. (1998). Accumulation and volatilization of different chemical species of selenium by plants, *Planta*, Vol, 206, pp. 284–292, ISSN 0032-0935

Zeng, H. & Combs, G. F. (2008). Selenium as an anticancer nutrient: roles in cell proliferation and tumor cell invasion, *Journal of Nutritional Biochemistry*, Vol 19, pp,1-7, ISSN 0955-2863

Zhao, F., McGrath, S., Gray, C. & Lopez-Bellido, J. (2007). Selenium concentrations in UK wheat and biofortification strategies. *Comparative Biochemistry and Physiology - Part A: Molecular & Integrative Physiology*, Vol.146, pp.S246, ISSN 1095-6433

Zhang, L.H., Byrne, P.F.& Pilon-Smits, E.A.H. (2006). Mapping quantitative trait loci associated with selenate Tolerance in *Arabidopsis thaliana*, *New Phytol*, Vol. 170, pp, 33–42, ISSN 1469-8137

Zhu, L., Jiang, C.J., Deng, W.W., Gao, X., Wang, R.J. and Wan, X.C. (2008). Cloning and expression of selenocysteine methyltransferase cDNA from *Camellia sinensis*. *Acta Physiol Plant*, Vol.30, pp, 167–174, ISSN 0137-5881

11

Significance of UV-C Hormesis and Its Relation to Some Phytochemicals in Ripening and Senescence Process

Maharaj Rohanie[1] and Mohammed Ayoub[2]
[1]*University of Trinidad and Tobago, Piarco*
[2]*National Agricultural Marketing Development Corporation, Piarco*
Trinidad

1. Introduction

Recently, the Food Agricultural Organization of the United Nations (FAO, 2009) predicted that the world population would top eight billion by the year 2030. Therefore, the demand for food would increase dramatically. Fruits and vegetables will play an important role in providing essential vitamins, minerals and dietary fibre to the world feeding populations in both developed and developing countries. Organizations such as FAO have also recommended increasing fruits and vegetables consumption to decrease the risk of cardiovascular diseases and cancer. Moreover, as far as health is concerned, quality and safety in foods are important criteria demanded by consumers.

Fruits and vegetables are perishable products with active metabolism during the postharvest period and the major limitations in their storability are senescence, fungal infection and water loss (Brady, 1987). Such crops are classified as being either climacteric or non climacteric based on their respiratory and ethylene production patterns during storage life. Climacteric fruits show a marked increase in respiration and increase ethylene (C_2H_4) production during their ripening whereas non-climacteric fruit and vegetables complete ripening without increases in respiration and independent of increase C_2H_4 production (Wills et al., 1981). There is extensive experimental evidence indicating that the increase in C_2H_4 production is linked to the main biochemical and physiological changes that occur during ripening, leading to a loss in skin resistance, increased water loss, senescence and ultimately postharvest diseases. Examples of postharvest diseases arising from quiescent infections include anthracnose of various tropical fruit caused by *Colletotrichum spp.* and grey mould of strawberry caused by *Botrytis cinerea*. Preservation of these crops in as fresh as possible state, with minimum loss in quality during handling and storage, will primarily involve retardation of the physiological and biochemical changes associated with ripening and senescence (Wills et al., 1981).

Methods aimed at improving postharvest shelf-life and quality particularly for tropical climacteric horticultural commodities should be addressed. Some of the current methods employed to delay ripening and senescence and increase shelf-life of crops include:-low temperature storage combined with high humidity, as well as the use of modified and controlled atmospheres, various coatings and waxes (Yahia, 1998). Certain fungicides have also been employed to control postharvest diseases and improve quality of horticultural crops but these have been met with resistance from consumers who want healthy, safe and nutritious foods. The benefits of these techniques have been reported for a wide myriad of fruits and vegetables however, all these techniques must be applied with refrigeration, as temperature is the dominant factor influencing all plant processes (Pearce, 1999). Developing newer techniques and technologies in order to improve postharvest longevity of horticultural crops has always been a challenge to researchers. With application of adequate technology to prevent deterioration after harvest and considering the biochemical characteristics of the produce, postharvest losses can be significantly reduced.

1.1 Phytochemicals in horticultural crops

Phytochemicals are plant components that have gained considerable attention as photoprotective agents in providing certain human health benefits and have been the subject of numerous investigations (Shahidi et al., 2011; Butt & Sultan, 2011; Heber, 2004; Ramana-Luximon et al., 2003). Phytochemicals belong to several classes that include polyphenols, flavonoids, isoflavonoids, phytoalexins, phenols, anthocyanidins and carotenoids. They are widely distributed with different structures at the tissue, cellular and sub-cellular levels (Shahidi & Naczk, 2004). Frequent consumption of fruits is associated with health promoting effects of plant phytochemicals in particular their antioxidant properties. Phytochemicals may work in promoting human health and disease prevention in different ways such as: by stimulating the immune response, by inducing gene suppression, by blocking oxidative damage to DNA, by detoxifying carcinogens and by initiating selected signaling pathways or by other mechanisms (Adhami et al., 2008; Schreiner & Huyskens-Keil, 2006).

Researchers have been able to isolate numerous bioactive phytochemicals which are well known for their powerful antioxidant and free radical scavenging potential (Jergensen et al., 1999). Moreover, specific postharvest elicitor treatments, such as low or high temperature treatments, ultraviolet and gamma irradiation, altered gas composition or application of signaling molecules may further enhance phytochemical content (Schreiner & Huyskens-Keil, 2006). Their possible impact on maintaining human health and prevention of diseases continue to be an active research (Ramana-Luximon et al., 2003).

2. UV-C hormesis

There is now considerable literature on the use of controlled abiotic stresses to delay postharvest senescence of horticultural crops. A treatment that could activate the mechanisms of the plant against the toll of senescence can be a useful method in the preservation of fresh tropical crops. One example of such technology, which is not expensive, simple and uses non-ionizing energy is ultraviolet (UV) radiation. UV-radiation has been classified as UV-C (200-280 nm), UV-B (280-320 nm) and UV-A (320-400 nm). Each band can induce significantly different biological effects in crops (Shama, 2007; Bintsis et al., 2000). UV-C wavelengths are absorbed in the stratosphere and are removed from the light reaching the earth's surface so long as there is ozone present. Most of the information on the biological effects of UV-radiation is derived from experiments using artificial UV-C, particularly 254 nm radiation (Bintsis et al., 2000).

Exposure to UV-radiation is well known to have deleterious effects of plant tissues. However, a biphasic dose response relationship in which low irradiation doses cause a stimulatory effect of beneficial responses while high doses cause detrimental (toxic) effects on plant tissues is termed Hormesis (Arul et al., 2001; Calabrese et al., 1987).

UV-C radiation is mainly used as a surface treatment because it penetrates only 5-30 microns of the tissue. It has been extensively used in the disinfection of equipment, glassware and air by food and medical industries for many years. Low pressure mercury (Hg) lamps are often called "germicidal" because most of their total radiation energy is at a wavelength of 253.7 nm, which is near the maximum for germicidal effectiveness, hence its usefulness in the control of microorganisms (Kowalski, 2009). While the application of hormesis is used in the context of this chapter on the benefits to plant tissues, hormic responses have also been shown in bacteria, fungi, animals and humans (Shama, 2007).

UV-light is absorbed by certain chemical groupings of molecules (chromophores) such as conjugated double bonds and results in photochemical reactions. The chromophores eg. nucleic acids, proteins, indoleacetic acid, flavoproteins and phytochrome have key roles in plant cell function and structure and any alterations of these compounds due to UV might be expected to cause physiological alterations in plants (Caldwell, 1981). DNA is one of the most important target molecules for photobiological effects. UV-C light can be (1) directly absorbed by DNA and (2) involved in photooxidation in plants via free radical production. Thus damage to plants as a result of UV-C can be classified into two categories: damage to DNA and damage to physiological processes (Stapleton, 1992). The intrusion of activated oxygen species into biological systems as well as changes to DNA results ultimately in deleterious effects for example, too much UV damages DNA in soyabeans, destroys their chlorophyll and disrupts photosynthesis which leads to poor yields (Vikhanski, 1989). However certain crops are resistant to such adverse effects by having DNA repair systems by gene activation and quenching systems which can undo the damage caused by UV-C.

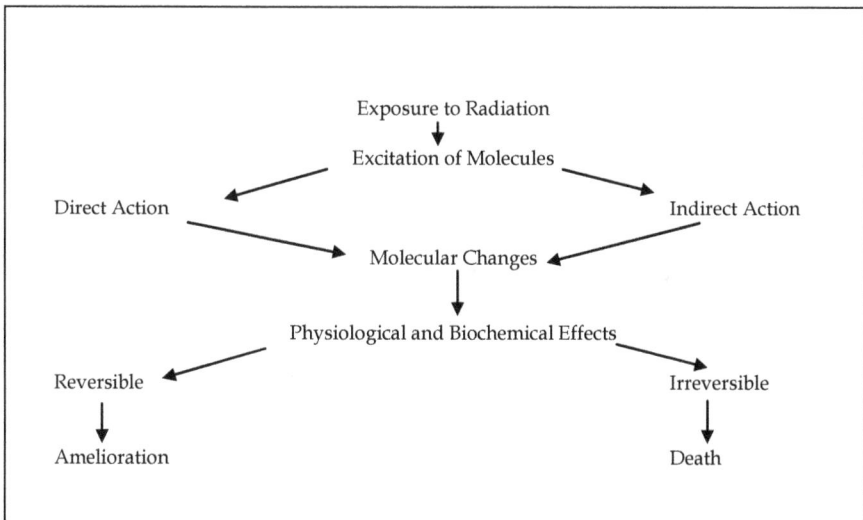

Fig. 1. Schematic representation of the photobiological effects of UV-C radiation in plants

2.1 Intense light pulses

In response to consumer preferences for high quality foods that are as close as possible to fresh products, athermal technologies are being developed to obtain products with high levels of organoleptic and nutritional quality but free of any health risks. Intense pulsed light (IPL) is a novel technology that rapidly inactivates pathogenic and food spoilage microorganisms. It appears to constitute a good alternative or a complement to conventional thermal or chemical decontamination processes (Elmnasser et al., 2007). Intense pulsed light (IPL) decontaminates food surfaces by killing microorganisms using short time high frequency pulses of an intense broad spectrum, rich in UV-C light. Different mechanisms have been proposed to explain the lethal effect of ILP, all of them related with the UV part of the spectrum and with its photochemical and/or photothermal effect (Anderson et al., 2000).

The decontamination effect of pulsed light on minimally processed vegetables have been reported, with log reductions between 0.56 and 2.04 achieved with mesophilic aerobic microorganisms (Gómez-López et al., 2005). On the other hand, exposure of pea plants to short pulses of UV-C radiation for 10, 14 and 21 consecutive days did not cause noticeable activation of the major antioxidant enzymes in the youngest leaves suggesting a different defence system for these plants (Katerova, 2009).

2.2 UV-C as an antimicrobial agent

The benefits of UV-C as a germicidal agent has found practical applications in a broad range of antimicrobial applications including inhibiting microorganisms from the surface of food products, destruction of microorganisms in air and sterilization of liquids (Bintsis et al., 2000). UV-C radiation can inactivate the pathogens that may be present at the surface of fruit (Artes-Hernandez et al., 2009). Exposure of fruits and vegetables to UV-C light at around 254 nm has been evaluated as a possible alternative to chemical fungicides for the control of postharvest diseases, rather than ripening. This is principally because of (1) its germicidal properties, and (2) UV-C activates the mechanism of defence of crops against infection by the *de novo* synthesis of antimicrobial compounds such as phytoalexins. The ultraviolet light acts as an antimicrobial agent directly due to DNA damage (Rame et al., 1997) and indirectly due to the induction of resistance mechanisms in different fruit and vegetables against pathogens (Mercier et al., 2001; Nigro et al., 1998; Liu et al., 1993). Shama (2007), cites several references on the use of UV-C treatment and its beneficial effects on reducing and/or preventing postharvest storage rots including onions, potatoes, sweet potatoes, carrots, tomato, peaches, mangoes and strawberries. Several studies have been conducted on the use of UV-C radiation and its benefit as a non chemical decontamination agent for human pathogens that may be found in food (Bintsis et al., 2000). Grapes and strawberries were UV-C irradiated for 24-48 h before being inoculated with *Botrytis cinerea* and the results compared with those inoculated immediately before irradiation. The results showed a reduction in the postharvest incidence of disease after UV-C irradiation (Nigro et al., 1998, 2000).

It was reported that the use of two sided UV-C radiation at the appropriate dose was effective in reducing the natural microflora and extending the shelf-life of minimally processed "Red Oak Leaf" lettuce (Allende et al., 2006) and in control of rots in sweet potatoes at certain doses (Stevens et al., 1990). UV-C treatment has been reported to induce resistance against pathogens in a number of species (El Ghaouth et al., 2003; Nigro et al., 2000; Stevens et al., 1996). The activity of the phenylalanine ammonia lyase (PAL; EC 4.3.1.5) enzyme was induced by UV-C radiation in several fruits (Charles et al., 2009; Nigro et al., 2000). This UV treatment also enhanced the accumulation of flavonoids, phytoalexins, and phenolic antifungal compounds (Liu et al., 2009).

The production of phytoalexins, antimicrobial compounds produced by plants in response to infection or physiological stimuli such as UV-radiation, is believed to be an important defence mechanism. One proposed hypothesis on the mechanism of induction of phytoalexin supposes that in normal tissue the genes involved are repressed. Agents which induce phytoalexin production are thought to depress these genes or promote their transcription by causing a conformational change in DNA (Langcake & Pryce, 1973).

UV-C light has been used in combination with other preservation techniques to preserve the quality of horticultural crops. Most of these studies showed the effectiveness of microbial reductions in fresh-cut fruits and vegetables by using chemical disinfection, low UV-C light doses (from 1 to 4 kJ/m^2) and storage under conventional MAP, without any detrimental effect on the organoleptic quality of the product. Additionally, UV-C light combined with other postharvest treatments such as mild thermal treatments and immersion in water at elevated temperatures showed improvements in keeping quality and reducing incidence of storage disease in various horticultural crops (Allende et al., 2006). The beneficial effects of a heatshock treatment to reduce browning in fresh-cut lettuce (e.g. 90 s at 45 °C) was due to the redirecting of protein synthesis away from the production of wound induced enzymes of phenolic metabolism, and toward the production of innocuous heat shock proteins (Saltveit, 2000 as cited in Allende et al., 2006). The efficacy of heat treatments and UV-C light for controlling postharvest decay of strawberries and sweet cherries were tested. In most of the cases, fungal inactivation was achieved for the treatments with the highest UV-C dose (10 kJ/m^2) combined with a long thermal treatment (15 min at 45 °C). The sequence of the treatments seemed to have an influence on microbial inactivation for strawberries. The fungal inactivation was greater when the ultraviolet treatment preceded the thermal treatment. The possibility of lowering the intensity of the heat treatment when preceded by an ultraviolet illumination resulted in a decrease of fruit damage caused by heating. Since less intense thermal conditions can be used, visual damage to the strawberries was also reduced (Marquenie et al., 2002).

2.3 UV-C and its role in plant tissue/organ senescence

One of the major factors in the process of plant tissue/organ senescence is free radical damage. As a result of oxidation inherent in aerobic respiration, free radicals are formed and their targets may be cell membranes, nucleic acids, enzymes and cell walls resulting in an acceleration of senescence and tissue softening (Brady, 1987). These oxidation stresses become severe with the physiological age of the tissue and it responds by generating an array of detoxifying mechanisms (antioxidants and enzymes) against free radical attack. It is documented that short wavelength radiation exerts two pronounced effects on plant metabolism viz: at low intensities, it may give rise to an enhancement of secondary stress metabolites which can protect the plant from free radical damage, while at high intensities, it can cause an inhibition of these substances often leading to detrimental effects on the plant (Kowalski, 2009). Certain plants have repair mechanisms which involve the production of secondary stress metabolites for example, antioxidant compounds such as carotenoids, phenols, flavonoids, and polyamines. The enzyme superoxide dismutase (SOD) is postulated to ameliorate the toxic effects of superoxide and this enzyme is ubiquitously present in aerobic organisms (Cunningham et al., 1985). These compounds, usually present in the plant are generally activated following stress when produced in sufficient abundance in pre-climacteric crops can counteract the effect of DNA and free radical damage (which becomes important with age) and stimulate increase longevity of harvested crops. By stimulating the natural defences of the crop against adverse stresses, one can improve the storability of such commodities. The UV doses reported to achieve beneficial effects in fruits and vegetables range from 0.5 kJ/m^2 for strawberries to 9.0 kJ/m^2 for oranges (Shama & Alderson, 2005). UV-C has also been documented to primarily control ripening and senescence in climacteric fruits (Maharaj et al., 2010; Hemmaty et al., 2006). On the other hand, prolonged exposure to UV irradiation has been found to accelerate ripening and senescence probably as a result of free radicals generated (Liu et al., 1993). Hyper UV-C doses have resulted in undesirable changes in skin colour, premature ripening, drying and infection in several crops as reported by Shama & Alderson (2005). Thus low levels or hormic doses of UV-C radiation can be an adjunct to refrigeration for preservation of horticultural crops. Effectiveness of UV-C treatments varies with plant species, stage of maturity, irradiation dose and duration.

3. Phytochemicals and UV-C hormesis

The impact of UV-C hormesis on fresh fruits and vegetables has been the subject of numerous studies over the last few years (Shama, 2007). Fruits contain a huge diversity of phytochemicals (antioxidant compounds). Of these, phenolics and carotenoids have been shown to protect the cellular systems from oxidative damage induced by free radicals. Promising results have been shown with abiotic stress treatment in positively impacting on phytochemicals, either by preserving its content in fruits and vegetables or increasing its contents following treatment. A number of factors affect the efficacy of treatment in maintaining or increasing photochemical content including: the type of horticultural product, the length of exposure and the class of phytochemicals. Non-enzymatic compounds consisting of lipid soluble membrane associated antioxidant (α-tocopherol, β-carotene) and water soluble reductants (glutathione, ascorbate, phenolics) and enzymatic antioxidants (SOD, catalase, peroxidases) are induced in response to oxidative stress (Jaleel et al., 2009). Changes in levels of polyamines have been linked to senescence suggesting that lowering of polyamine concentration is a step in triggering senescence or that exogenous application of polyamines could inhibit senescence (Shama & Alderson, 2005). The latter may be due to possible inhibition of ethylene synthesis and to stabilization and protection of membranes by associating with negatively charged phospholipids (Galston & Kaur-Sawhney, 1990).

3.1 Antioxidants: Carotenoids, lycopene and ascorbic acid

Colour development involves a decrease in chlorophyll pigments with a concomitant increase in pigments (e.g. carotenoids) and is a characteristic of many plant tissues during both ripening and senescence (Gong & Mattheis, 2003). Tomato fruit are a rich source of carotenoids, especially lycopene and β-carotene. UV-C has been shown to retard colour development, loss of chlorophyll and the development of lycopene which makes up the bulk of the carotenoid pigments in ripe tomato. Figure 2 illustrates the changes in lycopene content with storage time for both UV-treated and untreated tomato fruit stored under refrigeration. Lycopene is the major carotenoid in tomato accounting for more than 80% of the total carotenoids present in a fully ripe tomato fruit. The chloroplast seems to be one of the target sites for UV action and this offers evidence for reduction in the level of lycopene in irradiated tomatoes in comparison to the control fruits (Maharaj et al., 1999). However, production of total carotenoid pigments in tomato, were found to be higher in UV-treated fruits compared to the untreated ones. Carotenoids may be considered as an antioxidative phytochemical to counteract free radical damage with associated photosensitized reactions (Maharaj et al., 2010).

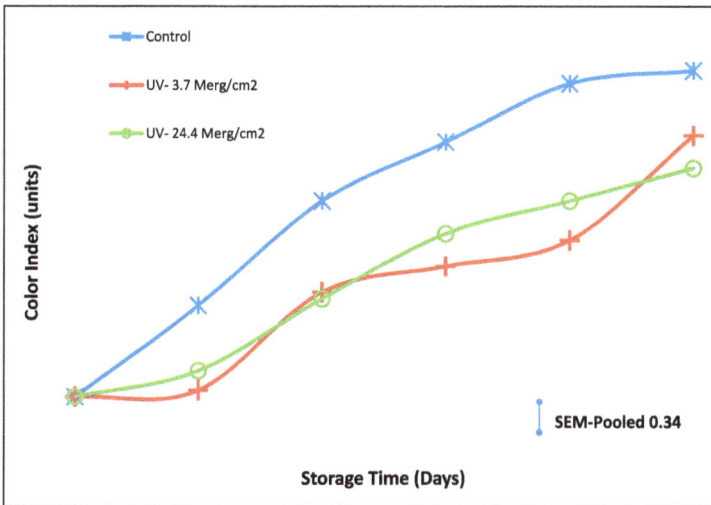

Fig. 2. Colour rating values (1=mature green, 6=red) of control and UV-C irradiated tomato stored at 16°C. Mean for n=4 (with permission Maharaj et al., 2010).

The impact of UV-C radiation on senescence of broccoli florets were investigated (Costa et al., 2006). One of the key symptoms of quality loss in broccoli is the loss of green colour of the sepals due to chlorophyll degradation. Apart from loss of green colour, lipid peroxidation, loss of antioxidant capacity, reduced nutritional value and increased tissue degradation are also well associated with quality loss in broccoli. Costa et al. (2006) subjected broccoli florets to four different UV-C dosages: 4, 7, 10 and 14 kJ m-2 as well as an untreated control. All samples were stored at 20 °C and evaluated after a six day storage period. The findings indicated that total chlorophyll loss was significantly higher in florets that were not radiated as compared to radiated florets having a mean chlorophyll content that was 53 % higher than the control. UV-C seemed to have had a positive impact in reducing the degradation of both chlorophyll-a and -b. It appears from the findings that UV-C treatment reduces the activity of chlorophyllase thus maintaining a greener colour than the control. Untreated florets had much higher chlorophyllase activity than UV-C treated florets. Pheophytin a by-product of chlorophyll degradation was also lower in UV-C irradiated broccoli with the exception of the highest (14 kJ m-2) dose where higher levels of pheophytin were detected. They suggested that at high UV-C dosage, the release of magnesium ions may be in part responsible for this observation. In *Shitake* mushrooms, treatment with UV-C radiation had higher level of flavonoids and ascorbic acid as well as reducing free radical scavenging potential and decreased peroxide (Jiang et al., 2010). Antioxidant enzyme activity of phenylalanine ammonia lyase (PAL), β-1,3-glucanase, superoxide dismustase, catalase and gluthione reductase were induced to high levels by UV-C treated Yali pears (Li et al., 2010).

Vitamin C content is considered to be a quality index for fruits and vegetables occurs as L-ascorbic acid and dehydroascorbic acid. Ascorbate is an electron donor and this property explains its function as an antioxidant or reducing agent. González-Aguilar et al. (2007) reported a reduction in total ascorbic acid content of UV-C irradiated fresh-cut mango fruits due to the oxidation of ascorbic acid by effect of the increment in UV-C exposure time.

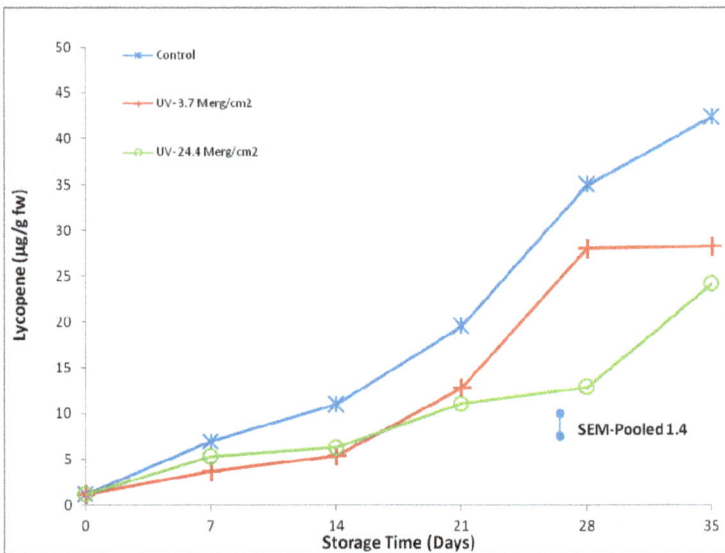

Fig. 3. Lycopene content of control and UV-C irradiated tomato stored at 16°C. Mean for n=4 (with permission Maharaj et al., 2010).

3.2 Phenolic compounds

The largest category of phytochemicals and most widely distributed in the plant kingdom are the phenolics. Bioactive phenolic compounds are plant secondary metabolites that are biosynthesized through the shikimic acid pathway (Tomás-Barberán & Espín, 2001). Phenolic compounds embrace a considerable range of substances which possess an aromatic ring bearing one or more hydroxyl groups. They may be classified into different groups as a function of the number of phenol rings that they contain and of the structural elements that bind these rings to each other. They exist in higher plants in many different forms including hydroxybenzoic derivatives, cinnamates, flavonoids (flavonols, flavones, flavanols, flavanones, isoflavones, proanthocyanidins), lignans, stilbenes, and which affect the quality characteristics of plants such as appearance, flavour and health-promoting properties. Plant phenolics are multifunctional and can act as reducing agents, metal chelators and singlet oxygen quenchers. Besides their role as antioxidants, phenolic compounds also possess antimicrobial properties and are involved in disease resistance by contributing to the healing of wounds by lignification of cell walls around wounded sites (Shahidi et al., 2011; Shahidi & Naczk, 2004; Tomás-Barberán & Espín, 2001). In addition, phenolics serve as substrates for browning reactions. The synthesis of certain phenols such as:- phytoalexins and lignins, that are all associated with local resistance processes, involves phenylalanine ammonia-lyase as a key step in the shikimic acid pathway (Shama & Alderson, 2005). The three amino acids phenylaline, tyrosine and tryptophan are the primary metabolites which serve as precursors for many secondary products. The enzymes phenylalanine ammonia-lyase (PAL; EC 4.3.1.5), polyphenol oxidases (PPO; EC 1.14.18.1) and peroxidises (POD; EC 1.11.1.7) are the main enzymes responsible for phenolic degradation that often leads to quality loss (Shahidi et al., 2011).

Light intensity and wavelength have an important effect on phenolic metabolism, as they affect flavonoid and anthocyanin biosynthesis. Some researchers point out that UV-radiation results in the accumulation flavonoids as they act as protective filter against (excessive) radiation and which may be implicated in the resistance of fruit and vegetables to microorganisms and senescence (Li et al., 2010; González–Aguilar et al., 2007; Shama & Alderson, 2005). Radiation is also known to cause depolymerisation of cell wall polysaccahrides resulting in higher extraction of phenols. UV irradiation during postharvest storage has been used to increase anthocyanin pigmentation in the skin of red apples, sweet cherries and strawberries leading to an increase in quality (Tomás-Barberán & Espín, 2001). In the case of apples, this increase correlated with an increase in both PAL and chalcone isomerase activities (Kataoka et al., 2003). On the other hand, no effect of postharvest UV irradiation on fruit pigmentation was detected in other crops such as nectarines, strawberries, grapes and plums showing that this treatment is not always useful for this purpose. In fact, UV irradiation produced necrosis and other quality losses in such irradiated fruits and these browning symptoms were attributed to the induced increased POD activity (Tomás-Barberán & Espín, 2001).

The effects of UV-C have also been investigated on fresh cut tropical fruits. Slices of bananas (cv. Pisang mas,) pineapple (cv. Honey pineapple) and Thai seedless guavas were irradiated with UV-C light and the results compared against untreated slices (control) (Alothman et al., 2009). In bananas and guava, exposure to UV-C radiation resulted in an increase in total phenols and flavonoids. In pineapples however, there was a significant increase in flavonoids but UV-C irradiation did not have any significant increase in total phenol content. In another study, mango slices were subjected to UV-C radiation at three different exposure times and its effect compared to untreated fruit slices (González-Aguilar et al., 2007). The study found that the length of exposure to radiation and storage time significantly affected phenols, flavonoids, β-Carotene and vitamin C. Fresh-Cut mango slices were radiated for 1, 3, 5 and 10 minutes or left untreated. After 3 days of storage, slices irradiated for 1, 3 and 5 minutes showed a sharp increase in total phenols followed by a plateau for the rest of the storage period. Mango fruit slices irradiated for 10 minutes, showed progressive increase in total phenols throughout the 15 day storage period. Slices used as control showed the smallest incremental increase after 3 days compared to all irradiated samples and also had a plateau for the rest of the 15 day storage period with the smallest quantities of total phenols. Flavonoid contents were also significantly affected by length of radiation exposure and storage time. Slices irradiated for 5 and 10 minutes showed a sharp continuous increase in flavonoids over the 15 day storage duration compared to those irradiated for shorter times (1 and 3 minutes), which also showed increased flavonoids but to a lesser extent than slices irradiated for longer times (5 and 10 minutes). Stimulation of flavonoid levels by UV-C radiation could be attributed to a defence mechanism in scavenging free radicals due to the reactivity of the hydroxyl groups. It has also been reported that UV light could be used to increase the content of health-promoting phenolics such as resveratrol in grapes and coumarins in grapefruit (Tomás-Barberán & Espín, 2001). The increase in total phenols and flavonoids in these studies may be attributed to their antioxidant and antimicrobial roles.

3.3 Polyamines

Polyamines (PA) are polyfunctional components present in plant and animal cells and are defined as "small polycationic biogenic amines" (Nambeesan et al., 2008). Polyamines (PA) are fairly ubiquitous, but of rather low concentrations ranging from micromolar to millimolar in mature fruit tissues (Galston & Kaur-Sawhney, 1990; Heby & Persson, 1990). They are implicated in a variety of regulatory processes ranging from regulation of growth and cell division, regulating the activity of ribonucleotides and proteinase to inhibition of C_2H_4 production and senescence (An et al., 2004; Valero et al., 2002; Pandey et al., 2000; Galston & Kaur–Sawhney, 1987). They have been shown to enhance the ability of plants to resist environmental stresses (An et al., 2004). Changes in PA biosynthesis in plant tissues have been correlated with various stresses such as K+ deficiency, cold acclimatization, chemical stress and controlled atmospheres.

Spermidine, spermine, and their precursor putrescine are the major polyamines in plants. In its free form, polyamines exist either as putrescine (diamine putresine), spermidine (triamine spermidine) and spermine (tetraamine spermine). Putrescine is the precursor for the synthesis of spermidine and spermine (Nambeesan et al., 2008). Apart from the three well known free forms of polyamines, there are some lesser known and studied polyamines such as 1,3-diaminopropane and homospermidine which have been detected in plants (Rodriguez-Garay et al., 1989), algae (Hamana & Matsuzuki, 1982), bacteria (Tait, 1985) and animals (Pandey et al., 2000).

At neutral (physiological) pH 7.0, PA are polycationic, they also occur in the free form or bound to phenolic acids. Conjugated amines have also detected in plant cells (Martin–Tanguy, 1997). They exist in either water soluble or forms which are insoluble in water (Pandey et al., 2000; Martin–Tanguy, 1997). While their exact roles are still uncertain, it is suggested that they act as reserve forms of amines which are released during growth at the point of synthesis or transported to other sites within plants as needed (Valero et al., 2002). In terms of the changes in polyamines in a number of crops, with few exceptions, the general trend is an increase in the concentration of free and bound polyamines especially spermine which coincides with the early stages of fruit growth at which time cell division is at its peak and a subsequent decrease at late fruit growth and at the time of ripening (Valero et al., 2002). However for crops such as avocado, pear and tomato fruit (cv. Rutgers), free PA levels declined during fruit development (Saftner & Baldi, 1990).

3.3.1 Biosynthesis and metabolism of the major polyamines

The biosynthetic pathways for polyamines have been delineated for mammals, fungi, bacteria plants (Nambeesan et al., 2008; Valero et al., 2002). Polyamine biosynthesis in plants involves two pathways. Putrescine, the primary precursor and first member in the PA sequence is derived from either ornithine (direct formation) or arginine (indirect formation). If putrescine is derived from ornithine, the pathway involves the decarboxylation of ornithine in the presence of ornithine decarboxylase (ODC). When putrescine is synthesized from arginine, it is first converted to agmatine by decarboxylation of arginine decarboxylase (ADC). The enzyme arginase can convert the amino acid arginine directly to ornithine with a loss of urea. The enzyme ODC which is located in the cystol and nucleus can convert ornithine into putrescine. Ornithine can be metabolised back to arginine via the ornithine cycle. With respect to the indirect formation, the enzyme ADC, found in the cytosol, decarboxylates arginine to agmatine. The agmatine is hydrolysed to an intermediate N-carbamoyl putrescine which then forms putrescine. Transformation of putrescine to spermidine (triamine) and spermine (tertaamine) requires stepwise transfer of aminopropyl groups from decarboxylated SAM (formed from methionine) to putrescine. The reaction is considered to be the rate-limiting step in PA biosynthesis. The PA 1,3-diaminopropane is formed by the oxidation of spermine and spermidine. Cadaverine another PA is formed via the decarboxylation of lysine (Slocom et al., 1984; Nambeesa et al., 2008).

Once formed, PA may be metabolized in various ways e.g. they can be conjugated or they can undergo oxidation. In plants, putrescine and spermidine are often bonded via their amino groups with the carboxyl groups of aromatic acids such as cinnamic and coumaric acids to form conjugates. PAs may be oxidized by a variety of enzymes in plants called amine oxidases. Diamine oxidases can convert putrescine to ammonia (NH_3), hydrogen peroxide (H_2O_2) and delta-pyrroline. Delta-pyrroline can be ultimately converted to succinic acid. PA oxidases can oxidize spermine and spermidine to 1,3-diaminopropane (Martin-Tanguy, 1997).

Polyamines are well known for their anti-senescence properties (Panday et al., 2000) whereas ethylene is known to initiate ripening and eventual senescence. The anti-senescent activity of polyamines may also be related to their ability to be effective free radical scavengers as well as stabilizing DNA and membranes by associating with negative charges on nucleic acids and phospholipids (Droplet et al., 1986). Their role in anti-senescence is believed to be related to their positively charged cations which are able to interact with negatively charged anions of membrane. Their cationic capability as well as their antioxidant property is believed to play a role preserving membrane integrity (Roberts et al., 1986). Ethylene, spermidine and sperminine share a common precursor S-adenosyl methionine (SAM) (Nambeesan et al., 2008; Pandey et al., 2000). Since they both exert opposite effects, it has been suggested that both ethylene and polyamines compete for this common precursor SAM. SAM is a substrate for 1-aminocyclopropane-1-carboxylic acid (ACC) in the synthesis of ethylene. SAM is also a substrate for SAM decarboxylase in a pathway that leads to the synthesis of PA. Thus the possibility exists that the biosynthesis of both may be closely linked (Panday et al., 2000; Saftner & Baldi, 1990). PA have shown inhibitory effects on ethylene synthesis in a variety of plant tissues as described by Nambeesan et al. 2008. Elevated PA (putrescine) levels have been correlated with prolonged storage characteristics in tomato fruit. From a postharvest physiological and quality perspective, free polyamines that are endogenously synthesized or exogenously applied have been found to exert its anti-senescence effect on a range of horticultural crops (Valero et al., 2002). These effects have been shown in retardation of colour changes, enhanced mechanical resistance, reduction in chilling injury, increase in fruit firmness and delayed respiration and ethylene production. The biosynthetic pathways of ethylene, spermidine and spermine are given in Figure 3.

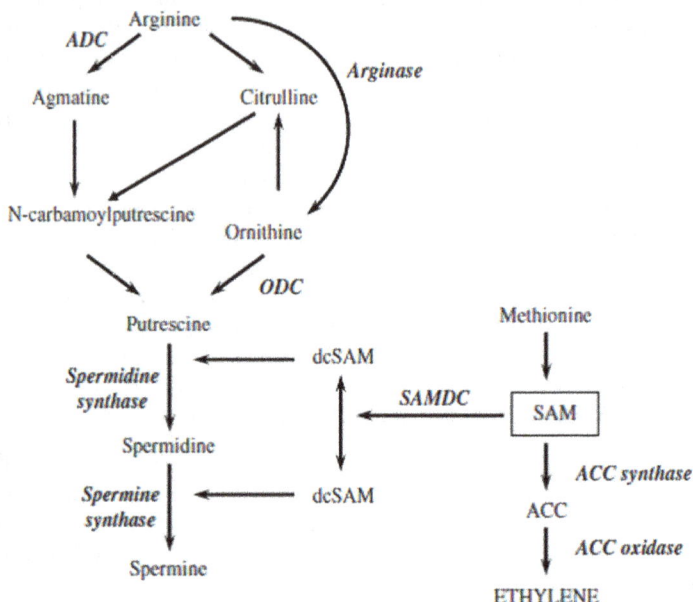

Fig. 4. Pathways (and related enzymes) of the ethylene and polyamine biosynthesis (adapted from Tassoni et al., 2006). ADC-arginine decarboxylase, ODC-ornithine decarboxylase, dcSAM-decarboxylated SAM, SAMDC-S-adenosyl methionine decaboxylase SAM- S-adenosyl methionine ACC- amino cyclopropane carboxylic acid

3.3.2 The effect of polyamines and UV-C stress

The intricate balance of polyamines and ethylene is suggested to play a central role in accelerating or retarding ripening and senescence (Pandey et al., 2000). A number of studies have been conducted to determine the possible effect of exogenous application of polyamines on ethylene production, senescence and other physio-chemical changes in fruit and vegetables and cut flowers. Four varieties of Japanese plums were treated with polyamines and their effects on ethylene production were measured (Serrano et al., 2003). Three varieties were typical climacteric types. In these varieties, treatment with polyamines resulted in a reduction and/or delay in the ethylene peak. One variety "Golden Japan" was known to behave in a non-climacteric fruit manner and therefore ethylene production remained low in treated and controlled fruits. Spermine and spemidine were more effective in inhibiting ethylene than putrescine in peaches when the application of polyamines to the canopy of the tree was studied (Breglio et al., 2002). In apricots, treatment with putrescine inhibited ethylene production during postharvest storage when compared to untreated fruit (Martinez-Romero et al., 2002). Application of exogenous polyamines on apple fruits however did not affect ethylene production when compared to untreated fruits (Wang et al., 1993). Valero et al. (2002) suggested that the difference in responses to exogenous application of ethylene production in climacteric fruits, may be explained by an understanding of the intrinsic ethylene production capacity of a particular horticultural commodity of a cultivar within a specific class of fruit. In commodities with low levels of ethylene production, treatment with polyamines result in the inhibition or delay in ethylene production while commodities with high levels of ethylene production, polyamines are not effective in delaying or inhibiting ethylene production.

However, in spite of well-documented protective role of PA in plants under damaging action of various stressors (high and low temperature, salinity, drought and others), molecular mechanisms of PA defensive effects and their biological significance in plant survival and adaptation remain obscure. Further only a few publications concerning PA oxidative degradation under the effect of UV-C irradiation have been reported. A high level of endogenous PA and plant tolerance to oxidative stress can be based not only on stress-induced but also on the constitutively high PA biosynthesis. It was reported that UV-C irradiation increased polyamine levels in the skin tissue of mango fruit. Higher levels of polyamines (putrescine and spermidine) were noted in mango fruits treated with UV-C for 10 minutes compared to controls and those irradiated for a longer period of 20 minutes (González-Aguilar et al., 2001). UV-C treated tomato fruits were firmer in texture and less red in colour indicating a delay in ripening (Liu et al., 1993). Similar results were noted in tomato fruit colour in addition, there was a delay in the appearance of the climacteric phase in irradiated fruit as well as reduced rates of CO_2 and C_2H_4 when compared to the controls. Further, optimal doses of UV-C produced higher levels of free and conjugated polyamines, particularly putrescine when compared to the control in tomato fruits during storage at 16°C (Maharaj et al., 1999). It has been reported that polyamines are not directly associated with a delay in tomato fruit ripening, but may prolong the fully-ripe stage before the fruit tissues undergo senescence (Tassoni et al., 2006). Thus the anti-senescent activity of polyamines may be related to their effectiveness as free radical scavengers more so than exerting opposite physiological effect to ethylene. Table 1 illustrates the increase in free PA in tomato pericarp subjected to UV-C hormic dose. In another study, UV-treated tomato during the first 5 days after irradiation, exhibited a significant induction of lipid peroxidation markers, suggesting that cell membrane was the primary target of UV-C irradiation. The levels then dropped lower than in control fruits suggesting the induction of a defence or repair mechanism. Treated fruits exhibited significantly less cell wall degrading enzymes (CWDE) activity compared to controls (Barka, et al., 2000a, 2000b).

Storage Time (days)	Free Putrescine in Tomato Pericarp (nmoles/g fw)		
	Control	UV-3.7 Merg/cm²	UV-24.4 Merg/cm²
0	123.8	125.0	120.0
7	306.7	418.1	386.4
14	374.5	513.0	400.7
21	39.0	452.0	292.8
SEM-Pooled	20.5		

Table 1. Effect of UV-C radiation on free putrescine levels in tomato pericarp during storage at 16°C. (n=3).

4. Conclusion

One of the primary objectives of postharvest interventions is to maintain the quality and safety of horticultural products along the value chain. Fruits and vegetables undergo a series of physiological changes from the point of harvest to final consumption. These processes eventually lead to senescence and ultimately death of the tissue. Over the last few decades, attention has shifted from quality maintenance of harvested crops to enhancement of health-promoting properties of phytochemicals. This has also shifted focus in developing technologies that may enhance shelf life and preserve the availability of such phytochemicals for the benefit of consumers. Non-ionizing, artifical, germicidial UV-C radiation is one such emerging technology. The role of UV-C hormesis (stimulation of beneficial responses by low levels of stressors which otherwise cause harmful responses) in altering the biotic relationship of higher plants as shown by changes in plant disease susceptibility, production of anti-fungal compounds, its relationship to polyamines, antioxidants and phenols will no doubt be a continued area of active research. This technology appears to be promising in the food and agriculture sector in order to minimize postharvest losses of horticultural crops by delaying ripening and senescence and controlling the incidence of decay which is also a significant and important development. In fact there is a view that UV-C is at the crossroads from laboratory research to commercial applications.

5. References

Adhami, V.M.; Syed, D.N.; Khan, N. & Afaq, F. (2008). Phytochemicals for prevention of solar ultraviolet radiation-induced damages. *Phytochemistry and Photobiology*, Vol.84, pp. 489-500.

Allende, A.; McEvoy, J.L.; Luo, Y.; Artes, F. & Wang, C.Y. (2006). Effectiveness of two-sided UV-C treatments in inhibiting natural microflora and extending the shelf-life of minimally processed "Red Oak Leaf" lettuce. *Food Microbiology*, Vol.23, No.3, (May 2006), pp. 241-249.

Allende, A.; Tomás-Barberán, F.A. & Gil, M.I. (2006). Minimal processing of healthy foods. *Trends in Food Science and Technology*, Vol.17, pp: 513-519.

Alothman, M.; Bhat, R. & Karim, A.A. (2009). UV-radiation-induced changes of antioxidant capacity of fresh-cut tropical fruits. Innovative *Food Science and Emerging Technologies*, Vol.10, No.4, pp. 512-516.

An I.N.; Liu G.X.; Zhang M.X.; Chen T.; Lin Y.H.; Feng H.Y.; Xu S.J.; Qiang, W.Y. & Wang, X.L. (2004). Effect of enhanced UV-B radiation on polyamine content and membrane permeability in cucumber leaves. *Russian Journal of Plant Physiology*, Vol. 51, No.5, pp. 658-662.

Anderson, J.G., Rowan, N.J., MacGregor, S.J., Fouracre, R.A. & Farish, O. (2000). Inactivation of food-borne enteropathogenic bacteria and spoilage fungi using pulsed-light. *IEEE Trans. Plasma Sci.* Vol.28, pp. 83-88.

Anderson Artes-Hernandez, F.; Escalona, VH.; Robles, PA.; Martinez-Hernandez, G.B. & Artes, F. (2009). Effect of UV-C radiation on quality of minimally processed spinach leaves. *J Sci Food Agric*, Vol. 89, pp. 414–21.

Arul, J.; Mercier, J.; Charles, M.T.; Baka, M. & Maharaj, R. (2001). Phytochemical treatment for control of post-harvest diseases in horticulture crops. In: *Physical Control Methods in Plant Protection*, C. Vincent, B. Panneton & F. Fleurat-Lessard (Eds.), pp. 146-161 INRA editions, Paris.

Barka, E.A.; Kalantari, S.; Makhlouf, J. & Arul, J. (2000a). Effects of UV-C irradiation on lipid peroxidation markers during ripening of tomato (*Lycopersicon esculentum* L.) fruits. *Australian Journal of Plant Physiology*, Vol.27, pp. 147-152.

Barka, E.A.; Kalantari, S.; Makhlouf, J. & Arul, J. (2000b). Impact of UV-C irradiation on the cell wall-degrading enzymes during ripening of tomato (*Lycopersicon esculentum* L.) fruit. *Journal of Agricultural and Food Chemistry*, Vol.48, pp. 667-671.

Bintsis, T.; Litopoulou-Tzanetaki, E. & Robinson R.K. (2000). Existing and potential application of ultraviolet light in the food industry – a critical review. *Journal of the Science of Food and Agriculture*, Vol.90, pp. 637-645.

Brady, C.J. (1987). Fruit Ripening. *Annual Review of Plant Physiology*, Vol.38, pp. 155-178.

Bregoli, A.M.; Scaramaglis, Costa G.; Sabatini, E.; Ziosi, V.; Stefania. B. & Torrigiani, T. (2002). Peach (Prunus persica) fruit ripening aminoethoxyvinyl glycine (AVG) and exogenous polyamines affect etlylene emission and flesh firmness *Physiologia Plantarum*, Vol.114, pp. 472 – 481.

Butt, M.S. & Sultan, M.T. (2011). Nutritional profile of vegetables and its significance to human health. In: *Handbook of Vegetables and Vegetable processing* N.K. Sinha (Ed.) pp.107-123 ISBN 978-0-8138-1541-1. Blackwell Publishing Ltd Ames Iowa.

Calabrese, M.M.; McCarthy.; M.E. & Kenyon, E. (1987). The occurrence of chemically induced hormesis. *Health Physics*, Vol. 52, pp. 531-541.

Caldwell, M.M. (1981). Plant response to solar ultraviolet radiation, In: *Physiological Plant Ecology 1-Responses to the Physical Environment*. O.L. Lange.;P.S. Nobel.; C.B. Osmond & H. Ziegler, (Eds.), Chpt 6, Springer-Verlac, ISBN, Berlin, Heidelberg.

Charles, M.T.; Tano, K.; Asselin, A. & Arul J. (2009). Physiological basis of UV-C induced resistance to *Botrytis cinerea* in tomato fruit. V. Constitutive defence enzymes and inducible pathogenesis related proteins. *Postharvest Biology and Technology*, Vol. 51, pp. 414–24.

Costa. L.; Vincente A.R.; Civello P.M.; Chaves A.R. & Martinez G.A. (2006). UV-C treatment delays postharvest senescence in broccoli floret. *Postharvest Biology and Technology*, Vol. 39, No. 2, pp. 204-210.

Cunningham, M.L.; Johnson, J.S.; Giovanazzi, S.M. & Peak, M.J. (1985). Photosensitized production of superoxide anion by monochromatic (290-405 nm) ultraviolet radiation of NADH and NADPH coenzymes. *Photochemistry and Photobiology*, Vol. 42, No.2, pp. 125-128.

Droplet, G.; Dumbroff, E.B.; Legge, R.L. & Thompson J.E. (1986). Radical scavenging of polyamines. *Phytochemistry*, Vol.25, pp. 367-37.

El Ghaouth, A.; Wilson, C.L. & Callahan, A.M. (2003). Induction of chitinase, beta-1,3-glucanase, and phenylalanine ammonia-lyase in peach fruit by UV-C treatment. *Phytopathology*, Vol.93, pp. 349–55.

Elmnasser, N.; Guillou, S.; Leroi, F.; Orange, N.; Bakhrouf, A. & Federighi, M. (2007). Pulsed-light system as a novel food decontamination technology: a review. *Canadian Journal of Microbiology*, Vol.53, No.7, pp.813-821.

FAO, (2009). Fruits and Vegetables: An Overview on Socio-Economical and Technical Issues, In: *Handling and preservation of fruits and vegetables by combined methods for rural areas*. Chapter 1, FAO Agricultural Services Bulletin 149.
http://www.fao.org/DOCREP/005/Y4358E/y4358e04.htm

Galston, A.W. & Sawhney, R.K. (1987). Polyamines as indigenous growth regulators. In: *Plant Hormones and Their Role in Plant Growth and Development 2nd Edition* . P.J Davies (Eds.), Dordrecht: Kluwer Academic Press, pp. 155-178.

Galston, A.W. & Sawhney, R.K. (1990). Polyamines in plant physiology. *Plant Physiology* Vol.94, pp. 406 – 410.

Gong, Y. & Mattheis, J.P. (2003). Effect of ethylene and 1-methyl-cyclopropane on chlorophyll catabolism of broccoli florets. *Plant Growth Regulation*, Vol. 40, pp. 33-38.

Gómez-López, V.M.; Devlieghere, F.; Bonduelle, V. & Debevere, J. (2005). Factors affecting the inactivation of micro-organisms by intense light pulses. *Journal of Applied Microbiology*, Vol.99, No.3, pp. 460-470.

González–Aguilar, G.A.; Wang, C.Y.; Buta, J.G. & Krizek, D.T. (2001). Use of UV-C irradiation to prevent decay and maintain postharvest quality of ripe "Tommy Atkins" mangoes. *International Journal of Food Science*, Vol.36, pp. 767-773.

González –Aguilar, G.A.; Villega-Ochroa M.A.; Martinez-Telez M.A.; Gardea A.A. & Ayala – Zavala J.F. (2007). Improving antioxidant capacity of fresh cut mangoes treated with UV- C. *Journal of Food Science*, Vol.36, pp. S197-S202.

Hamana, K. & Matsuzaki, S. (1982). Widespread occurance of non spermidine and non spermine in eukargolic algae *Journal of Biochemistry*, Vol.91, pp. 1321–1324.

Heby, O. & Persson, L. (1990). Motecular genetics of polyamine synthesis in eukaryotic cells. *Trends in Biochemical Science*, Vol.15, pp. 153 – 158.

Heber, D. (2004) Vegetables, fruits and phytoestorogen in prevention of diseases. *Journal of Postgraduate Medicine*, Vol.50, pp. 145-149.

Hemmaty, S.; Moallemi, N. & Naseri, L. (2006). Shelf-life and quality of apple fruits in response to postharvest application of UV-C radiation. *Journal of Applied Horticulture*,Vol.8, No.2, (July-December 2006), pp. 114-116.

Jaleel, C.A.; Riadh, K.; Gopi, R.; Manivannan, P.; Inès, J.; Al-Juburi, H.J.; Hong-Bo, Z.S. & Panneerselvam, R. (2009). Antioxidant defence responses: physiological plasticity in higher plants under abiotic constraints. *Acta Physiol Plant*, Vol.31, pp. 427-436.

Jiang,T.; Jahangir, M.M.; Jiang, Z. & Lu, X. (2010). Influence of UV-C treatment on antioxidant capacity, antioxidant enzyme activity and texture on postharvest Shitake (*Lentinus edodes*) mushrooms during storage. *Postharvest Biology and Technology*, Vol.56, No.3, pp. 209-215.

Kataoka, I.; Sugiyama, A. & Beppu, K. (2003). Role of ultraviolet radiation in accumulation of anthocyanin in berries of 'Gros Colman' grapes (*Vitis vinifera L.*). *Journal of the Japenese Society for Horticulture Science*, Vol.72, pp. 1–6.

Katerova, Z. (2009). Prolonged influence of short pulses ultraviolet-C radiation on young pea plant does not alter important antioxidant defence enzyme activities in young leaves. General and Applied Plant Physiology, Vol.35, No.3-4, pp. 134-139.

Kowalski, W. (2009). *Ultraviolet Germicidal Irradiation Handbook*, Springer-Verlag, Berlin Heidelberg pp.1-493.

Langcake, P. & Pryce, R. (1973). The production of resveratrol and viniferins by grapevines in response to ultraviolet irradiation. *Phytochemistry*, Vol.16, pp. 1193-1196.

Li, J.; Zhang, Q.; Cui, Y.; Cao, J.; Zhao, Y. & Jiang, W. (2010). Use of UV-C treatment to inhibit microbial growth and maintain the quality of Yali pear. *Journal of Food Science*, Vol.75, No.7, pp. M503-M507.

Liu, J.; Stevens, C.; Khan, V. A.; Lu, J. Y.; Wilson, C. L.; Adeyeye, O.; Kabwe, M.K; Pusey, P.L.; Chalutz, E.; Sultana, T. & Droby, S. (1993). Application of ultraviolet-C light on storage rots and ripening of tomatoes. *Journal of Food Protection*, Vol.56, (October 1993), pp. 868– 872.

Liu, L.H.; Zabaras, D.; Bennett, L.E.; Aguas, P. & Woonton, B.W. (2009). Effects of UV-C, red light and sun light on the carotenoid content and physical qualities of tomatoes during post-harvest storage. *Food Chemistry*, Vol.115, pp. 495–500.

Maharaj, R.; Arul, J. & Nadeau, P. (1999). Effect of photochemical treatment in the preservation of fresh tomato (*Lycopersicon esculentum* cv. Capello) by delaying senescence. *Postharvest Biology and Technology*, Vol.15, No.1, pp. 13-23.

Maharaj, R.; Arul, J. & Nadeau, P. (2010). UV-C Irradiation of Tomato and its Effects on Colour and Pigments. *Advances in Environmental Biology*, Vol.4, No.2, pp. 308-315, ISSN 1995-0756.

Marquenie, D.; Michiels, C.W.; Geeraerd, A.H.; Schenk, A.; Soontjens, C.; Van Impe, J.F. & Nicolaï, B.M. (2002). Using survival analysis to investigate the effect of UV-C and heat treatment on storage rot of strawberry and sweet cherry. *International Journal of Food Microbioogy*, Vol.73, pp. 191-200.

Martinez–Romero, D.; Serrano, M.; Carbonell, A.; Burgos L.; Riquelme, F. & Valero, D. (2002). Effects of post harvest putrescine treatment on extending shelf life and reducing mechanical damage in apricot. *Journal of Food Science*, Vol.67, No.5, pp. 1706 – 1712.

Martin-Tanguy, J. (1997). Conjugated polyamines and reproductive development: biochemical, molecular and physiological approaches. *Physiologia Plantarum*, Vol. 100, pp. 675-688.

Mercier, J.; Baka, M.; Reddy, B.; Corcuff, R. & Arul, J. (2001). Shortwave ultraviolet irradiation for control of decay caused by *Botrytis cinerea* in bell pepper: Induced resistance and germicidal effects. *J Am Soc Hort Sci*, Vol.126, pp. 128–33.

Nambeesam, S.; Handa, A.K. & Maltoo, A.K. (2008). Polyamines and regulation of ripening and senescence In : *Postharvest Biology and Technology of fruits, vegetables and flower seeds*. G. Paliyath, D.P Murr, A.K Handa and S.Larie.(Eds.) ISBN 978-0-8138-0408-8 Wiley, Black Well , Iowa U.S.A.

Nigro, F.; Ippolito, A. & Lima, G. (1998). Use of UV-C light to reduce Botrytis storage rot of table grapes. *Postharvest Biology and Technology*, Vol.13, pp. 171–181.

Nigro, F.; Ippolito, A.; Lattanzio, V.; Di Venere, D. & Salerno, M. (2000). Effect of ultraviolet-C light on postharvest decay of strawberry. *Journal of Plant Pathology*, Vol.82, No.1, pp. 29-37.

Pandey, S.; Ranade, S.A.; Nagar, P. K. & Kumar, N. (2000). Role of Polyamines and Ethylene as Modulators of plant senescence. *Journal of Bioscience*, Vol.25, No.3, pp. 291 – 298.

Pearce, R.S. (1999). Molecular analysis of acclimation to cold. *Plant Growth Regulation*, Vol. 29, pp. 47-76.

Ramana-Luximon, A.; Behorun, T. & Crozier, A. (2003). Antioxidant actions and phenolic and Vitamin C content of common Mauritian exotic fruits. *Journal of the Science of Food and Agriculture*, Vol.83, pp. 496-502.

Rame, J.; Chaloupecky, V.; Sojkova, N. & Bencko, V. (1997). An attempt to demonstrate the increased resistance of selected bacterial strains during repeated exposure to UV-radiation at 254 nm. *Central European Journal of Public Health*, Vol.4, pp. 30–31.

Roberts, D.R.; Dumdroff, E.B. & Thompson, J.E. (1986) Exogenous polyamines alter membrane fluidity – a potentail for misinterpretation of their physiological role *Planta*, Vol.167, pp. 395-401

Rodriguez–Garay, B.; Phillip, G.C. & Kuehn, G.D. (1989). Defection of norspermine in *Medicago sativa* L.(alfalfa). *Plant Physiology*, Vol.89, pp. 525–529.

Saftner, R.A. & Baldi, G.B. (1990). Polyamine levels and tomato fruit development possible interaction With Ethylene. *Plant Physiology*, Vol.92, No.2, pp. 547–550.

Schreiner, M. & Huyskens-Keil, S. 2006. Phytochemicals in fruit and vegetables: health promotion and postharvest elicitors, In: *Red Orbit*, 20.06.06, Available from http://www.redorbit.com/news/science/543640/phytochemicals_in_fruit_and_veg etables_health_promotion_and_postharvest_elicitors

Serrano, M.; Martinez-Romero, D.; Guillen, F. & Valero, D. (2003). Effects of exogenous putrescine on improving shelf life of four plum cultivars. *Postharvest Biology and Technology*, Vol.30, pp. 259-271.

Shahidi, F.; Chandrasekara, A. & Zhong ,Y. (2011). Bioactive phytochemicals in vegetables In: *Handbook of Vegatables and Vegetable processing*. N.K Sinha (Ed), pp. 125-158. Blackwell publishing Ltd. Ames Iowa.

Shahidi, F. & Naczk, M. (2004). *Phenolics in Food and Neutraceuticals*, pp. 1-82. CRC Press Boca Raton

Shama, G. (2007). Process challenges in applying low doses of ultra violet light to fresh produce for eliciting beneficial hermetic responses. *Postharvest Biology and Technology*, Vol.44, No.1, pp. 1-8.

Shama, G. & Alderson, P. (2005). UV hormesis in fruits: a concept ripe for commercialization. *Trends in Food Science and Technology*, Vol.16, pp. 128–136.

Slocum, R.D.; Kaur-Sawhney, R. & Galston, A.W. 1984. The physiology and biochemistry of polyamines in plants. *Archives of Biochemistry and Biophysics*, Vol.235, pp. 283-303.

Stapleton, A.E. (1992). Ultraviolet radiation and plants: burning questions. *The Plant Cell*, Vol.4, (November 1992), pp. 1353-1358.

Stevens, C.; Khan, Y.A.; Tang, A.Y. & Lu, Y.J. (1990). The effect of ultraviolet radiation on mold rots and nutrients of stored sweet potatoes. *Journal of Food Protection*, Vol.53, No.3, pp. 223-226.

Stevens, C.; Wilson, C.L.; Lu, Y.J.; Khan, V.A.; Chalutz, E.; Droby, S.; Kabwe, M.K; Haung, Z.; Wisniewski, M.E.; Adeyeye, O.; Pusey, L.P. & West, M. (1996). Plant hormesis induced by ultraviolet light-C for controlling postharvest diseases of tree fruits. *Crop Protection*, Vol.15, pp. 129-134.

Tait, G.H. (1985). Bacterial Polyamines, structure and biosynthesis: *Biochemical Society Transactions*, Vol.13, pp. 316 – 318.

Tassoni, A.; Watkins, C.B. & Davies, P.J. (2006). Inhibition of the ethylene response by 1-MCP in tomato suggest that polyamines are not involved in delaying ripening, but may moderate the rate of ripening or over-ripening. Journal of Experimental Botany, Vol.57, No.12, pp. 3313-3325.

Tomás-Barberán, F.A. & Espín, J.C. (2001). Phenolic compounds and related enzymes as determinants of quality in fruits and vegetables. Journal of the Science of Food and Agriculture, Vol.81, pp. 853-876.

Valero, D.; Romero, D.M. & Serrano, M. (2002). The role of polyamines in the improvement of shelf life of fruit. *Trends in Food Science and Technology*, Vol.13, pp. 228 – 234.

Vikhanski, L. (1989). Environmental Watch. *Discoverer*, Vol.9, pp 32.

Wang, C.Y.; Conway, W.S.; Abbott, J.A. & Kramer, G.F. (1993). Post harvest infiltration of polyamines and calcium influences ethylene production and technology changes in golden delicious apples. *Journal of the American Society of Horticultural Science*, Vol.188, No. 6, pp. 801 – 806.

Wills, R.B.H.; McGlasson, W.D.; Graham, D.; Lee, T.H. & Hall, E.C. (1981). *Postharvest: An Introduction to the Physiology and Handling of Fruits and Vegetables*, AVI Van Nostrand, ISBN, 9780870554025, New York.

Yahia, E. (1998). Modified and controlled atmospheres for tropical fruits. *Horticultural Reviews*, Vol. 22, pp. 123-183.

The Role of Root-Produced Volatile Secondary Metabolites in Mediating Soil Interactions

Sergio Rasmann[1*], Ivan Hiltpold[2] and Jared Ali[3]
[1]Department of Ecology and Evolution, UNIL Sorge, Le Biophore
University of Lausanne, Lausanne
[2]FARCE laboratory, University of Neuchâtel, Neuchâtel
[3]Department of Ecology and Evolutionary Biology
Cornell University, Corson Hall, Ithaca, NY
[1,2]Switzerland
[3]USA

1. Introduction

Since Darwin's suggestion that natural selection accounts for the diversity of plant morphological and chemical attributes, thousands of papers have been devoted to the ecology and evolution of plant secondary metabolites. Indeed, it is estimated that plants may produce over 200, 000 different compounds, the majority of which are classified as secondary metabolites (Pichersky and Gang 2000). The incredible diversity of particular classes of secondary metabolites is stunning. Terpenes, for example, comprise more than 30'000 described compounds (Hartmann 2007). Such incredible diversity of forms can be originated from various enzymes catalyzing the binding of different precursors (Wojciechowski 2003), promiscuity of enzymes (including multiple product and substrate enzyme specificity), changes in cellular compartmentalization patterns (Pichersky and Gang 2000; Bauer *et al.* 2010), or the matrix-like structure of pathways where natural products are formed by elaborate arrays of enzymes, concertedly controlled by the expression of their respective genes (Lewinsohn and Gijzen 2009).

Among early pioneers of plant secondary metabolites as mediators of ecological interactions was Jena botanist Ernst Stahl. Stahl, a fervent follower of Darwin's ideas, suggested some secondary plant metabolites might play protective roles, and thus herbivores may be a primary selecting force for specific biochemical compositions in plants (Stahl 1888). Some seventy years later, Fraenkel's (1959) paper in *Science* resurrected Stahl's pivotal ideas that plants produce an incredible diversity of secondary metabolites for mediating interactions with herbivores, thus providing a *raison d'être* for secondary plant substances. Fraenkel proposed that secondary metabolites have evolved to deter or kill herbivores. Additional coevolutionary frameworks such as Ehrlich and Raven's (1964) seminal paper on

* Corresponding Author

chemically-mediated coevolution between plants and butterflies, helped to foster a 'new wave' of research focused on plant organic chemistry and evolutionary ecology. Fifty years later, we have finally come to acknowledge the fact that there are ecological and evolutionary reasons for the fact that plants have become true "chemical factories" (Hartmann 2008).

Besides the production of toxic or anti-nutritive secondary metabolites, upon attack, individual plants rely on a matrix-like variety of defense mechanisms, involving physical barriers, and the attraction of "body-guards" to the plant (Schoonhoven, van Loon, and Dicke 2005; Agrawal and Fishbein 2006; Howe and Jander 2008; Karban and Baldwin 1997; Wink 2008; Kessler and Baldwin 2002). Indeed, an additional defence strategy available to plants when 'attacked' by herbivores is to attract mobile enemies of the herbivore. This phenomenon has been referred to as an "indirect plant defence" (Dicke and Sabelis 1988; Turlings, Tumlinson, and Lewis 1990). Natural enemies can be attracted to plants that harbour their host or prey by providing refuges or alternative food sources such as extrafloral nectars (i.e. resource-based indirect defences), or through various visual, chemical and tactile cues (i.e. information-based indirect defences) (Dicke and Baldwin 2010; Heil 2008; Kessler and Heil 2011). Particularly, plant volatile organic compounds (VOCs), which mainly comprise terpenoids, fatty acid derivatives, phenyl propanoids and benzenoids (Dudareva, Pichersky, and Gershenzon 2004) have been the center of intensive studies of plant-herbivore-predator interaction for more than two decades (Dicke and Sabelis 1988; Turlings, Tumlinson, and Lewis 1990). VOCs blends can be complex, comprising hundreds of compounds, some of which are not produced by intact or mechanically damaged plants and others of which are synthesized *de novo* in response to herbivore attack (Mumm and Dicke 2010; Turlings and Wäckers 2004).

Interestingly, the role of secondary metabolites as resistance factors has mainly been studied for aboveground plant parts and their associated communities. With the exception maybe of agricultural pests such as the larvae of various root flies feeding on cabbage, carrot and onion (Johnson and Gregory 2006; Blossey and Hunt-Joshi 2003), and root lesion nematodes *Pratylenchus* spp. (Potter *et al.* 1999), little attention has been paid to the role of secondary metabolites as defences against belowground feeding herbivores (Rasmann *et al.* 2011; van Dam 2009), and how this might shape soil communities (Wenke, Kai, and Piechulla 2010; Bais *et al.* 2006). Roots contain an equally rich variety of plant secondary metabolites as shoots do. Depending on the type of secondary metabolite that is analysed and the ontogenetic stage at which root and shoot levels are compared, the level of secondary plant compounds in the root may be even higher than in the shoot (Kaplan *et al.* 2008; Rasmann and Agrawal 2008). In a compilation of traditional Chinese pharmacopeia, it is surprising to note that more than one-quarter of the preparations are derived from roots and/or rhizomes (Bensky and Gamble 1986). Additionally, the past ten years have seen a surge of work on root exudation and their effect on soil communities. Similarly, as for aboveground herbivores, it has been found that root herbivores use specific chemicals emitted or exuded by the roots as cues to locate their host plant (Johnson and Gregory 2006; Perry and Moens 2006). This and the fact that belowground herbivores can do as much, or sometimes even more, damage to wild plants as aboveground feeding herbivores (Blossey and Hunt-Joshi 2003; Maron 1998; Hunter 2001) indicates that secondary plant metabolites fulfil similar roles belowground as they do aboveground. The soil surrounding and in close contact with roots of plants is generally termed "rhizosphere" a zone influenced by the plant with high biological activity, influencing many trophic levels (e.g. Bardgett 2005). Plant root exudates have been shown to mediate a wide variety of soil interactions in the rhizosphere (Bais *et al.* 2006). Plants can release up to 20% of the photosynthesized fixed carbon via root exudation (Barber and Martin 1976), affecting both biotic and abiotic soil conditions. Root exudates can chelate inorganic soil contaminants, change rhizosphere pH, and may increase degradation of organic contaminants by microbial metabolism. In addition to soil dwelling organisms, abiotic condition can also spur root exudations. For example, drought stress seem to stimulate the exudation of carbon-based molecules (Henry et al. 2007). Mucilage exudation under drought may favor water retention in the soil, and increase rhizosphere stabilization of soil particles (McCully and Boyer 1997). Among the wide variety of exuded molecules ranging from amino acids to complex polysaccharides and proteins, smaller, more volatile compounds have also been shown to directly or indirectly influence soil community of organisms. Thus, through the exudation of plant metabolites, plants may not only protect themselves against root herbivores and pathogens, but contribute to the regulation of the soil microbial community, inhibit growth of neighboring plants, promote the establishment of beneficial symbioses, and regulate soil physic-chemical properties. However, little is known about the biological significance of these metabolites for overall plant physiology and development, as well as their role in mediating soil food webs and interactions (Flores, Vivanco, and Loyola-Vargas 1999).

The role of root produced secondary metabolites in biotic interactions has been reviewed in recent seminal papers (e.g. Badri *et al.* 2009; Bais *et al.* 2001; Flores, Vivanco, and Loyola-Vargas 1999; Wenke, Kai, and Piechulla 2010)}. Here, we will focus our attention to the role of plant VOCs exudation and their effect on various soil organisms including other plants, microorganisms, herbivores and particularly predators of the herbivore. Particularly, volatile terpene production, physiology, emission, and effect on other organisms has been the concern of incredible work on aboveground interactions (Pichersky, Noel, and Dudareva 2006), and a recent trend in research has highlighting their role in belowground interactions and physiology (Degenhardt 2009; Degenhardt *et al.* 2009; Kollner *et al.* 2008). The evolution, detection, and manipulation of such compounds will finally be discussed as a potential tool for biotechnological improvement of resistance against agricultural root feeding pests (Turlings and Ton 2006).

2. Root volatile exudates and their effect on soil biota

Plant roots are known to emit an incredible variety of compounds from their roots, which are known to affect interactions between plants and other organism (Flores, Vivanco, and Loyola-Vargas 1999). Here we focus our attention on small lipophilic molecules that are volatile at ambient pressure and temperature. Such molecules have only recently seen a surge of interest as having potential to manipulate soil food-web and dynamics. With a general survey of the available literature, we will below discuss the role of volatile organic compounds (VOCs) in mediating plant-plant, plant-microbe, plant-herbivore, and plant-predator interactions (Fig. 1).

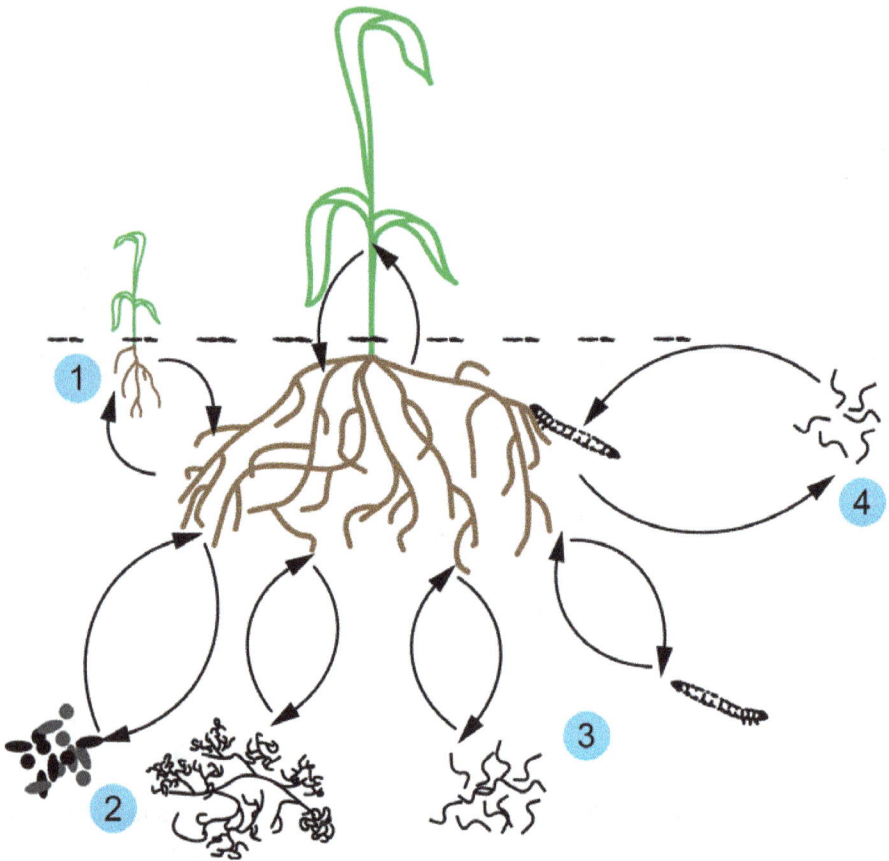

Fig. 1. Diagram showing different soil organisms under possible influence of root emissions of volatile organic compounds (VOCs). From left to right: root emissions of VOCs have been shown to influence 1) neighbouring plants, 2) microbes including bacteria and fungi, 3) herbivores including nematodes and insects, as well as 4) predators of the herbivores such as entomopathogenic nematodes seeking insect larvae for food. In text we highlight each of these interactions. Arrows represent emission of volatile secondary metabolites from roots, as well as animal or micro-organism movement toward roots.

2.1 Plant-plant interaction

Root volatiles can mediate allelopathic interaction between neighboring plants. After the first demonstration of seed germination and growth inhibition by plant volatiles (Muller and Muller 1964), several examples of belowground allelopathy have been described and their importance in plant biological invasion is seriously considered. Callaway and Ridenour (2004) proposed the 'Novel Weapon' hypothesis, which states that invading plants may possess novel biochemical weapons that function as unusually powerful allelopathic agents outside their native range, this including VOCs. In support of this, Barney *et al.* (2009) recently demonstrated that monoterpenes volatile emission from the perennial invasive plant *Artemisia vulgaris* could negatively impact native population of *Solidago canadensis*, hence helping in the invasion of this alien species. Root VOCs are also involved in the interaction between parasitic plants and their host plants (Bouwmeester *et al.* 2003). For instance, Zwanenburg and colleagues (2009) reviewed the bio-activity of the strigolactone family on the germination of parasitic weed germination. The use of such molecules could help in managing parasitic plants such as *Striga spp.* or *Orobanche spp.* population (Zwanenburg *et al.* 2009). Furthermore, root volatiles might also have positive effects on neighboring plants. Indeed, Zwahlen *et al.* (unpublished data) could measure a so-called priming effect of maize damaged roots on neighboring conspecifics. In the laboratory, they have shown that roots in contact with volatiles emitted by an insect induced plant would eventually be faster in responding and emit more volatiles when fed on by the same herbivore (Zwahlen *et al.*, unpublished data). Such examples are still scarce in the literature but may open unsuspected perspective in understanding belowground communities and evolution.

2.2 Plants-microbes interaction

Plants are also able to chemically interact with microbe populations using VOCs. Since microbe populations are often limited by carbon availability in soil, root volatiles, especially monoterpenes, might be a relevant carbon source and contribute to the belowground carbon cycle (Owen *et al.* 2007; Zak *et al.* 1994). Some organisms such as *Pseudomonase fluorescence* or *Alcaligenes xylosoxidans* have been able to develop on root monoterpenes as unique source of carbon (Kleinheinz *et al.* 1999). Simpler root VOCs such as carbon dioxide (CO_2) play an important generic role in plant belowground interactions with other organisms (Johnson and Gregory 2006). However, CO_2 for example, has also been shown to also mediate highly specific interactions. Indeed, Bécard and Piché (1989) could show that carbon dioxide was crucial in the growth of the vesicular-arbuscular fungus *Gigaspors margarita*. The authors showed a synergistic effect of CO_2 and root exudate factors in the hyphal growth of the obligate biotrophic symbiotic fungus. Carbon dioxide and root exudates taken alone had little or no effect, but when mixed together, hyphal growth was significantly stimulated (Bécard and Piché 1989). Further experimentations suggested that in this particular interaction carbon dioxide served as an essential source of carbon for the fungal growth (Bécard and Piché 1989). Since then, numerous plant volatile exudates, mainly belonging to the sesquiterpene lactone family, have been identified potentially mediating plant-microbe interactions. As an example, the strigolactone 5-desoxy-strigol has been isolated from *Lotus japonicus*. This volatile significantly triggered hyphal branching in *G. margarita* (Akiyama, Matsuzaki, and Hayashi 2005).

2.3 Plant-herbivore interaction

Two major groups of organisms utilize living roots of plants as their main diet: the insects and the nematodes. A recent survey of the literature indicated that about 17% of all insect families of North America contain species of root feeders (including chewers, sap suckers, and gall makers) (Rasmann and Agrawal 2008). During at least one life stage, a root-feeding phase is common to a variety of species belonging to the orders of crickets (Orthoptera), butterflies (Lepidoptera), flies (Diptera) , true bugs (Homoptera), beetles (Coleoptera) and ants or wasps (Hymenoptera) (Brown and Gange 1990). Root-feeding insects play an important role in both agricultural and natural ecosystems (Blossey and Hunt-Joshi 2003; Wardle *et al.* 2004). Indeed, through physiological and physical changes of roots, belowground herbivores have the potential to shape plant communities (De Deyn *et al.* 2003), belowground microorganism and macroorganism communities (Wardle 2006), as well as aboveground arthropod communities (Bezemer and van Dam 2005).

Volatile organic compounds have been commonly identified as arthropod attractants belowground. A nice review by Wenke *et al.* (2010) highlighted a whole range of compounds that are used by herbivores to locate the food source. Almost ubiquitous signals in the soils are the emissions of CO_2 by roots. A compilation of studies that looked at host attractants for root feeding arthropods underscored CO_2 as a major attractant for at least 20 studies examined (Johnson and Gregory 2006). Detection of CO_2 seems however to be dose-dependent, and soil insect are able to detect very small differences in the concentration of the CO_2 (Johnson and Gregory 2006). Whereas low concentrations are attractive, high concentrations of CO_2 may cause disorientation. Additionally, the orientation of insects within CO_2 gradients could be "masked" by other non-volatile or volatile gustatory and olfactory stimuli exuded by roots (Reinecke, Müller, and Hilker 2008). Besides CO_2, several disulfides and trisulfides have been identified as potent attractants for the root-feeding larvae of the fly *Delia antiqua* in *Allium cepa* (Carson and Wong 1961). Fatty acids in oaks (*Quercus* sp.) and monoterpenes in carrot (*Daucus carota* ssp. *sativus*) plants triggered the attraction of forest cockchafer larvae, *Melolontha hippocastani* (Weissteiner and Schütz 2006). Volatiles of fresh perennial ryegrass roots attracted larvae of *Costelytra zealandica* (Sutherland and Hillier 1972), and roots of *Medicago sativa* and *Trifolium pratense* attracted larvae of *Sitona hispidulus* (Wolfson 1987).

The second most important group of root feeders encompasses the plant parasitic nematodes. All species are obligate parasites, feeding exclusively on the cytoplasm of living plant cells. The most economically important groups of nematodes are the sedentary endoparasites, which include the genera *Heterodera* and *Globodera* (cyst nematodes) and *Meloidogyne* (root-knot nematodes). Cyst and root-knot nematodes differ in their parasitic life-cycle strategies. Cyst nematodes enter roots and move to the vascular cylinder, before establishing their feeding site; a multinucleate syncytium which results from the breakdown of the cell walls between the initial feeding site cell and its neighboring cells. In contrast to the cyst nematode, the juvenile of the root-knot nematode moves intercellularly after penetrating the root, migrating down the plant cortex towards the root tip. The juveniles then enter the base of the vascular cylinder and migrate up the root, where they establish a permanent feeding site in the differentiation zone of the root by inducing the formation of large, multinucleated cells. Then, the plant cells around the feeding site divide and swell, causing the formation of galls or 'root knots' (Williamson and Gleason 2003). A critical step in nematode life-cycle, host searching after hatching or molting into a new life stage, seems to also involve olfactory organs and other sensory organs, which allows sensing chemical gradients in soil (Robinson 2003), plant cell— specific surface determinants, as well as electrical signals (Riga 2004). Plant signals are essential for nematodes to locate hosts and feeding sites, however, besides the general signal furnished by carbon dioxide emissions, what particularly triggers and direct plant parasitic nematode attraction is still largely unexplored. Carbon dioxide was shown to attract *Meloidogyne incognita* (Dusenbery 1987; Pline and Dusenbery 1987), *Ditylenchus dipsaci* (Klinger 1963), and *Caenorhabditis elegans* nematodes (Dusenbery 1980). Aggregation and attraction have been demonstrated in plant parasitic nematodes, for example *M. juvanica* and *G. rostochiensis* juveniles respond to tomato (Prot 1980) and potato (Rolfe, Barrett, and Perry 2000) root diffusates, respectively (reviewed in Curtis, Robinson, and Perry 2009). Only very recently, it was shown that phytopathogenic nematodes can also follow gradients of herbivore induced terpene volatile organic compounds. *Tylenchulus semipenetrans* nematodes were more attracted to *Citrus* roots infested by weevil larvae compared to uninfested plants (Ali, Alborn, and Stelinski 2011). A series of terpene compounds were identified a possible attractants for the plant parasitic nematode, including α-pinene, β-pinene, limonene, geijerene, and pregeijerene (Ali, Alborn, and Stelinski 2011). Interestingly the same compounds were also responsible for the attraction of entomopathogenic nematodes, which function as bodyguards against insect root feeders (Ali, Alborn, and Stelinski 2010, 2011). It seems therefore that a plant parasitic nematode was able to exploit plant signals used for a mutualism.

2.4 Belowground tritrophic interactions

Mainly because of methodological constraints, most of the research on plant VOCs released after insect herbivory has so far been conducted mainly aboveground (reviewed in Kessler and Morrell 2010). However, an increasing number of studies are showing that herbivore induced belowground volatiles might also trigger predator attraction in the soil. For example, *Neoseiulus cucumeris* female predatory mites of rust mites (*Aceria tulipae*) responded to belowground volatiles signals of tulip bulbs infested by *A. tulipae* but not to volatiles of untreated or wounded bulbs (Aratchige, Lesna, and Sabelis 2004). Two inspiring papers demonstrated for the first time that unknown emissions of odorous cues were responsible for attracting entomopathogenic nematodes to insect damaged roots (Boff, Zoon, and Smits 2001; van Tol *et al.* 2001). To date, few additional tritrophic interactions implying belowground VOCs signaling have been described both in agricultural systems (Ali, Alborn, and Stelinski 2010, 2011; Rasmann *et al.* 2005) and in wild environment (Rasmann *et al.* 2011). Ali, Alborn, and Stelinski (2010) have demonstrated that citrus roots upon feeding by the root weevil *Diaprepes abbreviates* emit several terpenes in the surrounding soil. Using belowground olfactometers Ali *et al.* (2010) could show that the entomopathogenic nematode *Steinernema diaprepesi* was significantly more attracted by citrus roots induced by the insect pest larvae than by roots mechanically damaged or by control empty pots. However, Ali *et al.* (2011) recently pointed out that insect induced roots of citrus tree could also attract the phytopathogenic nematode *Tylenchulus semipenetrans* (see above section). Consequently, this may reduce the exploitation of citrus induced VOCs emission in biological control strategies targeting *Diaprepes abbreviates* where rootstocks are not resistant to *T. semipenetrans*.

Also recently, Rasmann *et al.* (2010) showed that the common milkweed *Asclepias syriaca*, which is generally fed by the specialist root herbivore larvae of the cerambycid beetle *Tetraopes tetraophthalmus,* can release volatiles in the soil. Increased emissions of VOCs after damage were correlated with increased entomopathogenic nematodes *Heterorhabidtis bacteriophora* nematodes in lab experiments. Subsequent field trials demonstrated that soil inoculation of nematodes benefitted the plants by restoring their biomass lost due to herbivory to control levels (Rasmann *et al.* 2010). Whether this was correlated with higher levels of emission in damaged plant was not however assessed. Root emission of *A. syriaca* plants are a very complex mixture of >30 compounds of which only few are described as being in the terpene family (Rasmann *et al.* 2010). Such complex blend, by itself, impedes the assessment of which particular compounds are really responsible for the attraction. A problem that is well known for above-ground systems, where, the emerging picture is that VOCs production in plants is the result of diffuse selection due to multiple players interacting with the plant (Kessler and Heil 2011). As aboveground, the functional role of belowground compounds, individual components or complex blends, would benefit our knowledge of organisms' intimate relationship with the plant, advanced metabolomics and multivariate statistical tools (van Dam and Poppy 2008).

Another example of highly complex volatile blends comes from the roots of cotton (*Gossypium herbaceum*). After feeding by the generalist root feeder larvae of the chrysomelid beetle *Diabrotica balteata*, cotton plants were scored to emit >10 compounds, among which the at least 7 terpenoid volatiles were observed (Rasmann and Turlings 2008). Among all cotton VOCs induced by the chrysomelid larva, the sesquiterpenoid aristolene was discussed as being a good candidate for playing a major role in *H. megidis* nematode attraction. This however remains to be confirmed in future studies (Rasmann and Turlings 2008). In the same study, among corn and cotton, nematode preference was also tested against damaged roots of cowpea (*Vigna unguiculata*) plants. In contrast to corn and cotton, cowpea plants emitted almost undetectable amounts of volatiles, which also resulted in lower nematode attractions (Rasmann and Turlings 2008).

Indeed, by far, corn system, first described by Rasmann *et al.* (2005) is today the best known belowground tritrophic interaction (Fig. 2). Upon attack of the voracious larvae of the western corn rootworm, *Diabrotica virgifera virgifera*, European maize varieties emit in soil the sesquiterpene (*E*)-β-caryophyllene (EβC) (Kollner *et al.* 2008; Rasmann *et al.* 2005), a highly attractive VOCs to entomopathogenic nematodes *Heterorhabidtis megidis* in the laboratory as well as in the field (Hiltpold, Toepfer, *et al.* 2010; Kollner *et al.* 2008; Rasmann *et al.* 2005) (Fig. 2). A series of experiments with various corn lines and synthetic compounds have shown that EβC is an ideal compound to diffuse through the complex belowground soil compartment, and that is among the less costly terpenoid that could be travelling within the soil matrix (Hiltpold and Turlings 2008), and that is under selection (Kollner *et al.* 2008; Kollner *et al.* 2004). Its production within the root system appears to be systemic even though the root area upon feeding emits more of the volatile and is more attractive to the *H. megidis* than distal regions (Hiltpold et al. 2011).

Recently, Hallem *et al.* (2011) reported positive chemotaxis of the two entomopathogenic nematode *H. bacteriophora* and *Steinernema carpocapsae* to several VOCs such as methyl salicylate, hexanol, heptanol, undecyl acetate or 4,5-dimethylthiazole. Interestingly, they showed that several volatiles repelled the nematodes. Similar effect of VOCs on the behavior of the entomopathogenic nematodes was already observed by (Hiltpold, Toepfer, *et al.* 2010; Rasmann and Turlings 2008) but no volatiles were identified yet. Most of the VOCs involved in belowground tritrophic interactions remain unknown but an increasing effort is invested in this field of research. Understanding more of these complex interactions would not only allow a better understanding of the rhizosphere but could also offer ecologically sound alternatives in pest management in agricultural systems.

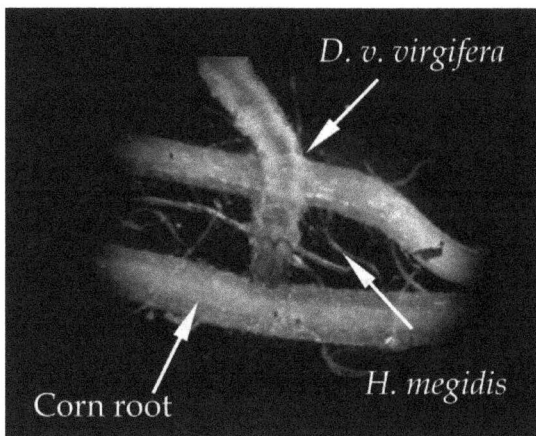

Fig. 2. Root of corn plants, when attacked by the larvae of the western corn rootworm *Diabrotica virgifera virgifera* emits a sesquiterpene volatile organic compound (*E*)-β-caryophyllene. This compound was then shown to increase the attractiveness of entomopathogenic nematodes *Heterorhabditis megidis* searching for the host larva (Rasmann *et al.* 2005).

3. Evolution of belowground volatile signaling

Similarly to any other adaptive plant traits, belowground emissions of volatiles may evolve if there is heritable variation in their production, which in turn affects fitness. This has barely been studied and little direct evidence indicates that natural selection has shaped root volatile production, and particularly as a defense (direct and indirect) against herbivores.

Nonetheless, root volatile chemistry is certainly heritable. Across 12 genotypes of *A. syriaca* constitutive emissions of total volatiles varied more than four-folds (Rasmann *et al.* 2010). In agricultural systems, corn (*Zea mays*) and *Citrus* plants also showed to display strong genotypic variability in insect-induced volatile production (Rasmann *et al.* 2005; Ali, Alborn, and Stelinski 2010, 2011). Root herbivores have been shown to display strong population level variation in their impact on plants, in turn having strong potential in impacting plant fitness (Maron 1998; Blossey and Hunt-Joshi 2003; Maron and Kauffman 2006). Higher herbivory might therefore lead to higher and more toxic compounds productions. Indeed, plants have been shown to produce some known nematicidal volatile organic compounds in their roots like benzaldehyde, thymol, limonene, neral, geranial, and carvacrol for defending themselves against the attacker in the underground (Bauske *et al.* 1994; Kokalis-Burelle *et al.* 2002; Oka *et al.* 2000; Rohloff 2002). This, however, has not been linked to plant fitness yet.

Allocation to belowground production of VOCs can also be costly for the plant, and classic theory would suggest trade-offs between growth and defense (Herms and Mattson 1992) or between defensive traits (Agrawal, Conner, and Rasmann 2010). Indeed, there is indication that faster growing bitter orange (*Citrus aurantium*) plants produce entomopathogenic nematodes attractive volatiles only after herbivory, whereas the slower growing trifoliate orange (*Poncirus trifoliata*) produce the same compounds also when undamaged (Ali,

Alborn, and Stelinski 2011), arguing in favor of a cost of producing high amount of VOCs constitutively. Also, in *A. syriaca*, levels of volatile production were negatively correlated with levels of toxic cardenolide production, across 12 genotypes (Rasmann *et al.* 2010), highlighting possible physiological trade-offs between direct and indirect defenses.

In corn, the production of (*E*)-β-caryophyllene has been tightly linked to reduction in herbivore performance and in subsequent reduction of root damage (Degenhardt *et al.* 2009) (Fig. 3). However, the actual fitness benefit for the plant has yet to be assessed. Most American cultivars, very likely through human breeding, are lacking the ability to produce this alarm signal (Kollner *et al.* 2008) even though the wild ancestor of maize, Teosinte, is able to synthetize and release this sesquiterpene upon attack of the chrysomelid larvae (Rasmann *et al.* 2005).

4. Methods for detecting volatile emissions in roots

The number of studies focused on belowground multitrophic interactions mediated by root volatile emissions 'pale in comparison' to the amount of research focused on their aboveground counterparts. This is largely due to technical difficulties associated with dynamics of the soil ecosystem. Soil is an opaque, tri-phasic medium making the analysis of individual factors and their interactions difficult. Most research has been based on *in vitro* analysis of individual factors.

Researchers hoping to study factors associated with roots signals are often directed to the study of root diffusates, root leachates, and or root exudates. Such terms are used interchangeably and can often be misleading. Leachate refers to a method of obtaining an extract from the roots, more than it does to the solution itself. Diffusate is used to convey non-volatile substances diffusing through the soil and establishing a gradient. Exudate is most often restricted to liquids that gradually 'ooze' from its source, but can be applied to volatiles as well. Approaches to evaluate roots volatiles have only recently been developed and applied in contexts of chemical and evolutionary ecology.

In 2005, Rasmann *et al.* evaluated the indirect volatile defences of maize roots using solid phase microextraction (SPME). SPME is a method of sampling volatiles without the use of solvents. In short, an adsorbent-coated fused silica fibre with properties similar to a gas chromatography column can collect volatile compound from the headspace of a sample. The volatile compounds once fixed to the SPME fibre can then be thermally desorbed in an injection port of a gas chromatograph and further analysed and/or identified when coupled with known standards or libraries of mass spectroscopy. In order to sample the effects root herbivory had on the plant produced VOCs Rasmann *et al.* (2005) crushed flash-frozen roots, either fed-upon or non-fed-upon into a fine powder. This powder was then exposed to the SPME fibre. This allowed for the volatiles that had accumulated in either treatment to be sampled and compared with GC-MS. SPME is a rapid and simple extraction method that doesn't require the use of solvents. Detection limits can reach parts per trillion (ppt) levels for certain compounds (Pawliszyn, 2009). Although this method is effective it is a destructive method of sampling root material. The plant and herbivore must be separated volatiles from this interaction can only be examined after harvesting and crushing the plant tissues.

Recently, Ali *et al.* (2010) was able to non-destructively sample belowground herbivore induced volatiles from citrus roots using a flow-through dynamic sampling technique coupled with an adsorbent traps. Volatiles can be collected and extracted by elution of an adsorbent with low boiling point solvents. Adsorbents traps are typically made of glass tubes filled with the granulated adsorbent, held in placed by stainless steel mesh, glass wool plugs, or Teflon fitted rings (in pre-made filters, ARS, Gainesville, FL, USA). The most common adsorbents are Hayesep Q® (HayeSeparations Inc., Bandera, TX, USA, the current alternative to Super Q® by Alltech Deerfield, IL, USA), and activated charcoal. By connecting the adsorbent trap to a vacuum pump and pulling air through glass chambers containing intact citrus plants either with or without feeding larvae Ali *et al.* (2010) was able to sample volatiles associated with belowground herbivory non-destructively and *in situ*. The volatiles collected on this trap are rinsed using solvent and analysed with GC-MS. This method allows for the sample to be retained in a solvent, which can be analysed more than once. The solvents containing root volatiles were also tested in sand-filled two-choice bioassays chambers. In this manner, Ali, Alborn, and Stelinski (2010, 2011) found evidence for entomopathogenic nematode attraction to volatiles from infested citrus roots.

Both techniques are effective and informative in different ways. The non-destructive sampling techniques are useful in evaluating belowground interactions *in situ* and may potentially prove useful in additional contexts. However, the properties of the surrounding soil may interfere and make resolution difficult with the potential for significant background. In this way SPME eliminates such background, but can introduce complications from tissue maceration were enzymes or oxidation can rapidly change the chemical profile and might not accurately represent the blend released from intact living organisms (Tollsten and Bergstrom 1988; Heath & Manukian 1992). Perhaps a combination of techniques and refinements of approaches will produce the best resolution for the dynamics of an individual system and the factors of concern (Rasmann, 2010).

5. Manipulation of root volatiles for agricultural improvement

The substantial advances in research on molecular mechanisms and ecological signaling of insect herbivore induced VOCs open promising prospects of manipulating the release of these compounds in order to enhance crop protection. Encouraging examples from laboratory and field experiments support this approach to develop novel ecologically sound crop protection strategies.

Manipulating the plant emitting VOCs appears as a first straightforward approach. Aboveground, for instance, Thaler and colleagues (1999) applied jasmonic acid on tomato plants in an experimental field. This treatment resulted in the emission of typical VOC blends that the plant would have produced upon herbivore attack and in a lepidopteran parasitism rate on plant treated with the phytohormone than on control plants (Thaler 1999). Supporting this plant manipulation approach, Rostàs and Turlings (2008) obtained a significant systemic resistance by treating maize plants with salicylic acid-mimic benzo-(1,2,3)-thiadiazole-7-carbothioic acid (S)-methyl ester resulting in an increased resistance against the fungal pythopathogen *Setosphaeria turcica*. Upon caterpillar attack, the treated plants were more attractive for the parasitoid wasp *Microplitis rufiventris* than plants damaged by the insect larvae only. Control experiments showed that these results were due to the plant-meditated effect rather than to the chemical treatment itself (Rostàs and Eggert 2008). So far there is no published example of such manipulation in the belowground compartment. Yet, preliminary results shows that root system treated with alginate, a polysaccharide extracted from brown algae, would emit faster and more VOCs than control plant upon belowground herbivory (Hiltpold, unpublished data). Ali *et al.* (in review) has recently increased mortality of root pests in the field by enhancing host location of naturally occurring entomopathogenic nematodes in Citrus and Blueberry crops with the application of the citrus root volatile, 1, 5-dimethylcyclodeca-1, 5, 7-triene (pregeijerene).

When the induced VOCs blend is clearly identified, it could be considered to genetically manipulate the plant in order to either (I) making the plant more attractive for beneficial predators or parasitoids or (II) to restore a new phenotype that was lost due to natural or human selection that is again attractive for the predators or parasitoids. (I) Again first examples of this approach are aboveground. Indeed, it has been first demonstrated in the model plant *Arabidopsis thaliana* in which a linalool/nerolidol synthase gene FaNES1 from strawberry was introduced. This resulted in the constitutive release of (3S)-(E)-nerolidol rendering the plants attractive to predatory mites *Phytoseiulus persimilis* (Kappers *et al.* 2005). Later, *Arabidopsis thaliana* was transformed with a TPS10 gene, coding for sesquiterpenes typically present in the blends emitted by plant upon caterpillar attacks (Schnee *et al.* 2006). The engineered plant was attractive to the parasitoid wasp *Cotesia marginiventris*, but only after they have learned to associate the HIPV blend to the presence of their hosts. (II) More recently, the terpene synthase gene TPS23 has been identified in maize (Kollner *et al.* 2008). TPS23 is responsible for the synthesis of EβC, a key attractant for some entomopathogenic nematodes (Rasmann *et al.* 2005). Most of the European maize varieties and Teosinte produce this sesquiterpene whereas American varieties doesn't (Kollner *et al.* 2008; Rasmann *et al.* 2005), indicating a shift in the gene activity through breeding selection (Kollner *et al.* 2008). In 2009, Degenhardt and colleagues were able to restore the ability of maize to recruit entomopathogenic nematodes by inserting a TPS23 gene from *Origanum vulgare* into a non-producing maize line (Degenhardt *et al.* 2009). In the field, the transformed maize line was significantly more attractive for the entomopathogenic nematode *Heterorhabditis megidis* compared to the wild type leading to a better protection for transformed plants (Fig. 3). Transformed plant received far less damage by the root pest *D. v. virgifera* and the transformation significantly reduced the beetle emergence, and overall this restored indirect defense resulted from the constitutive emission of EβC in the soil (Degenhardt *et al.* 2009). This was the first demonstration in the field that plant genotype engineering could enhance biological control. It has to be noticed that the variability in the emission of herbivore-induced volatiles in maize remains high (Degen *et al.* 2004) and therefore, because genetic modifications are still controversial, it should also be possible to incorporate highly attractive volatile compounds into new varieties using classical breeding programs.

In an inundative biological control strategy, the manipulation of the biological control agent can be considered as an option to enhance pest management (Hoy 1976). A classic idea that was put into practice with some relative success (Beckendorf and Hoy 1985; Hoy 2000). In that perspective, entomopathogenic nematodes appear as good candidates. Several studies have succeeded in selecting beneficial traits such as host finding (Gaugler and Campbell 1991; Gaugler, Campbell, and McGuire 1989), virulence (Peters and Ehlers 1998; Tomalak 1994) and tolerance to temperature (Ehlers *et al.* 2005; Grewal, Gaugler, and Wang 1996; Griffin and Downes 1994) or desiccation (Strauch *et al.* 2004).

For specific entomopathogenic nematodes, the knowledge of key attractants is now available (Hiltpold, Toepfer, et al. 2010; Rasmann et al. 2005). This prompted Hitlptold et al. (2010) to evaluate whether selection for enhanced responsiveness to the crucial root signal EβC could improve the efficiency of nematodes in controlling the larvae of the chrysomelid beetle D. v. virgifera. Using belowground six-arm olfactometers, a strain of the nematode Heterorhabditis bacteriophora was successfully selected. Originally, this nematode was not responding to EβC (Hiltpold, Baroni, et al. 2010) even though its effectiveness in controlling WCR larvae is high (Kurzt et al. 2009). The selected strain responded much better to EβC in laboratory experiments and was able to significantly control better the pest in the field, in presence of the belowground signal (Hiltpold, Baroni, et al. 2010). Because of the strong selective pressure, there were minor trade-offs in the infectiousness of the selected strain. Yet, the higher responsiveness to the HIPV overbalanced these weaknesses (Hiltpold, Baroni, et al. 2010). The establishment and the persistence in the field were not influenced by the selection process (Hiltpold, Baroni, et al. 2010). These results reflect the great potential of selecting beneficial organisms for a better and faster response, resulting in higher infection rates. Even though some constrains, such as knowledge of key compound/blends, and the laborious selection process, selecting for specific nematode strains could be coupled with selection of more attractive plant genotypes, making biological control of insect pests a success.

Beside from exploiting induced VOCs emitted while and after insect pest attack, luring the foraging insect with volatiles would eventually prevent any damage on the plant. Below ground examples of insect chemical luring are scarce. However, some promising attempts to control belowground pests with such a strategy have been published, notably in maize crops. Since Strnad and Bergman (1986) first reported that D. v. virgifera larvae were attracted to CO_2, disruption of pest foraging behavior with CO_2 has been demonstrated to have high potential method against this maize pest both in the laboratory and the field (Bernklau, Fromm, and Bjostad 2004). Beside this wide spread plant metabolite, more specific compounds, inducing a positive behavior of the beetle larvae, have been identified (Bernklau and Bjostad 2008; Bernklau et al. 2009; Bjostad and Hibbard 1992; Hibbard, Bernklau, and Bjostad 1994). Indeed, the authors described a combination of several sugars and fatty acids emitted by maize that serve as feeding stimulants for the D. v. virgifera neonate larvae (Bernklau and Bjostad 2008; Bernklau et al. 2009; Bjostad and Hibbard 1992; Hibbard, Bernklau, and Bjostad 1994). Recently, Bernklau and colleagues (2011) showed that D. v. virgifera larvae were feeding more and staying longer on pesticide-treated filter papers when the right blend of feeding stimulant chemicals was added compare to filter papers only treated with pesticides. The concentration of thiamethoxam required for equivalent kill without feeding stimulants was reduced 100,000 fold (Bernklau, Bjostad, and Hibbard 2011). Exploiting the same idea, capsules based on algae polymer and coated with D. v. virgifera feeding stimulants have been produced. When compared to maize root systems, the capsules were as much attractive for the pest larvae. These capsules could contain entomopathogenic nematodes and would then ease their application in the field as well as eventually lure D. v. virgifera larvae (Hiltpold et al., unpublished data). These belowground examples support the feasibility of such manipulative approach in the soil compartment even though more research and experiments are needed to achieve a good control of the targeted pest in the field.

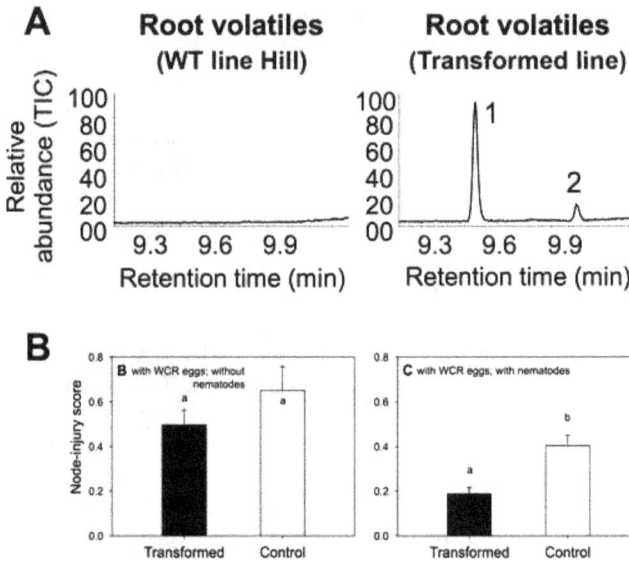

Fig. 3. Introduction of TPS23 genes into American lines of corn increases (E)-β-caryophyllene production (A) and better protect corn plants (B). Figure (A) shows a typical chromatogram obtained for the volatiles emitted by roots of the hybrid variety Hill line alongside a chromatogram for one of the transformed lines. Peak 1 is (E)-β-caryophyllene and peak 2 is α-humulene, a side-product of (E)-β-caryophyllene synthase. Figure (B) shows that in plants receiving WCR eggs and *H. megidis* nematodes, roots from transformed plants are significantly less damaged than roots from control lines. Modified from Degenhardt *et al.* (2009).

6. Acknowledgment

Funding sources for originate from the Swiss Science foundation Ambizione grant PZ00P3_131956 / 1 to SR.

7. References

Agrawal, A. A., J. K. Conner, and S. Rasmann. 2010. Tradeoffs and adaptive negative correlations in evolutionary ecology. In *Evolution After Darwin: the First 150 Years*, edited by M. A. Bell, D. J. Futuyma, W. F. Eanes and J. S. Levinton. Sunderland, MA, USA: Sinauer.

Agrawal, A. A., and M. Fishbein. 2006. Plant defense syndromes. *Ecology* 87 (7):S132-S149.

Akiyama, K., K. Matsuzaki, and H. Hayashi. 2005. Plant sesquiterpenes induce hyphal branching in arbuscular mycorrhizal fungi. *Nature* 435 (7043):824-827.

Ali, J. G., H. T. Alborn, and L. L. Stelinski. 2010. Subterranean herbivore-induced volatiles released by *Citrus* roots upon feeding by *Diaprepes abbreviatus* recruit entomopathogenic nematodes. *Journal of Chemical Ecology* 36 (4):361-368.

Ali, J. G., H. T. Alborn, and L. L. Stelinski. 2011. Constitutive and induced subterranean plant volatiles attract both entomopathogenic and plant parasitic nematodes. *Journal of Ecology* 99 (1):26-35.

Aratchige, N. S., I. Lesna, and M. W. Sabelis. 2004. Below-ground plant parts emit herbivore-induced volatiles: olfactory responses of a predatory mite to tulip bulbs infested by rust mites. *Experimental and Applied Acarology* 33 (1-2):21-30.

Badri, D. V., T. L. Weir, D. van der Lelie, and J. M. Vivanco. 2009. Rhizosphere chemical dialogues: plant-microbe interactions. *Current Opinion in Biotechnology* 20 (6):642-650.

Bais, H. P., V. M. Loyola-Vargas, H. E. Flores, and J. M. Vivanco. 2001. Invited review: Root-specific metabolism: The biology and biochemistry of underground organs. *In Vitro Cellular & Developmental Biology-Plant* 37 (6):730-741.

Bais, H. P., T. L. Weir, L. G. Perry, S. Gilroy, and J. M. Vivanco. 2006. The role of root exudates in rhizosphere interations with plants and other organisms. *Annual Review of Plant Biology* 57:233-266.

Barber, D. A., and J. K. Martin. 1976. The release of organic substances by cereal roots into soil. *New Phytologist* 76 (1):60-80.

Bardgett, R. D. 2005. *The biology of soil, Biology of habitats*. Oxford, New York: Oxford University Press.

Barney, J. N., J. P. Sparks, J. Greenberg, T. H. Whitlow, and A. Guenther. 2009. Biogenic volatile organic compounds from an invasive species: impacts on plant-plant interactions. *Plant Ecology* 203 (2):195-205.

Bauer, P., J. Munkert, M. Brydziun, E. Burda, F. Muller-Uri, H. Groger, Y. A. Muller, and W. Kreis. 2010. Highly conserved progesterone 5 beta-reductase genes (P5 beta R) from 5 beta-cardenolide-free and 5 beta-cardenolide-producing angiosperms. *Phytochemistry* 71 (13):1495-1505.

Bauske, E. M., R. Rodriguezkabana, V. Estaun, J. W. Kloepper, D. G. Robertson, C. F. Weaver, and P. S. King. 1994. Management of *Meloidogyne incognita* on contton by use of botanical aromatic compounds. *Nematropica* 24 (2):143-150.

Bécard, G., and Y. Piché. 1989. Fungal growth-stimulation by CO2 and root exudates in vesicular-arbusular mycorrhizal symbiosis. *Applied and Environmental Microbiology* 55 (9):2320-2325.

Beckendorf, S. K., and M. A. Hoy. 1985. Genetic improvement of arthropod naturel enemies through selection, hybridization or genetic engineering techniques. In *Biological Control in Agricultural IPM Systems*, edited by M. A. Hoy and D. C. Herzog: Academic, Orlando, FL.

Bensky, D., and A. . Gamble. 1986. *Chinese herbal medicine*. Seattle: Eastland Press.

Bernklau, E. J., and L. B. Bjostad. 2008. Identification of feeding stimulants in corn roots for western corn rootworm (Coleoptera: Chrysomelidae) larvae. *Journal of Economic Entomology* 101 (2):341-351.

Bernklau, E. J., L. B. Bjostad, and B. E. Hibbard. 2011. Synthetic feeding stimulants enhance insecticide activity against western corn rootworm larvae, Diabrotica virgifera virgifera (Coleoptera: Chrysomelidae). *Journal of Applied Entomology* 135 (1-2):47-54.

Bernklau, E. J., L. B. Bjostad, L. N. Meiiils, T. A. Coudron, E. Lim, and B. E. Hibbard. 2009. Localized Search Cues in Corn Roots for Western Corn Rootworm (Coleoptera: Chrysomelidae) Larvae. *Journal Of Economic Entomology* 102 (2):558-562.

Bernklau, E. J., E. A. Fromm, and L. B. Bjostad. 2004. Disruption of host location of western corn rootworm larvae (Coleoptera : Chrysomelidae) with carbon dioxide. *Journal Of Economic Entomology* 97 (2):330-339.

Bezemer, T. M., and N. M. van Dam. 2005. Linking aboveground and belowground interactions via induced plant defenses. *Trends in Ecology & Evolution* 20 (11):617-624.

Bjostad, L. B., and B. E. Hibbard. 1992. 6-Methoxy-2-benzoxazolinone - a semiochemical for host location by western corn rootworm larvae. *Journal of Chemical Ecology* 18 (7):931-944.

Blossey, B., and T. R. Hunt-Joshi. 2003. Belowground herbivory by insects: Influence on plants and aboveground herbivores. *Annual Review of Entomology* 48:521-547.

Boff, M. I. C., F. C. Zoon, and P. H. Smits. 2001. Orientation of *Heterorhabditis megidis* to insect hosts and plant roots in a Y-tube sand olfactometer. *Entomologia Experimentalis et Applicata* 98 (3):329-337.

Bouwmeester, H. J., R. Matusova, Z. K. Sun, and M. H. Beale. 2003. Secondary metabolite signalling in host-parasitic plant interactions. *Current Opinion in Plant Biology* 6 (4):358-364.

Brown, V K, and A C Gange. 1990. Insect herbivory below ground. *Advances in Ecological Research* 20:1-58.

Callaway, R. M., and W. M. Ridenour. 2004. Novel weapons: invasive success and the evolution of increased competitive ability. *Frontiers in Ecology and the Environment* 2 (8):436-443.

Carson, J. F., and F. F. Wong. 1961. Isolation of (+)S-methyl-L-cysteine sulfoxide and of (+)S-N-propyl-L-cysteine sulfoxide from onions as their N-2,4-dinitrophenil derivates. *Journal of Organic Chemistry* 26 (12):4997-&.

Carson, J. F., and F. F. Wong. 1961. Onion flavor and odor - Volatile flavor components of onions. *Journal of Agricultural and Food Chemistry* 9 (2):140-&.

Curtis, R .H.C. , A .F. Robinson, and R.N. Perry. 2009. Hatch and host location. In *Root-knot nematodes*, edited by R. Perry, M. Moens and J. Starr. CABI Publishing: Wallingford, UK.

De Deyn, G. B., C. E. Raaijmakers, H. R. Zoomer, M. P. Berg, P. C. de Ruiter, H. A. Verhoef, T. M. Bezemer, and W. H. van der Putten. 2003. Soil invertebrate fauna enhances grassland succession and diversity. *Nature* 422 (6933):711-713.

Degen, T., C. Dillmann, F. Marion-Poll, and T. C. J. Turlings. 2004. High genetic variability of herbivore-induced volatile emission within a broad range of maize inbred lines. *Plant Physiology* 135 (4):1928-1938.

Degenhardt, J. 2009. Indirect defense responses to herbivory in grasses. *Plant Physiology* 149 (1):96-102.

Degenhardt, J., I. Hiltpold, T. G. Kollner, M. Frey, A. Gierl, J. Gershenzon, B. E. Hibbard, M. R. Ellersieck, and T. C. J. Turlings. 2009. Restoring a maize root signal that attracts insect-killing nematodes to control a major pest. *Proceedings of the National Academy of Sciences of the United States of America* 106 (32):13213-13218.

Dicke, M., and I. T. Baldwin. 2010. The evolutionary context for herbivore-induced plant volatiles: beyond the 'cry for help'. *Trends in Plant Science* 15 (3):167-175.

Dicke, M., and M. W. Sabelis. 1988. How plants obtain predatory mites as bodyguards. *Netherlands Journal of Zoology* 38 (2-4):148-165.

Dudareva, N., E. Pichersky, and J. Gershenzon. 2004. Biochemistry of plant volatiles. *Plant Physiology* 135 (4):1893-1902.

Dusenbery, D. B. 1980. Responses of the nematode *Caenorhabditis elegans* to controlled chemical stimulation. *Journal of Comparative Physiology* 136 (4):327-331.

Dusenbery, D. B. 1987. Behavioral responses of *Meloidogyne incognita* to temperature and carbon dioxide. *Journal of Nematology* 19 (4):519-519.

Ehlers, R. U., J. Oestergaard, S. Hollmer, M. Wingen, and O. Strauch. 2005. Genetic selection for heat tolerance and low temperature activity of the entomopathogenic nematode-bacterium complex *Heterorhabditis bacteriophora-Photorhabdus luminescens*. *Biocontrol* 50 (5):699-716.

Ehrlich, P. R., and P. H. Raven. 1964. Butterflies and plants - a study in coevolution. *Evolution* 18 (4):586-608.

Firn, R. D., and C. G. Jones. 1996. An explanation of secondary product "redundancy". In *Phytochemical diversity and redundancy in ecological interactions*, edited by J. T. Romeo, I. A. Saunders and P. Barbosa. New York and London: Plenum Press.

Flores, H. E., J. M. Vivanco, and V. M. Loyola-Vargas. 1999. 'Radicle' biochemistry: the biology of root-specific metabolism. *Trends in Plant Science* 4 (6):220-226.

Fraenkel, G. S. 1959. The raison d'être of secondary plant substances. *Science* 129:1466-1470.

Gaugler, R., and J. F. Campbell. 1991. Selection for enhanced host-finding of scarab larvae (Coleoptera, Scarabaeidae) in an entomopathogenic nematode. *Environmental Entomology* 20 (2):700-706.

Gaugler, R., J. F. Campbell, and T. R. McGuire. 1989. Selection for host-finding in *Steinernema feltiae*. *Journal of Invertebrate Pathology* 54 (3):363-372.

Grewal, P. S., R. Gaugler, and Y. Wang. 1996. Enhanced cold tolerance of the entomopathogenic nematode Steinernema feltiae through genetic selection. *Annals of Applied Biology* 129 (2):335-341.

Griffin, C. T., and M. J. Downes. 1994. Selection of *Heterorhabditis* sp. for improved infectivity at low temperatures. In *Genetics of entomopathogenic nematode-bacterium complexes*, edited by A. M. Burnell, R. U. Ehlers and J. P. Masson: European Commission Publication EUR 15681 EN, Luxembourg.

Hallem, E. A., A. R. Dillman, A. V. Hong, Y. J. Zhang, J. M. Yano, S. F. DeMarco, and P. W. Sternberg. 2011. A sensory code for host seeking in parasitic nematodes. *Current Biology* 21 (5):377-383.

Hartmann, T. 2007. From waste products to ecochemicals: Fifty years research of plant secondary metabolism. *Phytochemistry* 68 (22-24):2831-2846.

Hartmann, Thomas. 2008. Chemical Ecology Special Feature: The lost origin of chemical ecology in the late 19th century.

Heil, M. 2008. Indirect defence via tritrophic interactions. *New Phytologist* 178 (1):41-61.

Henry, A., W. Doucette, J. Norton, and B. Bugbee. 2007. Changes in crested wheatgrass root exudation caused by flood, drought, and nutrient stress. *Journal of Environmental Quality* 36 (3):904-912.

Herms, D. A., and W. J. Mattson. 1992. The dilemma of plants - to grow or defend. *Quarterly Review of Biology* 67 (3):283-335.

Hibbard, B. E., E. J. Bernklau, and L. B. Bjostad. 1994. Long-chain free fatty-acids - Semiochemicals for host location by western corn rooworm lrvae. *Journal of Chemical Ecology* 20 (12):3335-3344.

Hiltpold, I., M. Baroni, S. Toepfer, U. Kuhlmann, and T. C. J. Turlings. 2010. Selection of entomopathogenic nematodes for enhanced responsiveness to a volatile root signal helps to control a major root pest. *Journal of Experimental Biology* 213 (14):2417-2423.

Hiltpold, I., M. Baroni, S. Toepfer, U. Kuhlmann, and T. C. J. Turlings. 2010. Selective breeding of entomopathogenic nematodes for enhanced attraction to a root signal did not reduce their establishment or persistence after field release. *Plant Signaling and Behavior* 5 (11):1450-1452.

Hiltpold, I., M. Erb, C. A. M. Robert, and T. C. J. Turlings. 2011. Systemic root signalling in a belowground, volatile-mediated tritrophic interaction. *Plant Cell and Environment* 34 (8):1267-1275.

Hiltpold, I., S. Toepfer, U. Kuhlmann, and T. C. J. Turlings. 2010. How maize root volatiles affect the efficacy of entomopathogenic nematodes in controlling the western corn rootworm? *Chemoecology* 20 (2):155-162.

Hiltpold, I., and T. C. J. Turlings. 2008. Belowground chemical signaling in maize: When simplicity rhymes with efficiency. *Journal of Chemical Ecology* 34 (5):628-635.

Howe, G. A., and G. Jander. 2008. Plant immunity to insect herbivores. *Annual Review of Plant Biology* 59:41-66.

Hoy, M. A. 1976. Genetic Improvement of Insects - Fact or Fantasy. *Environmental Entomology* 5 (5):833-839.

Hoy, M. A. 2000. Transgenic arthropods for pest management programs: Risks and realities. *Experimental and Applied Acarology* 24 (5-6):463-495.

Hunter, Mark. D. . 2001. Out of sight, out of mind: the impacts of root-feeding insects in natural and managed systems. *Agricultural and Forest Entomology* 3 (1):3-9.

Johnson, S. N., and P. J. Gregory. 2006. Chemically-mediated host-plant location and selection by root-feeding insects. *Physiological Entomology* 31 (1):1-13.

Kaplan, I., R. Halitschke, A. Kessler, S. Sardanelli, and R. F. Denno. 2008. Constitutive and induced defenses to herbivory in above- and belowground plant tissues. *Ecology* 89 (2):392-406.

Kappers, I. F., A. Aharoni, Twjm van Herpen, L. L. P. Luckerhoff, M. Dicke, and H. J. Bouwmeester. 2005. Genetic engineering of terpenoid metabolism attracts bodyguards to Arabidopsis. *Science* 309 (5743):2070-2072.

Karban, Richard, and Ian T Baldwin. 1997. *Induced Responses to Herbivory*. 1st edition ed. Chicago: The University of Chicago Press.

Kessler, A., and I. T. Baldwin. 2002. Plant responses to insect herbivory: The emerging molecular analysis. *Annual Review of Plant Biology* 53:299-328.

Kessler, A., and M. Heil. 2011. The multiple faces of indirect defences and their agents of natural selection. *Functional Ecology* 25 (2):348-357.

Kessler, A., and K. Morrell. 2010. Plant Volatile Signalling: Multitrophic Interactions in the Headspace. In *The Chemistry and Biology of Volatiles*, edited by A. Herrmann. Chichester: Wiley.

Kleinheinz, G. T., S. T. Bagley, W. P. St John, J. R. Rughani, and G. D. McGinnis. 1999. Characterization of alpha-pinene-degrading microorganisms and application to a bench-scale biofiltration system for VOC degradation. *Archives of Environmental Contamination and Toxicology* 37 (2):151-157.

Klinger, J. 1963. Die orientierung von *Ditylenchus dipsaci* in gemessenen künstlichen und biologischen CO_2 gradienten. *Nematologica* 9 (185-199).

Kokalis-Burelle, N., N. Martinez-Ochoa, R. Rodriguez-Kabana, and J. W. Kloepper. 2002. Development of multi-component transplant mixes for suppression of *Meloidogyne incognita* on tomato (*Lycopersicon esculentum*). *Journal of Nematology* 34 (4):362-369.

Kollner, T. G., M. Held, C. Lenk, I. Hiltpold, T. C. J. Turlings, J. Gershenzon, and J. Degenhardt. 2008. A maize (E)-beta-caryophyllene synthase implicated in indirect defense responses against herbivores is not expressed in most American maize varieties. *Plant Cell* 20 (2):482-494.

Kollner, T. G., C. Schnee, J. Gershenzon, and J. Degenhardt. 2004. The variability of sesquiterpenes cultivars is controlled by allelic emitted from two *Zea mays* variation of two terpene synthase genes encoding stereoselective multiple product enzymes. *Plant Cell* 16 (5):1115-1131.

Kurzt, B., I. Hiltpold, T. C. J. Turlings, U. Kuhlmann, and S. Toepfer. 2009. Comparative susceptibility of larval instars and pupae of the western corn rootworm to infection by three entomopathogenic nematodes. *Biocontrol* 54 (2):255-262.

Lewinsohn, Efraim, and Mark Gijzen. 2009. Phytochemical diversity: The sounds of silent metabolism. *Plant Science* 176 (2):161-169.

Maron, J. L. 1998. Insect herbivory above- and belowground: Individual and joint effects on plant fitness. *Ecology* 79 (4):1281-1293.

Maron, J. L., and M. J. Kauffman. 2006. Habitat-specific impacts of multiple consumers on plant population dynamics. *Ecology* 87 (1):113-124.

McCully, M. E., and J. S. Boyer. 1997. The expansion of maize root-cap mucilage during hydration .3. Changes in water potential and water content. *Physiologia Plantarum* 99 (1):169-177.

Muller, W. H., and C. H. Muller. 1964. Volatile growth inhibitors produced by *Salvia* species. *Bull Torrey Bot Club* 91:327-330.

Mumm, R., and M. Dicke. 2010. Variation in natural plant products and the attraction of bodyguards involved in indirect plant defense. *Canadian Journal of Zoology-Revue Canadienne De Zoologie* 88 (7):628-667.

Oka, Y., S. Nacar, E. Putievsky, U. Ravid, Z. Yaniv, and Y. Spiegel. 2000. Nematicidal activity of essential oils and their components against the root-knot nematode. *Phytopathology* 90 (7):710-715.

Owen, S. M., S. Clark, M. Pompe, and K. T. Semple. 2007. Biogenic volatile organic compounds as potential carbon sources for microbial communities in soil from the rhizosphere of Populus tremula. *Fems Microbiology Letters* 268 (1):34-39.

Perry, R. N., and M. Moens. 2006. *Plant Nematology*. Wallingford, UK: CABI Publishing.

Peters, A., and R. U. Ehlers. 1998. Evaluation and selection for enhanced nematode pathogenicity against *Tipula* spp. In *Pathogenicity of entomopathogenic nematodes versus insect defense mechanisms: impact on selection of virulent strains*, edited by N. Simoes, N. Boemare and R. U. Ehlers: European Commission Publication COST819, Brussels.

Pichersky, E., and D. R. Gang. 2000. Genetics and biochemistry of secondary metabolites in plants: an evolutionary perspective. *Trends in Plant Science* 5 (10):439-445.

Pichersky, E., J. P. Noel, and N. Dudareva. 2006. Biosynthesis of plant volatiles: Nature's diversity and ingenuity. *Science* 311 (5762):808-811.

Pline, M., and D. B. Dusenbery. 1987. Responses of plant-parasitic nematode *Meloidogyne incognita* to carbon dioxide determined by video camera-computer tracking. *Journal of Chemical Ecology* 13 (4):873-888.

Potter, M. J., V. A. Vanstone, K. A. Davies, J. A. Kirkegaard, and A. J. Rathjen. 1999. Reduced susceptibility of Brassica napus to Pratylenchus neglectus in plants with elevated root levels of 2-phenylethyl glucosinolate. *Journal of Nematology* 31 (3):291-298.

Prot, J-C. 1980. Migration of plant-parasitic nematodes towards plant roots. *Revue de Nématologie* 3 (2):305-318.

Rasmann, S., and A. A. Agrawal. 2008. In defense of roots: A research agenda for studying plant resistance to belowground herbivory. *Plant Physiology* 146 (3):875-880.

Rasmann, S., T. L. Bauerle, K. Poveda, and R. Vannette. 2011. Predicting root defence against herbivores during succession. *Functional Ecology* 25 (2):368-379.

Rasmann, S., A. C. Erwin, R. Halitschke, and A. A. Agrawal. 2010. Direct and indirect root defences of milkweed (*Asclepias syriaca*): trophic cascades, trade-offs and novel methods for studying subterranean herbivory. *Journal of Ecology* 99 (1):16-25.

Rasmann, S., T. G. Kollner, J. Degenhardt, I. Hiltpold, S. Toepfer, U. Kuhlmann, J. Gershenzon, and T. C. J. Turlings. 2005. Recruitment of entomopathogenic nematodes by insect-damaged maize roots. *Nature* 434 (7034):732-737.

Rasmann, S., and T. C. J. Turlings. 2008. First insights into specificity of belowground tritrophic interactions. *Oikos* 117 (3):362-369.

Reinecke, Andreas, Frank Müller, and Monika Hilker. 2008. Attractiveness of CO2 released by root respiration fades on the background of root exudates. *Basic and Applied Ecology* 9 (5):568-576.

Riga, E. . 2004. Orientation behavior. In *Nematode behaviour*, edited by R. Gaugler and A. C. Bilgrami. Wallingford,: CABI.

Robinson, A.F. 2003. Nematode behaviour and migrations through soil and host tissue. In *Nematology advances and perspectives.*, edited by X. Zhongxiao, S. Chen and D. W. Dickson. Wallingford: CABI.

Rohloff, J. 2002. Volatiles from rhizomes of *Rhodiola rosea* L. *Phytochemistry* 59 (6):655-661.

Rolfe, R.N., J. Barrett, and R.N. Perry. 2000. Analysis of chemosensory responses of second stage juveniles of Globodera rostochiensis using electrophysiological techniques. *Nematology* 2:523-533.

Rostas, M., and K. Eggert. 2008. Ontogenetic and spatio-temporal patterns of induced volatiles in Glycine max in the light of the optimal defence hypothesis. *Chemoecology* 18 (1):29-38.

Schnee, C. , T.G. Köllner, M. Held, T.C.J. Turlings, J. Gershenzon, and J. Degenhardt. 2006. The products of a single maize sesquiterpene synthase form a volatile defense signal that attracts natural enemies of maize herbivores. *Proceedings of the National Academy of Sciences USA* 103 (4):1129-1134.

Schoonhoven, L. M., J. J. A. van Loon, and M. Dicke. 2005. *Insect-Plant Biology*. Oxford: Oxford University Press.

Stahl, E. 1888. Pflanzen und Schnecken: Eine biologische Studie über die Schutzmittel der Pflanzen gegen Schneckenfrass. *Jenaer Zeitschr. Medizin Naturwissenschaften* (22):557-684.

Strauch, O., J. Oestergaard, S. Hollmer, and R. U. Ehlers. 2004. Genetic improvement of the desiccation tolerance of the entomopathogenic nematode *Heterorhabditis bacteriophora* through selective breeding. *Biological Control* 31 (2):218-226.

Strnad, S. P., and M. K. Bergman. 1986. Movement of thirst-instar western corn rootworm (Coleoptera: Chrysomelidae) in soil. *Environmental Entomology* 16 (4):975-978.

Sutherland, O. R. W., and J. R. Hillier. 1972. Olfactory responses of *Costelytra zealandica* (Coleoptera: Melolonthinae) larvae to grass root odours. *New Zealand Journal of Science* 15 (2):165-172.

Thaler, J. S. 1999. Jasmonate-inducible plant defences cause increased parasitism of herbivores. *Nature* 399 (6737):686-688.

Tomalak, M. 1994. Selective breeding of *Steinernema feltiae* Filipjev (Nematoda, Steinernematidae) for improved efficacy in control of a mushroom fly, *Lycoriella solani* Winnertz (Diptera, Sciaridae). *Biocontrol Science and Technology* 4 (2):187-198.

Turlings, T. C. J, and F Wäckers. 2004. Recruitment of predators and parasitoids by herbivore-injured plants. In *Advances in Insect Chemical Ecology*, edited by R. T. Cardé and J. G. Millar: Cambridge University Press.

Turlings, T. C. J., and J. Ton. 2006. Exploiting scents of distress: the prospect of manipulating herbivore-induced plant odours to enhance the control of agricultural pests. *Current Opinion in Plant Biology* 9 (4):421-427.

Turlings, T. C. J., J. H. Tumlinson, and W. J. Lewis. 1990. Exploitation of herbivore-induced plant odors by host-seeking parasitic wasps. *Science* 250:1251-1253.

van Dam, N. M. 2009. Belowground herbivory and plant defenses. *Annual Review of Ecology Evolution and Systematics* 40:373-391.

van Dam, N. M., and G. M. Poppy. 2008. Why plant volatile analysis needs bioinformatics - detecting signal from noise in increasingly complex profiles. *Plant Biology* 10 (1):29-37.

van Tol, R. W. H. M., A. T. C. van der Sommen, M. I. C. Boff, J. van Bezooijen, M. W. Sabelis, and P. H. Smits. 2001. Plants protect their roots by alerting the enemies of grubs. *Ecology Letters* 4 (4):292-294.

Wardle, D. A. 2006. The influence of biotic interactions on soil biodiversity. *Ecology Letters* 9 (7):870-886.

Wardle, D. A., R. D. Bardgett, J. N. Klironomos, H. Setala, W. H. van der Putten, and D. H. Wall. 2004. Ecological linkages between aboveground and belowground biota. *Science* 304 (5677):1629-1633.

Weissteiner, S., and S. Schütz. 2006. Are different volatile pattern infuencing host plant choice of belowground living insects. *Mitt. Dtsch. Ges. Allg. Angew. Entomol.* 15:51-55.

Wenke, K., M. Kai, and B. Piechulla. 2010. Belowground volatiles facilitate interactions between plant roots and soil organisms. *Planta* 231 (3):499-506.

Williamson, Valerie M., and Cynthia A. Gleason. 2003. Plant-nematode interactions. *Current Opinion in Plant Biology* 6 (4):327-333.

Wink, Michael. 2008. Plant secondary metabolism: diversity, function and its evolution. *Natural Product Communications* 3 (8):1205-1216.

Wojciechowski, Zdzislaw A. . 2003. Biosynthesis of plant steroid glycosides. In *Advances in Phytochemistry*, edited by F. Imperato. Kerala, India: Research Signpost.

Wolfson, J. L. 1987. Impact of *Rhizobium nodules* on *Sitona hispidulus*, the clover root curculio. *Entomologia Experimentalis Et Applicata* 43 (3):237-243.

Zak, J. C., M. R. Willig, D. L. Moorhead, and H. G. Wildman. 1994. Functional diversity of microbial communities - a quantitative approach. *Soil Biology & Biochemistry* 26 (9):1101-1108.

Zwanenburg, B., A. S. Mwakaboko, A. Reizelman, G. Anilkumar, and D. Sethumadhavan. 2009. Structure and function of natural and synthetic signalling molecules in parasitic weed germination. *Pest Management Science* 65 (5):478-491.

Section 4

Reproduction

13

Nutritional and Proteomic Profiles in Developing Olive Inflorescence

Christina K. Kitsaki, Nikos Maragos and Dimitris L. Bouranis
Laboratory of Plant Physiology, Department of Agricultural Biotechnology
Agricultural University of Athens, Athens
Greece

1. Introduction

Flowering is a crucial phase of the plant life cycle hardly programmed through floral time genes, cadastral genes and floral organ identity genes (Jack, 2004). Four pathways control flowering time, floral meristem identity genes and floral organs genes: light and photoperiod, temperature, energy supplies (mainly through carbohydrates) and hormones (Taiz and Zeiger, 2010).

Plant nutrient status and especially nitrogen is determinant for flowering and completion of the plant life cycle. Tree nutrient status, as well as its seasonal nutrient fluctuations have been based almost exclusively on leaf analysis. Flower analysis has been used for the study of the distribution of different nutrients through the various parts of the fruit tree (Drossopoulos et al., 1996, Sanchez et al., 1992), as well as for the detection and handling of deficiencies during this early stage of growth (Sanz and Montanes, 1995).

Olive tree (*Olea europaea L.*) is a very important indeterminate long lived, evergreen fruit tree, cultivated mainly in the Mediterranean basin, since experimental evidences have demonstrated both fruits and leaves to be important organs for synthesis of several interesting biological compounds. The tree has been cultivated for approximately 6000 years. Some specimens have been reported to live for thousand, and it may be in the spirit of peace that olive branches appear on the flag of the United Nation Organization (Rugini and Fedeli, 1990; Martin, 2009).

Olive tree has a plentiful bloom but a low percentage of normal fruit set. (Reale et al., 2006). In plants nutrient status affects flowering, and play a crucial role for the completion of the life cycle. Except from nutrients, proteins play a special role in inflorescence development, as they play crucial role in flower organ development (Taiz and Zeiger, 2010). In olive tree, nitrogen status affects flower quality and ovule longevity (Fernandez-Escobar et al., 2008).

Proteomic analyses in trees have also gained little interest, as they are difficult as plant materials. Especialy, olive tree has major difficulty, because of the presence of many phenolic compounds. In spite of the importance of olive tree for human diet, no interest has been focused on olive inflorescence development. In this work we studied the profile of nutrients, as well as the proteomic profile in olive inflorescence cv "Konservolia" in order to add data on this interest subject.

2. Materials and methods

2.1 Plant material

Three about 30-year-old fruit bearing olive trees (*Olea europaea* L. cv "Konservolia"), were selected. The experiment started about one week after floral bud burst, nine weeks before full bloom, in early March, and lasted up to one week after full bloom, in late May. Hanging or horizontal reproductive shoots 15-25 cm long with ten or more inflorescence bearing nodes were cut from positions symmetrically distributed around the crown of each tree at weekly intervals. The inflorescences from the central region of each reproductive shoot (3rd to 6th node measuring from the top of the shoot) were used for further analysis. Three groups of samples were collected each sampling date at 09:00 h i.e. (i) for measurements of developmental parameters of inflorescences (length, FW, DW, and morphological observations), (ii) for nutritional profile analyses and (iii) for proteomic analyses.

2.2 Developmental parameters of inflorescence development

Twenty inflorescences per tree were used for measurements of the inflorescence length and fresh and dry weight of the organ. For detailed morphological and anatomical observations two representative inflorescences with four lateral branches were selected from four reproductive shoots of each of the three trees and fixed in FAA.

Anatomy (median longitudinal section of perfect flowers) were observed on flowers from the first branch (counting from the apex of the inflorescence). Photographs were taken with an Olympus SZX12 stereoscope, equipped with an Olympus DP71 P-30N camera.

2.3 Nutritional analyses

Three samples per tree were separately analysed. Fresh weight per sample was recorded, samples were oven-dried at 80 °C, dry weight was recorded and the samples were ground to pass a 40-mesh screen using an analytical mill (IKA, model A10) prior to chemical analysis (Mills and Jones 1996). N was analyzed by micro-Kjeldahl digestion followed by distillation (Jones, 1991). P, K, Mg, Ca, Cu, Fe, Zn and Mn were determined following a wet acid digestion procedure based on the combination of HNO_3 and 30% H_2O_2 (Mills and Jones 1996). Phosphorus quantitative analysis in the diluted digests was carried out colorimetrically by determining the absorption of the blue phosphomolybdate complex at 660 nm, using the ammonium molybdate and stannus chloride procedure (Peach and Tracey 1956). The concentrations of all other nutrients were determined in the diluted digests by atomic absorption spectrophotometry using a GBC Avanta spectrophotometer. For the determination of Ca and Mg, 1% (w/v) lanthanum was added in the digests.

2.4 Proteomic analyses

Three separate replicates were used for each of the proteomic analyses we followed. A modified protocol based on those suggested by Garcia et al. (2000) Süle et al. (2004), and Wang et al. (2003, 2004) for olive was followed for protein extraction and purification. Proteins were extracted using a 10fold volume/weight buffer (50 mM sodium borate, 50 mM ascorbic acid pH 9.0, 1% β-mercaptoethanol, 1% soluble PVP, 1% insoluble PVP, 10 mM PMSF) under continuous vortexing for 60 min at 4 ºC. The homogenate was centrifuged at 4 ºC for 30 min at 35,000 g and the pellet was discarded. Proteins in the supernatant were precipitated by adding equal volume of cold 10% TCA, 0.07% β-mercaptoethanol in acetone. The precipitation was carried out overnight at -20 °C. A 10 min centrifugation in cold at 10,000 g resulted in a three phase system, an upper aqueous, an intermediate protein phase, and a lower organic phase. The upper aqueous phase was discarded and the remained proteins were homogenized with the lower organic phase by vortexing. Proteins were then rinsed three times by adding a twofold volume of 10% aqueous TCA, 0.07% β-mercaptoethanol, vortexing and centrifugation for 5 min in cold at 10,000 g. Purification of proteins completed by washing once with 5fold volume ddH$_2$O. Finally, the protein pellet was washed twice with ice cold acetone, 0.07% β-mercaptoethanol and centrifuged for 5 min in cold at 10,000 g. The pellet was dried under a gentle stream of air. Total water extractable proteins (WSP) were quantified using bovine serum albunin as standard (Bradford, 1976; Bearden,1978).

The pellets for SDS-PAGE analysis (about 50 μg) were solubilised in sample buffer 62,5 mM Tris-HCl, pH 6.8, 2% SDS, 5% β-mercaptoethanol, 10% glycerol and traces of bromophenol blue (Garcia et al. 2000) enhanced with 8.0 M urea, and denatured by heating in boiling water for 5 min. Then cooled to room temperature and centrifuged for 5 min at 14,000g. Electrophoresis was carried out on a Perfect Blue M 14x16 cm (Peqlab Biotechnologie GmbH) vertical electrophoresis system using 12,5% continuous pH 8,3 SDS running gel (1,5 mm thick), with a 4% stacking gel parallel with Sigma markers (14 to 205 kDa), according to Hoefer (1994) at 30 mA for about 1.5 hours. The gels were stained with Coomassie Brilliant Blue R-250, according to Sigma (1994) protocol.

For 2-DE about 300 μg of purified proteins were diluted with rehydration buffer (9.5 M urea, 2% CHAPS, 2% IPG buffer pH 4-7 and 20 mM DTT (Amersham Biosciences). The samples were then put on Immobiline Drystrip 13 cm with a linear pH range 4.0-7.0, the strips were covered with cover fluid and left to rehydrate for at least 10 h. Isoelectric focusing was carried out on a horizontal Multiphor II Amersham Biosciences system under a four phases program (300 V/1 Vh, 300 V/1800 Vh, 3500 V/9500 Vh and 3500 V/19250 Vh). Prior to running the second dimension the proteins were reduced with 65 mM DTT followed by alkylation with 135 mM iodoacetamide and traces of bromophenol blue, both diluted in equilibration buffer (50 mM Tris-HCl, pH 8.8, 6 M urea, 30% glycerol and 2% SDS). The second dimension was carried out as in SDS-PAGE analysis but with 0.8 mm thick gels, without stacking gel. Sigma markers (14-205 KDa) were run parallel to samples. Proteins were stained with silver nitrate and data were assessed by mean of Photoshop CS3 program.

3. Results and discussion

Flower and inflorescence development is one of the most complicated and well regulated plant process (Rolland-Lagan et al., 2003). Flowering time genes, floral meristem identity genes, and floral organ identity genes have well studied in *Arabidopsis, Pisum and Antirrhinum*, as well as in other annual plants. Tetramers of MADS proteins are thought to determine floral organ identity (Jack, 2004; Taiz and Zeiger, 2010). The size and the shape of petals as well as of the rest whorls is genetically determined (Laitinen et al., 2007). However little data are available until now concerning flower proteomic (Dafny-Yelin et al., 2005; Moccia et al., 2009; Sun et al., 2009; Logacheva et al. 2011; Zhu et al., 2011).

3.1 Olive inflorescence development

Olive inflorescence is a panicle which arises with a central axis terminated by a flower, with lateral axes branching from the peduncle. The lateral axes may in turn branch, resulting in a tertiary branched panicle (Weis et al., 1988; Seifi et al., 2008; Ganinoa et al., 2011).

Under normal conditions, olive inflorescence development last a prolonged period starting in March and full bloom occurs in May or June (Cirik, 1989; Weis et al., 1991).

The inflorescence development in "Konservolia" (Fig. 1, photos above plate a) seemed to constitute of three stages, as was also reported for "Kalamon" inflorescence (Bouranis et al., 2010) and lasted about 10 weeks until full bloom. Nine to seven weeks before full bloom inflorescence looked as a small berry covered with bracts and the elongation of the main axis prevailed, obviously due to the elongation of its cells. Six to four weeks before full bloom lateral axes having compact floral meristems on them were developed on the inflorescence, which suggest mitotic activity, as well as cell enlargement in the organs. At this stage the development of floral meristems and their whorls on the lateral axes indicates the implication of proteins derived from floral organ identity genes transcription. The next three weeks up to full bloom the development of flower whorls tended to complete. The week after full bloom only fertilized ovaries were remained on inflorescence, while lateral axes, unfertilized flowers and bracts were massively abscised during this week.

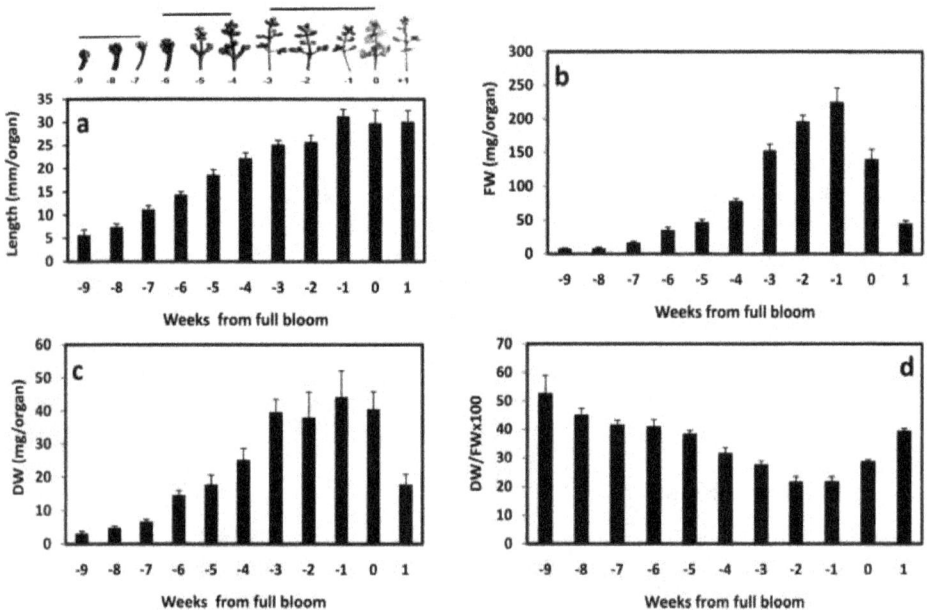

Fig. 1. The developmental stages of olive cv "Konservolia" inflorescence from floral bud burst 9 week before full bloom (plate a) up to one week after full bloom as well as the profiles of the changes of the length (plate b), of the fresh weights (plate c), dry weights (plate d) and the DW/FW ratio (plate d) (full bloom week 0). Vertical bars represent SE.

The increase of inflorescence length (Fig. 1, plate a) followed a sigmoid type of development and reached a maximum one week before full bloom. The fresh weight of inflorescence (Fig. 1, plate b) was progressively increased for a period of five weeks, with an intensive increase up to one week before full bloom, decreasing rapidly afterwards. Dry weight presented a similar but milder progress in comparison to that of FW during the same period (Fig. 1, plate c). The DW/FW ratio (plate d) showed a decreasing trend for a period of seven weeks before full bloom, stabilized on the lower level the two weeks before full bloom, increased at full bloom when floral whorls were separated by each other, and increased sharply the week after full bloom. This decreasing trend is obviously due to the enlargement of the floral organ cells and the contemporary hydration of them. The opening of petals and thereafter the higher transpiration rate, as well as the abscission of floral organs even before full bloom (Weis et al., 1988) and mainly after full bloom leads to a recovery of the DW/FW ratio at full bloom and the next week.

In Fig.2 are shown representative developmental stages of perfect flowers of the first lateral branch of four-branching inflorescence on median longitudinal sections five weeks before full bloom up to full bloom. Bracts which covered individual flowers were already shed five weeks before full bloom (plate -5). The stamen sac development found to be completed about two weeks before full bloom (plate -2). Pistil development followed that of the stamen and continued until full bloom in perfect flowers. Nucellus was already visible in the ovary five weeks before full bloom, while two well developed embryosacs were found on sections two and one week before full bloom (plates –2, –1). At full bloom, the opened petals revealed the flattened anthers and a global-like ovary (plate 0).

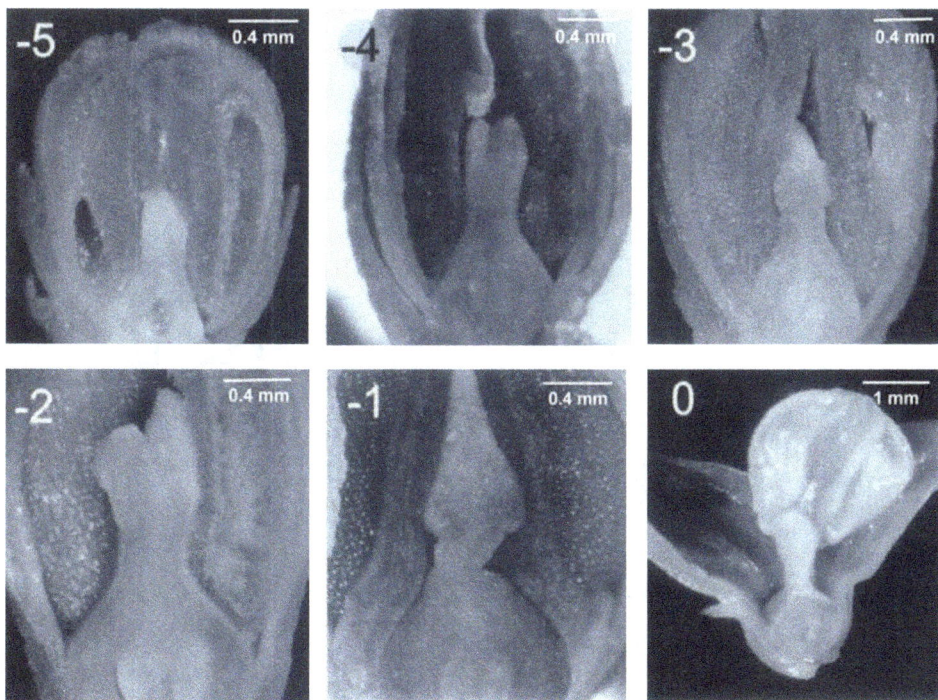

Fig. 2. Median longitudinal sections of olive perfect flowers (*Olea europaea* L. cv "Konservolia") showing the developmental stages of the organ at weekly intervals from 5 weeks before full bloom (plate -5) to full bloom (plate 0).

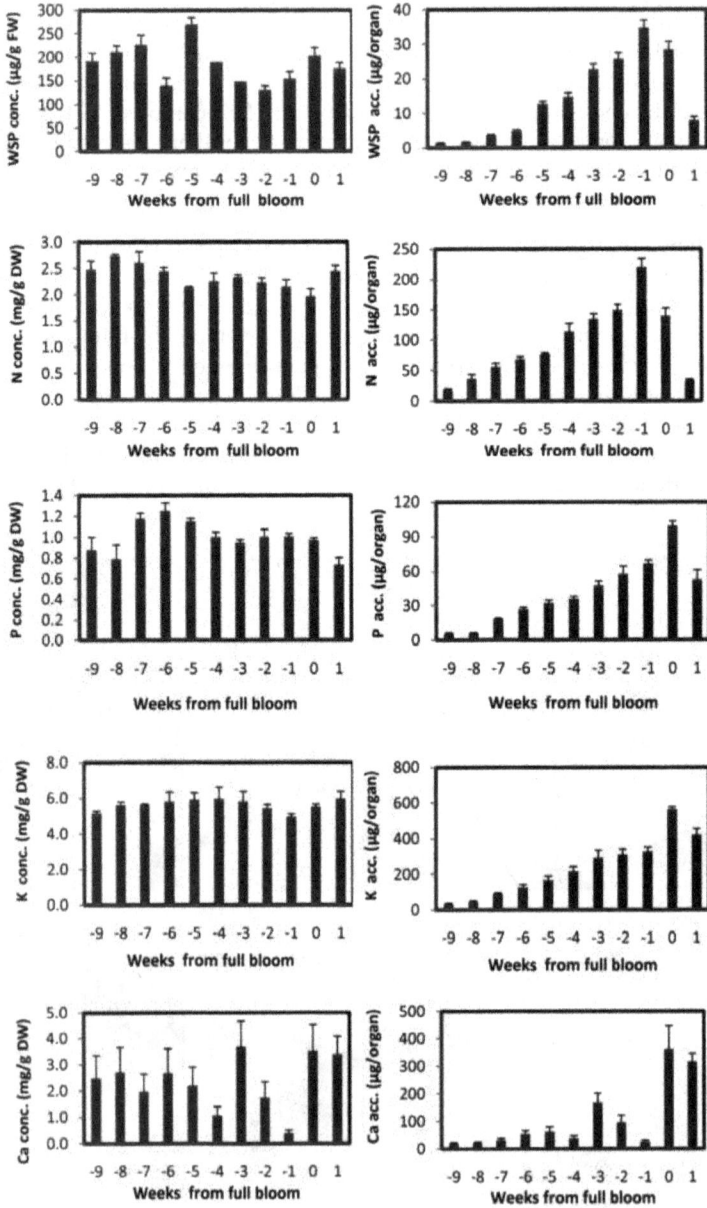

Fig. 3a. The profiles of concentration and accumulation of water soluble proteins, total nitrogen, phosphorus, potassium and calcium in olive inflorescence (*Olea europaea* L. cv "Konservolia") from floral bud burst (nine weeks before full bloom) to one week after full bloom.

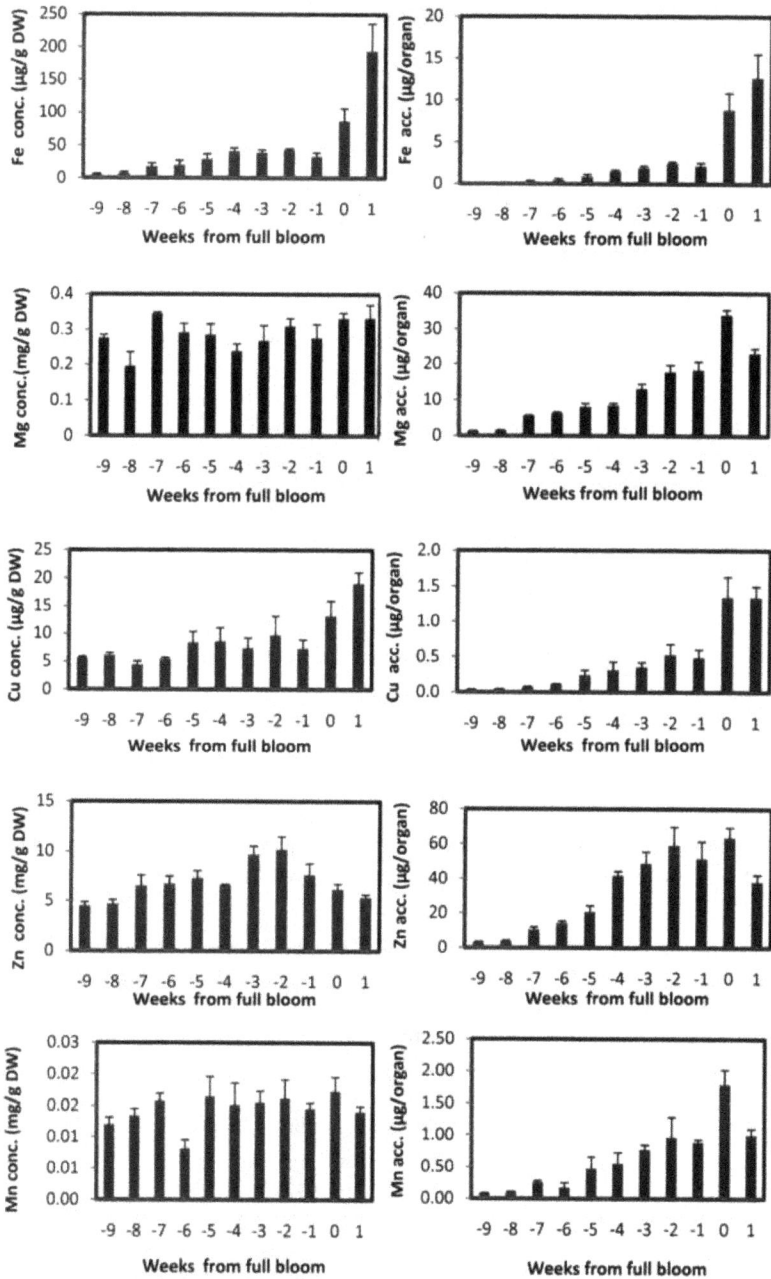

Fig. 3b. The profiles of concentration and accumulation of iron, manganese, copper, zink, and manganese in olive inflorescence (*Olea europaea* L. cv "Konservolia") from floral bud burst (nine weeks before full bloom) to one week after full bloom.

3.2 Profiles in total water extractable proteins and macro- and micro-nutrients

A pulse like profile found in total water extractable proteins in olive inflorescence which seemed to be consisted of three phases. The two lower levels at water soluble proteins found six and 2 weeks before full bloom. This profile showed to be similar to that of the progress of inflorescence development as discussed above. Water soluble protein accumulation showed a sigmoid like profile in which the phase of the higher rate of accumulation localized in the period of five to one week before full bloom, which corresponds to the period of flower and floral whorls development on inflorescence. On the other hand, the profile of nitrogen concentration showed to be more uniform, having a decreasing trend. Nitrogen accumulation showed an increasing trend up to one week before full bloom. In spite the similar trend, the rate of increase of nitrogen accumulation was lower to that of DW accumulation, resulting in a progressive dilution of nitrogen level in the developing inflorescence. A pulse like profile on water soluble proteins was found in differentiating olive floral buds during late winter and thought to be due to a need of a relaxing phase of the dividing cells (Kitsaki et al., 2010). Proteins may be enzymes for the different metabolic processes among which mitosis, structural proteins, membrane proteins, or storage proteins. In flowers tetramers of proteins which determine floral organ identity and development, as well as those implicated to meiosis have a special role (Jack, 2004; Luu and Maurel 2005; Ahsan and Komatsu, 2009; Malumbres et al., 2009; Echalier et al., 2010; Taiz and Zeiger, 2010). As different physiological processes take place during floral organ development, among them meiosis, a fluctuated profile of water soluble proteins during this period could be expected. The observed fluctuations in the water soluble proteins could also be attributed to interchanges between soluble and insoluble protein fractions in the organ.

In plant cells except from proteins nitrogen is a constituent of free amino acids, amides, nucleic acids nucleotides, coenzymes, amines and others substances Therefore, nitrogen deficiency rapidly inhibits plant growth and if persists transition of plant from vegetative to reproductive phase may not be completed (Marschner, 1995; Taiz and Zeiger, 2010). Differences in the level of nitrogenous compounds between vegetative and reproductive shoots on the same plant have been reported for the majority of plant, among them for olive (Bouranis et al., 2004). In olive tree, nitrogen status affects flower quality and ovule longevity (Fernandez-Escobar et al., 2008).

The higher levels of phosphorus concentration observed during the early stages of flower development (-7 to -5 weeks), while the nutrient was accumulated progressively up to full bloom in the inflorescence. Potassium also accumulated in the inflorescence up to full bloom but changes in the concentration showed a rather mild profile. The accumulation of calcium remained at low levels up to full bloom when a huge increase was observed. On the other hand the fluctuations in calcium concentration were the most dramatic (Fig. 3a). Except from the special role of P in energy storage, of K in osmoregulation and of Ca in the structure of cell wall and as a second messenger, as well as their participation in many enzymes and coenzymes, they occupy a special role in the pollen grain germination as components of phytate salts (Fu et al., 2008; Taiz and Zeiger, 2010). The decreasing trend in Ca up to full bloom is possibly the result of the immobile character of this nutrient (Marschner, 1995), while the rapid increase in its accumulation after full bloom may be due to the dehydration of inflorescence at this stage and the consequent increase of percentage of cell walls in the organ.

Profiles of iron and copper concentration showed an increasing trend during inflorescence development, zink concentration also tended to increase up to two weeks before full bloom, while magnesium and manganese showed not considerable fluctuations during inflorescence development. On the other hand accumulation of the above nutrients showed a slow or medium increasing rate up to full bloom, when a rapid increase was observed in the accumulations (except from that of Cu). All the above nutrients are constituents of enzymes which implicate in the basic metabolic activity and especially Mg and Fe. Iron and copper seemed to have special interest during the massive flower organ abscission one week after full bloom, as their concentrations was increased considerably. On the other hand three to two weeks before full bloom (late stages of ovary and stamens development) zink concentration showed to be of increased interest. The strong positive linear regression among DW and water soluble proteins, total N, as well as macro- and micro-nutrients accumulation during inflorescence development (Table 1) reinforces the concept that olive inflorescence is a strong sink organ.

	WSP	N	P	K	Mg	Ca	Cu	Fe	Zn	Mn
R^2	0.78**	0.98**	0.97**	0.98**	0.97**	0.64**	0.81**	0.67**	0.76**	0.90**
SE	14.2	7.0	6.3	18.6	1.8	70.6	0.25	3.1	0.5	0.4
n	11	11	11	11	11	11	11	11	11	11

Table 1. The R^2 values among dry weight accumulation and of total water soluble proteins (WSP), as well as mineral nutrients (N, P, K, Mg, Ca, Cu, Fe, Zn and Mn) in developing olive inflorescence (*Olea europaea* L. cv "Konservolia").

3.3 Proteomic changes in developing olive inflorescence

In Fig. 4 is shown the profile of proteins revealed by SDS-PAGE analysis in developing olive inflorescence nine weeks before full bloom up to one week after full bloom. In table 2 are summarized the number and the kDa of the detected protein bands as well as the most abundant and the common protein bands in developing olive inflorescence by SDS-PAGE analysis.

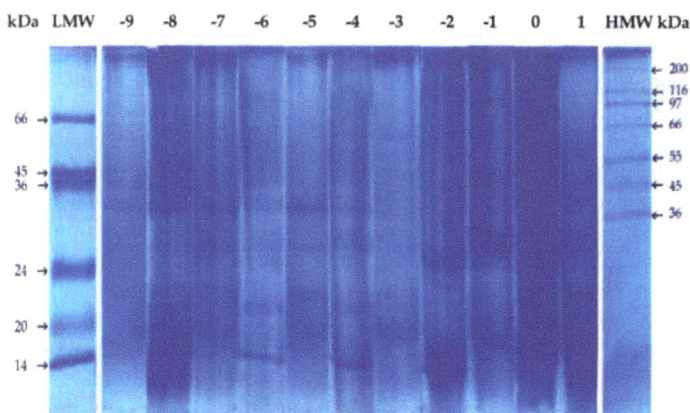

Fig. 4. The protein profile in developing olive inflorescence (*Olea europaea* L. cv "Konservolia") nine weeks (-9) before full bloom up to one week (1) after full bloom. Proteins (50 µg) were resolved in 12,5% SDS/acrylamide gels (1,5 mm thick) parallel with Sigma markers (14 to 66 and 36 to 205 kDa) and gels were stained with Coomassie Brilliant Blue R-250, according to Sigma protocol.

W	-9	-8	-7	-6	-5	-4	-3	-2	-1	0	1
1								121.5			
2									**101.6**	101.1	
3							80.26		**79.74**		
4								77.36		77.36	
5		65.56								**67.25**	
5					53.81			54.85		**54.51**	
6	51.06	**52.80**	50.73								
7									49.04		
8				45.90							
9	42.78	**43.62**			43.34	43.89				43.89	44.20
10			**41.39**		39.02		39.54	40.15	40.59		40.41
11											
12				33.18						**33.60**	32.27
13		31.93	31.22		31.22	32.33	31.43		**31.85**		
14								29.58			
15				24.44				24.28			
16					20.13	20.45	20.66				**20.78**
17	19.48	19.07	19.86	19.43		19.06					
18		15.28			15.90		15.08			**15.00**	
19	14.31			**13.40**					13.94	13.87	13.58
20		**12.53**				12.37				12.45	
21								**11.83**			
22		**10.61**									10.52

Table 2. The kDa of the detected protein bands as well as the most abundant (bold numbers) and the common (shadowed cells) protein groups in developing olive inflorescence by SDS PAGE analysis nine weeks before (-9) full bloom to one week after full bloom (1).

Protein bands molecular weights were fluctuated between 10 and about 100 kDa. Protein bands of 44, 40, 32, 19, 15 and 14 kDa were the most frequently appeared, while the groups of 14 and 12 kDa were the most abundant. The number of protein bands was fluctuated between 5 and 9 during the three developmental stages of inflorescence. Most of protein groups were visualized at full bloom (9 groups), at the middle of the phase of inflorescence elongation (-8 week), as well as at the late stage of floral organ development (-2 week), 7 protein groups. As full bloom was coming, new groups of high molecular weights were appeared (3 weeks before full bloom to full bloom). These groups were disappeared the week after full bloom. In this week the range of molecular weights of protein groups was reduced to 10 to 44 kDa (Table 2). The fluctuation of the number of protein bands may be related with the intensity of meristematic activity, as well as cell enlargement processes, which are correlated with enzyme systems relative to mitosis, as cyclin-dependent-kinases, structural proteins for dauter cells and others, as mentioned above (Dembinsky et al., 2007; Malumbres et al., 2009; Echalier et al., 2010; Nafati et al., 2011). This concept is reinforced by the pulse like data we found in the profile of water extractable proteins. At full bloom new physiological processes take place e.g pollen tube growth, pollination and fertilization which need specific enzymatic activity (Chen et al., 2009). Proteome analysis of soybean flowers and leaves at various developmental stages revealed organ specific functional differentiation of proteins (Ahsan and Komatsu, 2009). Moreover, Dafny-Yelin et al. (2005) through expression analysis of resolved proteins from rose petals at advanced stage development found about 30% of them to be stage specific. Recently Zhu et al. (2011) identified differentially expressed proteins in mature and germinated maize pollen and found 26 proteins (among about 470) to be changed between the two stages, 13 of them (up-regulated) were mainly involved in tube wall modification, actin cytoskeleton organization and energy metabolism. However, although genetic control of flowering and floral organ development has been extensively studied (Jack, 2004; Leitanen at al., 2007) much remains to be done concerning proteomic approach. Three groups of proteins (about 20 kDa, 31 kDa and 40 kDa) which were present in a wide range of developmental stages of inflorescence may be related to the basic cell metabolic activity or may be storage proteins. Ahsan and Komatsu (2009) referred an abundance of a glycoprotein of 31 kDa in soybean petals, which may be similar to that we also realized. Except from any relation to pollen germination, pollination and fertilization, the presence of the high molecular weight proteins exclusively at the late stage of inflorescence development, may also be related to genes that regulate synthesis of secondary metabolites, as flavonoids, carotenoids and other aromatic substances, which have been reported that appear at the late stage of floral organ development (Ben-Meir et al., 2002).

Further proteome analyses with 2-D electrophoresis carried out in three sampling dates distributed in the three phases of inflorescence development: (a) eight weeks before full bloom (middle of the phase of inflorescence axis elongation), (b) four weeks before full bloom (middle of the phase of floral organ development and (c) at full bloom. In Fig. 5a, 5b, 5c are shown the correspond to the above sampling dates profiles of proteins revealed by 2-DE analysis. In table 3 are summarized the kDa and the pIs of the detected proteins by 2-DE analyses. Proteins were distributed all over the range of 4.0 to 7.0 pH, while molecular weights fluctuated between about 10 to 55 kDa the -8 and the -3 weeks and between 10 to 105 kDa at full bloom (Table 3). Most of the abundant proteins were found at full bloom. Two protein groups (42.5 kDa/4.12-4.60 pIs and 28.2 kDa/4.64-5.83 pIs) were visualized in the three sampling dates. Some of proteins were common in the three stages of development, while other were present in one or two of the three phases. The most abundant protein was that of 44.3 kDa/4.67-5.19 pIs, especially at -8 week and at -4 week. Proteins with high molecular weights (over 60 kDa) were monitored exclusively at full bloom.

Early tree proteome studies were focused on the process of xylogenesis mainly in *Populus* and *Pinus* (Costa et al., 1999; Mijnsbruge et al., 2000), but some fruit trees have recently gained interest for proteome analysis (Cartu et al., 2008; Paiva et al., 2008). However, little effort have been done on olive proteomic, probably because of the difficulty on sample handling, due to the presence of many phenolic compounds (Garcia et al. 2000, Wang et al., 2003, Wang et al., 2004, Wang et al., 2010,).

The pIs of most of the proteins we visualized by 2-DE in developing olive inflorescence were localized in the low pH region. Similar data have been reported for olive leaf (Wang et al., 2003), as well as for other plants (Dafny-Yelin et al., 2005; Ahsan and Komatsu, 2009; Chen et al., 2009). The fluctuation in abundance in the three stages of inflorescence development may be related to developmental stages specific functional proteins (Ahsan and Komatsu, 2009). The majority of the abundance found at full bloom is in accordance with the recent data of Ahsan and Komatsu (2009) and Zhu et al. (2011), concerning stage specific functional proteins.

-8 week (middle of the phase of inflorescence axis elongation)									
SN	kDa/pI	SN	kDa/pI	SN	kDa/pI	SN	kDa/pI	SN	kDa/pI
1	55.6/4.61	10	25.6/4.20	20	13.6/4.92	25	**11.8/6.28**	36	10.5/4.80
2	**46.8/4.96**	11	24.7/6.91	21	13.6/5.20	32	**11.5/5.34**	38	**10.0/4.40**
3	**44.3/4.95**	12	23.8/6.90	22	**12.9/6.03**	33	11.6/5.49	40	10.0/4.29
4	**42.7/4.96**	13	22.7/4.54	26	12.9/4.03	16	11.5/5.88	41	10.0/4.88
5	**40.9/4.96**	15	18.9/4.40	28	12.5/4.41	29	11.5/4.30	43	**10.0/5.81**
6	34.8/4.21	14	18.5/4.23	27	12.5/4.76	31	10.9/5.20	39	10.0/4.05
7	29.7/4.43	17	16.6/4.35	23	**12.3/5.95**	37	10.9/4.30	42	**10.0/5.29**
8	29.4/4.69	18	15.7/4.37	24	**12.3/6.27**	34	**10.5/5.77**	44	**10.0/6.03**
9	28.4/5.11	19	14.8/4.92	30	**12.3/4.84**	35	**10.5/5.93**	45	**10.0/6.10**
-4 week (middle of the phase of floral organ development)									
SN	kDa/pI	SN	kDa/pI	SN	kDa/pI	SN	kDa/pI	SN	kDa/pI
1	**52.8/5.2**	7	**42.3/4.21**	16	28.0/4.64	26	17.3/4.27	31	**14.4/4.84**
2	**50.5/5.19**	13	**35.4/5.51**	15	27.2/4.20	25	17.5/4.27	29	14.4/4.47
3	**48.7/5.19**	10	**33.5/4.20**	22	23.7/5.46	27	17.5/4.68	28	14.4/4.28
4	**47.5/5.19**	11	33.5/4.83	21	22.6/5.40	35	15.3/5.33	36	14.0/6.43
5	**44.8/5.19**	12	33.5/5.16	20	22.6/4.69	34	**15.3/5.27**	37	14.0/6.48
9	**44.7/4.67**	14	31.7/6.72	19	21.3/4.28	33	15.3/5.20	38	10.7/4.27
8	**43.5/4.38**	17	29.4/4.91	23	19.7/4.27	32	15.3/4.95	39	10.0/4.47
6	**43.5/5.19**	18	29.4/5.15	24	19.7/4.70	30	15.3/4.72	40	10.0/4.48
0 week (full bloom)									
SN	kDa/pI	SN	kDa/pI	SN	kDa/pI	SN	kDa/pI	SN	kDa/pI
8	106.5/6.29	6	93.5/5.34	17	**42.9/4.12**	25	32.3/4.84	32	16.5/4.39
10	105.2/6.47	4	92.0/5.19	19	**42.9/4.47**	26	**28.4/5.83**	34	**16.5/5.39**
9	104.3/6.45	12	68.7/4.57	20	**42.9/4.60**	27	**28.4/5.19**	35	**15.2/5.36**
2	**104.3/4.35**	11	67.3/4.38	21	**41.3/5.20**	28	26.3/4.50	36	**14.3/4.57**
3	**103.2/4.74**	13	**56.6/5.19**	18	**41.3/4.31**	29	**26.3/4.23**	37	14.3/4.36
7	103.2/5.86	14	**49.7/5.47**	22	**36.6/5.41**	30	20.7/4.70	38	13.2/4.16
1	**102.5/4.10**	15	**48.3/5.27**	23	**32.3/5.41**	31	20.7/4.34	39	**10.0/4.45**
5	93.5/5.34	16	**46.6/5.20**	24	**32.3/5.20**	33	17.2/4.59	40	**10.0/4.08**

Table 3. The kDa and the pIs of the detected proteins by 2-DE analyses in developing olive inflorescence (*Olea europaea* L. cv "Konservolia") eight weeks before full bloom (-8 week), three weeks before full bloom (-3 week) and at full bloom (0 week). The most abundant proteins are indicated with bold numbers (SN: spot number).

pH 4.0 pH 7.0

KDa

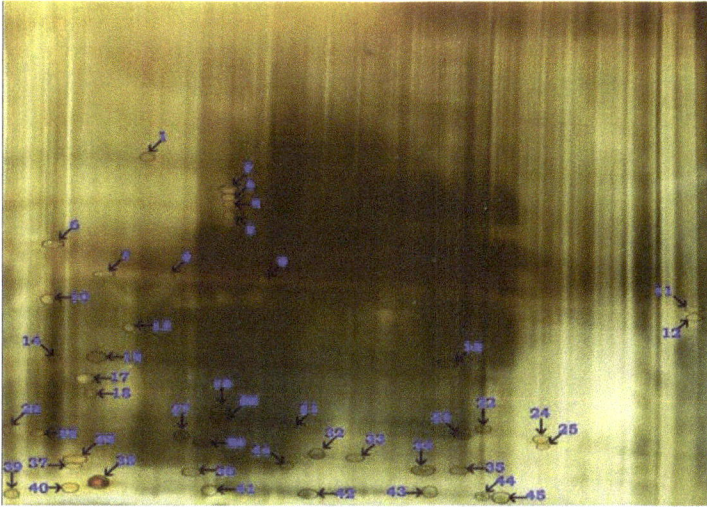

Fig. 5a. Protein profile in developing olive inflorescence eight weeks before full bloom. Proteins (300 µg) were resolved in 13 cm linear immobilized dry strips pH 4.0-7.0 and in the second dimension in 12.5 % SDS/acrylamide 0.8 mm thick gels.

Ph 4.0 pH 7.0

KDa

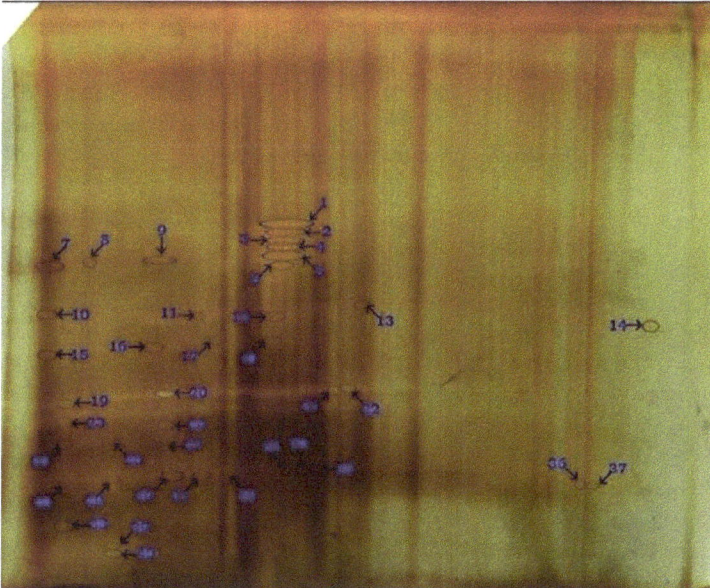

Fig. 5b. Protein profile in developing olive inflorescence four weeks before full bloom. Proteins (300 µg) were resolved in 13 cm linear immobilized dry strips pH 4.0-7.0 and in the second dimension in 12.5 % SDS/acrylamide 0.8 mm thick gels.

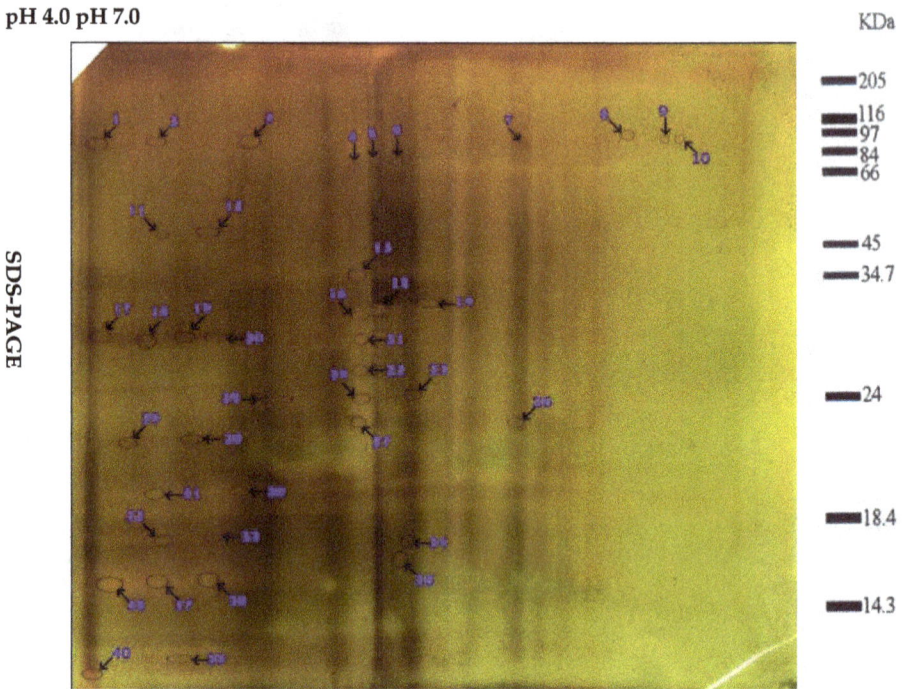

Fig. 5c. Protein profile in developing olive inflorescence at full bloom . Proteins (300 µg) were resolved in 13 cm linear immobilized dry strips pH 4.0-7.0 and in the second dimension in 12.5 % SDS/acrylamide 0.8 mm thick gels.

4. Conclusions

Olive inflorescence development seems to be separated in three stages: a first one corresponding to the inflorescence axis elongation, a second corresponding to the development of flowers and floral whorls development and the last one which corresponds to the full bloom and flower fertilization. Changes in DW/FW ratio revealed intense hydration of the organ two weeks before full bloom. Strong positive linear regression found among DW and water soluble proteins, as well as macro- and micronutrients. A pulse like profile found in total water extractable proteins in olive inflorescence which seemed to be consisted of three phases. This profile showed to be similar to that of the progress of inflorescence development. Most of protein groups were visualized at full bloom, at the middle of the phase of inflorescence elongation, as well as at the late stage of floral organ development. As full bloom was coming, new groups of high molecular weights were appeared. Proteins with high molecular weights (over 60 kDa) were monitored exclusively at full bloom. The pIs of most of the proteins we visualized by 2-DE in developing olive inflorescence were localized in the low pH region.

5. Acknowledgments

This work was supported by the Ministry of National Education Affairs. The authors are grateful to the staff of the Laboratory of Pomology of Agricultural University of Athens, which take care of the olive orchard.

6. References

Ahsan, N., Komatsu, S. (2009). Comparative analysis of the proteomes of leaves and flowers at various stages of development reveal organ-specific functional differentiation of proteins in soybean. *Proteomics* 9, pp 4889-4907.

Bearden, J. C. (1978). Quantitation of submicrogram quantities of protein by an improved protein-dye binding assay. *Biochim. Biophys. Acta* 533 pp 525-529.

Ben-Meir, H., Zuker, A., Weiss, D.,Vainstain, A., in: Veinstain, A., (Ed) (2002). Molecular control of flower pigmentation: Anthocyanins, Kluwer, Dordrecht. pp 253-272.

Bouranis D. L., Kitsaki, C. K., Chorianopoulou, S. N., Papadimitriou, A., Chondroyianni, V. (2010). Does the Nutritional Status of Reproductive Shoots of Olive Tree (cv. 'Kalamon') Differ from that of Vegetative Shoots during Inflorescence Development? *European Journal of Plant Science and Biotechnology* pp 137-144

Bouranis, D. L., Kitsaki, C. K., Tzakosta A. (2004). Differences in nitrate and ammonium homeostasis of reproductive and vegetative shoots of olive tree cv Kalamon during inflorescence development. *Journal of Plant Nutrition 27(5)* pp 797-813.

Bradford, M. M., (1976). A rapid and sensitive method for the quantitation of microgram quantities of protein utilizing the principle of protein-dye binding. *Anal. Biochem.* 72, pp 248-254.

Cantú M. D., Mariano, A. G., Palma, M. S., Carrilho, E., Wulff, N. A. (2008). Proteomic analysis reveals suppression of bark chitinases and proteinase inhibitors in *Citrus* plants affected by the citrus sudden death disease. *Biochem. Cell Biol.* 98 (10) pp 1084-1092.

Chen, S., Liu, H., Chen, W., Xie, D., Zheng, S., (2009). Proteomic analysis of differentially expressed proteins in longan flowering reversion buds. *Sci. Hortic.* 122 pp 275-280.

Cirik, N. 1989. Factors influencing olive flower bud formation. *Olivae VI* Year, No 27, pp 25-27.

Costa, P., Pionneau, C., Bauw, G., Dubos, C., Bahrmann, N., Kremer, A., Frigerio, J.M., Plomion, C., (1999). Separation and characterization of needle and xylem maritime pine proteins. *Electrophoresis* 20(4-5) pp 1098-1108.

Dafny-Yelin, M., Guterman, I., Menda, N., Ovadis, M., Shalit, M., Pichersky, E., Zamir, D., Lewinsohn, E., Adam, Z., Weiss, D., Vainstein, A. (2005). Flower proteome: changes in protein spectrum during the advanced stages of rose petal development. *Planta* 222 pp 37–46.

Dembinsky, D., Woll, K., Saleem, M., Liu, Y., Fu,Y., Borsuk, L.A., Lamkemeyer, T., Fladerer, C., Madlung, J., Barbazuk, B., Nordheim, A., Nettleton, D., Schnable, P.S., Hochholdinger, F. (2007). Transcriptomic and proteomic analyses of pericycle cells of the maize primary root. *Plant Physiol.* 145 pp 575-588.

Drossoloulos, J. B., Kouchaji, G.,G., Bouranis, D., L. (1996). Seasonal dynamics of mineral nutrients by walnut tree reproductive organs. *J. Plant Nutr.* 19(2) pp 421-434.

Echalier, A., Endicott, J.A., Nobl, M. E. M. (2010). Recent developments in cyclin-dependent kinase biochemical and structural studies. *Biochim. Biophys. Acta* 1804 pp 511-519.

Fernández-Escobar R, Ortiz-Urquiza, A., Prado, M., Rapopport, H.F. (2008). Nitrogen status influence on olive tree flower quality and ovule longevity. *Environmental and experimental Botany* (64): 113-119.

Fu, J., Peterson, K., Guttieri, M., Souza, E.,Raboy, V. (2008). Barley (*Hordeum vulgare* L.) inositol monophosphatase: Gene structure and enzyme characteristics. *Plant Molecular Biology* 67 (6) pp 629-642.

Ganinoa, T., H., F., Rapoport, H., F., Fabbri, A. (2011). Anatomy of the olive inflorescence axis at flowering and fruiting. *Scientia Horticulturae* 129 pp 213–219.

Garcia, J., L., Avidan, N., Troncoso, R., Sarminto, S., Lavee, S., (2000). Possible juvenile-related proteins in olive tree tissues. *Sci. Hortic.* 85, pp 271-284.

Jack, T. (2004). Molecular and Genetic Mechanisms of Floral Control. *The Plant Cell*, 16 (Suppl), pp 1-17.

Jones, J. B., Kjeldahl, Jr. (1991). *Method for Nitrogen Determination:* MicroMacro Publishing , Inc.: Athens, GA, pp 56.

Kitsaki, C. K., Andreadis, E., Bouranis, D. L., (2010). Developmental events in differentiating floral buds of four olive (*Olea europaea* L.) cultivars during late winter to early spring. *Flora* 205(8), 599-607.

Laitinen, R. A., Pöllänen, E., Teeri, T. H., Elomaa, P., Kotilainen, M. (2007). Transcriptional analysis of petal organogenesis in *Gerbera hybrid*. *Planta* 226 pp 347–360.

Logacheva, M., Artem, D., Kasianov, S., Vinogradov, D. V., Samigullin, T. H., Gelfand, M. S. (2011). De novo sequencing and characterization of floral transcriptome in two species of buckwheat (Fagopyrum). *Genomics* pp 1471-1485.

Luu, D.T., Maurel, C. (2005). Aquaporins in a challenging environment: molecular gears for adjusting plant water status. *Plant Cell Environ.* 28 pp 85-96.

Malumbres, M., Harlow, E., Hunt, T., Hunter, T., Lahti, J.M., Manning, G., Morgan, D.O., Tsai, L.H., Wolgemuth, D.J. (2009). Cyclin-dependent kinases: a family portrait. *Nat. Cell Biol.* (11) pp1275-1276.

Martin, G.C., (2009). http://diegoromerosellanes.files.wordpress.com/2009/11/8-olea-europaea-l-george-c-martin-professor-of-pomology-at-the-university-of-california-davis-california2.

Marschner, H. (1995). Mineral nutrition of higher plants, 2nd Ed. ISBN 0-12-473543-6. Academic press. London

Mills, H. A., Jones, J. B. Jr. (1996). *Plant Analysis Handbook II*, Micro Macro Publishing, Inc, Athens, GA.

Mijnsbrugge, K.V., Meyermans, H., Van Montagu, M., Bauw, Guy., Boerjan, W., 2000. Wood formation in poplar: identification, characterization, and seasonal variation of xylem proteins. Planta 210, 589-598.

Moccia, M. D., Oger-Desfeux, C., Marais, G. A. B., Widmer, A. (2009). A White Campion (*Silene latifolia*) floral expressed sequence tag (EST) library: annotation, EST-SSR characterization, transferability, and utility for comparative mapping *BMC Genomics* 10 pp 243-251.

Nafati, M., Cheniclet, C., Hernould, M., Do, P.T., Fernie, A.R., Chevalier, C., Gévaudant, F. (2011). The specific overexpression of a cyclin-dependent kinase inhibitor in tomato fruit mesocarp cells uncouples endoreduplication and cell growth. *Plant J.* 65(4) pp 543-556.

Paiva, J. A.P., Garcés, M., Alves, A., Garnier-Géré, P., Rodrigues, J.C., Lalanne, C., Porcon, S., Le Provost, da Silva Perez, D., Brach, J., Frigerio, J.-M., Claverol, S., Barré, A., Fevereiro, P., Plomio C., (2008). Molecular and phenotypic profiling from the base to the crown in maritime pine wood-forming tissue. New Phytol. 178 pp 283-301.

Peach, K., Tracey, M.,V. (1956). *Moderne Methoden der Pflanzenanalyse* (Vol 1), Springer-Verlag, Berlin, Germany, pp 487-488.

Reale, L., Sgromo, C., Bonofiglio, T., Orlandi, F., Fornaciari, M., Ferranti, F., Romano, B. (2006). Reproductive biology of olive (Olea europaea L.) DOP Umbria cultivars. *Sexual Plant Reproduction* 19 pp 151-161.

Rolland-Lagan, A-G., Bangham, J. A., Coen, E. (2003). Growth dynamics underlying petal shape and asymmetry. *Nature* 422 pp 161-163.

Rugini, E., Fedeli., E. (1990). Olive as an oilseed pp. 593-641.In Y.P.S. Bajaji (ed). Biotechnology in Agriculture and Forestry. Vol. 10 *Legumes and Oilseed* Crops I. Springer-Verlag, Berlin, Germany.

Sanchez, E., Righetti, T. L., Sugar, D., Lombard, P., B. (1992). Effects of timing of nitrogen application on nitrogen partitioning between vegetative, reproductive and structural components of mature "comice" pears. *J Hort. Sci.* 67 pp 51-58.

Sanchez, E., Righetti, T. L., Sugar, D., Lombard, P., B. (1992). Effects of timing of nitrogen application on nitrogen partitioning between vegetative, reproductive and structural components of mature "comice" pears. *J Hort. Sci.* 67 pp 51-58.

Sanz, M., Montanes, L. (1995). Floral analysis as a new approach to diagnosing the nutritional status of the peach tree. *J. Plant Nutr.* 18(8):1667-1675.

Seifi, E., Guerin, J., Kaiser, B., Sedgley, M. (2008). Inflorescence architecture of olive. *Scientia Horticulturae* 116 pp 273–279.

Sun, Q., Chaofeng, Hu, C., Hu, J., Li, S., Zhu, Y. (2009). Quantitative Proteomic Analysis of CMS-Related Changes in Honglian CMS Rice. *Anther Protein J* 28: pp 341–348.

Süle, A., Vanrobaeys, F., Hajos, G., Beeumen, V., J., Devreese, B. (2004). Proteomic analysis of small heat shock protein isoforms in barley shoots. *Phytochemistry* 65, pp 1853-1863.

Taiz, L., Zeiger, E., (2010). *Plant Physiology*. 5th ed, ISBN 978-0-87893-511-6. Sinauer Associates Inc. Publishers, Sunderland.

Wang, W., Scali, M., Vignani, R., Padafora, A., Sensi, E., Mazzuca, S., Cesti, M., (2003). Protein extraction for two-dimensional electrophoresis from olive leaf, a plant tissue containing high levels of interfering compounds. *Electrophoresis.* 24, pp 2369-2375.

Wang, W., Vignani, R., Scali, M., Sensi, E., Tiberi, P., Cresti, M. (2004). Removal of lipids contaminants by organic solvents from oilseed protein extract prior to electrophoresis. *Analytical. Biochemistry.* 329, pp 139-141.

Wang, W., Tai, F., Hu, X. (2010). Current initiatives in proteomics of olive tree. In Preedy, V.R., Watson, R. (Eds), *Olives and Olive Oil in Health and Desease Prevention.* Academic Press, UK, pp 25-32.

Weis, K., G., Goren, R. Martin, G.,C., Webster, B., D. (1988). Leaf and inflorescence abscission in olive I. Regulation by ethylene and ethephon. *Bot. Gaz.*149 pp 391-397.

Weis, K., G., Webster, B., D., Goren, R., Martin, G., C. (1991). Inflorescence abscission in olive: Anatomy and Histochemistry in response to ethylene and ethephon. . *Bot. Gaz.* 152 pp 51-58.

Zhu, Y., Zhao, P., Wu, X., Wang, W., Scali, M., Cresti, M. (2011). Proteomic identification of differentially expressed proteins in mature and germinated maize pollen *Acta Physiol Plant* 33 pp 1467-1474.

14

Regulatory Mechanism in Sexual and Asexual Cycles of *Dictyostelium*

Aiko Amagai

Graduate School of Life Sciences, Tohoku University

Japan

1. Introduction

Cellular slime molds exhibit dimorphism in development: sorocarp formation as an asexual cycle and macrocyst formation as a sexual cycle. These two developmental forms are regulated by environmental conditions, such as light and water. *Dictyostelium mucoroides 7* (Dm 7), a species of cellular slime molds forms sorocarps in the light, while forming macrocysts in the dark or water (Fig.1).

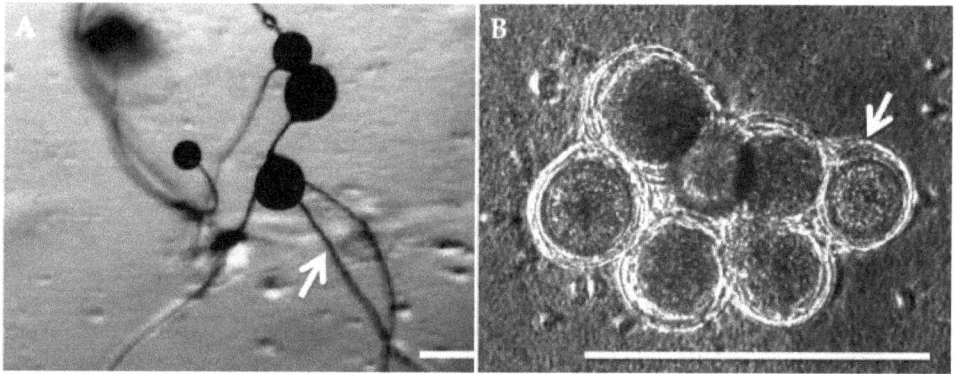

Fig. 1. Two developmental forms in Dm7. A: Starved Dm7 cells were developed on 1.5 % agar in the light. After 48 hours of incubation, they formed sorocarps. A globular spore mass is supported by a stalk (arrow). B: When starved Dm7 cells were developed under submerged conditions, they formed macrocysts. Each macrocyst is surrounded by a thick wall (arrow). Bars: 200 μm.

In an asexual cycle, amoeboid cells grow and multiply feeding on bacteria. Upon exhaustion of the bacterial food supply (starvation), starving cells stop growing and start the differentiation process. They gather together to form cell aggregates. A tip is formed on the top of each cell aggregate, which then migrates as a slug-shaped mass. After migration, the slug changes its shape dramatically to form a sorocarp consisting of a stalk with an apical mass of spores.

Macrocyst formation as a sexual cycle is characterized by the formation of large aggregates after starvation. Large aggregates are subdivided into smaller masses (precysts), each of which is surrounded by a fibrillar sheath. At the center of each precyst, a cytophagic cell (a giant cell) arises, which in turn engulfs all the other cells in the precyst. The engulfed cells (endocytes) are eventually broken down into granular remnants. The enlarged cytophagic cell finally becomes surrounded by a thick wall to form the mature macrocyst (Fig. 2) (Filosa & Dengler, 1972). After a resting period, the macrocyst germinates to release several amoeboid cells and initiates a new life cycle (Nickerson & Raper, 1973). Cytophagic cells (giant cells) formed during macrocyst formation have been proved to be zygotes that are produced by cell fusion and subsequent nuclear fusion (Amagai, 1989).

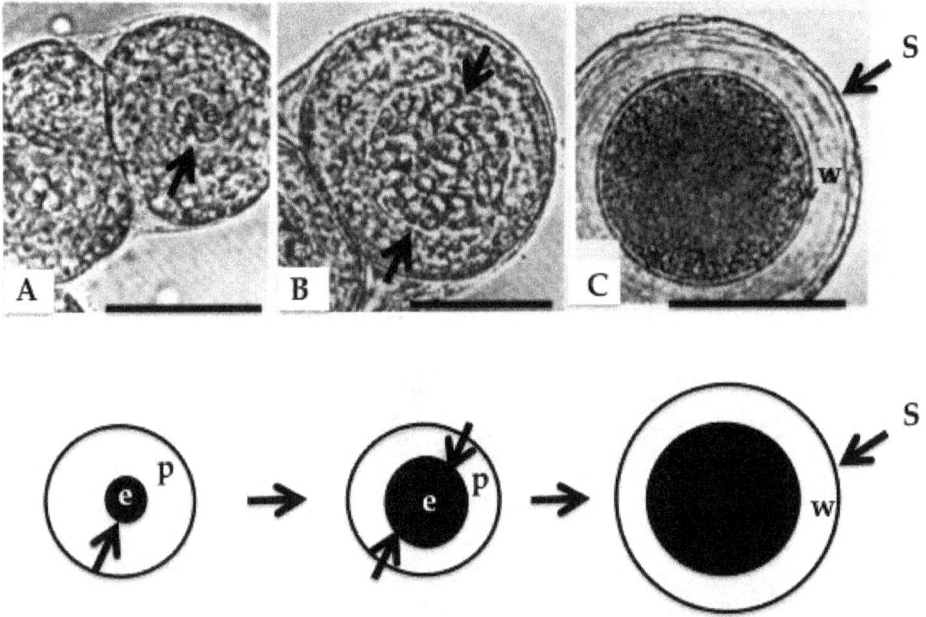

Fig. 2. The developmental process of macrocyst formation. Starved Dm7 cells were developed by shaking. A: 21 hours development after starvation. The first few endocytes have formed at the center of a precyst. A boundary delimits the cluster of endocytes (e) from the surrounding peripheral cells (p) (arrow). B: 24 hours of development. The number of endocytes has increased as more peripheral cells have been engulfed by the cytophagic cell (delimited by arrows). C: 6 day-old cyst filled with granules in Brown motion. After all peripheral cells have been transformed into endocytes, endocytes are broken down into granular remnants. S: fibrillar sheath, W: macrocyst wall. Drawings show the same stages as photographs shown at the top (modified data cited from Filosa & Dengler, 1972). Bars: 35μm.

There are two kinds of mating systems in the macrocyst formation; homothallic and heterothallic (Clark et al., 1973; Erdos et al., 1973). Dm 7 forms macrocysts without mating types as a homothallic strain. On the other hand, *Dictyostelium discoideum* (*D. discoideum*), a heterothallic strain, undergoes mating with an opposite mating type, V12M2 cells.

In this chapter, the regulatory mechanism for determining one of two cycles in Dm 7 and *D. discoideum* is reviewed, particularly focusing on the function of ethylene as a potent regulator. The signal transduction pathways involved in zygote formation are also noted, specifically paying attention to the function of a novel protein ZYG1, whose expression is augmented by ethylene.

2. Advantages of using cellular slime molds for cellular and developmental studies

Cellular slime molds are known as model organisms and have a lot of advantages as materials in the fields of cellular and developmental biology. As the differentiation phase of the cellular slime molds starts by removal of nutrients (starvation), it is temporarily separated from the growth phase. The pattern of cell differentiation is also relatively simple: *Dictyostelium* cells eventually differentiate into mainly two cell types, spore and stalk cells in an asexual cycle. These characters are quite suitable for studies of differentiation including the mechanisms of the transition from growth to the differentiation phase. As they are usually haploid, it is easier to manipulate genes, such as clone, knockout genes and so on. In fact, many transformants were cloned and used in this chapter. Transformants described here are summarized in Table 1.

Name of transformants	Organisms transformed	Name of genes	Origin of genes
(Over expression)			
ACO^{OE}	Dm 7	*Dd-aco*	*D. discoideum*
Dm-ACO^{OE}	Dm 7	*Dm-aco*	Dm 7
OH10	Dm 7	*zyg1*	Dm 7
GFP/ZYG1^{OE}	*D. discoideum* (Ax2)	*gfp/zyg1*	*Aequorea Victoria*/Dm 7
GFP^{OE}	*D. discoideum* (Ax2)	*gfp*	*Aequorea victoria*
HA/ZYG1	myoblasts (C2C12)	*ha/mzyg1*	*Influenza virus*/Dm 7
(Under expression)			
ACO-RNAi	Dm 7	*Dd-aco*	*D. discoideum*

Table 1. Transformants used in this study. *Dd-aco*: an acc oxidase homologue gene isolated from *D. discoideum*. *Dm-aco* : an acc oxidase homologue gene isolated from Dm 7. *zyg1*: a novel gene isolated from Dm 7. *gfp*: a green fluorescent protein gene isolated from *Aequorea Victoria*. *gfp/zyg1*: a fusion gene of *gfp* and *zyg1*. *ha/mzyg1*: a fusion gene of a hemagglutinin gene (*ha*) and a humanized *zyg1* gene (*mzyg1*). Introduction of the vector constructs into *Dictyostelium* cells was performed by electroporation as described by Howard et al. (1988). In case of C2C12 cells (myoblasts), they were placed on coverslips and transfected with a *ha/mzyg1* fusion gene, using Lipofectamine™2000 (Invitrogen) according to the manufacturer's instruction. After 6 hours of transfection in Opti-MEM (Invitrogen), samples were washed twice with phosphate buffered saline (PBS) and incubated with Dulbecco's Modified Eagle Medium (DMEM, Invitrogen) containing 10% fetal bovine serum (FBS) and 80 µg/ml of Kanamycin for 24 h at 37°C in 5% carbon dioxide (CO_2) at 95% humidity. Transformants underexpressing *Dd-aco* (ACO-RNAi) were obtained by means of the RNAi method. For this purpose, a vector construct of the stem-loop RNA directed against the full length of Dd-aco was infected into Dm 7 cells (Amagai et al., 2007).

3. Cyclic AMP and ethylene are potent regulators required for the choice of two life cycles

3.1 Identification of ethylene as an inducer of macrocyst formation

Two life cycles, sorocarp and macrocyst formation, are regulated by several environmental conditions, such as light and water. These environmental conditions affect the synthesis of chemical regulators in cells. The two chemicals, ethylene and cyclic AMP (cAMP), have been demonstrated as regulators for the choice of life cycles in Dm7 (Amagai & Filosa, 1984; Amagai, 1984). It had already been reported that a volatile substance (CAG) might be involved in the macrocyst formation (Filosa, 1979). Based on the knowledge that the volatile gas is hydrophobic and acts antagonistically to carbon dioxide (CO_2), ethylene was speculated as a potent candidate of CAG. The possibility that ethylene might act as CAG was tested using MF1 cells, a spontaneous mutant isolated from Dm7. As the mutant MF1 cells are able to form macrocysts even in the light depending upon cell densities plated, the use of MF1 cells makes it easier to analyse results by excluding the factor of light. Considering the effects of cell densities, CAG was suggested to have the threshold concentration to induce macrocyst formation. In fact, MF1 cells failed to form macrocysts even at higher cell densities when they were allowed to develop in a larger incubation chamber. It was supposed that the concentration of CAG produced by the cells would be insufficient for induction of macrocysts by being diluted in the larger chamber. However, MF1 cells changed their developmental forms from sorocarps to macrocysts even in the larger chamber by the addition of ethylene. As was expected, inhibitors of ethylene biosynthesis, such as aminooxyacetic acid (AOA) and aminoethoxyvinyl glycine (AVG) greatly inhibited macrocyst formation. Ethylene production by MF1 and Dm7 cells was confirmed by gas chromatography (Amagai, 1984). From these results, ethylene was concluded to be a CAG, an inducer of macrocyst formation.

3.2 Determinination of life cycles by the balance of cAMP and ethylene amounts

Strangely, Dm7 cells formed sorocarps in the light, even though a significant amount of ethylene was produced from them. Why did the Dm7 cells not form macrocysts in the light, though they produced ethylene? This question was resolved by finding cAMP as a negative regulator of macrocyst formation. Developmental fate was actually changed from macrocyst to sorocarp formation in the presence of cAMP (Amagai & Filosa, 1984). To conclude, it was proposed that the choice of developmental pathways was determined by the balance of cAMP and ethylene amounts at the aggregation stage when developmental fate was determined. This was confirmed by determining the amounts of two regulators produced at the aggregation stage. The amount of cAMP was higher in the sorocarp than in the macrocyst formation, whereas the amount of ethylene production was actually decreased when sorocarps were formed in the presence of AOA (Amagai, 1987).

3.3 Relation between ethylene amounts and macrocyst formation confirmed using transformants

The close relationship between the amount of ethylene and the induction of macrocyst formation was confirmed directly, using two kinds of transformants, over- and under-producing ethylene (Amagai et al., 2007). *Dd-aco*, an 1-aminocyclopropane-1-carboxylic acid (ACC) oxidase homologue gene, isolated from *D. discoideum* (DDBJ, EMBL and GenBank databases, accession no. AB105858) was introduced into Dm7 cells to obtain the transformant overproducing ethylene. As ACC oxidase (ACO) catalyzes the last step in the biosynthesis of ethylene, the transformant overexpressing *Dd-aco* (ACO[OE]) was expected to increase the production of ethylene (Fig. 3). When the production of ethylene was determined by gas chromatography, ACO[OE] actually produced a larger amount of ethylene as compared with wild type, Dm7. On the other hand, the transformant underexpressing *Dd-aco* (ACO-RNAi) which had been isolated by the method of RNAi, produced a smaller amount of ethylene as compared with Dm7. Depending upon the amount of ethylene produced, ACO[OE] cells formed macrocysts, while ACO-RNAi cells failed to form them, regardless of culture conditions. The relationship between the amount of ethylene produced and the developmental forms in the transformants is summarized in Table 2.

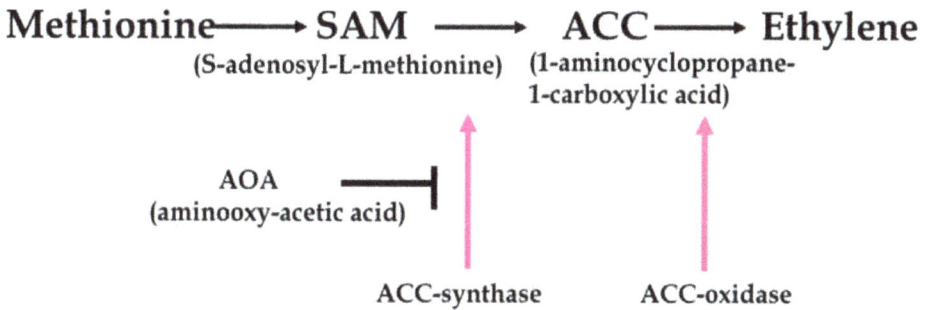

Methionine⟶ SAM ⟶ ACC⟶ Ethylene
(S-adenosyl-L-methionine) (1-aminocyclopropane-
1-carboxylic acid)

AOA ⊣
(aminooxy-acetic acid)

ACC-synthase   ACC-oxidase

Fig. 3. Ethylene biosynthesis in the cellular slime molds. Ethylene is synthesized from methionine through S-adenosyl-L-methionine (SAM) and 1-aminocyclopropane-1-carboxylic acid (ACC) in *Dictyostelium* as the case in higher plants. ACC synthase catalyzes the production of the ethylene precursor ACC from SAM (S-adenosylmethionine). ACC oxidase catalyzes the last step in the ethylene biosynthesis. AOA (amino-oxy acetic acid) prevents ethylene synthesis by blocking ACC synthase activity.

3.4 How to induce macrocyt formation by ethylene

How does ethylene induce macrocyst formation? 1-methylcyclopropene (1-MCP) is known as an inhibitor of ethylene by binding specifically to the cellular ethylene receptors. When starved Dm7 cells were incubated in air containing 224-2,240 ppm of 1-MCP, macrocyst formation in the dark was inhibited (Amagai et al., 2007). A family of ethylene receptors such as ETR1 has been first reported as a two-component histidine kinase in *Arabidopsis* (Bleecker et al., 1988) and then in many kinds of plants such as rice (Goff et al., 2002; Yu et al., 2002), carnation (Shibuya et al., 2002) and tomato (Tieman et al., 2002). Though the ethylene receptors in *Dictyostelium* have not been identified yet, the effect of MCP suggests the existence of ethylene receptors in *Dictyostelium* cells. The binding of ethylene with its receptor possibly triggers the signaling pathways leading to macrocyst formation.

4. Similarities of ethylene biosynthesis and functions between cellular slime molds and higher plants

Ethylene is well known to regulate many aspects of the life cycles in plants, including seed germination, root initiation, flower development, fruit ripening, senescence, and responses to several stresses (Abeles et al., 1992). In general, ethylene is synthesized from methionine through S-adenosyl-L-methionine (SAM) and 1-aminocyclopropane-1-carbo-xylic acid(ACC) (Adams & Yang, 1979). It has been suggested that ethylene may be biosynthesized in *Dictyostelium* cells through the same pathway as that in plants (Fig.3) (Amagai & Maeda, 1992). In fact, *Dictyostelium* homologue genes of ACC synthase (*Dd-acs*) and ACC oxidase (*Dd-aco*) have been isolated from *D. discoideum* by the Japanese *Dictyostelium* cDNA project and the genome project of *Dictyostelium* (Eichinger et al., 2005). The ACC oxidase homologue gene was also isolated from Dm7 (*Dm-aco*) (DDBJ, EMBL and GenBank databases, accession no. AB291210). When the nucleotide sequence of *Dm-aco* is compared with that of *Dd-aco*, only one nucleotide, thymidine at 672nd in *Dd-aco* is replaced with cytosine in *Dm-aco*. The deduced amino acid sequences encoded by the two genes are completely identical (Amagai et al., 2007). Dd-ACO shows 24.3% of identity and 40.0% of similarity, in comparison with the ACC-oxidase of *Arabidopsis* (accession number

Cells	The amounts of ethylene production (%)	Developmental forms		zygote formation
		light	dark	
Dm 7	100	SC	MC	+
ACO[OE]	167	MC	MC	nd
Dm-ACO[OE]	183.9	nd	nd	++
ACO-RNAi	82.7	SC	SC	−

Table 2. The relation of ethylene production to developmental forms. The determination of ethylene production from Dm7 and transformants was performed by gas chromatography. The amounts of ethylene produced from transformants are shown as relative values to that from Dm7 cells. Developmental forms: Starved cells were plated separately at 5×10^5 cells/cm² on agar in a glass dish (diameter: 3.5 cm) and incubated at 22°C either in the light or in the dark. After 48 hours of incubation, Dm 7 cells form sorocarps in the light and macrocysts in the dark (See Fig.1). ACO[OE] cells form macrocysts even in the light. ACO-RNAi cells form sorocarps in the light. Even in the dark, ACO-RNAi cells form tiny sorocarps and loose aggregates instead of macrocysts (Amagai et al., 2007). Zygote formation: Dm7 and a transformant overexpressing *Dm-aco* (Dm-ACO[OE]) were developed by shaking for 1.5 hours after starvation, fixed with methanol, and stained with DAPI. The binucleate cells caused by cell fusion were formed in Dm 7 cells. On the other hand, in Dm-ACO[OE], huge cells containing a large number of nuclei were frequently formed due to the enhanced cell fusion (Amagai, 2011) (See Fig.5). When ACO-RNAi cells were developed under the submerged conditions and stained with DAPI, such binucleate cells were scarcely formed (Amagai et al, 2007). SC: sorocarp, MC: macrocyst, nd: not determined. The number of "+", the degree of zygote formation: "− ", no zygote formation.

AAC27484). These are not high. However, by the Pfam homology search, Dd-ACO was found to contain a conserved 2-oxoglutalate (2OG) and Fe (II) dependent oxygenase superfamily domain (210-307th amino acids from the initiation site), which is characteristic of enzymes catalyzing oxidation of the organic substrate such as ACC. It also contained two histidine sites at 229th and 287th for binding iron which are necessary for activation of the enzyme (Aravind & Koonin, 2001). Determining the ethylene production by gas chromatography, it was confirmed that ethylene was actually biosynthesized by *Dd-aco* and *Dm-aco* (Table 2) (Amagai et al., 2007; Amagai, 2011). In addition, it is important to note the existence of ethylene receptor(s) in *Dictyostelium* cells as suggested before (Amagai et al., 2007). Taken together, these results suggest high similarities in the biosynthetic pathway and functions of ethylene between higher plants and *Dictyostelium*. Therefore, the study using *Dictyostelium* will contribute to obtaining new insights into the functions of ethylene beyond species.

5. Two chemicals regulate zygote formation at cellular level

5.1 Determination of developmental stages requiring ethylene for macrocyst formation

There are several cellular events occurring in the process of macrocyst formation, such as zygote formation and engulfment of the other cells by a zygote in a precyst, etc. Which events are controlled by ethylene? In order to determine the developmental stages when ethylene is effective to induce macrocyst formation, ethylene gas produced by Dm7 and MF1 cultures was removed by the use of charcoal at various stages in development. As a result, the developmental fate was shifted from macrocyst to sorocarp formation by removal of ethylene at the early developmental stages. This means that ethylene is necessary for macrocyst formation at the early stages of development. However, ethylene becomes unnecessary for macrocyst formation at the later stages of development. The timing when ethylene becomes unnecessary for macrocyst formation is consistent with the timing of the appearance of binucleate cells. When cells were stained with 4'-6-diamidino-2-phenylindole (DAPI), binucleate cells formed by cell fusion were recognized. The appearance of zygotes during development was determined by counting the number of binucleate cells as a marker of zygotes. As a result, binucleate cells were noticed at the aggregation stage (Fig. 4).

5.2 Control of zygote formation by two regulators

The developmental time requiring non-necessity of ethylene for macrocyst formation was delayed due to below the threshold concentration of ethylene when the cell densities plated were lowered. However, when ethylene was applied in the culture dishes at the beginning of this culture, the time requiring non-necessity of ethylene for macrocyst formation was advanced in concert with the advanced and increased formation of binucleate cells (Amagai, 1989). These results strongly suggest that ethylene may directly induce zygote formation. As was expected, cAMP, the second regulator, exhibits an inhibitory effect on the process of zygote formation (Suzuki et al., 1992). When the number of binucleate cells stained with DAPI was counted, the number was decreased by the addition of cAMP, while it was increased by the addition of phosphodiesterase (PDE), which hydrolyzed cAMP to 5' AMP. In conclusion, two regulators, ethylene and cAMP, directly regulate zygote formation.

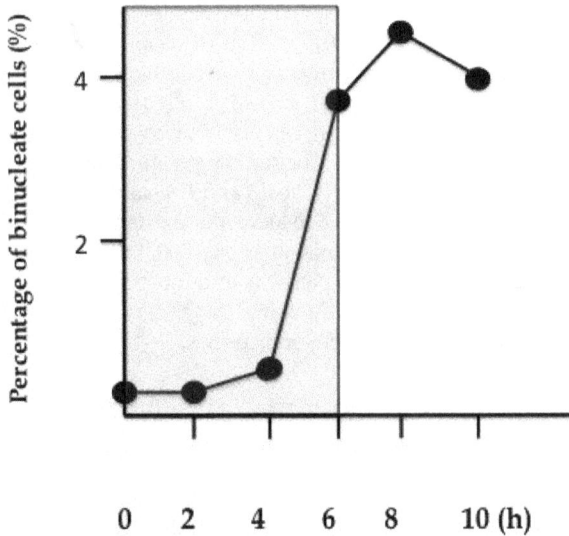

Fig. 4. The formation of binucleate cells during macrocyst formation in Dm7 cells. Starved Dm7 cells at 1 x10^6 cells/cm^2 were allowed to develop for 2 hours at 22°C in the dark, transferred to 4°C and kept overnight. Subsequently, the temperature was again shifted from 4°C to 22°C and incubated for the designated hours (h). This was followed by fixation with methanol and staining with DAPI. The percentages of binucleate cells were determined by counting the number of cells under a fluorescence microscope using UV excitation. Cell aggregates are formed at 6 hours, and subdivided into smaller cell masses at 10 hours. Each of the cell masses eventually develops to macrocysts. The low temperature treatment was performed to synchronize cellular development. The periods during which ethylene is necessary for macrocyst formation are shown by a grey zone. The timing when ethylene becomes unnecessary for macrocyst formation is consistent with the timing of the appearance of binucleate cells (basically from Amagai, 1989).

5.3 Relation between ethylene amounts and zygote formation confirmed using transformants

The close relationship between the amount of ethylene produced and zygote formation was also shown clearly using two types of transformants: Cells overexpressing *Dm-aco* (Dm-ACO^OE) formed huge cells containing a large number of nuclei as a result of enhanced cell fusion (Fig. 5), accompanying the expression of *Dm-aco* mRNA and the larger amount of ethylene production as compared with Dm 7 (Table 2) (Amagai, 2011). In contrast, ACO-RNAi cells never formed zygotes as was expected (Table 2) (Amagai et al., 2007). In heterothallic strains, zygote formation is also regulated by ethylene and cAMP. It was induced by ethylene, while it was inhibited by cAMP (Amagai, 1992; O'Day & Lydan, 1989). Incidentally, *D. discoideum* and its mating type, V12M2 cells also produce ethylene (Amagai & Maeda, 1992). Accordingly, ethylene and cAMP may act on the regulation of zygote formation beyond the difference of mating systems. Ethylene functions in other cellular events occurring during macrocyst formation remain to be elucidated.

Besides ethylene and cAMP, Ca^{2+} has been shown to induce zygote formation. That is, the percentage of zygotes was elevated by the presence of extracellular Ca^{2+} in *D. discoideum* and Dm 7 and by the increase of intracellular Ca^{2+} in *D. discoideum* (Chagla et al., 1980; Lydan & O'Day, 1988a; Suzuki et al., 1992; Szabo et al.,1982).

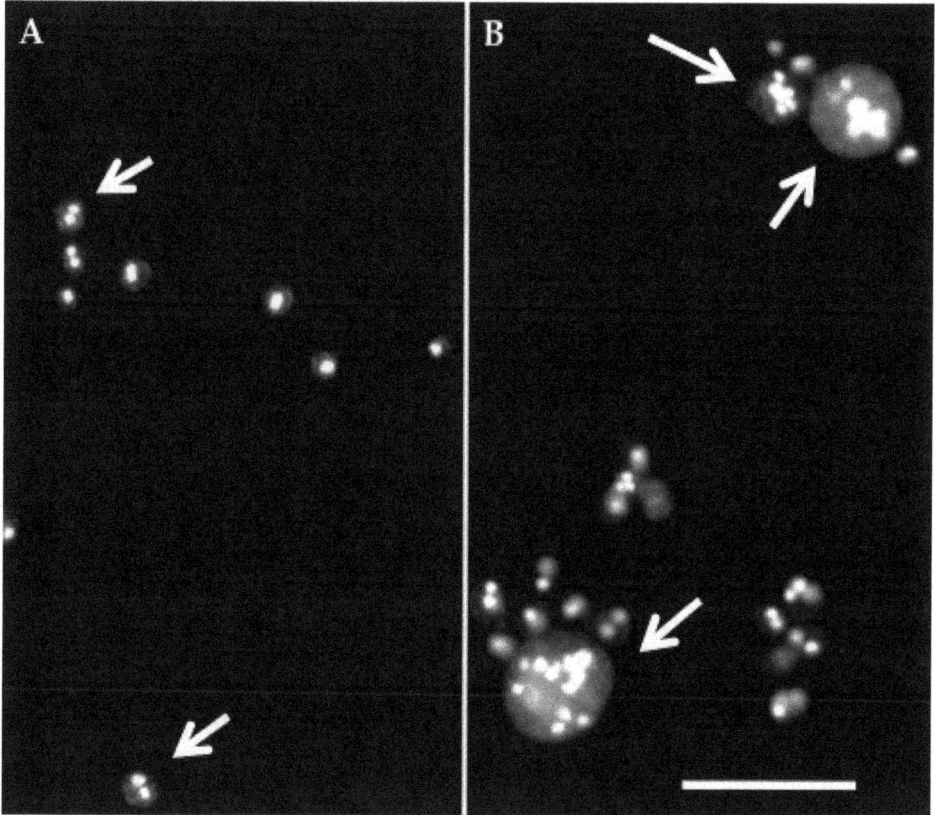

Fig. 5. Induction of multinucleate giant cells by enforced expression of the *Dm-aco* gene and the larger amount of ethylene production. Dm7 and their transformant overexpressing *Dm-aco* (Dm-ACO[OE]) were developed separately by shaking for 1.5 hours after starvation, fixed with methanol and stained with DAPI. Some binucleate cells (arrows) caused by cell fusion are observed in Dm 7 cells (A). In Dm-ACO[OE] cells (B), huge cells containing multi-nuclei are formed, indicating that cell fusion is markedly enhanced (Amagai, 2011). Bar: 50 μm.

6. Ethylene induces the expression of a novel gene, *zyg1*

6.1 Description of *zyg1*

As a gene involved in zygote formation, a novel gene *zyg1* cDNA (DDBJ/EMBL/ GenBank, accession no. AB006956) expressed predominantly during macrocyst formation was isolated by the differential screening method (Amagai, 2002). The *zyg1* gene was isolated also from genomic DNA of Dm7 and *D. discoideum* (Ax2) cells by PCR and sequenced (accession number: AB479506 for *Dm-zyg1*, AB479507 for *Dd-zyg1*). The comparison of sequences of *zyg1* between genomic DNA of Dm7 and Ax2 cells shows 100% identity besides the existence of intron (578-648) in Dm7 genome (in preparation). The predicted protein, ZYG1, consists of 268 amino acids with a molecular mass of 29.4 kDa. After BLAST (Altschul et al., 1990) and FASTA (Pearson, 1990) searches, the amino acid sequence as a whole shows no convincing similarity to known proteins. Although the ZYG1 protein is predicted to have several sites phosphorylated by protein kinase C (PKC), it has neither transmembrane domains nor specific signal sequences (Fig. 6). The expression of the *zyg1* gene began 2 hours after starvation, reached maximum level at 8 hours, and then decreased, when Dm7 cells were cultured under submerged conditions (Amagai, 2002). Such an expression pattern is quite similar to the developmental kinetics of zygote formation with about 1 hour precedence. The number of zygotes began to increase 5 hours after starvation, reached the maximum level at 9 hours and then gradually decreased (Kawai et al., 1993). From these observations, the *zyg1* gene was predicted to be closely involved in zygote formation. As was expected, a transformant overexpressing a *zyg1* gene (OH10) was found to form macrocysts on agar even in the light. In addition, they formed a number of zygotic giant cells besides macrocysts. These results suggested that the *zyg1* gene might be involved in the induction of zygote formation (Amagai, 2002).

```
M E I D S K I T N F E D A G T I N L N L H N F V S E K F A N K P
K V L N V A S L A S N S V D E A G D S E Q K V S F R I N Q T G
N I F Y S T T T P E L T L E S K K L F N S V T V L F A A M T K A
L G E K G L N L F N Y E A V A S L I Q K S G Y F V E V Q K F Q K
N L S I K S G S L S I D T Q I I Q Q L I P G L T S G A S L D I A K
G V L G A L N G E F S A S S S D E K V K I A H L L F I C E E L F
G A P S V T V R L F Y A T K E T H K T L T S S P C H K S S S V S
F E L N Q E A S T F L F V S P D T I A E F S Q K F E T Q P E E Y
K N L I E K L K G Y L P
```

Fig. 6. The predicted amino acid sequence of ZYG1. The nucleotide sequence of *zyg1* is deposited in the DDBJ, EMBL and GenBank databases with the accession number AB06956. Amino acid sequences of ZYG1 shown as red color indicate the position of predicted PKC-phosphorylation sites (Amagai, 2002).

6.2 Relation between ethylene and *zyg1*

It is quite possible that the action of ethylene may be realized through an enhanced ZYG1 expression. In order to examine this possibility, the expression of the *zyg1* gene during development was examined and compared, using two kinds of transformants, ACO[OE] and ACO-RNAi. Depending upon ethylene production, Dm7 cells and ACO[OE] cells exhibited higher levels of *zyg1* expression, while ACO-RNAi cells had a significantly lower level of *zyg1* expression. Incidentally, the *zyg1* expression was decreased by application of AOA, an inhibitor of ethylene biosynthesis (Fig. 7). From this, it was certified that the expression of *zyg1* was induced by ethylene. However, the mechanism of how ethylene induces *zyg1* expression has not been resolved yet. Taken together these results indicate that ethylene induces zygote formation through an enhanced expression of *zyg1* (Amagai et al., 2007).

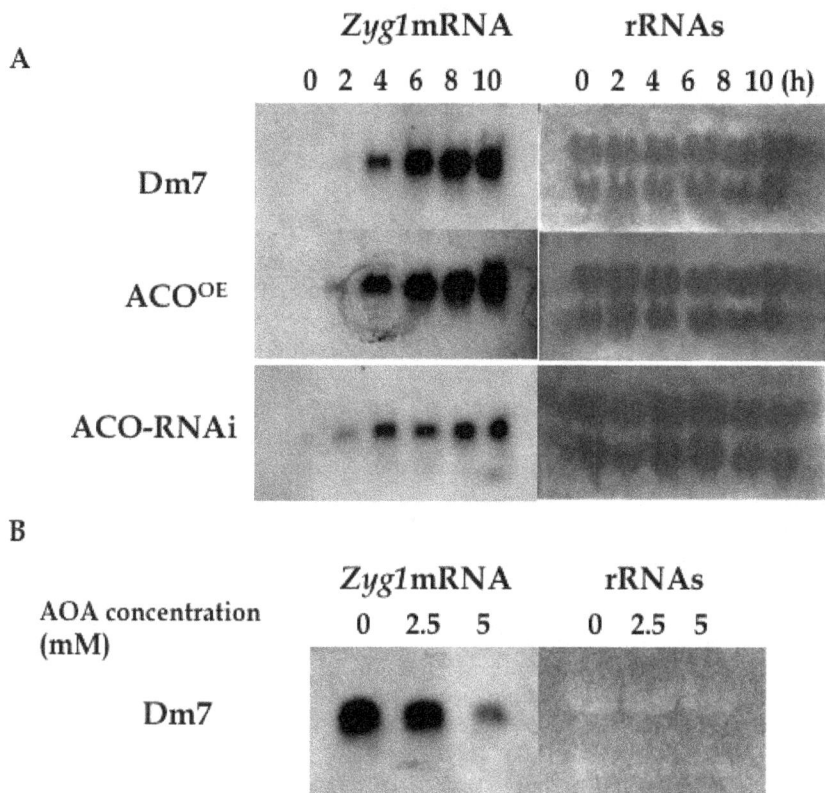

Fig. 7. A. Expression patterns of *zyg1* in Dm7, ACO^{OE} and ACO-RNAi cells during development. Starved Dm7, ACO^{OE} and ACO-RNAi cells were placed at 5 x 10^5 cells/cm^2 in glass dishes (diameter: 9cm) and developed under submerged conditions (Bonner's salt solution), respectively. Total RNAs were extracted at 2 hours intervals during development. 30 μg of total RNA was loaded on each lane, electrophoresed and transferred to membranes. The expression of *zyg1* was detected using the full length of *zyg1* cDNA as a probe. The membranes were exposed on X-ray films for a week. The large (26S) and small (17S) subunits of ribosomal RNAs blotted on the membranes are shown as loading controls in the right panels. It is clear that *zyg1* mRNA is expressed earlier in ACO^{OE} cells than in Dm7 cells, but that *zyg1* expression is significantly suppressed in ACO-RNAi cells. B. Effect of AOA on *zyg1* expression. Starved Dm7 cells were placed at 1 x 10^6 cells/cm^2 in glass dishes (diameter: 3.5 cm) under the submerged conditions (20 mM MES, pH 7.0). Subsequently, the indicated concentrations (final concentrations) of AOA were added to culture dishes, respectively. After 8 hours of incubation, total RNAs were extracted from cells. 12.6 μg of total RNA was loaded on each lane and probed using the full length of *zyg1* cDNA, as described above. The membranes were exposed on X-ray films for 1 day. It is evident that *zyg1* expression is decreased by AOA in a dose-dependent manner. The large (26S) and small (17S) subunits of ribosomal RNAs stained with EtBr are shown as loading controls in the right panel. These results indicate that ethylene induces the expression of *zyg1* (cited from Amagai et al., 2007).

7. The possible mechanisms of zygote induction by ethylene

As described above, ethylene, $zyg1$, Ca^{2+}, and cAMP regulate zygote formation. How are these regulators involved in signal transduction pathways for zygote formation? There are inductive and inhibitory signal transduction pathways for zygote formation. Inductive signal transduction pathways are described first, and inhibitory signal transduction pathways next.

7.1 Inductive signal transduction pathways for zygote formation

Intracellular Ca^{2+} binds to calmodulin and PKC. Ca^{2+}-calmodulin complex binds the target proteins, acting positively on zygote formation (Lydan & O'Day, 1988b). Concerning PKC, phorbol esters such as 12O-tetradecanoylphorbol-13-acetate (TPA), potent activators of PKC, have been reported to enhance the formation of zygotes in both $D.$ $discoideum$ and Dm 7. In contrast, staurosporine, an inhibitor of protein kinases including PKC, inhibited zygote formation in both $D.$ $discoideum$ and Dm 7 (Amagai, 2011; Gunther et al., 1995) and macrocyst formation in Dm7 (Kawai et al., 1993). From these results, PKC activated by the increase of intracellular Ca^{2+} possibly participates in the induction of zygote formation. However, the target substrate of PKC had not been identified. Since ZYG1 has several predicted sites phosphorylated by PKC, ZYG1 was a likely candidate for the substrate of PKC. Therefore, whether ZYG1 was phosphrylated by PKC or not was examined, using a transformant overexpressing a $gfp/zyg1$ fusion gene. The localization of ZYG1 was monitored by GFP, and phosphorylation of ZYG1 was detected by the anti-phosphoserine antibody which specifically recognized serine residues phosphorylated by PKC and then the Rhodamine conjugated secondary anti-rabbit IgG antibody. When the green color of GFP (GFP/ZYG1 fusion protein) was merged with the red color of Rhodamine (proteins phosphorylated by cPKC), the color changed to yellow in GFP/ZYG1[OE] cells at the regions of cell-to-cell contacts (Fig. 8H, arrow). This showed that GFP/ZYG1 fusion protein was co-localized with the proteins phosphorylated by cPKC there. Since the co-localization of GFP and the protein phosphrylated by PKC was not observed in GFP[CONT] cells (Fig. 8G, arrow), the localization of ZYG1 at the region of cell-to cell contact was shown clearly (Amagai et al., 2012). This suggests a strong implication of ZYG1 phosphorylated by PKC in zygote formation. However, the phorphorylated sites of ZYG1 protein have not been identified yet.

7.2 Inhibitory signal transduction pathways for zygote formation

Inhibitory signal transduction pathways are described next. As 2'-deoxy-cAMP (a cAMP analog) with a high affinity for the surface cAMP receptor inhibited zygote formation, it was suggested that cAMP operates by binding to a surface receptor to mediate downstream signaling events (Suzuki et al., 1992). The inhibitory effect of cAMP on zygote formation was nullified, when K252a, a potent inhibitor of protein kinases (PKA, PKC, and PKG) was co-applied with cAMP. In addition, when KT5720 (a specific inhibitor of PKA) and W7 (calmodulin inhibitor) were co-applied, zygote formation was enhanced (Kawai et al., 1993). From these results, it is evident that the activation of PKA triggered by binding cAMP with its receptor and calmodulin dependent protein kinase are involved in the inhibition of zygote formation. However, target proteins for PKA and calmodulin dependent protein kinase presently remain to be identified. Since these protein kinase inhibitors are effective at the developmental stage when the cells acquired fusion competence, it is most likely that they may inhibit the process of gamete formation. In this way, calmodulin dependent protein kinases are involved in both inductive and inhibitory signal transduction pathways

(Lydan & O'Day, 1988b). The inductive and inhibitory signal transduction pathways involved in zygote formation are summarized in Fig. 9.

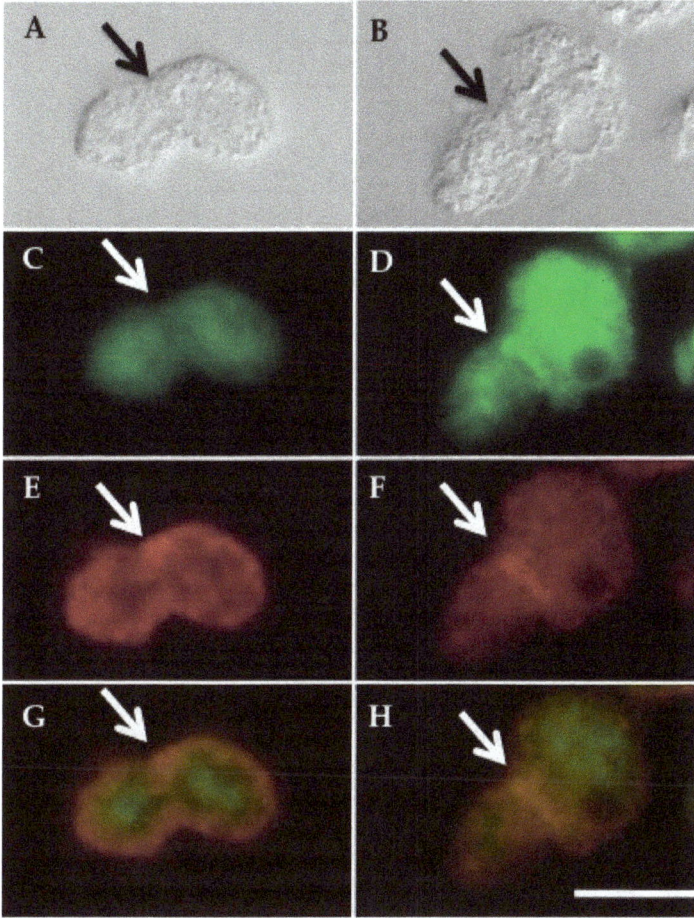

Fig. 8. Immunocytochemical detection of the PKC-mediated ZYG1 phosphorylation. Starved GFP^CONT and GFP/ZYG1^OE cells were developed for 2 h, fixed with 4% paraformaldehyde and stained by the Phospho-(Ser) PKC Substrate antibody. This was followed by the rhodamine-conjugated anti-rabbit secondary antibody to detect the proteins phosphorylated by cPKC. Photographs were taken under DIC (A and B) and fluorescence microscopes (GFP, C and D; Rhodamine, E and F; GFP and Rhodamine merged, G and H). Photographs represent the same fields of GFP^OE cells (A, C, E, G) and of GFP/ZYG1^OE cells (B, D, F, H). In GFP/ZYG1^OE cells, the green color of GFP (GFP/ZYG1 fusion protein) merged with the red color of Rhodamine (proteins phosphorylated by cPKC) shows yellow color at the region of cell-to-cell contacts (H, arrow). This shows that GFP/ZYG1 and the protein phosphorylated by PKC are co-localized particularly there. While, in GFP^CONT cells, the co-localization of GFP and the protein phosphorylated by PKC is not observed (G, arrow). This shows that ZYG1 itself is located at the region of cell-to-cell contact and phosphorylated by PKC (Amagai et al., 2012). Bar, 25 μm.

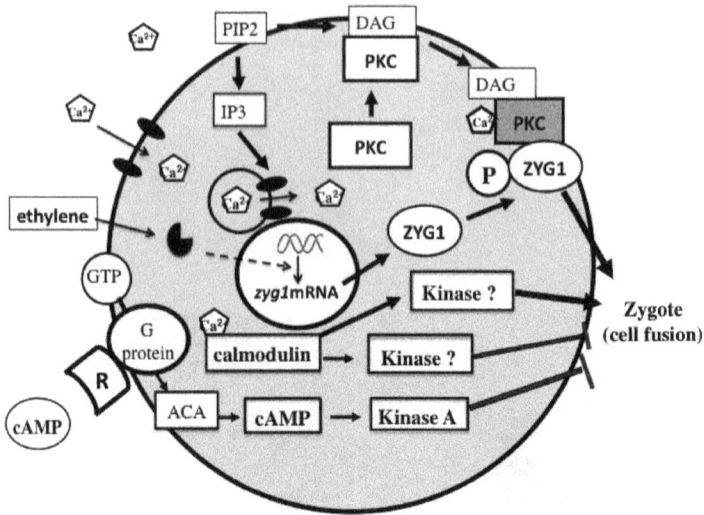

Fig. 9. Schematic representation of the proposed signal transduction pathways for the induction and inhibition of sexual cell fusion (zygote formation). The binding ethylene with its receptor (⬤) and the increase of intracellular calcium begin the inductive signal transduction pathway. The increase of intracellular Ca^{2+} activates PKC (gray color shows activated PKC). Ethylene induces the expression of *zyg1*. ZYG1 phosphorylated by activated PKC causes cell fusion. For the inhibition of zygote formation, cAMP binds to its receptor (R) and then activates PKA. Calmodulin dependent protein kinases are involved in both inductive and inhibitory signal transduction pathways. GTP, guanosine-5′-triphosphate; PIP2, phosphatidylinositol 4,5-bisphosphate; IP3, inositol triphosphate; DAG, diacylglycerol. →: induction; ⊥: inhibition.

7.3 Induction of myoblast fusion by *zyg1*

Ca^{2+} -and PKC- mediated signal transduction pathways are known to be involved in myoblast fusion during myogenesis (David et al., 1990; Paterson & Strohman, 1972; Shainberg et al., 1969). Therefore, it is possible that there may be common signal transduction pathways for cell fusion between myoblasts and *Dictyostelium* cells. To test this possibility, a humanized version of the *zyg1* gene (*mzyg1*) was synthesized in which each amino acid codon was replaced by that most commonly found in mammalian cells (DNA 2.0 Inc.). The construct containing a fusion gene of *ha* and *mzyg1* (*ha/mzyg1*) was introduced into mouse C2C12 cells. At 24 hours of incubation after transfection, cells were fixed with 4 % paraformaldehyde, followed by immunostaining with the anti-HA antibody (mouse) and then the FITC-conjugated secondary anti-mouse IgG antibody. Their nuclei were stained with DAPI. As a result, cells expressing the HA/ZYG1 fusion protein (HA/ZYG1) were recognized as the green color of FITC. Quite interestingly, green colored cells expressing the HA/ZYG1 fusion protein were found to form giant multinucleate cells frequently (Fig.10). They exhibited fibroblast-like morphology. However, the highly elongated cells as observed during typical myoblast fusion, in which nuclei were arranged in a single file, were not observed. Though the cell shape was different from normal multinucleate cells in myotube, it is clear that cell fusion is actually induced by ZYG protein (Amagai et al., 2012). Since ZYG1 has not a mammalian ortholog, ZYG1 could take place of the protein phosphorylated by PKC in mammal cells. It is supposed that the protein phosphorylated by PKC during myoblast fusion might be functionally similar, though the target protein of PKC in myoblasts has not been identified yet.

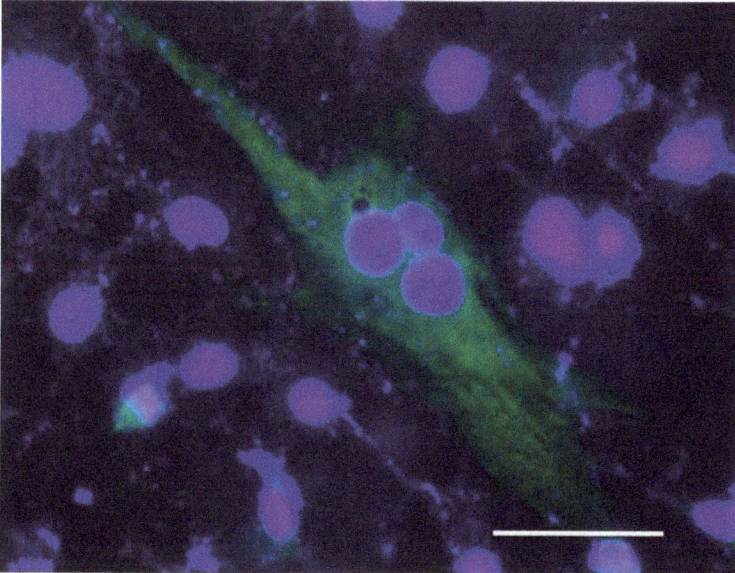

Fig. 10. Formation of multinucleate cells in mouse C2C12 cells transfected with a *ha/mzyg1* fusion gene. C2C12 cells were transfected with a *ha/mzyg1* fusion gene. After 24 hours of incubation, cells were fixed with 4 % paraformaldehyde, followed by immunostaining with the anti-HA antibody (mouse) and then the FITC-conjugated secondary anti-mouse IgG antibody. The samples were also stained with DAPI. The photograph shows a FITC image merged with a DAPI image under a fluorescence microscope. A cell expressing HA/ZYG1 fusion protein shows the green color of FITC. Nuclei show the purple color of DAPI. Since a large-sized green cell expressing HA/ZYG1 contains three nuclei, it is most likely that cell fusion may be induced by ZYG1. *mzyg1*: a humanized version of the *zyg1* gene (cited from Amagai et al., 2012). Bar, 50 μm.

8. Conclusion

It is evident that the regulatory mechanisms to determine the reproduction cycles are common between homothallic and heterothallic strains, regardless of different mating systems. Ethylene induces macrocyst formation as a sexual cycle, while cAMP induces sorocarp formation as an asexual cycle. At the cellular level, ethylene induces zygote formation, which is formed by cell fusion and subsequent nuclear fusion during macrocyst formation, through an enhanced expression of a novel gene, *zyg1*. ZYG1 may be a likely substrate for PKC in the pathway of Ca^{2+} -and PKC- mediated signal transduction. Inversely, cAMP inhibits zygote formation, possibly through a PKA-related signaling pathway. In this chapter, the functions of ethylene in macrocyst formation have been proved at the levels of individuality (macrocyst formation), cell (zygote formation), and molecule (*zyg1* expression). It is of interest to note that the enforced expression of ZYG1 induces cell fusion in myoblast as well as in *Dictyostelium* cells. Thus the data presented here must give us a principal insight into the mechanisms of developmental regulation in a wide range of organisms including *Dictyostelium*. The functions of ethylene and ZYG1 in the cell fusion process will be particularly memorable.

9. Acknowledgment

I am grateful to Prof. Yasuo Maeda (Tohoku University) and Prof. Michael MacManus (Miyagi University of Education) for their critical reading and insightful comments of the manuscript. I would like to thank Dr. Michael F. Filosa, and all people who support these works.

10. References

Abeles, F. B., Morgan, P. W., Saltveit, M. E. J. (1992). *Ethylene in plant biology*, 2nd (ed.), New York: Academic Press.

Adams, D. O. & Yang, S. F. (1979). Ethylene biosynthesis: identification of 1-aminocyclopropane-1-carboxylic acid as an intermediate in the conversion of methionine, *Pro.Natl. Acad. Sci. USA* 76: 170-174.

Altschul, S. F., Gish, W., Miller, W., Myers, E. W. & Lipman, D. J. (1990). Basic local alignment search tool, *J Mol Biol* 215: 403-10.

Amagai, A. (1984). Induction by ethylene of macrocyst formation in the cellular slime mould *Dictyostelium mucoroides*, *J Gen Microbiol* 130: 2961-2965.

Amagai, A. & Filosa, M. F. (1984). The possible involvement of cyclic AMP and volatile substance(s) in the development of a macrocyst-forming strain of *Dictyostelium mucoroides*, *Dev Growth Differ* 26: 583-589.

Amagai, A. (1987). Regulation of the developmental modes in *Dictyostelium mucoroides* by cAMP and ethylene, *Differentiation* 36: 111-115.

Amagai, A. (1989). Induction of zygote formation by ethylene during the sexual development of the cellular slime mold *Dictyostelium mucoroides*, *Differentiation* 41: 176-183.

Amagai, A., Soramoto, S., Saito, S. & Maeda, Y. (2007). Ethylene induces zygote formation through an enhanced expression of zyg1 in *Dictyostelium mucoroides*, *Exp Cell Res* 313: 2493-2503.

Amagai A. (1992). Induction of heterothallic and homothallic zygotes in *Dictyostelium discoideum* by ethylene, *Dev Growth Differ* 34: 293-299.

Amagai, A. & Maeda Y. (1992). The ethylene action in the development of cellular slime molds: an analogy to higher plants, *Protoplasma* 167: 159-168.

Amagai A. (2002). Involvement of a novel gene, zyg1, in zygote formation of *Dityostelium mucoroides*, *J Muscle Res Cell Motility* 23: 867-74.

Amagai, A. (2011). Ethylene as a potent inducer of sexual development, *Dev Growth Differ* 53: 617-623,

Amagai, A., MacWilliams, H., Isono, T., Omatsu-Kanbe, M., Urano, S., Yamamoto, K. & Maeda, Y. (2012). PKC-mediated ZYG1 phosphorylation induces fusion of myoblasts as well as of *Dictyostelium* cells, *Int J Cell Biol* 2012: 11 pages, ArticleID 657423, doi:10.1155/2012/657423

Aravind, L. & Koonin, E. V. (2001). The DNA-repair protein AlkB, EGL-9, and leprecan define new families of 2-oxoglutarete- and iron-dependent dioxygenases, *Genome Biol* 2: research0007.

Bleecker, A. B., Estelle, M. A.. Somerville, C. & Kende, H. (1988). Insensitivity to ethylene conferred by a dominant mutation in Arabidopsis thaliana, *Science* 241: 1086-1089.

Brefeld, O. (1869) *Dictyostelium mucoroides*. Ein neuer Organismus aus der Verwandshaft der Myxomyceten. *Abhandl. Senckenbergish Naturf. Ges* 7: 85-107.

Chagla, A. H., Lewis, K. E. & O'Day, D. H. (1980). Ca^{2+} and cell fusion during sexual development in liquid cultures of *Dictyostelium discoideum*, *Exp Cell Res* 126: 501-505.

Clark, M. A., Francis, D. & Eisenberg, R. (1973). Mating types in cellular slime molds, *Biochem Biophys Res Com* 52: 672-678.

David, J. D., Faser, C. R. & Perrot, G. P. (1990). Role of protein kinase C in chick embryo skeletal myoblast fusion, *Devel Biol* 139: 89-99.

Eichinger, L., Pachebat, J. A., Glockner, G., Rajandream, M. A., Sucgang, R., Berriman, M., Song, J., Olsen, R., Szafranski, K., Xu, Q., Tunggal, B., Kummerfeld, S., Madera, M., Konfortov, B. A., Rivero, F., Bankier, A. T., Lehmann, R., Hamlin, N., Davies, R., Gaudet, P., Fey, P., Pilcher, K., Chen, G., Saunders, D., Sodergren, E., Davis, P., Kerhornou, A., Nie, X., Hall, N., Anjard, C., Hemphill, L., Bason, N., Farbrother, P., Desany, B., Just, E., Morio, T., Rost, R., Churcher, C., Cooper, J., Haydock, S., van Driessche, N., Cronin, A., Goodhead, I., Muzny, D., Mourier, T., Pain, A., Lu, M., Harper, D., Lindsay, R., Hauser, H., James, K., Quiles, M., Madan Babu, M., Saito, T., Buchrieser, C., Wardroper, A., Felder, M., Thangavelu, M., Johnson, D., Knights, A., Loulseged, H., Mungall, K., Oliver, K., Price, C., Quail, M. A., Urushihara, H., Hernandez, J., Rabbinowitsch, E., Steffen, D., Sanders, M., Ma, J., Kohara, Y., Sharp, S., Simmonds, M., Spiegler, S., Tivey, A., Sugano, S., White, B., Walker, D., Woodward, J., Winckler, T., Tanaka, Y., Shaulsky, G., Schleicher, M., Weinstock, G., Rosenthal, A., Cox, E. C., Chisholm, R. L., Gibbs, R., Loomis, W. F., Platzer, M., Kay, R. R., Williams, J., Dear, P. H., Noegel, A. A., Barrell, B. & Kuspa, A. (2002). The genome of the social amoeba *Dictyostelium discoideum*, *Nature* 435: 43-57.

Erdos, G. W., Raper, K. B. & Vogen, L. K. (1973). Mating types and macrocyst formation in *Dictyostelium discoideum*, *Proc Natl Acad Sci USA* 70:1828-1830.

Filosa, M. F. & Dengler, R.E. (1972). Ultrastructure of macrocyst formation in the cellular slime mold *Dictyostelium mucoroides*: Extensive phagocytosis of amoebae by a specialized cell, *Dev Biol* 29: 1-16.

Filosa, M.F. (1979). Macrocyst formation in the cellular slime mold *Dictyostelium mucoroides*: Involvement of light and volatile morphogenetic substance(s), *J Exp Zool* 207: 491-495.

Goff, S. A., Ricke, D., Lan, T. H., Presting, G., Wang, R., Dunn, M., Glazebrook, J., Sessions, A., Oeller, P., Varma, H., Hadley, D., Hutchison, D., Martin, C., Katagiri, F., Lange, B. M., Moughamer, T., Xia, Y., Budworth, P., Zhong, J., Miguel, T.,. Paszkowski, U., Zhang, S., Colbert, M., Sun, W. L., Chen, L., Cooper, B., Park, S., Wood, T. C., Mao, L., Quail, P., Wing, R., Dean, Ryu. Y., Zharkikh, A.,. Shen, R., Sahasrabudhe, S.,A. Thomas, A.,. Cannings, R., Gutin, A., Pruss, D., Reid, J., Tavtigian, S., Mitchell, J., Eldredge, G., Scholl, T., Miller, R. M., Bhatnagar, S., Adey, N., Rubano, T., Tusneem, N., Robinson, R., Feldhaus, J., Macalma, T., Oliphant, A. & Briggs, S. (2002). A draft sequence of the rice genome (*Oryza sativa* L. ssp. *japonica*), *Science* 296: 92-100.

Gunther, K. E., Ramkisson, H., Lydan, M. A. & O'Day, D. H. (1995). Fertilization in *Dictyostelium discoideum*: pharmacological analysis and the presence of a substrate protein suggest protein kinase C is essential for gamete fusion, *Exp Cell Res* 220: 325-31.

Howard, P. K., Aher, K. G. & Firtel, R. A. (1988). Establishment of a transient expression system for *Dictyostelium discoideum*, *Nucleic Acids Research* 16: 2613–2633.

Kawai, S., Maeda, Y. & Amagai, A. (1993). Promotion of zygote formation by protein kinase inhibitors during the sexual development of *Dictyostelium mucoroides*, *Dev Growth Differ* 35: 601-607.

Lydan, M. A. & O'Day, D. H. (1988a). The role of intracellular Ca2+ during early sexual development in *Dictyostelium disocideum*: effects of LaCl3, Ins(1,4,5)P3, TMB8, chlortetracycline and A23187 on cell fusion, *J Cell Sci* 90: 465-473.

Lydan, M. A. & O'Day, D. H. (1988b). Different developmental functions for calmodulin in *Dictyostelium*: Trifluoperazine and R24571 both inhibit cell and pronuclear fusion but enhance gamete formation, *Exp Cell Res* 178: 51-6.

Nickerson, A. W. & Raper, K. B. (1973). Macrocysts in the life cycle of the Dictyosteliaceae. II. Germination of the macrocyst, *Amer J Bot* 60: 247–253.

O'Day, D. H. & Lydan, M. A. (1989). The regulation of membrane fusion during sexual development in *Dictyostelium discoideum*, *Biochem Cell Biol* 67: 321-6. Review.

Paterson, B. & Strohman, R. C. (1972). Myosin synthesis in cultures of differentiating chicken skeletal muscle, *Dev Biol* 29: 113-138.

Pearson, W. R. (1990). Rapid and sensitive sequence comparison with FASTP and FASTA, *Method Enzymol* 183: 63-98.

Shainberg, A., Yagil, G. & Yaffe, D. (1969). Control of myogenesis in vitro by Ca2+ concentration in nutritional medium, *Exp Cell Res* 58: 163-167.

Shibuya, K., Nagata, M., Tanikawa, N., Yoshioka, T., Hashiba, T. & Satoh, S. (2002). Comparison of mRNA levels of three ethylene receptors in senescing flowers of carnation (*Dianthus caryophyllus*), *J Exp Bot* 53: 399-406.

Suzuki, T., Amagai, A. & Maeda, Y. (1992). Cyclic AMP and Ca2+ as regulators of zygote formation in the cellular slime mold *Dictyostelium mucoroides*, *Differentiation* 49:127-32.

Szabo, S. P., O'Day, D. H. & Chagla, A. H. (1982). Cell fusion, nuclear fusion, and zygote differentiation during sexual development of *Dictyostelium discoideum*, *Dev Biol* 90: 375-82.

Tieman, D. M., Taylor, M. G., Ciardi, J. A. & Klee, H. J. (2000). The tomato ethylene receptors NR and LeETR4 are negative regulators of ethylene response and exhibit functional compensation within a multigene family, *Proc Natl Acad Aci USA* 97: 5663-5668.

Yu, J.,. Hu, S.,J. Wang, J., Wong, G. K., Li, S., Liu, B., Deng, Y., Dai, L., Zhou, Y., Zhang, X., Cao, M., Liu, J., Sun, J., Tang, J., Chen, Y., Huang, X., Lin, W., Ye, C., Tong, W., Cong, L., Geng, J., Han, Y., Li, L., Li, W., Hu, G., Li, J., Liu, Z., Qi, Q., Li, T., Wang, X., Lu, H., Wu, T., Zhu, M., Ni, P., Han, H., Dong, W., Ren, X. Feng, X., Cui, P., Li, X., Wang, H., Xu, X., Zhai, W., Xu, Z., Zhang, J., He, S., Xu, J., Zhang, K., Zheng, X., Dong, J., Zeng, W., Tao, L., Ye, J., Tan, J., Chen, X., He, J., Liu, D., Tian, W., Tian, C., Xia, H., Bao, Q., Li, G., Gao, H., Cao, T., Zhao, W., Li, P., Chen, W., Zhang, Y., Hu, J., Liu, S., Yang, J., Zhang, G., Xiong, Y., Li, Z., Mao, L., Zhou, C., Zhu, Z., Chen, R., Hao, B., Zheng, W., Chen, S., Guo, W.,.Tao, M., Zhu, L., Yuan, L., & Yan, H. (2002). A draftsequence of the rice genome (*Oryza sativa* L. ssp. *indica*), *Science* 296: 79-92.

15

Cytokinins and Their Possible Role in Seed Size and Seed Mass Determination in Maize

Tomaž Rijavec[1], Qin-Bao Li[2],
Marina Dermastia[3,*] and Prem S. Chourey[2,4,*]
[1]Institute of Physical Biology
[2]Unites States Department of Agriculture –
Agricultural Research Service, Gainesville
[3]National Institute of Biology
[4]Agronomy and Plant Pathology Department
University of Florida, Gainesville, FL
[1,3]Slovenia
[2,4]USA

1. Introduction

Cytokinins (CKs) are plant hormones promoting cell division and differentiation, morphogenesis in tissue culture, leaf expansion, bud formation, delay of leaf senescence and chloroplast development (reviewed in Rijavec & Dermastia, 2010). Natural CKs are adenine derivatives and based on the side chain moiety, they can be divided into two subgroups: isoprenoid and aromatic cytokinins.

One of the richest sources of CKs in various plants species are developing seeds. In fact, zeatin, a major natural CK, was first discovered in developing seeds of maize (Letham, 1963). Not surprisingly, developing seeds of maize, rice and *Lupinus albus* have played a major role in CK research (reviewed in Emery & Atkins, 2005). A major focus of these studies has been on the possible role of CKs in sink strength of developing seeds. The role of CKs in controlling sink strength is inferred largely from (a) the well known role of CK in stimulating cell division that may lead to increased organ size of the sink tissue, and (b) the coincidence of CK accumulation with seed cell division profiles in several plant species (Dietrich et al., 1995; Arnau et al., 1999; Yang et al., 2002).

Nearly all CK metabolites identified thus far have been reported to be present in developing seeds of different plants. In maize, the zeatin type of isoprenoid cytokinins is most common (Brugière et al. 2003; Veach et al., 2003; Rijavec et al., 2009, 2011). Developmental profiles of various CKs in maize caryopsis, a single seeded fruit of plants from the grass family, have shown high levels of zeatin riboside (Brugière et al., 2008; Rijavec et al., 2009, 2011) during the early stages of endosperm development (Dietrich et al., 1995; Brugière et al., 2003; Rijavec et al., 2011). In maize caryopsis there are also many reports on biochemical and

---

* These authors have contributed equally to this work.

molecular aspects of genes and proteins involved in CK biosynthesis, conjugation, and degradation. Enzyme activity for isopentenyl transferase (IPT) that marks the first committed step in CK biosynthesis was reported in immature maize caryopsis by Blackwell and Horgan (1994). Recently, Brugière et al., (2008) described a small family of *IPT* genes from maize based on the similarity with *Arabidopsis* IPT proteins. The *Ipt* genes *ZmIpt1*, *ZmIpt2* and *ZmIpt10* are expressed in the pedicel, endosperm and embryo (Brugière et al., 2008; Šmehilová et al., 2009; Vyroubalová et al., 2009; Rijavec et al., 2011), indicating local biosynthesis of CKs in these parts of the caryopsis. Degradation and reversible/irreversible inactivation are key regulators of CK levels. Several maize CK dehydrogenase genes (*Ckx*) are expressed in the caryopsis pedicel region, endosperm and embryo (Massonneau et al., 2004; Šmehilová et al., 2009; Vyroubalová et al., 2009), and their corresponding protein products irreversibly cleave the N^6-side chain from the main purine ring (Massonneau et al., 2004). Recently, the importance of conjugation has also been described. In maize caryopsis *cis*-zeatin riboside-*O*-glucoside was shown to be a major CK metabolite in caryopsis (Veach et al., 2003), and similarly, increased levels of zeatin-*O*-glucoside have been reported in roots and leaves of maize transformants harboring the *Zog1* gene encoding a zeatin-*O*-glucosyltransferase from *Phaseolus lunatus* (Rodó et al., 2008). Although the exact role of the conjugates CK-*O*-glucosides is not clear, they are generally assumed to be the storage products. Finally, information on the molecular biology of CK-*N*-glucosides, a group of metabolically inactive CKs (Brzobohatý et al., 1994) is scarce (Hou et al., 2004; Rijavec et al., 2011). However, in maize caryopsis there is a notably high concentration of *trans*-zeatin-9-glucoside (Z-9-G) recorded in the developmental phase following intense mitotic activity (Rijavec et al., 2009, 2011).

As in other eukaryotes the plant cell cycle is governed by cyclin-dependent kinases (CDKs). While the catalytic CDK subunits are responsible for recognizing the target motif, which is present in substrate protein, the regulatory proteins — cyclins play a role in discriminating between distinct protein substrates and thus regulate different cell cycle transitions. Plants contain many cyclins. In *Arabidopsis*, for example, at least 32 cyclins in seven classes have been described (reviewed in Inzé and De Veylder 2006). In particular, cyclins from the class D were proposed to have a role as external growth factor sensor that integrates the external signals with the cell cycle machinery (Sherr and Roberts 1999; Planchais et al., 2004). In *Arabidopsis*, the expression of *CycD2* and *CycD3* during the G1 phase is controlled by the availability of sugars (Riou-Khamlichi et al., 2000). In addition, *CycD3*, but not *CycD2*, expression responds to CKs. *CycD3* is elevated in an *Arabidopsis* mutant, exhibiting high CK levels, and is rapidly induced by CK application in both cell cultures and whole plants (Riou-Khamlichi et al., 1999). On the other hand, in maize CKs also stimulated the expression of *CycD2* at the late stages of germination (Gutiérrez et al., 2005). In *Arabidopsis* developing leaves *CycD3* function contributes to the control of cell number by regulating the duration of the mitotic phase and timing of the transition to the endoreduplication stage (Dewitte et al., 2007), in which nuclear DNA content is increased by rounds of full genome duplication without intervening mitoses, giving rise to cells with higher ploidy levels (Joubès and Chevalier 2000). It has also been suggested that cellular expansion and its accompanying endoreduplication are inhibited in *Arabidopsis* plants overexpressing the *CycD3;1* gene (Dewitte et al., 2003).

In this study, the role of various CK metabolites in the control of seed mass in maize was investigated using an allelic series of *miniature 1* (*mn1*), a loss-of-function mutation in the

Mn1 gene that codes for an endosperm-specific cell wall invertase 2 (INCW2) (Miller & Chourey, 1992, Cheng et al., 1996). The *mn1* seed mutation is associated with a conspicuous loss of seed mass at maturity and small seed size, related to the decreased cell number and size in developing endosperm (Vilhar et al., 2002). Several lines of correlative evidence from various plant species are available to show that cell wall invertase plays a major role in controlling sink strength through source-to-sink unloading of sucrose (Chourey et al., 2006). As expected from the invertase-deficiency, the *mn1* mutation also exhibits much altered sugar metabolism relative to *Mn1* (LeClere et al., 2010). The *mn1* mutant accumulates higher levels of sucrose and lower levels of hexoses in the basal region of the caryopsis, while its upper regions accumulate either similar or even increased concentrations of sugars compared to the upper regions of the wild type caryopsis. Given the importance of hexose signaling, these changes may also be associated with the levels of CKs and auxin (Rolland et al., 2002). Indeed, CK metabolism in the *mn1* mutant differs from that of the wild type (Rijavec et al., 2009, 2011) and also has greatly reduced levels of the auxin indole-3-acetic acid (IAA) throughout caryopsis development (LeClere et al., 2008). Given these relationships, it was reasoned that the four lineage-related genotypes of the *mn1* allelic series are an ideal genetic system in which to analyze possible relationship among various CK metabolites, RNA level expression of a few selected genes of CK metabolism and sink strength in developing seeds. In this regard, a previously described *mn1-89* mutation in maize is of special interest. It is an EMS-induced point mutation representing a single amino acid change from the conserved proline to leucine at the position that is critical for either efficient translation or for stabilization of the protein *in planta* (Carlson et al., 2000). The *mn1-89* allele encodes normal or higher levels of *Incw2* RNA compared to the wild type, but exceedingly low levels, ~ 6% of the *Mn1*, of the INCW2 protein and enzyme activity (Cheng et al., 1996). More importantly for the studies here, genetic analyses showed that the *mn1-89* allele is semi-dominant in seed phenotype (Cheng et al., 1996) when crossed to the standard null allele, *mn1-1*, which has approximately ~1% of the wild type levels of invertase activity encoded by the non-allelic gene, *Inc1*, a paralog of the *Mn1* locus (Chourey et al., 2006). The reciprocal hybrids from such crosses yield gene-dose dependent invertase activity levels of ~4 and 2% encoded by two and one *mn1-89* alleles, respectively, present in the triploid endosperms. Thus, the invertase activity is strictly gene-dose dependent in that the number of allele copies determines the amount of enzyme activity.

It has recently been shown that both *Mn1* and *mn1* genotypes have extremely high, but similar CK levels during the very early stages of development from 6 to 8 days after pollination (DAP), which are followed by a marked and genotype-specific reduction (Rijavec et al., 2009). While the decrease of CKs in *Mn1* was associated with their deactivation by 9-glucosylation, the absolute CK concentrations as well as concentrations of the biologically active CKs remained higher in the mutant until 16 DAP. Based on the correlative results from different studies, a developmental model showing possible crosstalk among CKs, cyclins CycD2 and CycD3, and INCW as causal to increase of cell number and sink strength of the *Mn1* developing endosperm has been proposed (Fig. 1) (Rijavec et al., 2009). In the present work the proposed hypotheses that *CycD2* and *CycD3* genes are temporally differentially expressed in the mutant *mn1* caryopsis and that their transcript abundance is associated with the elevated CK level or CK activation were further explored.

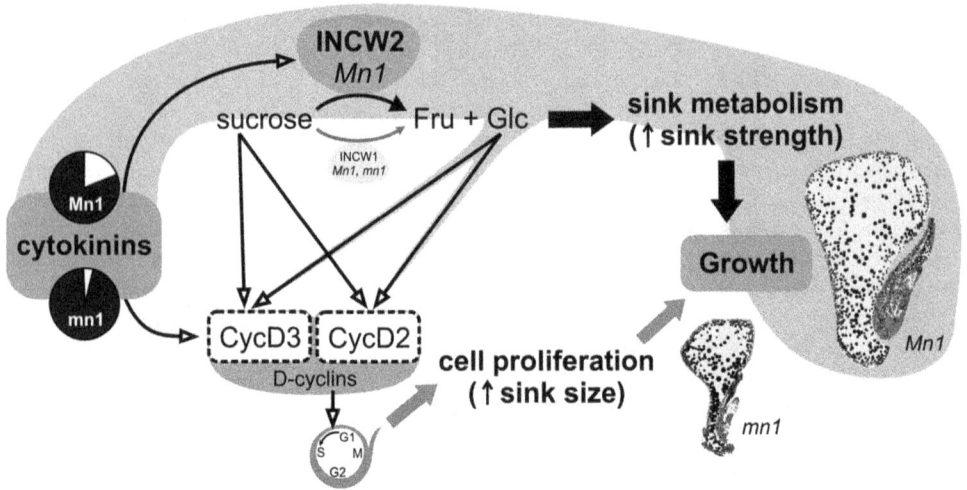

Fig. 1. Model of two alternative developmental pathways induced by cytokinins that lead to different sizes of filial tissues (endosperm and embryo) in wild-type (*Mn1*) and mutant *miniature1* (homozygous for the *mn1* allele) maize caryopsis. Cytokinin pie charts indicate the ratio of metabolically active cytokinins (dark) versus metabolically inactive *trans*-zeatin-9-glucoside (Z-9-G) (white) in both genotypes. The pathway that utilizes INCW2 for increasing cell number and sink strength is shaded gray. The images of *Mn1* and *mn1* caryopsis are bubble graph presentations of their actual longitudinal sections at 16 DAP. Steps in the model presented by dotted boxes and arrows with open heads have been proven in experimental systems other than maize. (From Rijavec et al., 2009; with the permission of the Journal of Integrative Plant Biology).

2. Caryopsis dry mass and *ZmIncw2* transcript level correlate with the gene dose of mutant alleles

A series of mutant maize caryopses (i.e. endosperm genotypes *mn1-89/mn1-89/mn1-89*, *mn1-89/mn1-89/mn1-1*, *mn1-1/mn1-1/mn1-89*, *mn1-1/mn1-1/mn1-1*) were obtained by reciprocal crossing of maize plants homozygous for *mn1-89* and *mn1-1* mutant alleles. The mass of developing caryopses rose from 6 to 28 DAP and was gene-dose dependent (Fig. 2). In mature caryopses dry mass of the wild type *Mn1* was significantly higher than that of the mutants. Seed mass of the homozygous *mn1-89* seed, which was ~80 % that of the *Mn1*, was in accordance with reported phenotypical similarity between this mutant and the wild type (Cheng et al., 1996). Additionally, the caryopsis with slightly more reduced seed mass (i.e. 64 % of the size of *Mn1*) had two copies of the *mn1-89* allele and one of the *mn-1* allele in the triploid endosperm. Seed masses of a heterozygous genotype with two copies of the *mn-1* allele and the *mn1* homozygous genotype were almost indistinguishable in the developmental period from 6 to 28 DAP (Fig. 2), confirming the similarity of caryopsis phenotypes (Cheng et al., 1996). However, the differences among the genotypes were more pronounced in mature caryopses (Fig. 2), in which the dry mass of *mn1-1/mn1-1/mn1-89* represented 41 % of the wild-type (*Mn1*) and that of *mn1-1/mn1-1/mn1-1* only 25.6 %.

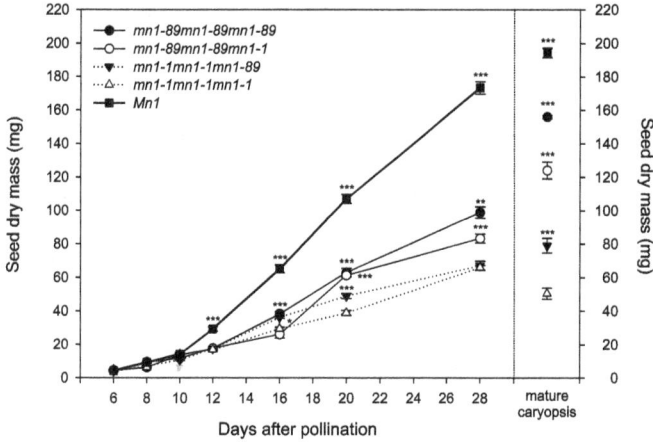

Fig. 2. Dry mass of developing caryopsis. The results are an average mass of 10 dry caryopses ±SE. Maize plants, homozygous for the *mn1-89* and *mn1-1* allele (both in the W22 genetic background) were cross-pollinated to produce four different endosperm genotypes (*mn1-89/mn1-89/mn1-89*, *mn1-89/mn1-89/mn1-1*, *mn1-1/mn1-1/mn1-89* and *mn1-1/mn1-1/mn1-1*). Student's t-test between the wild type Mn1 and each mutant yielded P-values < 0.05 (*), 0.01 (**), or <0.001 (***).

The general temporal profiles of the *ZmIncw2* transcript levels measured with quantitative real-time PCR were similar for all endosperm genotypes (Fig. 3).

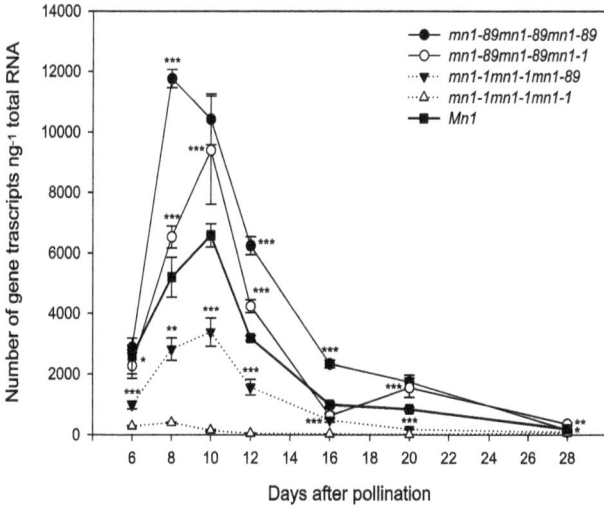

Fig. 3. Temporal expression of *ZmIncw2*. Data represents average transcript abundance ±SE for 6-9 real-time PCR reaction prepared from two individual RNA isolations and 3 subsequent reverse transcription reactions. Gene pecific primers based on the sequences with the GenBank Accesion number AF165179. Student's t-test between the wild type Mn1 and each mutant yielded p-values < 0.05 (*), 0.01 (**), or <0.001 (***).

Transcript levels were low at 6 DAP and between 16 and 28 DAP, while showing a distinct peak in transcript abundance at 8 or 10 DAP (Fig. 3). The highest abundance of the transcript preceded the peak of INCW2 activity at 12 DAP (Cheng et al., 1996). In spite of the similar profiles, *ZmIncw2* transcript levels showed a gene dose dependent trend in the *mn1-89/mn1-1* mutant series. Accordingly, genotypes *mn1-89/mn1-89/mn1-89* and *mn1-89/mn1-89/mn1-1* showed the highest transcript levels and even exceeded those of the wild type, confirming previous observations of high *Incw2* transcript levels in the *mn1-89* homozygous mutant using Northern blotting (Carlson et al., 2000). On the other hand, *Incw2* transcript levels were low in the heterozygous mutant with one *mn1-89* allele and almost undetectable in the homozygous *mn1-1* mutant genotype (Fig. 3). It is noteworthy that there is a discrepancy between protein and enzyme activity levels and *Incw2* transcript levels in *Mn1* and *mn1* endosperms (Chourey et al., 2006), since high RNA level in the homozygous *mn1-89* mutant results in only ~6 % of the *Mn1* invertase activity (Cheng et al., 1996). It has been suggested that a large proportion of INCW2 is dispensable without a significant change in seed phenotype (Cheng et al., 1996).

3. Cytokinin profiles and transcript levels of CK metabolism related genes

The total concentration of nine isoprenoid CKs (*trans*-zeatin, tZ; dihydrozeatin, DHZ; N^6-isopentenyl adenine, iP; *trans*-zeatin riboside, ZR; dihydrozeatin riboside, DHZR; N^6-isopentenyl adenosine, iPA; *trans*-zeatin-9-glucoside, Z-9-G; dihydrozeatin-9-glucoside, DHZ-9-G; N^6-isopentenyladenine-9-glucoside, iP-9-G) detected in maize caryopses with applied immunoaffinity chromatography followed by HPLC analyses (Rijavec et al., 2009, 2011) was similar in all examined genotypes (Fig. 4a). Depending on the mutant, a CK peak was recorded in the period between 8 and 12 DAP, followed by a drop of cytokinin content at 16 DAP and a slight increase at 20 DAP (Fig. 4a). Similar CK patterns have been reported in maize caryopses before (Dietrrich et al., 1995; Brugière et al., 2008; Rijavec et al., 2009, 2011). Specific CK concentrations at peak stages were 40.87 ± 5.35, 81.7 ±33.3 and 90,68 ±23.21 pmol CKs per caryopsis from *mn1-89/mn1-89/mn1-1*, *Mn1* and *mn1-1/mn1-1/mn1-89*, respectively at 8 DAP; 93,54 ± 13.91 pmol per caryopsis from *mn1-1/mn1-1/mn1-1* caryopsis at 10 DAP; and 88.62 ±17.99 of total CKs per caryopsis from *mn1-89/mn1-89/mn1-1* at 12 DAP (Fig. 4a). The peak of CK concentration before 12 DAP has been previously suggested to be associated with the induction of programmed cell death in the placento-chalazal region of the caryopsis pedicel (Rijavec et al., 2011), as its essential developmental phase (Kladnik et al., 2004). The proper activity of the placento-chalazal region is crucial for the transport of photosynthates from the maternal to the filial tissues of the caryopsis. An interesting observation in the present research was the lowest detected amount of total CKs at 10 DAP in *mn1-89/mn1-89/mn1-1*, in which the total content never exceeded 41 pmol per caryopsis and differed substantially from the CK concentrations in *mn1-89/mn1-89/mn1-89* (Fig. 4a). However, the overall appearance of caryopses remained similar. It has been shown before that the absolute concentration of CKs in plants is not the crucial factor controlling their development, but that local changes in cytokinin level can have global consequences for the plant (Dermastia & Ravnikar, 1996; Dermastia et al., 1996). Indeed, specific distribution of CK metabolites in various caryopsis tissues has been demonstrated (Rijavec et al., 2009, 2011).

The observed CK distribution was associated with a temporal expression of several CK metabolic genes (Fig. 5). The transcript abundance of isopentenyl transferase 2 (*ZmIpt2*), cytokinin dehydrogenase 1 (*ZmCkx1*), *cis*-zeatin-*O*-glucosyl transferase (*ZmCzog*), histidine kinase 1 (*ZmHK1*) and a putative *N*-glucosyl transferase (*ZmCngt*; Rijavec et al., 2011) were determined in the period from 6 DAP to 28 DAP using quantitative real-time PCR. The temporal expression profiles of genes *ZmIpt2*, *ZmCkx*, *ZmCzog* and *ZmcHK1* roughly clustered into two groups related to the genotype; the first group consisted of *Mn1*, *mn1-89/mn1-89/mn1-89* and *mn1-89/mn1-89/mn1-1*, and the second group of *mn1-1/mn1-1/mn1-89* and *mn1-1/mn1-1/mn1-1* (Fig. 5).

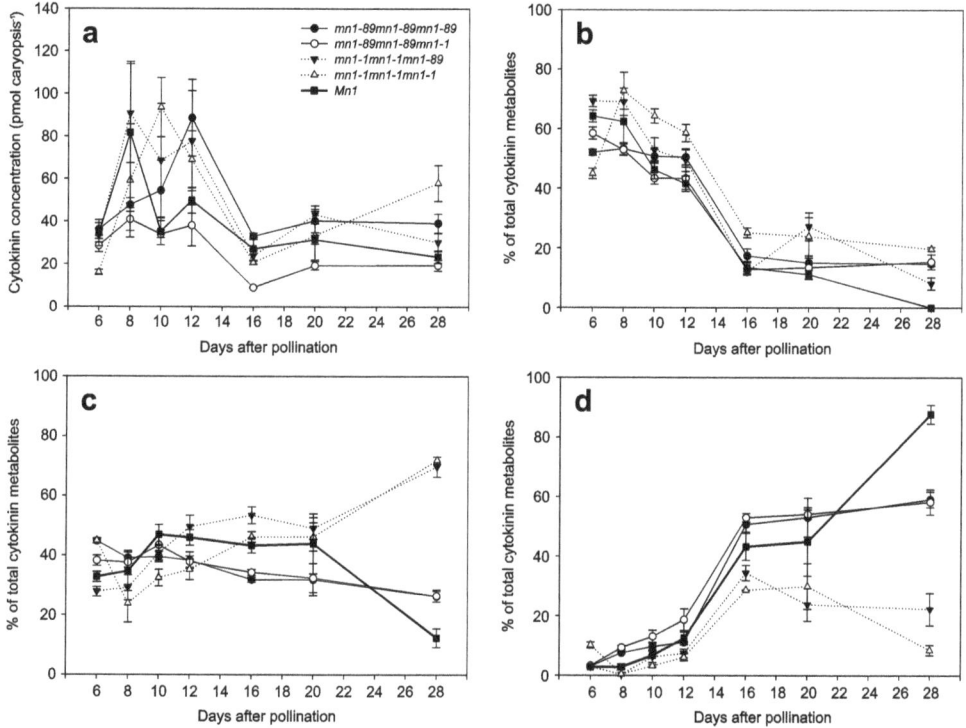

Fig. 4. Temporal profiles of cytokinins in caryopses with genotypes *Mn1*, *mn1-89/mn1-89/mn1-89*, *mn1-89/mn1-89/mn1-1*, *mn1-1/mn1-1/mn1-89* and *mn1-1/mn1-1/mn1-1* between 6 and 28 days after pollination. Data represent the average relative amount of cytokinin metabolites ± SE for 3 separate quantifications. a) Total CK concentration per caryopsis of CKs *t*Z, ZR, Z-9-G, DHZ, DHZR, DHZ-9-G, iP, iPA, and iP-9-G; b) percent of CK free bases *t*Z, DHZ and iP from total detected CKs; c) percent of CK ribosides ZR, DHZR and iPA from total detected CKs; d) percent of CK 9-glucosides Z-9-G, DHZ-9-G, iP-9-G from total detected CKs.

The *ZmIpt2* transcript profile of the first group of endosperm genotypes was low at 6 DAP, peaked between 8 DAP and 12 DAP and dropped after 12 DAP. The transcript abundance was similar in the two mutant genotypes and was significantly higher than in *Mn1*. In the second group the transcript profile showed level amount of the *ZmIpt2* transcript between 6 DAP and 12 DAP and, again, steep decrease after 12 DAP (Fig. 5a). The broader expression peak of the CK biosynthetic gene *ZmIpt2* in the wild type, in the homozygous *mn-1* mutant and in *mn1-1/mn1-1/mn1-89* (Fig. 5a) preceded the elevated CK concentration through several DAP (Fig. 4a). On the contrary, a narrow increase and subsequent drop in *ZmIpt2* transcript in *mn1-89/mn1-89/mn1-89* was associated with a

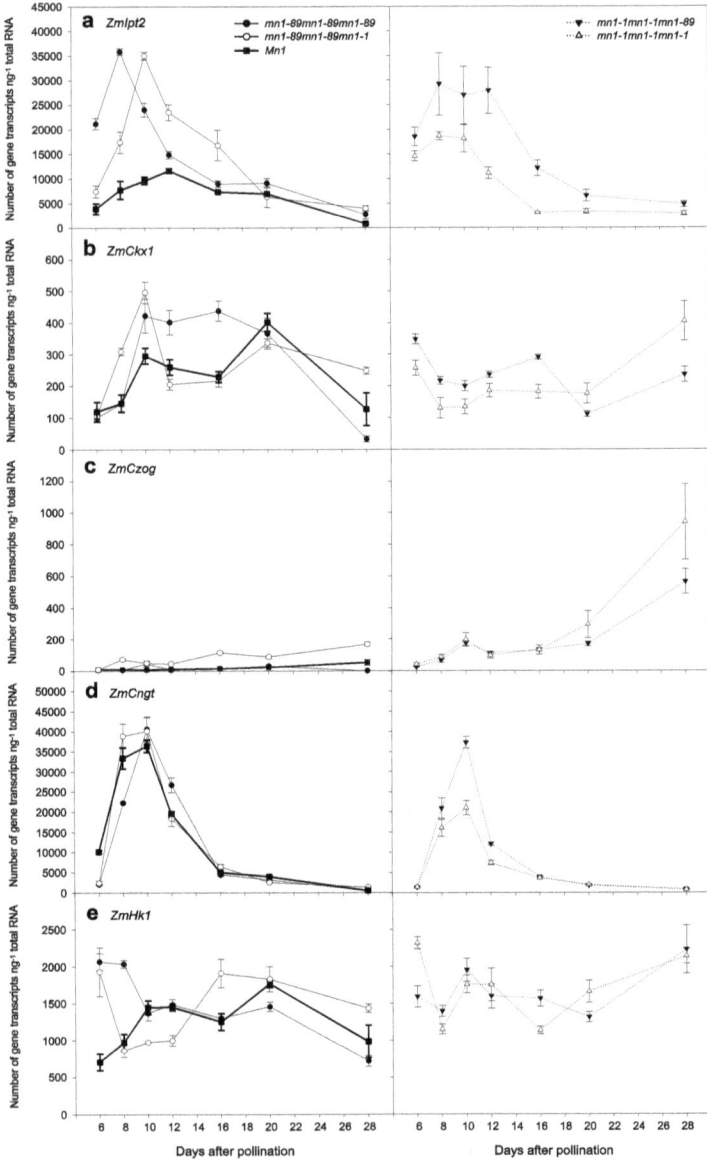

Fig. 5. Temporal expression profiles of cytokinin metabolism related genes. Gene expression of (a) *ZmIpt2*, (b) *ZmCkx1*, (c) *ZmCzog* d) *ZmCngt*, and (e) *ZmHk1*, genes was examined in developing caryopsis 6 to 28 DAP in four *miniature1* endosperm mutants with different relative numbers of *mn1-89* and *mn1-1* mutant alleles. Data represents average transcript abundance ±SE for 6-9 real-time quantification reaction prepared from two individual RNA isolations and three subsequent reverse transcription reactions. Gene-specific primers were based on the sequences with the following GenBank Accesion numbers: *ZmCkx1*, Y18377; *ZmIpt2*, DV527975; *ZmCngt*, BT016809; *ZmHk1*, AB042270; *ZmCzog*, AF318075.

narrow peak of CK concentration at 12 DAP. However, *ZmIpt2* transcript abundance was not related to the quantity of detected CKs. Interestingly, in a heterozygote with two alleles of *mn1-89* no clear increase in CKs was observed (Fig. 4a), but a high and sharp increase of *ZmIpt2* transcript was detected at 10 DAP (Fig. 5a).

The transcript abundance of *ZmCkx1* in the first group of genotypes peaked at 10 DAP. In *mn1-89/mn1-89/mn1-89* stayed approximately at the same level until 20 DAP and then decreased (Fig. 5b). However, in *Mn1* and *mn1-89/mn1-89/mn1-1* the transcript level decreased after 10 DAP, increased again between16 DAP and 20 DAP and declined after that. The absence of a clear CK concentration peak in *mn1-89/mn1-89/mn1-1* (Fig. 4a) might be explained because the highest detected level of the *ZmCkx* transcript peak corresponds with the expression peak of *ZmIpt2* (Fig. 5a), suggesting enhanced regulation of CKs during this time period.. In the group of *mn1-1/mn1-1/mn1-89* and *mn1-1/mn1-1/mn1-1* the transcript abundance was similarly high at 6 DAP, decreased in the period between 6 DAP and 8 DAP, slightly increased after 10 DAP, decrease again between 16 DAP and 20 DAP and at 28 DAP reached the level of 6 DAP (Fig. 5b).

The expression profile of the *ZmCzog* gene was low between 6 DAP and 16 DAP. It remained low until 28 DAP in the first group of genotypes, but it steeply increased after 20 DAP in *mn1-1/mn1-1/mn1-89* and *mn1-1/mn1-1/mn1-1* (Fig. 5c).

Regardless of the genotype, the profile of the transcript *ZmCngt*, encoding the corresponding enzyme *N*-glucosyl transferase, was very low at 6 DAP, followed by a clear expression peak at 8 DAP and rapid drop afterward (Fig. 5d). However, its abundance was similar in all mutants except the homozygous *mn1* mutant, where it was two-fold lower (Fig. 5d). Previously reported high concentration of the putative product of the *N*-glucosyl transferase reaction, Z-9-G, in the developmental phase following intense mitotic activity in maize endosperm (Rijavec et al., 2009, 2011) was confirmed in mutant hybrids. The concentration of this metabolically inactive CK-*N*-glucoside (Brzobohatý et al., 1994) in mutants with the *mn1-89/mn1-89/mn1-89* genotype was close to the concentration previously reported for the wild type (Fig. 4d) (Rijavec et al., 2011). On the other hand, its concentrations were lower in the other examined mutant hybrids, especially at 28 DAP (Fig. 4d), but the peaks corresponded to the gene expression profile of *ZmCngt*.

In addition, the expression of the *ZmHk1* gene was evaluated in caryopses with different genotypes (Fig. 5e). Histidine kinases (HK) are transmembrane receptors that bind cytokinin on the outer side of the cell's membrane and transducer the signal across the membrane to the inner side (reviewed in Rijavec & Dermastia, 2010). When cytokinin is not bound to the receptor, the latter is not active, suppressing downstream CK signaling. The oscillating expression patterns of *ZmHk1* were similar for all genotypes until 20 DAP, but by 28 DAP a distribution into two groups was evident again (Fig. 5e). Whether this higher *ZmHk1* expression was related to more active CK forms in *mn1-1/mn1-1/mn1-89* and *mn1-1/mn1-1/mn1-1* at this time point is currently not known. However, it has been suggested that the cellular cytokinin level modulates signaling and that mutual control mechanisms exists between metabolism and signaling, which may contribute to fine-tuning of the cytokinin response (Riefler et al., 2006). Study of cytokinin receptor mutants has also led to the conclusion that the seed size is a direct consequence of loss of receptor functions and their role in growth control (Riefler et al., 2006).

Although absolute CK concentrations (Fig. 4a) did not cluster into two groups as did expression levels of CK metabolic genes (Fig. 5), specific groups of CKs, specifically ribosides and 9-glucosides, clearly followed this pattern (Fig. 4c, d). Specifically, in *Mn1* and the hybrids with two or three *mn1-89* alleles the share of CK ribosides continuously decreased after 10 DAP, while in the hybrids with one *mn1-89* allele or in the homozygous hybrid with three alleles of *mn1* their share increased and was 2.7-fold higher at 28 DAP. The share of 9-glucosides, which in maize caryopses were represented solely by Z9G, did not exceed 17% from 6 DAP to 12 DAP. However, in *Mn1*, *mn1-89/mn1-89/mn1-89* and *mn1-89/mn1-89/mn1-1* it continuously rose and represented at 28 DAP about 57% in the hybrids and even 88% in the wild type (Fig. 4d). On the other hand, the share of Z-9-G in *mn1-1/mn1-1/mn1-89* and *mn1-1/mn1-1/mn1-1* increased to 34 % and 39 %, respectively, at 16 DAP but subsequently decreased, and in the homozygous genotype represented only 8 % of total cytokinin metabolites at 28 DAP (Fig. 4d).

4. Differential expression of type D2 and type D3 cyclin genes in the *Mn1* and *mn1* caryopsis supports the alternative development of *mn1*

The degree of change in transcript abundance of putative maize cycling genes *ZmCycD2*, *ZmCycD3-1* and *ZmCycD3-2* recovered by homology searches against all translated maize sequences (TBLASTN) with known protein sequences from *Arabidopsis* and rice was quantified by a real-time PCR in *Mn1* and *mn1* caryopsis samples at developing stages from 0 to 32 DAP. The *ZmCycD2* expression profile was similar in both, wild type and mutant caryopsis. The gene was highly expressed at pollination, down-regulated in the period between 0 and 12 DAP, up-regulated from 12 to 20 DAP, and in the wild type the transcript abundance afterward decreased. However, the the *ZmCycD2* transcript in *mn1* mutant caryopses was slightly more abundant from 0 to 8 DAP than in *Mn1*; the gene was more heavily down-regulated at 12 DAP and was up-regulated in the late stages of caryopsis development (Fig. 6a).

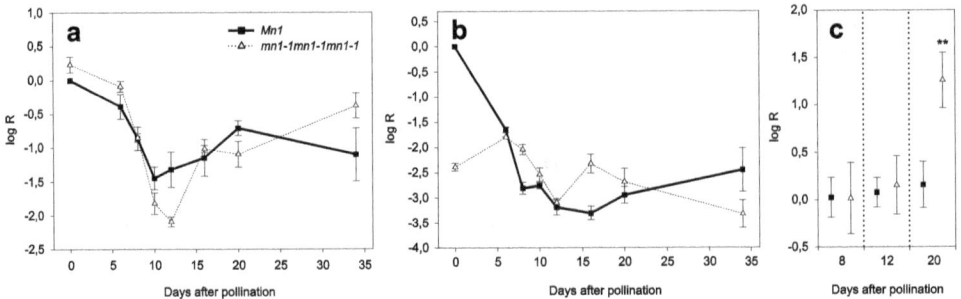

Fig. 6. The expression profiles of a) *ZmCycD2*, b) *ZmCycD3-1* and c) *ZmCycD3-2* genes in caryopsis from 0 DAP to 28 DAP. Three independent RT preparations from three separate RNA isolations were used for transcript quantification each in three replicate real-time PCR reactions. Final data represented the mean ±SE of nine quantification reactions; ** $P < 0.01$. The relative quantification of gene transcripts was done after normalization of all samples to a geometric average of the expression levels of cytochrome oxidase (COX), elongation factor 1α (EF-1α) and cyclophilin (CyP) genes. Gene specific primers based on the sequences with the following GenBank Accesion numbers: *ZmCycD2* (NM_001111578), *ZmCycD3-1* (NM_001158642) and *ZmCycD3-2* (NM_001196691).

The expression level of *ZmCycD3-1*was 2- to 4- fold lower in the mutant caryopsis compared to the wild type at 0 DAP. It increased during 0 - 5 DAP and was higher in comparison to the wild type until 20 DAP, with an additional peak of expression at 16 DAP. However, at 34 DAP the expression of *ZmCycD3-1*in the mature caryopsis was lower than that of the wild type (Fig. 6b).

The expression profile of the gene *CycD3-2* was similar to that of *CycD3-1*, but more differentially expressed (data not shown). Fig. 6c shows its relative expression analysed in detail at three time points. It is evident that its expression at 20 DAP was significantly (2.6-fold) higher in the *mn1* caryopsis in comparison with *Mn1*.

The expression of *CycD2* and *CycD3* has been previously characterized in *Arabidopsis*. It has been shown to be induced by sucrose and glucose, but the increase in *CycD2* levels is more responsive to glucose (Riou-Khamlichi et al., 2000). It has been recently demonstrated in maize caryopsis that in the period between 8 and 12 DAP sucrose levels remain constant in the basal part and only slightly decrease in the upper part of the wild type caryopsis, but increase in both parts of the mutant maize caryopsis (LeClere et al., 2010). On the other hand, in the same time period glucose decrease in both caryopses (LeClere et al., 2010). In rough accordance with those results was a down-regulation of *ZmCycD2* between 8 and 12 DAP (Fig. 6a). The subsequent up-regulation of *ZmCycD2* between 12 and 16 DAP is consistent with the constant concentration of sucrose in the whole *Mn1* caryopsis, the slight decrease of glucose in its upper part and strong increase of glucose in its basal part. Similar patterns of sugars and *ZmCycD2* expression were observed in the *mn1* caryopsis. The differences in the expression profiles of the *ZmCycD3* genes in the *Mn1* and *mn1* caryopsis in comparison with *ZmCycD2* might be at least partially explained by the levels of cytokinins present, immunolocalized as described (Rijavec et al., 2011) (Fig. 7). The immunosignal was similarly high in the pedicel regions, slightly stronger in the mutant embryo but very abundant throughout *mn1* endosperm development. While the signal was relatively strong in the aleurone layer of both caryopses, it was very faint in the upper part of the *Mn1* endosperm. On the other hand it was very strong in the upper endosperm of *mn1*. It has been shown, that *CycD3* expression is induced by cytokinins when sucrose remains present (Riou-Khamlichi et al., 1999).

Another interesting observation from this study is the increased concentration of the *ZmCycD3-2* transcript, which was shown to be significantly higher at 20 DAP (Fig. 6c) in the *mn1* caryopsis in comparison with the wild type. The peak of *ZmCycD3-1* in *mn1* at 16 DAP (Fig.6b) corresponds to the peak of endosperm endoreduplication. In the *mn1* mutant at the endoreduplication developmental stage endosperm comprise 55 % fewer cells with reduced cell size (Vilhar et al., 2002). Despite this, the progress of endoreduplication was not affected in the *mn1* endosperm in comparison with the wild type. In the model of caryopsis growth (Fig. 1) (Rijavec et al., 2009) it has been proposed that growth of *mn1* caryopsis that is highly reduced in size due to the absence of the invertase *Incw2* gene may be partially compensated by *CycD2* and *CycD3* induced cell division regulated by sugars and cytokinins. The results of this and other studies support the model. Specifically, in transgenic tobacco lines expressing *Arath-CycD3* gene the ploidy levels of mature stem tissues were not affected, suggesting no effect on the extent of endoreduplication. Moreover, the *Arath-CycD3* overexpressing lines have an increased cell number together with a reduced cell size (Boucheron et al., 2005). Similar effects were also observed in *Arabidopsis* that expressed very high levels of *CycD3;1* (Dewitte et al., 2003). However, in *Arabidopsis* endoreduplication was also inhibited in plants overexpressing *CycD3;1*.

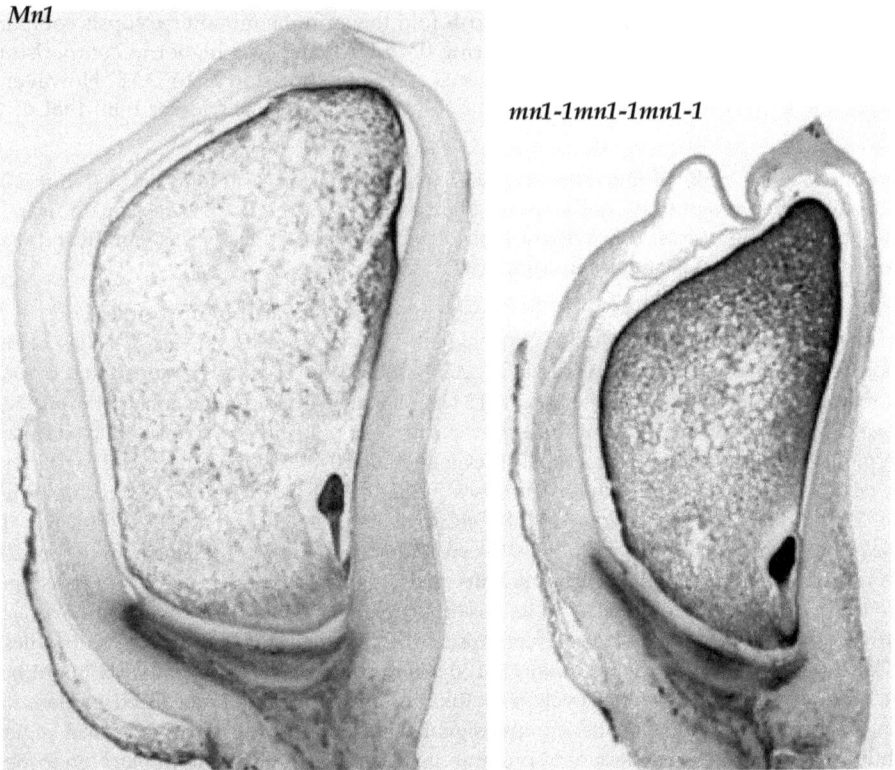

Fig. 7. Immunolocalization of cytokinins on the central longitudinal section of wild type *Mn1* and *mn1-1/ mn1-1/ mn1-1* caryopsis at 12 DAP.

5. Conclusions

The results here confirm the previously demonstrated linear gene-dose relationship of the *Incw2* transcripts with the number of *mn1-89* copies. Additionally we show that the seed mass was markedly different between the homozygous *Mn1* and the *mn1-89* series beginning at 12 to 16 DAP. Quantitative measurements of nine CK metabolites showed that the genotypes with reduced levels of invertase activity have reduced concentrations of the metabolically inactive CK form zeatin-9-glucoside and increased concentrations of active CK ribosides. Quantitative real-time PCR analyses for the CK genes showed that the *cytokinin-N-glucosyl transferase* gene (*ZmCngt1*) was expressed at the highest level at 10 DAP and was greatly reduced in expression in the *mn1* mutant. Additionally, this study supports our model (Rijavec et al., 2009) which proposes that high amounts of CKs and a higher ratio of their active forms may partially compensate the INCW2-deficiency in regulating CycD3 and CycD2 induced cell divisions by CKs and sucrose in the *mn1* endosperm. Consistent with the model, the sink strength is greater in caryopses with the *Mn1* genotype. In contrast, the levels of glucose, fructose and sucrose were higher in the upper part of the mutant caryopsis, due to as yet unknown mechanisms, and may contribute to the increased expression of the CycD3 gene and to normal, although substantially slower, development of *mn1* caryopses.

6. Acknowledgement

The authors thank the Slovenian Research Agency (grant 1000-05-310055 to T.R.) and the USA-Slovenia Cooperation in Science and Technology (grant BI-US/06-07-031). The authors also thank Drs. Aleš Kladnik for his help with data visualization, S. Chen and Mukesh Jain for critical review of the manuscript and Elizabeth Covington for proof-reading. M. Dermastia and P.S. Chourey contributed equally to this work.

7. References

Arnau, J. A, Tadeo, F. R., Guerri, J., & Primo-Millo E. (1999). Cytokinins in peach: Endogenous levels during early fruit development. *Plant Physiology and Biochemistry* 37, 10 (October):, p.p. 741-750, ISSN 0981-9428

Blackwell, J. R., Horgan, R. (1994). cytokinin biosynthesis by extracts of *Zea mays*. *Phytochemistry* 35, p.p. 339 342, ISSN 0031-9422

Brugière, N., Jiao, S., Hantke, S., Zinselmeier, C.m, Roessler, J. A., Niu, X., Jones, R. J., & Habben, J. E. (2003). Cytokinin oxidase gene expression in maize is localized to the vasculature, and is induced by cytokinins, abscisic acid, and abiotic stress. *Plant Physiology* 132, p.p. 1228-1240, ISSN 0032-0889

Brugière, N., Humbert, S., Rizzo, N., Bohn, J., & Habben, J. (2008). A member of the maize isopentenyl transferase gene family, *Zea mays isopentenyl transferase 2 (ZmIPT2)*, encodes a cytokinin biosynthetic enzyme expressed during kernel development. *Plant Molecular Biology* 67, no. 3 (June 1), p.p. 215-229, ISSN 0167-4412.

Brzobohatý, B., Moore, I.,& Palme, K. (1994). CK metabolism: implications for regulation of plant growth and development. *Plant Molecular Biology* 26, p.p. 1483-1497, ISSN 0167-4412

Boucheron, E.,. Healy, J. H. S., Bajon , C., Sauvanet, A., Rembur, J., Noin, M., Sekine, M., Riou Khamlichi, C., Murray, J. A. H., Van Onckelen, H., & Chriqu, D.(2005). Ectopic expression of *Arabidopsis* CYCD2 and CYCD3 in tobacco has distinct effects on the structural organization of the shoot apical meristem. *Journal of Experimental Botany* 56, p.p. 123–134, ISSN 1460-2431

Carlson, S.J., Shanker, S. & Chourey, P.S. (2000.) A point mutation at the *Miniature1* seed locus reduces levels of the encoded protein, but not its mRNA, in maize. *Molecular and General Genetics* 263, p.p. 367-373, ISSN 0026-8925

Cheng, W.H., Taliercio, E.W., & Chourey, P. S. (1996). The *Miniature1* seed locus of maize encodes a cell wall invertase required for normal development of endosperm and maternal cells in the pedicel. *The Plant Cell* 8, no. 6, p.p. 971-983, ISSN 1040-4651

Chourey, P. S., Jain, M., Li, Q.-B., & Carlson, S. J. (2006). Genetic control of cell wall inveratses in developing endosperm of maize. *Planta* 223, p.p. 159-167, ISSN 0032-0889

Dermastia, M., & Ravnikar, M. (1996) Altered cytokinin pattern and enhanced tolerance to potato virus Y[NTN] in the susceptible poptao cultivar (*Solanum tuberosum* L. cv. Igor) grown *in vitro*. *Physiological and Molecular Plant Pathology* 48, p.p. 65-71, ISSN 0885-5765

Dermastia, M., Ravnikar, M., & Kovač, M. (1996) Morhology of potato (*Solanum tuberosum* L. cv. Sante) stem node cultures in relation to the level of endogenous cytokinins. *Journal of Plant Growth Regulation* 15, p.p. 105-108, ISSN 0721-7595

Dewitte, W., & Murray, J. A. H. (2003) The plant cell cycle. *Annual Reviews of Plant Biology* 54, p.p. 235-264; ISSN 1543-5008

Dewitte, W., Scofield, S., Alcasabas, A. A., Maughan, S. C., Menges, M., Braun, N., Collins, C., Nieuwland, J., Prinsen, E., Sundaresan, V., & Murray, J. A. H. (2007). *Arabidopsis* CYCD3 D-type cyclins link cell proliferation and endocycles and are rate-limiting for cytokinin responses. *Proceedings of the National Academy of Sciences of the United States of America* 104, no. 36, p.p. 14537-14542, ISSN 1091-6490

Dietrich, J.T, Kaminek, M., Blevins, D.G., Reinbott, T.M., & Morris, R.O. (1995). Changes in cytokinins and cytokinin oxidase activity in developingmaize kernels and the effects of exogenous cytokinin on kernel development. *Plant Physiology and Biochemistry* 33, no. 3, p.p. 327-336, ISSN 0981-9428

Emery, N., & Atkins, C. (2006). Cytokinins and seed development, In: *Handbook of seed science and technology*, A. S. Basra (Ed.), pp. 63-93, The Haworth Press, Inc., ISBN1560223146, Binghamton

Gutiérrez, R., Quiroz-Figueroa, F., & Vázquez-Ramos, J. M. (2005). Maize cyclin D2 expression, associated kinase activity and effect of phytohormones during germination. *Plant and Cell Physiology* 46, p.p. 166-173, ISSN 0032-0781

Hou, B., Lim, E.K., Higgins, G.S., Bowles, D.J. (2004). N-glucosylation of CKs by glycosyltransferases of *Arabidopsis thaliana*. *The Journal of Biological Chemistry* 279, p.p. 47822–47832, ISSN 0021-9259

Inzé. D., & De Veylder L. (2006). Cell cycle regulation in plant development. *Annual Review of Genetics* 40, p.p. 77-105, ISSN 0066-4197

Joubès, J., & Chevalier, C. (2000). Endoreduplication in higher plants. *Plant Molecular Biology* 43,735–745, ISSN 0167-4412

Kladnik, A., Chamusco, K., dermastia, M., Chourey, P. S. (2004). Evidence of programmed cell death in post-phloem transport cells of the maternal pedicel tissue in developing caryopsis of maize. *Plant Physiology* 136, p.p. 3572-3581, ISSN 0032-0889

LeClere, S., Schmelz, E. A., & Chourey, P. S. (2008). Cell wall invertase-deficient miniature1 kernels have altered phytohormone levels. *Phytochemistry* 69, no. 3 (February), p.p. 692-699, ISSN 0031-9422

LeClere, S., Schmelz, E. A., & Chourey, P. S. (2010). Sugar levels regulate tryptophane-dependent auxin biosynthseis in developpink maize kernels. *Plant Physiology* 153, p.p. 306-318, ISSN 0032-0889

Letham, D. S. (1963). Zeatin, a factor inducing cell division from *Zea mays*. *Life Sciences* 8, p.p. 569–573, ISSN 0024-3205

Massonneau, A., Houba-Herin, N., Pethe, C., Madzak, C., Falque , M., Mercy M, Kopecny, D., Majira, A., Rogowsky, P., Laloue, M. (2004). Maize CK oxidase genes: differential expression and cloning of two new cDNAs. *Journal of Experimental Botany* 55, p.p. 2549-2557, ISSN 1460-2431.

Miller, M.E., Chourey, P.S. (1992). The maize invertase-deficient miniature1 seed mutation is associated with aberrant pedicel and endosperm development. *The Plant Cell* 4, p.p. 297–305, ISSN 1040-4651

Planchais, S., Samland, A. K., & Murray J. A. H. (2004). Differential stability of *Arabidopsis* D-type cyclins: CYCD3;1 is a highly unstable protein degraded by a proteasome-dependent mechanism . *The Plant Journal* 38, p.p. 616-625, ISSN 1365-313X

Riefler., M., Nova, O., Strnad, M., & Schmüling, T. (2006). *Arabidopsis* cytokinin receptor mutants revela functions in shoot growth, leaf seenscence, seed size, germination, root development, and cytokinin metabolism. *The Plant Cell* 18, p.p. 40-54, ISSN 1040-4651

Rijavec, T., & Dermastia, M. (2010). Cytokinins and their function in developing seeds. *Acta Chimica Slovenica*, 57,3, p.p. 617-629, ISSN 1318-0207

Rijavec, T., Kovač, M., Kladnik, A., Chourey, P.S . & Dermastia, M. (2009). A comparative study on the role of cytokinins in caryopsis development in the maize miniature1 seed mutant and its wild type. *Journal of Integrative Plant Biology* 51, no. 9, p.p. 840-849, ISSN 1672-9072

Rijavec, T., Jain, M., Dermastia, M., Chourey, P. S. (2011). Spatial and temporal profiles of cytokinin biosynthesis and accumulation in developing caryopsis of maize. *Annals of Botany* 107, p.p. 1235-1245, ISSN 0305-7364

Riou-Khamlichi, C., Menges, M., Healy, S. J. M., & Murray, J. A. H. (2000). Sugarc ontrol of the plant cell cycle: Differential regulation of *Arabidopsis* D-type cyclin gene expression. *Molecular and Cellular Biology* 20, no. 13 (July 1), p.p. 4513-4521, ISSN 1098-5549

Riou-Khamlichi, C., Huntley, R., Jacqmard, A., & Murray, J. A. H. (1999). Cytokinin activation of *Arabidopsis* cell division through a D-type cyclin. *Science* 283, no. 5407, p.p. 1541-1544, ISSN 0036-8075

Rodó, A.P., Brugière, N., Vankova, R., Malbeck, J., Olson, J.M., Haines, S.C., Martin , R.C., Habben, J.E., Mok, D.W.S., & Mok, M.C. (2008). Over-expression of a zeatin O-glucosylation gene in maize leads to growth retardation and tasselseed formation. *Journal of Experimental Botany* 59, p.p. 2673 - 2686, ISSN 1460-2431

Rolland, F., Moore, B., & Sheen, J. (2002). Sugar sensing and signaling in plants. *The Plant Cell* 14(suppl.), S 185-S205, ISSN 1040-4651

Sherr, C.J. & Roberts, J.M. (1999). CDK inhibitors: positive and negative regulators of G1-phase progression. *Genes & Development* 13, p.p. 1501–1512, ISSN 0890-9369

Šmehilová , M., Galuszka, P., Bilyeu, K.D., Jaworek, P., Kowalska, M., Šebela, M., Massonneau, A, Houba-Herin, N., Pethe, C., Madzak, C., Falque, M., Mercy, M,. Kopecny, D., Majira, A., Rogowsky, P., & Laloue, M. (2004). Maize CK oxidase genes: differential expression and cloning of two new cDNAs. *Journal of Experimental Botany* 55, p.p. 2549-2557, ISSN 1460-2431

Veach, Y.K., Martin, R.C., Mok, D.W.S., Malbeck, J., Vankova, R., & Mok, M.C. (2003). O-glucosylation of *cis*-zeatin in maize. Characterization of genes, enzymes, and endogenous cytokinins. *Plant Physiology* 131, p.p. 1374- 1380, ISSN 0032-0889

Vilhar, B., Kladnik, A., Blejec, A., Chourey, P. S., & Dermastia, M. (2002). Cytometrical evidence that theloss of seed weight in the *miniature1* seed mutant of maize is associated with reduced mitotic activity in the developing endosperm. *Plant Physiology* 129, no. 1 (May 1), p.p. 23-30, ISSN 0032-0889

Vyroubalova, S., Vaclavikova, K., Tureckova, V., Novak, O., Šmehilová , M., Hluska, T., Ohnoutkova, L., Frebort, I., & Galuszka. P. (2009). Characterization of new maize genes putatively involved in cytokinin metabolism and their expression during osmotic stress in relation to cytokinin levels. *Plant Physiology*, no. 1 (September 1), p.p. 433-447, ISSN 0032-0889

Yang, J., Zhang, J., Huang, Z., Wang, Z., Zhu, Q., & Liu, L. (2002). Correlation of cytokinin levels in the endosperms and roots with cell number and cell division activity during endosperm development in rice. *Annals of Botany* 90, 3, p.p. 369 -377, ISSN 0305-7364

16

Terpenoids and Gibberellic Acids Interaction in Plants

Zahra Asrar

Department of Biology, Shahid Bahonar University of Kerman, Kerman
Iran

1. Introduction

Plants synthesize an astonishing diversity of isoprenoids, some of which play essential roles in photosynthesis, respiration, and the regulation of growth and development. In spite of economic significance of the terpenoids and their many essential functions, relatively little is known about terpenoid metabolism and its regulation in plants (Mansouri et al, 2009). However, two independent pathways for the biosynthesis of isoprenoid precursors coexist within the plant cell: the cytosolic mevalonic acid (MVA) pathway and the plastidial methyerythritol phosphate (MEP) pathway. *Cannabis* is a diecious species that is a source of fiber, food, oil and medicine. Cannabinoids represent a distinctive class of compounds belong to the chemical class of natural terpenophenols. Among others, Δ^9-tetrahycannabinol (THC) and cannabidiol (CBD) are the most important of these compounds. The experiments with labeling patterns showed that the cannabinoids are derived entirely or predominantly (98%) from the deoxyxylulose pathway (Fellemeier et al., 2001). They are produced by glandular trichomes that occur on most aerial surfaces of the plant (Hilling 2004). Cannabis is used in modern medicine for the treatment of emesis in chemotherapy. As well as being useful anti-emetics, cannabinoids appear to have therapeutic value as antispasmodics, analgesics, and appetite stimulants and have potential in the treatment of epilepsy, glaucoma, and asthma (Agurell and Nilsson 1972; Guzman 2003; Howlett et al., 2004).

However little is known about the effects of plant hormones on the regulation of these pathways. Thus, we investigated the effect of gibberellic acid (GA₃) on changes in the amount of many produced terpenoids and the activity of the key enzymes, 1-deoxy-D-xylulose 5-phosphate synthase (DXS) and 3-hydroxy-3-methy glutaryl coenzyme A reductase (HMGR) in these pathways. To understand the role of gibberellic acid (GA₃) in the regulation of two terpenoid biosynthesis pathways, we gained experience on the response of the main end products of these pathways and cannabinoids under GA₃ treatment in cannabis plants.

This chapter will provide focus on recent development of the interaction effects of terpenoids and gibberellic acid in *Cannabis sativa*.

2. Biosynthesis of terpenoids

Isoprenoids (also known as terpenoides) made from assembly of isoprene units. Isoprenes are flexible five carbon units – $(CH_2)_2C=CH-CH_2-$, and they comprise a diverse class of plant metabolites; most of them are known as defense-related compounds, flavors or scents (Chris et al., 2007). Terpenoids constitute the largest family of natural plant products, with over 30,000 members (Dewick, 2002). Members of this diverse group of natural products are found in all organisms. In primary metabolism, they function as photosynthetic pigments (chlorophylls and carotenoids), electron transport (ubiquinone and plastoquinone), plant hormones (abscisic acid and gibberellins) and membrane fluidity such as sterols (Lang and Ghassemian, 2003; Mcgarvey and Croteau, 1995). The plant produced isoprenoids β-carotene (provitamine A) and α-tocopherol (VitE) are required for the maintenance of human health (Shintani and Dellapenna, 1998; Hirschberg, 1999). Industrial uses of isoprenoids include products such as colorants, fragrance, and flavorings (Lang and Croteau, 1999).

Otto Wallach who in 1910 recognized that isoprene is the basic constituent of terpens (Fig. 1) and Leopold Ruzicha found that isoprene is the basic element for the synthesis of many natural compounds including steroids. He postulated the biogenic isoprene rule, according to which all terpenoids (derivatives of terpens) are synthesized via a hypothetical precursor which he named active isoprene. This speculation was verified by Feodor Lynen in 1964, when he identified isopentenyl pyrophosphate to be "the active isoprene" (Heldt, 2005).

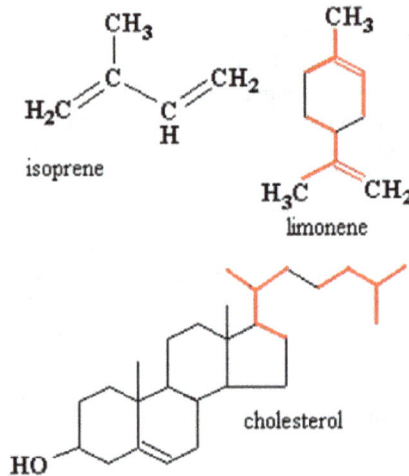

Fig. 1. Isoprene unit and Terpens

2.1 Terpenoids have two different biosynthetic pathways

All isoprenoides are derived from the ubiquitous C5 building blocks isopentenyl diphosphate (isopentenyl pyrophosphate) (IPP) and dimethylallyl diphosphate (DMAPP). These precursors can be synthesized by independent pathways in higher plants and algae (Lichtenthaler, 1999), the classical mevalonate pathway in the cytoplasm or the alternative non-mevalonate pathway in plastids (Arigoni et al., 1997; Rohmer, 1999). The plastidial pathway, now known as the 2-C-methy-D-erythriol-4-phosphate (MEP) pathway, has been fully elucidated by a combination of biochemical and genomic approaches (Rodriguez and Boronat, 2002); it provides the precursors for monoterpenes, diterpenes, carotenoids, tocopherols, and the prenyl moiety to chlorophyll (Eisenreich et al., 2001).

2.1.1 Mevalonate pathway

Konrad Bloch in 1964 discovered that acetyl CoA is a precursor for the biosynthesis of steroids. In fact, the mevalonate pathway (MVA pathway) has long been assumed to be the exclusive biosynthesis route used in all organisms for the synthesis of Isopentenyl pyrophosphate (IPP); particularly in mammalian where it is responsible for the biosynthesis of cholesterols. Fig. 2 shows the synthesis of the intermediary product IPP: two molecules of acetyl CoA react to produce acetoacety CoA and with another acety CoA yielding β-hydroxy-β methyl glutaryl CoA (HMG CoA). In plants a single enzyme, HMG CoA synthase catalyzes both reactions (Heldt, 2005). Several studies have emphasized the regulatory role of 3-hydroxy 3-methy glutaryl CoA enzyme reductase (HMGR) in the mevalonate pathway. HMG R is considered to be the most heavily regulated enzyme in mammalian metabolism (Goldstein and Brown, 1990) and there are numerous studies demonstrating that it is also closely regulated in plants, particularly in the case of sterol and phytoalexine synthesis (Weissenborn, 1995; Stermer et al., 1994).

The esterified carboxyl group of HMG Co A is reduced by two molecules of NADPH to a hydroxyl group accompanied by hydrolysis of the energy-rich thioester bond. Thus mevalonic acid is formed. A pyrophosphate ester is formed in two successive phosphorylation steps, catalyzed by two different kinases. Third molecule of ATP, then involving the transitory formation of a phosphate ester, a carbon-carbon double bond is generated and the remaining carboxyl group is removed (Heldt, 2005). Isopentenyl pyrophosphate is the basic unit for the formation of an isoprenoid chain and the cytosolic pathway, which starts from acetyl-CoA and proceeds through the intermediate mevalonate (MVA), provides the precursors for sterols and ubiquinons (Laule et al., 2003).

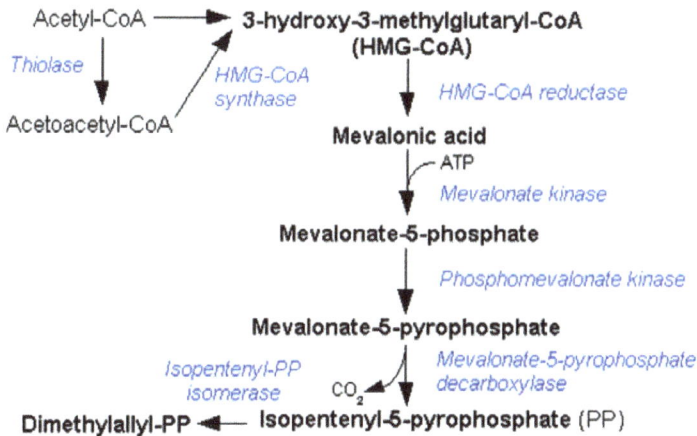

Fig. 2. Scheme of Mevalonate (MVA) pathway from acetyl CoA to Isopenteny phyrophosphate (IPP)

2.1.2 The DOXP/ MEP pathway

Experiments with plants led to the discovery that synthesize of isopentenyl pyrophosphate in the plastids follow a different pathway (Fig. 3). The precursors for this pathway are

DOXP/MEP Pathway

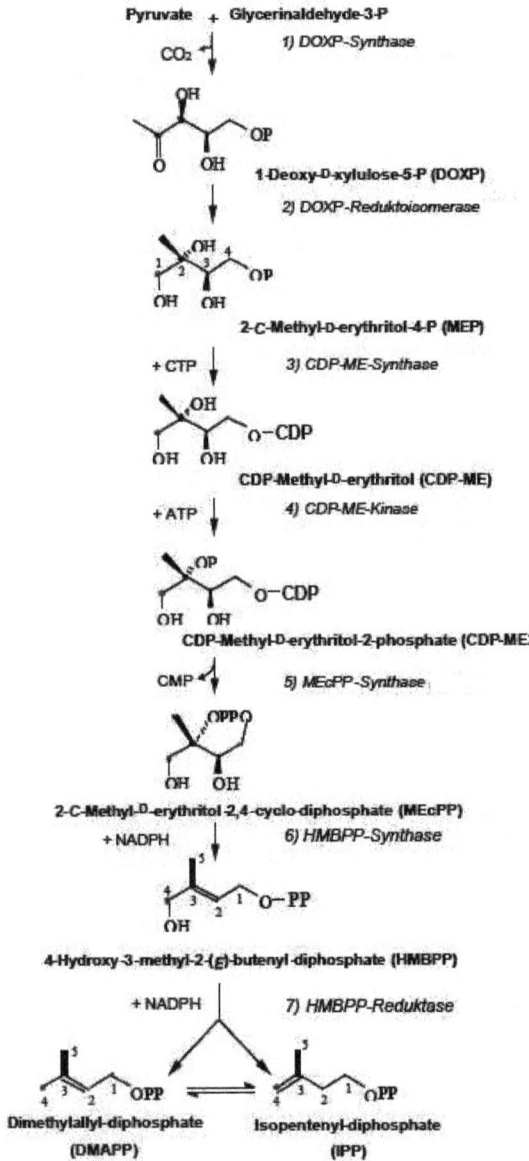

Fig. 3. Scheme of 1-deoxy-D-xylulose-5-phosphate/methyl-D-erythrithol (DOXP/MEP pathway). The enzymes of both pathways are numbered. MEP/DOXP pathway for isopentenyl diphosphate (IPP) and isoprenoid biosynthesis

pyruvate and D-glyceraldehyde-3-phosphate. Pyruvate is decarboxylated via thiamine pyrophosphate (TPP) and then is transferred to D-glyceraldehyde-3-phosphate to yield 1-deoxy-D-xylulose-5-phosphate (DOXP). After isomerization and reduction by NADPH, 2−C-methyl erythriol-4-phosphate (MEP) is synthesized. MEP is then activated by reacting with CTP to yield CDP methyl erythriol. Two further reduction steps, following dehydration and phosphorylation. Finally yield isopentenyl pyrophosphate (Heldt, 2005).

The MEP-synthase pathway for isoprenids is present in bacteria, algae, and plants, but not in animals. A large part of plant isoprenoids, including the hemiterpene, monoterpenes like limnone, phytohormones such as gibberellins (Crozier et al., 2000), brasinoesteroids, and phytoalexine (Lang and Gassemian, 2003; Kanno et al., 2006).

The production and accumulation of cannabinoids in plants of *Cannabis sativa* follow non-mevalonate pathway (Fellermeier et al., 2001) synthesized via the MEP pathway located in plastids (Rohmer, 1999).

1-deoxy-D-xylulose-erythriol phosphate synthase (DXS) catalyzes the first step in the methy-erythriol phosphate (MEP) pathway and is hypothesized to be an important control step within MEP pathway (Lichtenthaler, 1999). The role of the DXS in plants can be seen in *Arabidopsis*, the over expression of DXS led to an increase in terpenoids concentrations while the repression of DXS decreased terpenoid concentrations (Estevez et al., 2001).

3. Biosynthesis of gibberellins

Farmers in Asia were aware of a disease of rice plants called bakanae (foolish seedling) disease. Infected plants would grow excessively taller than normal healthy plants, and being fall over and be unharvestable. This disease was found to be caused by a fungus known as *Gibberella fujikuroi* (Yabuta, T., 1935). Thus, Work on this special disease in Japan occurred by Japanese scientists who isolated a growth promoting substances from cultures of a fungus that parasitizes rice plants in the 1930s. They called it gibberellin. Scientists in the 1950s rediscovered this work and extracted a range of chemical that elicit the growth response in rice seedlings. Since that time, more than 100 slightly different gibberellins have been identified chemically.

Gibberellins (GAs) are diterpene plant hormones or plant growth regulator of vascular plants, fungi, and bacteria that are biosynthesized through complex pathways and control diverse aspects of growth and development, including seed germination, shoot elongation, leaf expansion, pollen-tube growth, trichome, flower and fruit development, cell division, cell elongation and floral transition (Olszewski et al., 2002; Razem et al., 2006). Among more than a hundered GAs identified from plants (http://www.plant-hormones.info/gibberelline-nomenclature.htm) , only a small number of them, such as GA_1, GA_3, GA_4, and GA_7 are thought to function as bioactive (biologically active gibberellins) hormones. Therefore, many non bioactive gibberellins exist in plants as precursors for the bioactive forms or deactivated metabolism (Yamaguchi, 2008). The major bioactive GAs, commonly have a hydroxyl group on C-3β, a carboxyl group on C-6 and a lactone between C-4 and C-10 (Fig. 4). GAs has been identified frequently in a variety of plant species, implying that it acts as a widespread bioactive hormone.

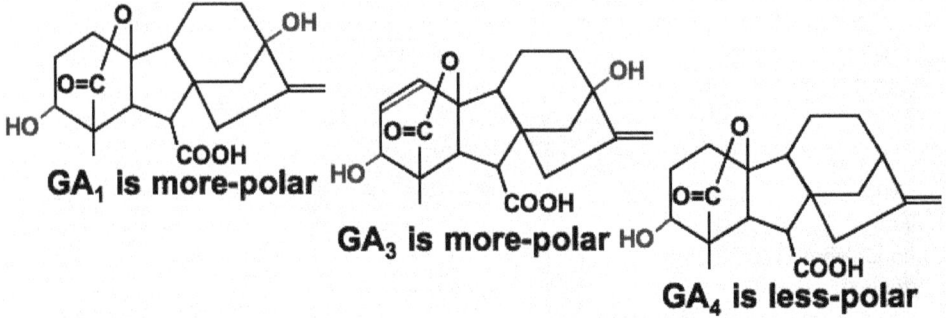

GA₁ is more-polar

GA₃ is more-polar

GA₄ is less-polar

Fig. 4. Active Gibberellins

According to the nature of the enzymes involved, gibberellins biosynthesis pathways are usually divided into three parts. Each of the three parts consists of several steps:

I- Biosynthesis of ent-Kaurene

Recently, a second, MVA- independent pathway to isopentenyl pyrophosphate (IPP) via glyceraldehydes 3-phosphate and pyruvate was discovered in a green alga (Schwender et al., 1996). Isopentenyl pyrophosphate is further converted to gerany geranyl pyrophosphate (GGPP) by two enzymes, IPP-isomerase and GGPP-synthase in plastids of higher plants (Dogbo & Camara, 1987).

Gerany gerany pyrophosphate is cyclized to ent-copalyl pyrophosphate (CPP) and finally to ent-kaurene. In higher plants these reactions are catalyzed by two enzymes, CPP synthase formerly Kaurene synthase A and ent-Kaurene (K) synthase formerly Kaurene synthase B, renaming was suggested by Mcmillan 1979. Both enzymes were localized in isolated proplastids of meristematic shoot tissues, but not in mature chloroplasts of pea and wheat (Aachet al., 1995, 1997) or of pumpkins endosperm (Simcox et al., 1975; Aach et al., 1995).

II- From ent-Kaurene to GA₁₂

In the second stage of GA biosynthesis (Fig. 5), the C-19 methy (CH₃) group of Kaurene is oxidized in three steps to give ent-kaurenoic acid (KA). These oxidations are catalyzed by ent-kaurene oxidase (KO), which is multifunctional because it can catalyze all three reactions. In Arabidopsis, KO is encoded by GA₃, and mutation in this gene will again produce severly dwarfed plant (Shinjiro, 2008).

The intermediate, second, part of the pathway is catalyzed by microsomal NADPH-dependent cytochrome P-450 monooxygenases at the endoplasmic reticulum (Lange, 1998). Thus, the precursor ent-kaurene must be translocated from the proplastids to the endoplasmic reticulum, although nothing is known about the transport mechanism (Lange, 1998). Ent-kaurene is oxidized into GA₁₂, via ent-kaurenol, ent kaurenal, entkaurenoic acid, and ent-7α-hydroxy kaurenoic acid, and GA12-aldehyde. Certain biosynthesis steps, Such as 7-oxidation, 12α-hydroxylation, and 13-hydroxylation are catalyzed by both monooxygenases and soluble dioxygenases, which occasionally occur together within the same species or even within the same tissue (Lange & Graebe, 1993).

III- Gibberellins from GA12

Some initial steps of the final part of pathway might overlap with the second part. However, the third part starts with GA12-aldehyde is oxidized by soluble 2-oxoglutarate- dependent dioxygenases (Fig. 5). The GA dioxygenases are multifunctional with broad substrate specificity, resulting in many side reactions and numerous gibberellines found in higher plants (Lange & Graebe, 1993; Hedden & Kamiya, 1997).

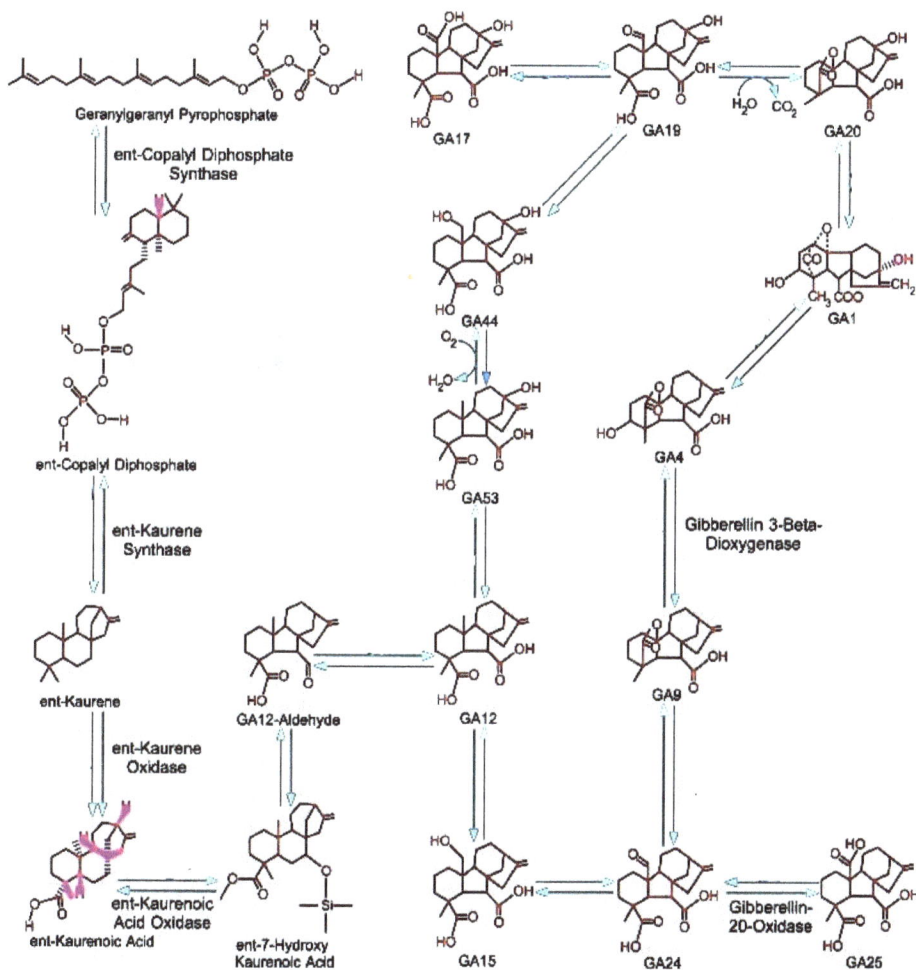

Fig. 5. Scheme of gibberellins biosynthesis (DXS) in the leaves of cannabis plants.

4. The interaction effect of terpenoids and GA₃ at vegetative stage

The influence of gibberellic acid (GA₃) on plastidic and cytosolic terpenoids and on two key enzymes, 1-deoxy-D-xylulose-5-phosphate synthase (DXS) and 3-hydroxy-3-methylglutaryl coenzyme A reductase (HMGR), for terpenoid biosynthesis was compared in vegetative cannabis plants. However, the regulation of the biosynthesis of terpenoids in plant cell is poorly understood. We do not know how the metabolite fluxes between the primary and secondary metabolisms are orchestrated. Plant growth regulators have essential roles in growth and development of plants and plant resposes to environment. In order to understand how interact with terpenoid biosynthesis; we focused on the role of GA3 in the regulation of primary and secondary terpenoid production in *Cannabis sativum* at vegetative stage.

Treatment with exogenous GA3 resulted in a decrease of DXS activity in comparison with the control plants (Fig. 6). Although, there is no report on the effect of GA3 on DXS activity, many reports support a regulatory role of DXS for the production of MEP-derived isoprenoids in plants (Rodriguez-concepcion, 2006). Estevez et al (2001) reported that transgene mediated up regulation or down regulation of DXS levels in Arabidopsis were correlated with concomitant changes in the levels of MEP derived isoprenoid end product.

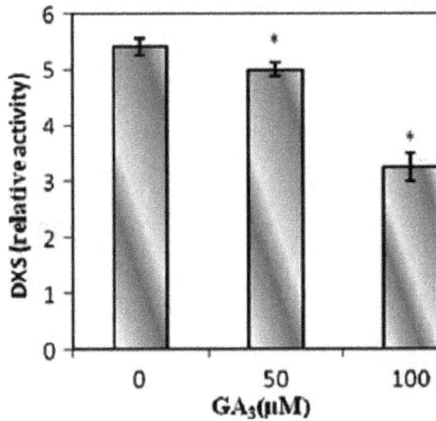

Fig. 6. Effects of gibberellic acid (GA3) on 1-deoxy-D-xylulose 5-phosphate synthase (DXS) in the leaves of cannabis plants.

The effects of the concentration of GA3 on the chlorophyll and carotenoid contents of cannabis leaf are present in Fig. 7 & 8. Reduction in chlorophyll content after GA3 application has been reported in wheat (Misra & biswal, 1980), rice seedlings (Yim et al., 1997) and pea (Bora & Sarma, 2006) which are consistence with our results (Mansouri et al., 2011).The variation in the level of the cartenoids were similar to those observed for the chlorophylls. Apparently carotenoids and chlorophylls accumulation are controlled through a similar mechanism, because both of them are reduced by GA3. The changes in chlorophyll and carotenoid were parallel with changes in DXS activity. It can show the limiting role of DXS activity in chlorophyll and carotenoid synthesis.

The tocopherol is lipophilic antioxidants that are synthesized by photosynthetic organisms, occurring mainly in leaves and seeds. Literature sources indicate that the major tocopherol form in leaf tissues is α− tocopherol (Szymanska and Kruk, 2008). Abdul Jaleel et al. (2007) reported that GA3 treatment stimulated α− tocopherol accumulation in *Catharanthus roseus*.The α− tocopherolcontent of cannabis plants decrease in 50 µM GA3 treatment and increased with increasing GA3 at vegetative stage (Fig. 9) which is consistence with other reports.

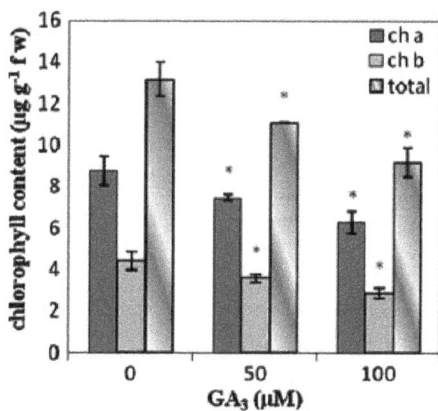

Fig. 7. Effects of gibberellic acid (GA3) on chlorophylls a, b and total in the leaves of cannabis plants.

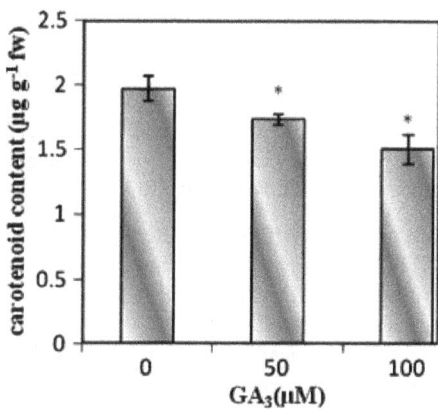

Fig. 8. Effects of gibberellic acid (GA3) on carotenoids in the leaves of cannabis plants.

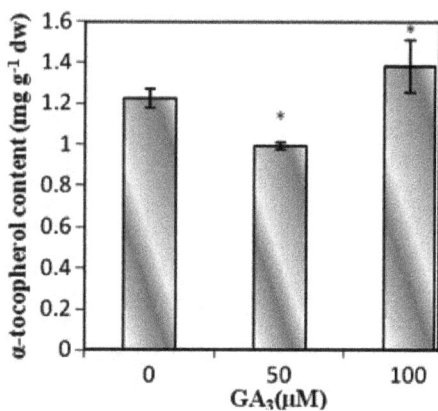

Fig. 9. Effects of gibberellic acid (GA3) on a-tocopherol content in the leaves of cannabis plants

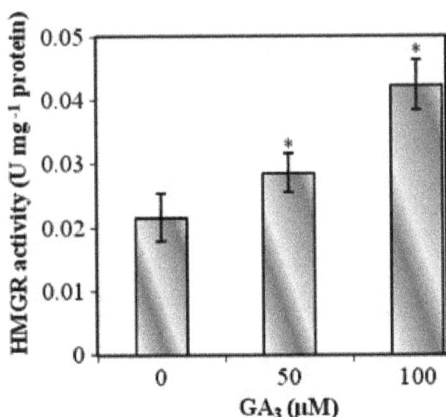

Fig. 10. Effects of gibberellic acid (GA3) on 3-hydroxy-3-methylglutaryl coenzyme A reductase (HMGR) activities in the leaves of cannabis plants.

Exogenous GA_3 caused an increasing in HMGR activity (Fig. 10). Since HMGR is an important control point for the MVA pathway in plants (Kato-Emori et al., 2001), the increase in HMGR activity should result in increasing the supply of phytosterols. Squalene and phytosterol contents showed similar changes and increased concomitant with HMGR activity in cannabis plants (Fig.11, 12). Consistent with our results, pea seedlings treated with GA_3 showed an increase in the HMGR activity (Russell & Davidson, 1982).

The amount of THC and CBD increase with increasing GA_3 treatment in comparison with the control plants (Fig. 13).The THC content was higher than those of CBD content (Mansouri et al., 2011). Perhaps, the increase observed in the THC and CBD content at high level of GA_3 is not a direct effect of GA_3 treatment and could reflect the GA_3 interaction with other plant hormones. As has been shown that exogenous application of GA3 caused a clear increase in ACC content. ACC oxidase activity and ethylene biosynthesis occur during the breaking of dormancy and onset of germination in *Fagus sylvatica* L. seeds (Calvo et al., 2004). Furthermore, it is possible that ethylene caused the increase observed in THC and CBD content.

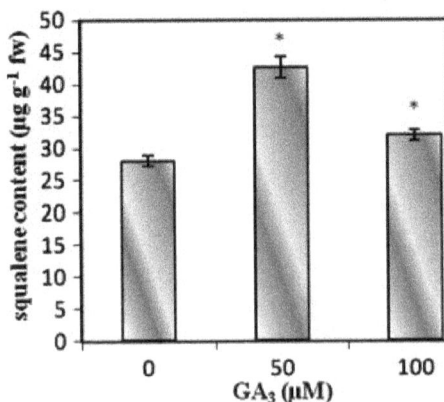

Fig. 11. Effects of gibberellic acid (GA3) on squalene content in the leaves of cannabis plants.

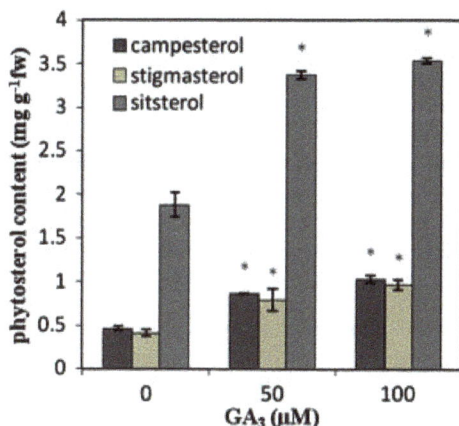

Fig. 12. Effects of gibberellic acid (GA3) on phytosterol content in the leaves of cannabis plants.

On the whole, these results showed that GA_3 treatment had opposite effect on primary terpenoid biosynthesis by MVA and MEP pathways. GA_3 treatment caused a decrease in DXS activity and biosynthesized primary terpenoids from MEP pathway, but this treatment increased HMGR activity and phytosterols from MVA pathway. Whereas, secondary terpenoids showed different response to GA_3 treatment and it could be because of interference of two biosynthetic pathways in their formation.

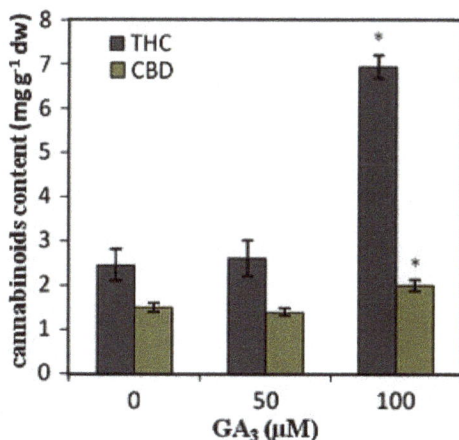

Fig. 13. Effects of gibberellic acid (GA3) on THC and CBD content in the leaves of cannabis plants.

5. The effect of GA3 and terpenoids at flowering stage

In spite of economic importance of the terpenoids and their many essential functions, relatively little is known about terpenoid metabolism and its regulation in plants. To understand the role of gibberellic acid (GA_3) in the regulation of two terpenoid biosynthesis pathways, we studied the response of the main end products of these pathways and cannabinoids under GA₃ treatment in cannabis plants.

Effects of different concentrations of GA₃ on levels of chlorophyll, carotenoids and α— tocopherol in leaves of male and female cannabis plants were investigated. Male plants treated with 50uM GA₃ had lower chlorophyll a and total contents compared to the control (Fig. 14A). Treatment of female leaves with 50 and 100uM GA₃ significantly decreased chlorophyll a, b and total contents in a dose-dependent pattern (Fig. 14B). The carotenoid contents of treated plants were lower compared with the control (Fig. 15A). Low concentrations of GA₃ were more effective on decreasing carotenoid contents in male plants.

The phytyl (C_{20}) conjugates chlorophylls and tocopheroles, and carotenoids (C_{40}) are produced by the MEP pathway. GA₃ treatment decreased chlorophyll contents in *C.sativa* plants. Reduction in chlorophyll content after GA₃ application has been reported in, wheat (Misra & Biswal 1980), peach trees (Monge et al. 1994), rice seedlings (Yim et al. 1997) and pea (Bora & Sarma 2006). Perez et al. (1974) have shown that a mutant of tomato with increased levels of GA₃ contains less chlorophyll. Our results showed that GA₃ caused a decrease in carotenoid contents (Fig. 14). Apparently carotenoid and chlorophyll accumulationis controlled through a similar mechanism, because both of them are reduced by GA₃. The previous studies indicate that GA₃ treatments delayed the chloroplast – chromoplast conversion of colored fruit (Goldshmidt, 1998; Pfander, 1992). The development of chromoplasts is accompanied by the accumulation of carotenoids (Vainstein et al., 1994). Also by Rodrigo and Zacarias (2007) reported that GA₃ reduced the ethylene induced expression of early carotenoid biosynthetic genes and the accumulation of phytoene in orange.

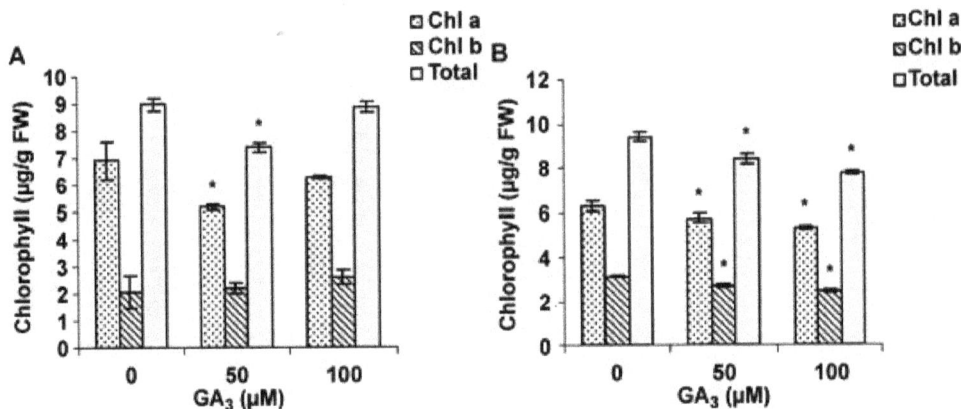

Values are means of four replications ± standard deviation (*SD*).

Fig. 14. Effects of gibberellic acid (GA3) on chlorophyll *a*, *b* and total in leaves of male (**A**) and female (**B**) cannabis plants.

Fig. 15B shows the effect of different concentrations of GA₃ on α— tocopherol content all concentrations of GA₃ sprayed on leaves caused an increase in α— tocopherol content in both sexes. The highest increase was detected with 50uM GA₃ in male plants. Apparently, the pattern of changes in α− tocopherol content was an invert of that in chlorophyll and carotenoid contents.

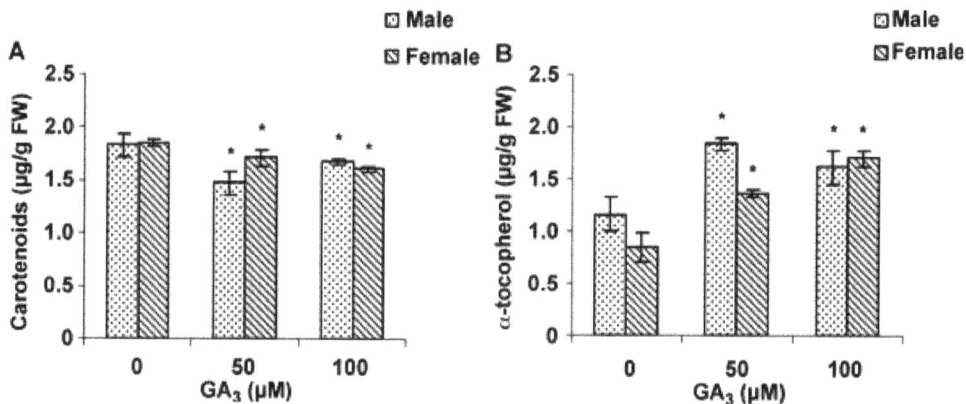

Values are means of four replications ± standard deviation (SD).

Fig. 15. Effects of gibberellic acid (GA3) on (A) carotenoids and (B) a-tocopherol in leaves of female and male cannabis plants.

Tocopherols (a, β, γ, and δ − tocopherol) are lipophilic antioxidants that collectively constitute vitamin E (Crowell et al, 2008). Therefore, we measured the changes in the amounts of α − tocopherol in response to GA$_3$ treatment. Our results showed that GA$_3$ increased α− tocopherol content in cannabis plants (Mansouri et al., 2009). A stimulatory effect of applied GA$_3$ on α− tocopherol content was also absorbed in Catharanthus roseus (Abdul Jaleel et al., 2007). On the other hand, we found a reversed relationship between photosynthetic pigments (chlorophyll and carotenoids) and α− tocopherol contents in treated plants with GA$_3$. Both α− tocopherol and chlorophyll contain a phytol moiety as part of their molecule. The substrate used for their biosynthesis may be derived from a common pool. It is possible that the decrease in chlorophyll and carotenoid biosynthesis caused an increase in the substrate accumulation and therefore biosynthesis. Also Rise et al. (1989) reported a transient increase in α− tocopherol accompanied with a decrease in chlorophyll during the course of senescence in several plant species. Their results indicated that the phytol released by chlorophylase during the initial stages of chlorophyll breakdown, perhaps used for the biosynthesis of α− tocopherol during senescence.

The effects of GA$_3$ on key enzyme activity of terpenoid biosynthetic pathways (DXS and HMGR) were investigated. As shown in Fig. 16 male plants had more DXS activity than female plants and a significant decrease was observed over DXS activity in treated plants (male and female) by GA$_3$ which was liner with increasing GA$_3$ concentration. GA$_3$ treatment had a stronger effect on decreasing DXS activity in male plants.

GA$_3$ treatment caused a decrease in the 1-deoxy-D-xylulose5-phosphate synthase (DXS) activity in male and female of Cannabis sativa plants (Mansouri et al., 2009).This is the first report about the effect of GA$_3$ on DXS activity. Many reports support a regulatory role of DXS for the production of MEP- derived isoprenoids in plants (Rodriguez-Concepcion 2006). Estevez et al. (2001) reported that transgene-mediated up regulation or down regulation of DXS levels in Arabidopsis were correlated with concomitant changes in the levels of MEP-derived isoprenoid end products. However, in our investigation, the decrease in the entire MEP pathway end product. The results showed an increase in $\alpha-$ tocopherol content and a decrease in chlorophyll and carotenids. These data support this hypothesis that several enzymes share control over the metabolic flux through the MEP pathway. The MEP pathway might be regulated at several control points to compensate for fluctuations in precursor/product equilibrium and redistribute the balance of control within the pathway (Enfissi et al., 2005). Another possibility for this uncoordinated activity between the DXS and the plastidic isoprenoids is the exchange or precursors between the cytosol and the plastid. Strong biochemical evidence for such exchange of precursors was reported by Kasahara et al., (2002), Nagata et al. (2002) and Hemmerlin et al. (2003).

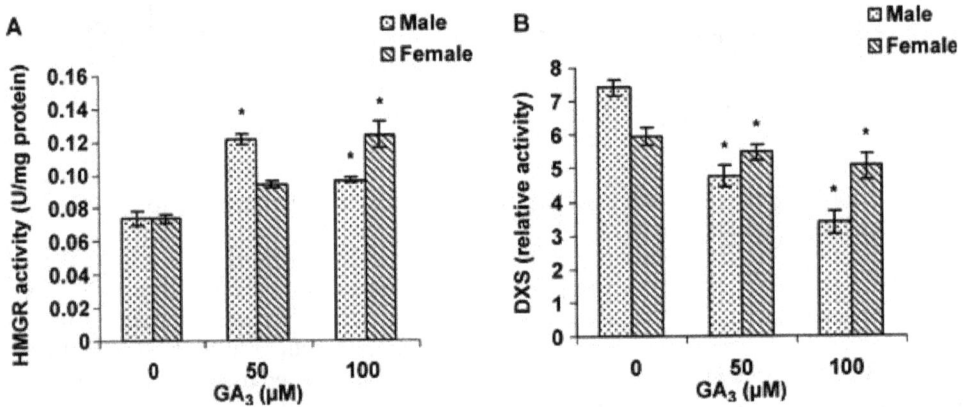

(HMGR) activities in the leaves of female and male cannabis plants.
Values are means of four replications ± standard deviation (SD).

Fig. 16. Effects of gibberellic acid (GA3) on 1-deoxy-D-xylulose 5-phosphate synthase (DXS) and 3-hydroxy-3-methylglutaryl coenzyme A reductase.

Gibberellic acid treatment increased HMGR activity in the treated plants (Fig. 16). However some differences are obvious between the two groups of plants of different sexes (Fig.16). The male individuals had a greater HMGR activity at lower GA3 concentrations than female plants.

Activity of 3-hydroxy-3-methyglutaryl coenzyme A reductase (HMGR) in cannabis plants demonstrated that HMGR activity was greater in plants treated with GA3. Consistent with our result, Russell and Davidson (1982) reported that GA3 increased HMGR activity in pea seedlings. It was shown that higher levels of HMGR activity were usually associated with rapidly growing parts of plants (Brooker and Russell 1975). On the othe words, Ga is widely regarded as a growth-promoting compound. Thus, it can be a candidate for stimulating of the HMGR activity. The lower HMGR activity was seen in the male plants treated with 100 μM GA3. The decrease observed in the HMGR activity could reflect the GA3 interaction in high concentration with other plant hormones. As it has been shown that exogenous application of GA3 caused a clear increase in ACC content, ACC oxidase activity and ethylene biosynthesis occur during the breaking of dormancy and onset of germination in *Fagus sylvatica* seeds (Calvo et al. 2004). In addition, it is possible that ethylene caused the decrease observed in HMGR activity.

The mature leaves of male and female cannabis plants were used to determine the effect of GA3 on squalene (biosynthetic precursor to all steroids), campestrol, stigmasterol, and sitosterol (the most representative phytosteroids of the MVA pathway). A significant increase in squalene (Fig. 17), stigmasterol and sitosterol accumulation occurred in female and male plants treated with gibberellic acid (Fig. 18). The changes in the amount of squalene and sitosterol were coordinate with the changes in HMGR activity. In addition, GA3 had a stimulatory effect on campestrol accumulation in male plants.

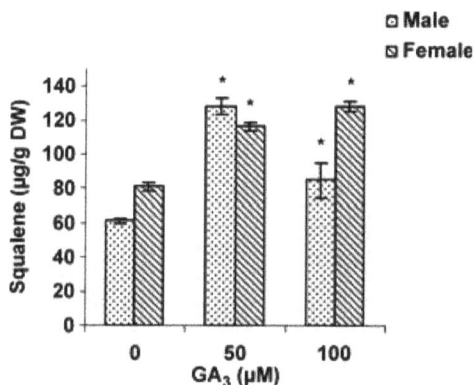

(HMGR) activities in the leaves of female and male cannabis plants.
Values are means of four replications ± standard deviation (SD).

Fig. 17. Effects of gibberellic acid (GA3) on squalene content in leaves of female and male cannabis plants.

The data in our study showed that GA_3 treatment increased squalene and phytostrol contents in a pattern similar to the changes in the HMGR acitivity. Since HMGR is an important control point for the MVA pathway in plants (Kato-Emori et al. 2001), the increase in HMGR activity should result in increasing the supply of phytostrols. Free sterols are found predominantly in cell membranes and are thought to contribute to the proper functioning of membranes by controlling the fluidity characteristics of the membrane (Devarenne et al. 2002). Douglas and paleg (1974) demonstrated that the inhibitors of GAs biosynthesis caused a decrease in sterol accumulation. Huttly and Philips (1995) suggested that GA_3 causes an increase in cell number and size to produce a significant effect. Since these processes need the production of cell membrane. It can be assumed that GA_3 should induce the phytostrol biosynthesis to influence its effects on growth in plants.

A comparison between male and female plants showed that females had higher amounts of THC; especially in the flowers (Fig. 19). THC content of the leaves was slightly lower than that of flowers. The application of GA_3 to cannabis plants resulted in a decrease in THC content. Gibberellic acid treatment had a stronger effect in decreasing THC content in the flowers of male plants in comparison with that of female plants. However, the leaves of the two sexes indicated similar responses to GA_3 treatment.

Δ^9-tetrahydrocannabinol (THC) is the cannabinoid responsible for the main psychoactive effects of most Cannabis drug preparations (Mecoulam 1970). Factors that control biosynthesis and distribution of cannabinoids within the plant are unknown. We investigated the impact of altered GA3 levels on this secondary metabolite in *C.sativa*, and the results indicated that apart from the influence of GA3, female plants had more THC than male plants in leaves and flowers. The treated plants with GA3 had lower THC content in comparison with that in control plants. It is demonstrated that cannabinoids are synthesized from the DXP pathway (Fellemeier et al. 2001). Furthermore, our results showed that GA3 decreased THC content by decreasing the DXS activity and necessary precursors for THC biosynthesis.

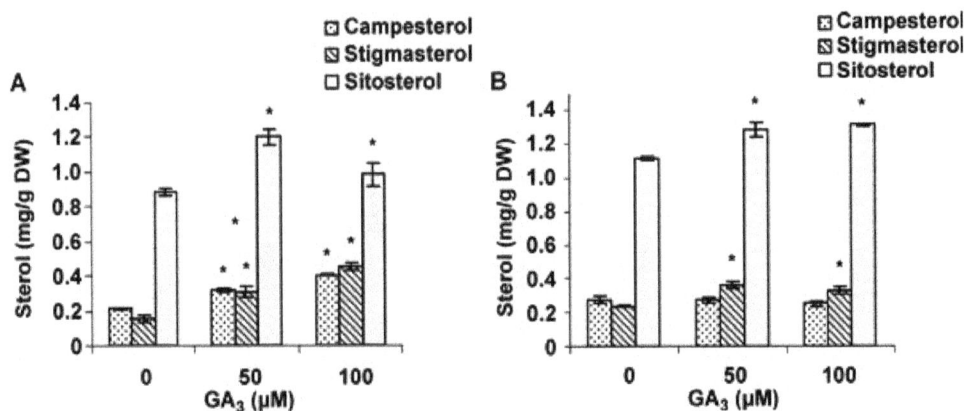

(HMGR) activities in the leaves of female and male cannabis plants.
Values are means of four replications ± standard deviation (SD).

Fig. 18. Effects of gibberellic acid (GA3) on campesterol, stigmasterol and sitosterol content in leaves of (**A**) male and (**B**) female cannabis plants.

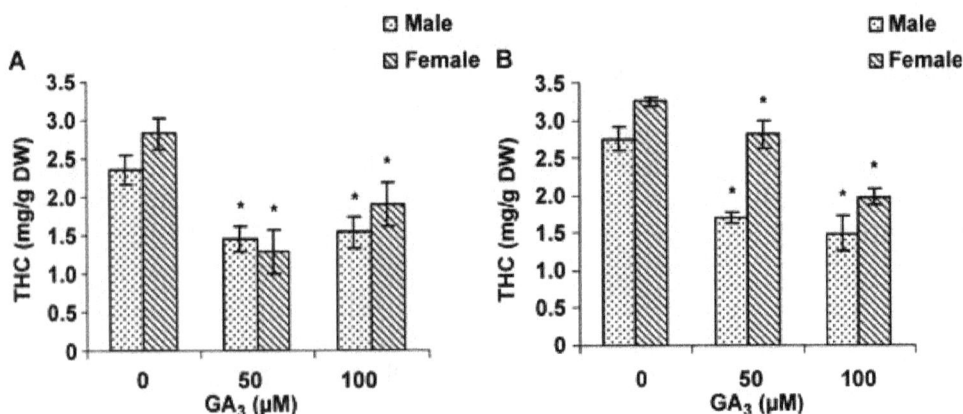

(HMGR) activities in the leaves of female and male cannabis plants.
Values are means of four replications ± standard deviation (SD).

Fig. 19. Effect of gibberellic acid (GA3) on _9-tetrahydrocannabinol (THC) content in (**A**) leaves and (**B**) flowers of the male and female cannabis plants.

6. Conclusion

In conclusion, the pattern of changes in the amounts of primary terpenoids (chlorophyll, carotenoids, and phytosterols) in *Cannabis sativa* suggest that GA3 have opposite effects on the primary terpenoid biosynthesis of MEP and MVA pathways. In addition, the appearance of the direct relationship between DXS and HMGR activity, and their main end products confirm an important role for these enzymes in the MEP and MVA pathways regulation. However, to understand the role of GA3 in terpenoid biosynthesis regulation, we need further investigation.

7. Acknowledgement

The autor wishes to thank Mr. Hossein Mozafari phD student of Shahid Bahonar university of Kerman for his critical comments and help of the manuscript.

8. References

Aach, H.; Bode, H.; Robinson, DG. & Graebe, JE. (1997). *ent*-Kaurene synthase is located in proplastids of meristematic shoot tissues, *Planta*, Vol.202, pp.211-219.

Aach, H.; Bose, G. & Graebe, JE. (1995).*ent*-Kaurene biosynthesis in a cell-free system from wheat (*Triticum aestivum* l.) seedlings and the localization of *ent*-Kaurene synthase in plastids of three species, *Planta*, Vol.197, pp. 333-342

Abdul Jaleel, c.; Gopi, R.; Manivannan, P.; Sankar, B.; Kishorekumar, A. & Panneerselvam, R. (2007). Antioxidant potentials and ajmalicine accumulation in *Catharanthus roseus* after treatment with gibberellic acid. *Colloids Surf B Biointerf.* Vol. 60, pp.195-200.

Agurell, S. & Nilsson, LG. (1972). The chemistry and biological activity of cannabis. *Bullet. Narcotic.* pp 35-37.

Arigoni, D.; Sagner, S.; Latzel, C.; Eisenrich, W.; Bacher , A. & Zenk, MH. (1997). Terpenoidbiosynthesis from 1-deoxy-D-xylulose in higher plants by intermolecular skeletal rearrangement, *Proc Natl Acad Sci*, Vol. 94, pp.10600-10605.

Bora, RK. & Sarma, CM. (2006). Effect of gibberellic acid and Cycoel on growth , yield and protein content of pea. *Asi J plant Sci.* Vol.5, pp.324-330.

Brooker, JD. Russell, DW. (1975). Properties of microsomal 3-hydroxy-3-methylglutaryl coenzyme A reductase from Pisum sativum seedings, *Arch. Biochem. Biophys.* Vol. 167, pp723-729

Calvo, Ap.; Nicolas, C.; Lorenzo, O.; Nicolas, G. & Rodriguez, D. (2004). Evidence for positive regulation by gibberellins and ethylene of ACC oxidase expression and activity during transition from dormancy to germination in Fagus sylvatica L. seeds,*J. Plant Growth Regul.* Vol. 23, pp.44-53.

Crowell, EF., McGrath, JM.& Douches, DS,. (2008). Accomulation of vitamin E in potato (Solanum toberosum) tubers, *Transgenic Res.* Vol.17, pp.205-217.

Crowell, EF.; Mcgrath, Jm.& Douches, DS. (2008). Accumulation of vitE in potato (Salanum tuberosum) tubers, Transgenic Res. Vol. 17 pp.205-217.

Crozier, A.; Kamiya, Y.; Bishop, G. & Yokota, T. (2000). Biosynthesis of hormones and elicitor molecules. In: Buchanan BB, Gruissem W, Jones RL (eds) Biochemistry and molecular biology of plants, *Amer Soc Plant Physio*, Roclville, pp.850-929.

Devarenne, TP.; Ghosh, A. & Chappell, J. (2002). Regulation of squalene synthase, a key enzyme of sterol biosynthesis, in tobacco. Plant Physiol. Vol 129, pp 181-222.

Dewick, PM. (2002). The biosynthesis of C5-C25 terpenoid compounds, *Nat Prod Rep*, Vol.19, pp.181-222.

Dogbo O. & Camara, B. (1987). Purification of isopentenyl pyrophosphate isomerase and geranylgeranyl pyrophosphate synthase from *Capsicum* chrooplasts by affinity chromatography. *Biochim Biophys Acta*. Vol. 920, pp.140-148.

Douglas, TG.; Paleg, LG. (1974). *Plant Physiol*. Vol 54, pp 238-245.

Eisenreich, W.; Rojdich, F. & Bacher, A. (2001). Deoxyxylulose phosphate pathway to terpenoids, Trends Plant Sci., Vol.6, pp.78-84.

Enfissi, MA.; Fraser, PD.; Lois, LM.; Boronat, A.; Schuch, W. & Bramley, PM. (2005). Metabolic engineering of the mevalonate and non-mevalonate isopentenyl diphosphate-forming pathways for the production of health-promoting isoprenoids in tomato, *Plant Biotech J*, Vol.3, pp.17-27.

Estevez, JM.; Cantero, A.; Reindl, A.; Reichler, S. & Leon, P. (2001) 1-deoxy-D-Xylulose 5-phosphate synthase, limiting enzyme for plastidic isoprenoid biosynthesis in plants, *J Bio Chem* Vol. 276, pp. 22901-22909.

Fellermeier, M.; Eisenreich,W.; Bacher, A. & Zenk, MH. (2001). Biosynthesis of cannabinoids: incorporation experiment with 13C-labeled glucoses, *Eur . J. Biochem*. Vol. 268, pp1596-1604.

Fellermeier, M.; Eisereich, W.; Bacher, A. & Zenk, MH. (2001). Biosynthsis of cannabinoids: incorporation experiments with 13C- labeled glucose. *Eur J Biochem* Vol.268, pp.1596-1604.

Goldschmidt, EE. (1988). Regulatory aspects of chloro-chromoplast interconversion in seeding citrus fruit peel, Isr. *J. Bot*. Vol. 37, pp.23-130.

Gusman, M. (2003). Cannabinoids: potential anticancer agents, *Nat. Rev*. Vol. 3, pp. 745-755.

Hedden, P. & Kamiya, Y. (1997). Gibberellin biosynthesis enzymes, genes and their regulation, AnnRev Plant Physiol Plant Mol Biol, Vol.48, pp.431-460.

Heldt, HW., (2005). A large diversity of isoprenoids has multiple functions in plant metabolism, *Plant Biochemistry*. 3rd Edition ISBN: 0-12-088391-0

Hemmerlin, A.; Hoeffler, JF.; Meyer, O.; Tritsch, D.; Kagan, IA. & Grosdemanage-Billiard C. (2003). Cross talk between the cytosolic mevalonate and the plastidial methlerythritol phosphate pathways in tobacco Bright Yellow-2 cells, *J. BIol. Chem*. Vol.278, pp.26666-26676

Hilling, KW. (2004). A chemotaxonomic analysis of terpenoid variation in cannabis, *Biochem. Systems Ecol*. Vol. 32, pp. 875-891

Hischberg, J. (1999). Production of high value compounds: carotenoids and vitamin E, *Curr. Opin. Biotechnol*, Vol. 20, pp.186-191.

Howlett, AC.; Breivogel, CS.; Cholders, SR.; Deadwyler, SA. Hampson, Re. & Porrino LY (2004). d physiology and pharmacology: 30 year progress. *Neuropharmacol*. Vol. 47, pp. 345-358.

Huttly, AK. & Phillips, AL. (1995). Gibberellin regulated plant genes. *Physol. Plant*. Vol. 95, pp. 310-317.

Kanno, Y.; Otomo, K.; Kenmoku, H.; Mitsuhashi, W.; Yamane, H.; Oikawa, H.; Toshima, H.; Matsuoka, M.; Sassa, T. & Toyomasu, T. (2006). Characterization of a rice gene family encoding type A diterpene cyclases, *Biosci Biotechnol Biochem*, Vol. 70, pp.1702-1710.

Kasahara, H, Hanada A, Kuzuyama T, Takagi M, Kamiya Y, Yamaguchi S. (2002).Contribution of the mevalonate and methylerythritol phosphate pathways to the biosynthesis of gibberellin in Arabidopsis, *J. Biol. Chem.* Vol. 277, pp. 45188-45194.

Kato-Emori, S, Higashi K, Hosoga K, Kobayashi T, Ezura H (2001). Cloning and characterizing of the gene encoding 3-hydroxy-3-methylglutaryl coenzyme A reductase in melon (Cucumic melo L. reticulatus), *Mol. Genet. Genomics* Vol. 265, pp. 135-142.

Lang, Bm. & Ghassemian, M. (2003). Genome organization in *Arabidopsis thaliana* : a survey for gene involved in isoprenoid and chlorophyll metabolism. *Plant Mol Biol,* vol. 51 pp. 925-948.

Lang, H. (1998). Molecular biology of gibberelline synthesis, *Plant,* Vol.204 pp. 409-419.

Lang, T. & Graebe, JE. (1993). Enzymes of gibberellins synthesis. In: Lea PJ (ed) Method in plant biochemistry, Vol.9, *Academic Press,* London, pp.403-430.

Lange, BM. & Croteau, R. (1999). Genetic engineering of essential oil production in mint, *Curr Opin Pant Biol,* Vol.2, pp.139-144.

Laule, O.; Futholz, A.; Chang, HS.; Zhu, T.; Wang, X. & Heifetz, PB. (2003). Cross talk between cytosolic and plastidial pathways of isoprenoid biosynthesis inArabidopsis thaliana, Plant Biol, Vol.100, pp.6866-6871.

Lichtenthaler, HK. (1999). The 1-deoxy-D-xylulose-5-phosphate pathway of isoprenoid biosynthesis in plants. *Annual Review of plant physiology and plant molecular biology,* Vol. 50, pp.47-65.

Lichtenthaler, HK.; Rohmer, M. & Schwender, J. (1997) Two independent biochemical pathways for isopentenyl diphosphate and isoprenoid biosynthesis in higher plants. *Pysiol plant,* vol. 101 pp. 643-652.

MaGarvey, DJ. & Croteau, R. (1995). Terpenoid metabolism. *The plant cells* Vol. 7, pp. 1015-1026.

Mansouri, H.; Asrar, Z. & Amarowicz, R. (2011). The respose of terpenoids to exogenous gibberellic acid in Cannabis sativa at vegetative stage, *Acta physiol plant,* Vol.33, pp.1085-1091.

Mansouri, H.; Asrar, Z. & Mehrabani, M. (2009) Effect of gibberellic acid on primary terpenoids and Tetrahydrocannabinol in *Cannabis sativa* at flowerin stage. *Journal of integrative plant Biology* Vol. 51, pp. 553-561.

McCaskill, D. & Croteau, R. (1998). Some caveats for bioengineering terpenoid metabolism in plants, *TIBITECH,* Vol.16, pp.349-355.

Mechoulam, R. (1970). Marijuana chemistry. *Science,* Vol. 168, pp.1159-1166.

Misra, AN. Biswal, UC. (1980). Effect of photohormons on chlorophyll degradation during again of chloroplasts in vivo and in vitro, *Protoplasma* Vol. 105, pp. 1-8.

Monge, E.; Aguirre. R. & Blanco, A. (1994). Application of paclobutrazol and GA3 to adult peach trees: effects on nutritional status and photosynthetic pigments, J. *Plant Growth Regul.* Vol. 13, pp. 15-19.

Negata, N.; Suzuki, M.; Yoshida, S. & Muranaka, T. (2002). Mevalonic acid partially restres chloroplast and etioplast development in Arabidopsis lacing the non-mevalonate pathway, Planta, Vol.216, pp.345-350

Olszewski, N.; Sun, T. & Gubler, T. (2002). Gibberellin signaling: Biosynthesis, catabolism, and response pathways, *Plan Cell*, Vol.14, pp.61-80.

Perez, AT.; Marsh, HV. & Lachman, WH. (1974). Physiology of the the yellow –green 6 gene in tomato, *Plant Physiol*, Vol.53, pp.192-197.

Pfander, H. (1992). Carotenoids: an overview, *Meth Enzymol*, Vol.213, pp.3-13.

Razem, FA.; Baron, K. & Hill, RD. (2006). Turning on gibberellins and abscisic acid signaling, *Curr Opin Plant Bio*, Vol.9, pp.454-459.

Rise, M.; Conjocaru, M.; Gottlieb, HE. & Goldschmidt, EE. (1989). Accumulation of tocopherol in sensscing organs as related to chlorophyll degradation, *Plant Physiol*, Vol.89, pp.1028-1030.

Rodrigo, MJ. & Zacarias, L. (2007). Effect of postharvested ethylene treatment on carotenoid accumulation and the expression of carotenoid biosynthetic genes in the flavedo of orange (*Citrus sinensis* L.), *Postharvest Biol Technol*, Vol.43, pp.14-22.

Rodriguez-concecion, M. (2006). Early steps in isoprenoid biosynthesis: Multilevel regulation of the supply of common precursors in plant cells, Phytochem Rev, Vol.5, pp.1-15.

Rodriguez-concepcion, M. & Boronat, A. (2002). Elucidation of the methylerythritol phosphate pathwayfor isoprenoids biosynthesis in bacteria and plastids, A metabolic milestone achieved through genomics, *Plant Physiol*, Vol. 130, pp.1079-1089.

Russell, DW. & Davidson, H. (1982). Regulationof cytosolic HMG-CoA reductase activity in pea seedlings: contrasting responses to different hormones, and hormone product interaction, suggest hormonal modulation of activity, *Biochem. Biophys. Res. Commun.* Vol. 104pp. 1537-1543.

Shinjiro, Y. (2008). Gibberellin Metaboilism and its regulation, *Anne. Rev. plant Biol.* Vol. 59, pp. 225-251.

Shintani, D. & Della Penna, D. (1998). Elevation the Vit E content of plants through metabolic engineering, Science, Vol.282, pp.2098-2100.

Szymanska, R. & Kruk, J. (2008). α-tocopherol content and isomers composition in selected plant species, Plant Physiol Biochem, Vol.46, pp.29-33.

Yabuta, T. (1935). Biochemistry of the "Bakabae'fungus of rice". *Agr. Hort.* Vol. 10, pp. 17-22.

Yim, KD.; Kwon, YW. & Bayer, DE. (1997). Growth response and allocation of assimilates of rice seedlings by paclobutrazol and gibberellins treatment, J Plant Growth Regul, Vol.16, pp.35-41.

17

Lipotubuloids – Structure and Function

Maria Kwiatkowska, Katarzyna Popłońska, Dariusz Stępiński,
Agnieszka Wojtczak, Justyna Teresa Polit and Katarzyna Paszak
University of Łódź, Department of Cytophysiology
Poland

1. Introduction

In the 19th and 20th centuries these structures were called „elaioplasts" according to Wakker (1888) who introduced this term with regard to nucleus-size lipid bodies present in *Vanilla planifolia* leaf epidermis which strongly refracted light. Since they are clearly visible under light microscope (Fig. 1A,B) they became the focus of interest in those days and were described in about 120 mono- and dicotyledonous plant species. Also in our laboratory "elaioplasts" were for the first time observed in 12 *Gentiana* species (Kwiatkowska, 1959, 1961) and in *Dahlia variabilis* (Kwiatkowska, 1963).

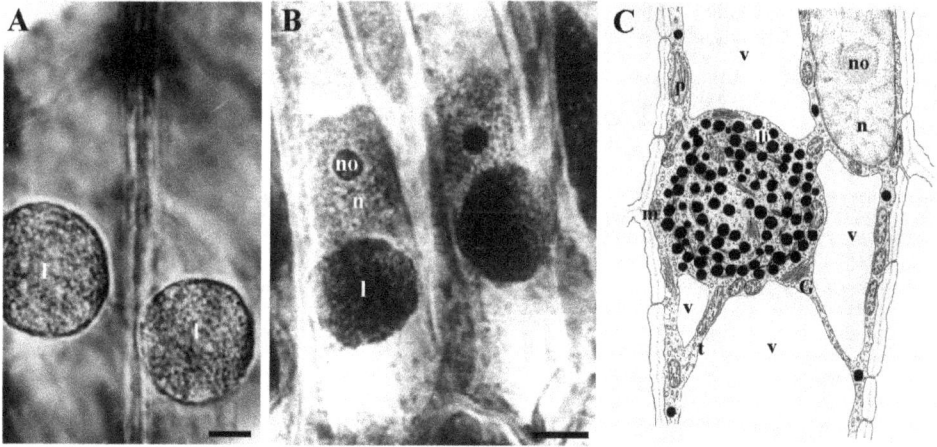

Fig. 1. *Ornithogalum umbellatum* "elaioplast" (lipotubuloid); A – in a living cell; B - after OsO₄ fixation; C – a scheme of epidermis cell with lipotubuloid; G – Golgi apparatus, l – lipotubuloid, lb – lipid bodies, m – mitochondrion, n – nucleus, no – nucleolus, p – plastid, t – tonoplast, v - vacuole; bars: 10 μm.

In 1883 Schimper introduced the term plastids which has been widely accepted since then. Among plastids, which are cell organelles containing double phospholipid bilayer, there are those producing lipids, i.e. elaioplasts which this term actually means. However, they are totally different from Wakker's "elaioplasts", the latter not being plastids in the contemporary meaning of this term. This was unequivocally proved by EM observations of "elaioplasts" in *Ornithogalum umbellatum* ovary and stipule epidermis described by Raciborski (1895) and Kwiatkowska (1966). Their ultrastructure indicates that they are specific cytoplasm domains (Fig. 2) (Kwiatkowska, 1971a, 1971b, 1972a). They do not have their own membrane but invaginating into vacuoles are surrounded by a tonoplast (Fig. 1C).

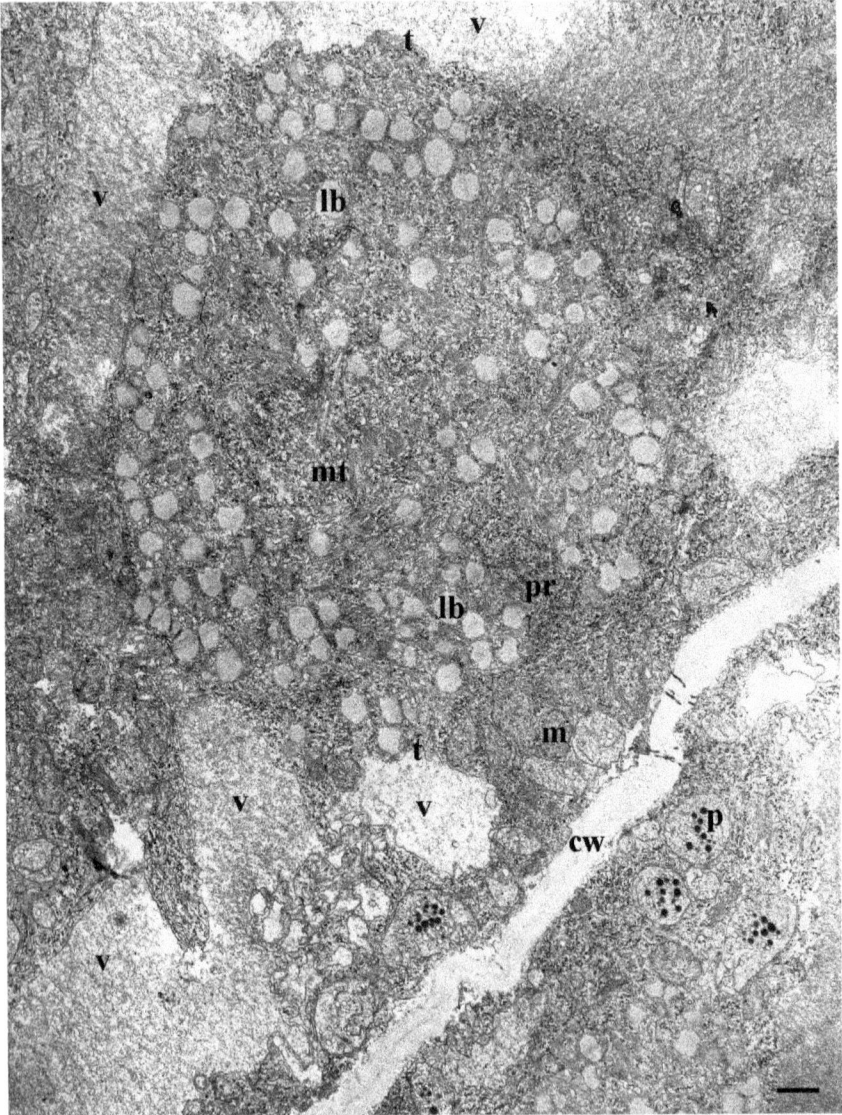

Fig. 2. *O. umbellatum* lipotubuloid. Electronogram from the ovary epidermis fixed in freshly prepared mixture of OsO$_4$ and glutaraldehyde; cw – cell wall, lb – lipid body, m – mitochondrion, mt – microtubules, p – plastid, pr – polyribosomes, t – tonoplast; v – vacuole; bar: 0.3 μm.

Elaioplasts are mainly filled with aggregates of osmiophilic granules identified as lipid bodies (Kwiatkowska, 1973a) and surrounded by a phospholipid half unit membrane characteristic only of these structures (Yatsu & Jacks 1972). Differences in the structure and thickness of lipid mono- and bilayer were revealed only after fixation of *O. umbellatum* epidermis in freshly prepared mixture of OsO_4 and glutaraldehyde (Kwiatkowska, 1973a).

Lipid bodies of plants are covered with proteins, oleosins, which prevent their merging during oil seed germination (Huang, 1996; Hsieh & Huang, 2004). It was shown that the amount of oleosins in a cell determined the size of lipid bodies (Siloto et al., 2006).

O. umbellatum lipid bodies are entwined by microtubules adhering to their surfaces and running in different directions forming a basket-like structure around them (Fig. 3).

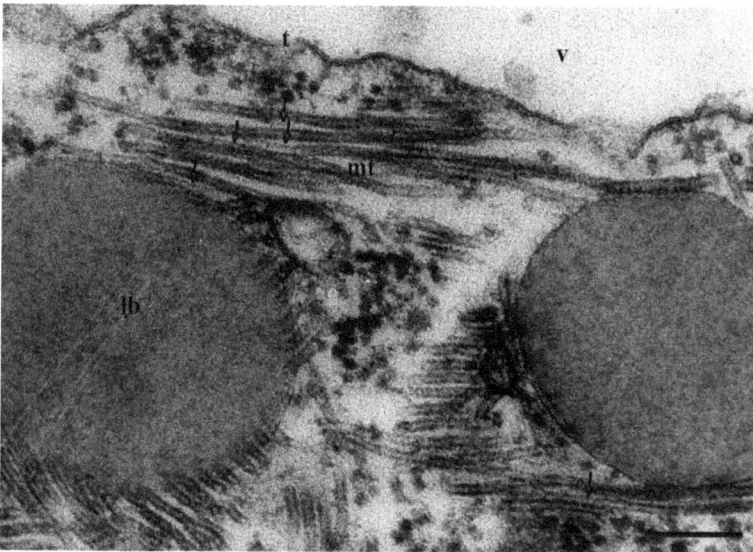

Fig. 3. *O. umbellatum* lipid bodies surrounded by a half unit membrane, entwined by microtubules; arrows indicate actin filaments; lb – lipid body, mt – microtubules, t – tonoplast, v – vacuole; bar: 0.1 μm (Kwiatkowska et al., 2005).

Beside lipid bodies and microtubules in this domain there are numerous ribosomes, ER cisternae and vesicles as well as few mitochondria, microbodies and Golgi structures and also autolytic vacuoles later during development.

In order to emphasis that the structures observed in *O. umbellatum* are not plastids a term lipotubuloids has been coined, it reflects the fact that they are rich in lipids and microtubules (Kwiatkowska, 1971a, 1971b, 1972a). Lipotubuloids were first reported in *O. umbellatum*, however recently they have also been identified in *Haemanthus albiflos* (Fig. 4) (Kwiatkowska et al., 2010), where they were previously described as "elaioplasts" by Politis (1911) and Tourte (1964, 1966). These structures are also present in *Vanilla planifolia, Funkia Sieboldiana* and *Athaea rosea* (Kwiatkowska et al., 2011a) in which they were previously called "elaioplasts" by Wakker (1888), Zimmermann (1893), Wałek-Czernecka & Kwiatkowska (1961), respectively.

All the above mentioned lipotubuloids contain the same organelle (Fig. 4) which implies that their presence there is not accidental but results from a functional relationship between them and lipid bodies. The main difference consists in stability of microtubules, those in *O. umbellatum* are markedly more stable, probably due to the polysaccharide layer covering them, than the others (Kwiatkowska, 1973b).

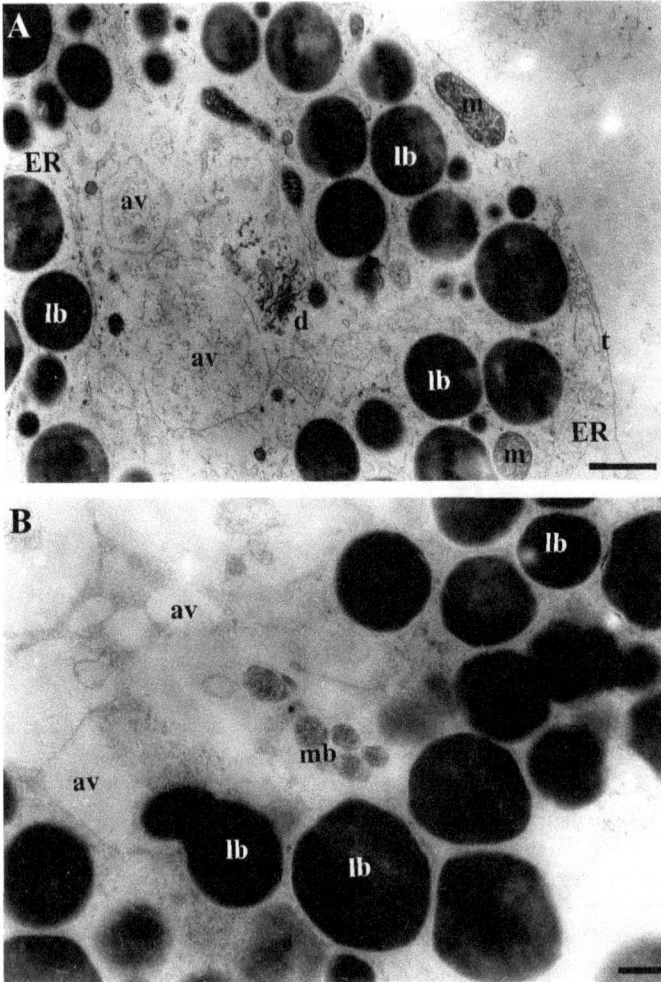

Fig. 4. Fragments of *Haemanthus albiflos* lipotubuloid with various organelles (A) and moreover with microbodies (B); av – autolytic vacuole, d – dictyosome, ER – endoplasmic reticulum, lb – lipid bodies, m – mitochondria, mb – microbodies, t – tonoplast; bars: 1 μm (A), 0.6 μm (B) (Kwiatkowska et al., 2010).

In the case of *H. albiflos* lipotubuloids become visible only after preincubation in 15 μm taxol with microtubule stabilizing buffer (100 mM PIPES, 1 mM MgCl₂, 5 mM EGTA) pH 7.2 for 2 h prior to fixation in the freshly prepared mixture of 2.5% glutaraldehyde and 1%

OsO_4, followed by postfixation in 1% OsO_4 (Fig. 5). The presence of tubulin around lipid bodies was confirmed by the immunogold method with the α anti-tubulin antibody (Kwiatkowska et al., 2010).

In the case of *V. planifolia, F. Sieboldiana and A. rosea* lipotubuloids were fixed according to the same procedure but without taxol preincubation. α anti-tubulin antibody application also gave positive results: gold grains were localized around lipid bodies of lipotubuloids in these plants.

Contrary to *O. umbellatum* lipotubuloids, microtubules in the other species were mostly observed in cross-sections near lipid bodies, no microtubules between lipid bodies, characteristic of *O. umbellatum*, were noted. This might be the result of poorer stability of these microtubules as varying stability of microtubules even in the same cell (Kwiatkowska et al., 2011c) is a well known phenomenon resulting, among others, from post-translational tubulin modifications (PTM) such as acetylation, tyrosination, polyglutamylation, polyglycylation, phosphorylation, palmitoylation (Verhey & Gaertig, 2007; Hammond et al., 2008; Fukushima et al., 2009). Nevertheless, we think that it is fully justified to call the structures present in *H. albiflos, V. planifolia, F. Sieboldiana A. rosea* lipotubuloids.

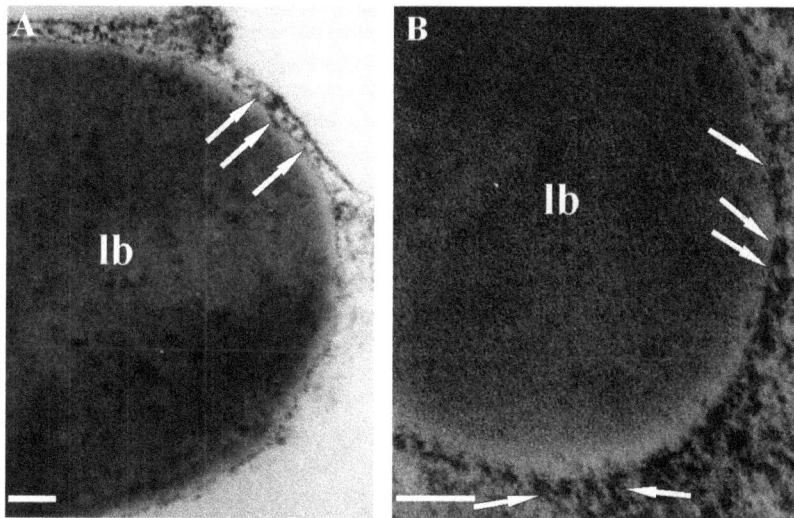

Fig. 5. *H. albiflos* lipid bodies entwined by microtubules (A,B) in cross and longitudinal sections (white arrows); lb – lipid bodies; bars: 0.1 μm (Kwiatkowska et al., 2010).

It is highly probable that among "elaioplasts" which are not plastids described in other species further lipotubuloids will be identified. Due to that old papers concerning "elaioplasts" are listed in literature and a table presenting families in which "elaioplasts" were described is included (Table. 1)

The main subject of our work have been lipid bodies and microtubules in *O. umbellatum* lipotubuloids which are most stable during fixation for EM. We returned to them after 30 years (during which we concentrated on *Chara* spermiogenesis) due to the development of new methods and new ideas concerning lipid bodies.

Family	References
Monocotyledoneae	
Dioscoreaceae	Politis, 1911
Iridaceae	Politis, 1911; Faull, 1935
Liliaceae	Zimmermann, 1893; Raciborski, 1895; Politis, 1911; Guillermond, 1922; Wóycicki, 1929; Weber, 1955; Thaler, 1956; Tourte, 1964, 1966; Kwiatkowska, 1966
Orchidaceae	Wakker, 1888; Zimmermann, 1893; Raciborski, 1895; Politis, 1911; Faull, 1935
Dicotyledoneae	
Anacardiaceae	Guttenberg, 1907 in Linsbauer, 1930
Campanulaceae	Górska-Brylass, 1962
Compositae	Beer, 1909
Cucurbitaceae	Riss, 1918
Gentianaceae	Kwiatkowska, 1959, 1961
Malvaceae	Politis, 1911; Luxenburgowa, 1928; Wałek-Czernecka & Kwiatkowska, 1961
Orobanchaceae	Bidermann, 1920
Pirolaceae	Bidermann, 1920

Table 1. Examples of plant family of mono- and dicotyledonous class in which "elaioplasts" have been identified.

2. Lipid bodies – Biogenesis and lipid synthesis

Lipid bodies (lipid droplets, oil bodies, spherosomes) are organelles commonly present in plants, animals and some Prokaryote. For many years they were treated only as the source of energy, however during the last decade they were shown to play a role in lipid homeostasis and to contain numerous proteins involved in signal transductions as well as proteins transiently stored or degraded (Martin & Parton, 2005; Cermelli et al., 2006; Welte, 2007; Fujimoto et al., 2008). Moreover, they became the focus of interest because disturbances in their functioning cause many serious dieses in humans and animals such as obesity, diabetes, atherosclerosis, cardiovascular disease, allergic inflammation, arthritis, mycobacterial infections, bacterial sepsis, acute respiratory distress syndrome (Bozza et al., 2009). Lipid bodies are also important for plant productivity and biotechnology (Baud & Lepiniec, 2010; Lu et al., 2011).

Cytological studies with the use of light and electron microscopy combined with autoradiography and immunocytochemistry might significantly elucidate lipid bodies biogenesis and function as well as their interactions with other cell organelles. Lipotubuloids which contain an enormous number of lipid bodies and microtubules are especially suitable experimental material.

Biogenesis of lipid bodies is one of cytological controversies. Some scientists believe that they are formed in cytoplasm as naked droplets (Lung & Weselake, 2006) which are later surrounded by a half unit membrane, others suggest that both ER membranes take part in their biogenesis (Robenek et al., 2011). However, the most common hypothesis is based on old ultrastructural observations (Frey-Wyssling et al., 1963) of plant cells which indicate that ER contributes to lipid bodies formation resulting from the accumulation of lipids between leaflets of a phospholipid bilayer (Fujimoto et al., 2008; Ducharme & Bickel, 2008; Guo et al., 2009; Ohsaki et al., 2009).

In *O. umbellatum* lipotubuloids lipid bodies are formed in the latter way. This process can be observed in the lipotubuloids enlarging due to *de novo* formation of lipid bodies during intensive ovary epidermis growth (Kwiatkowska, 1971b). Ovary epidermis cells are characterized by intensive growth without mitotic divisions leading to their 30-fold (by 3000%) enlargement (Fig. 6). It is also closely correlated with lipotubuloid growth (correlation coefficient 0.98) which indicates the important role played by these structures in cells (Kwiatkowska et al., in prep.).

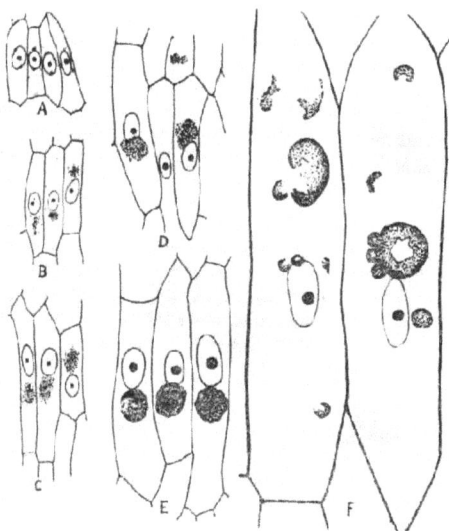

Fig. 6. *O. umbellatum* ovary epidermis and lipotubuloid development (Kwiatkowska, 1971b)

Forming lipid bodies first appear as a gap (Fig. 7A) then change into a lens-like structure grey in color (Fig. 7B,C) . Later it enlarges and becomes greatly osmiophilic, due to that the phospholipid half unit membrane which surrounds it is poorly visible as a thinner and lighter structure than a bilayer membrane (Fig. 7E-G). It appears very clearly later on around mature spherical lipid bodies which are filled with less osmiophilic and more homogenic substance (Fig. 7I,J) as well as in structures which are transient forms between forming and mature lipid bodies (Fig. 7H); these transient forms are connected with ER. However, mature lipid bodies are not linked with ER directly, but through microtubules which are adjacent to ER at one end and to lipid bodies at the other.

The immunogold technique revealed single or aggregated gold grains indicating the presence of diacylglicerol acyltransferase 1 and 2 (DGAT1 and DGAT2) in the specific rough ER regions and which means that these enzymes are synthesized in ER bound ribosomes (Fig. 8) (Kwiatkowska et al., 2011b).

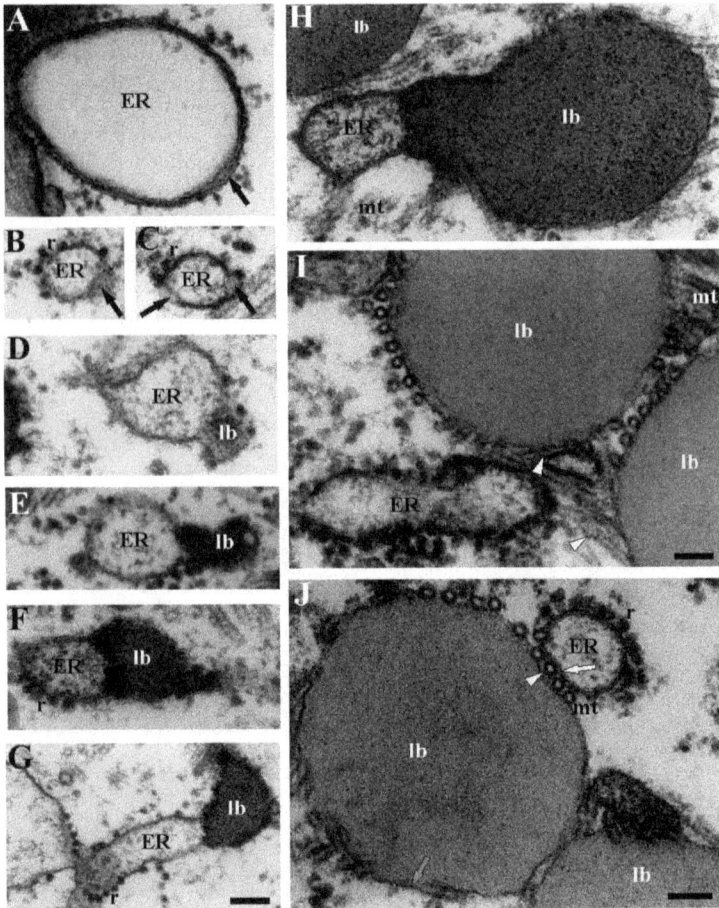

Fig. 7. Biogenesis of lipid bodies in *O. umbellatum* lipotubuloids; A - early step of lens-like structure formation (arrow); B,C - accumulation of electron-clear substance between two leaflets of ER membrane (arrows); D-H - enlargement of nascent lipid body; I,J - matured lipid bodies. Connection between lipid bodies and ER by microtubules (white arrow heads); ER - endoplasmic reticulum, lb - lipid body, mt – microtubules, r – ribosomes, white arrow - ER membrane bilayer, gray arrow - lipid body monolayer; bars: 50 nm (A-F, H,I), 100 nm (G), 50 nm (J), (Kwiatkowska et al., 2011b).

More gold grains indicating DGAT1 and DGAT2 were observed in characteristic swollen regions surrounded by thinner membranes at the ends or sides of ER cisternae (Fig. 9).

These pictures seem to present lipid bodies *in statu nascendi*. They suggest that during their formation enzymes are transported from ER to the surface of lipid bodies. These pictures correspond to gold grains localization indicating DGAT1 and DGAT2 situated at the outside zone of mature lipid bodies (Fig. 10). Thus it seems possible that the last stage of lipid synthesis, transformation of DAG into TAG, can take place in mature lipid bodies similarly as in adipocytes and COS7 fibroblast (Kuerschner et al., 2008).

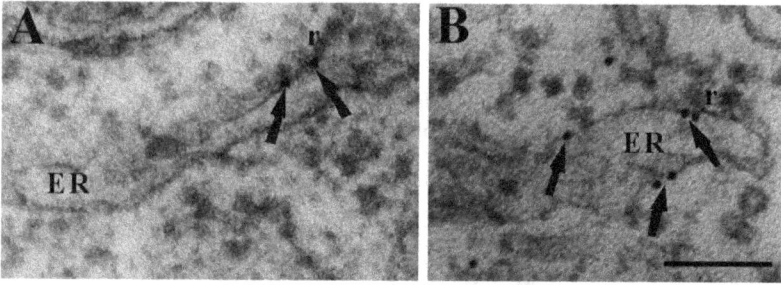

Fig. 8. 10 nm gold grains (arrows) indicating DGAT2 presence near ribosomes on ER in
O. umbellatum lipotubuloid (A,B); ER – endoplasmic reticulum, r – ribosomes; bar: 100 nm,
(Kwiatkowska et al., 2011b).

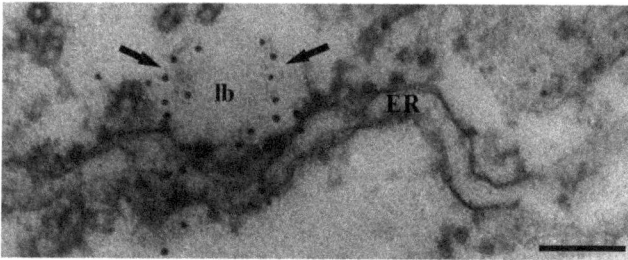

Fig. 9. Swelling ER cisternae surrounded with a membrane thinner that ER bilayer with gold
grains indicating DGAT2 presence (arrows) in *O. umbellatum* lipotubuloid; ER –
endoplasmic reticulum, lb – lipid body; bar: 100 nm, (Kwiatkowska et al., 2011b).

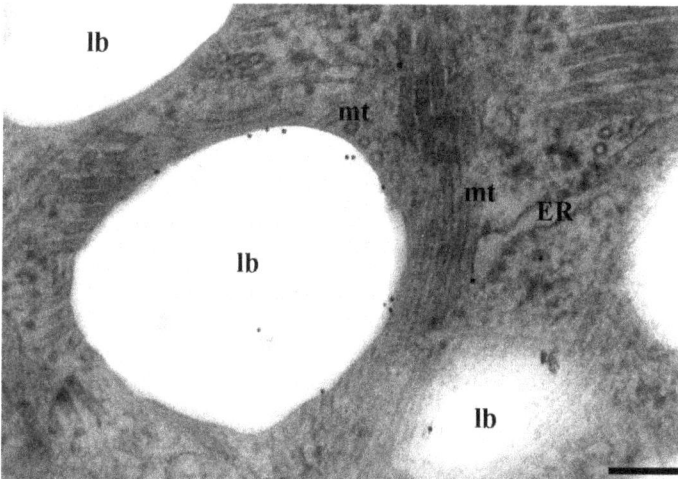

Fig. 10. Mature lipid bodies in *O. umbellatum* lipotubuloid; 10 nm gold grains indicating
DGAT2 presence; sections after hydrogen peroxide treatment to remove osmium; ER –
endoplasmic reticulum, lb – lipid body, mt - microtubules; bar: 100 nm (Kwiatkowska et
al., 2011b).

Buers et al. (2009) also observed that in macrophages freeze-fracture replica immunogold method revealed DGAT2 on lipid bodies surfaces. To the best of our knowledge our results of immunogold research revealing DGAT1 and DGAT2 on the surfaces of lipid bodies in *O. umbellatum* lipotubuloids are the only such results to date in plants. In the case of *O. umbellatum* lipotubuloids autoradiographic studies with ^{3}H-palmitic acid at the ultrastructural level (Kwiatkowska et al., 2011b) directly prove that lipid synthesis does take place at the outside of mature lipid bodies (Fig. 11B-J).

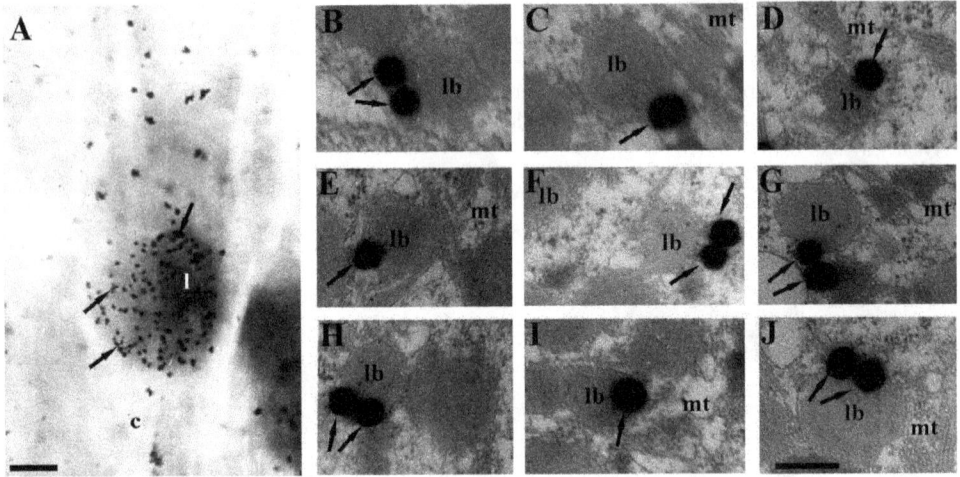

Fig. 11. *O. umbellatum* lipotubuloid autoradiograms after 2 h ^{3}H-palmitic acid incorporation; A - light microscope picture of a lipotubuloid labeled with silver grains (arrows); B-J – electronograms of silver grains localized at the half unit membrane/core border of lipid bodies (arrows); c – cytoplasm, l – lipotubuloid, lb – lipid bodies, mt – microtubules; bars: 10 µm (A), 0.2 µm (B-J), (Kwiatkowska et al., in prep.).

Autoradiographic labeling disappears after lipid extraction with lipid solvents which shows that this precursor becomes incorporated into lipids. It is most clearly seen on light microscope pictures of autoradiograms where lipotubuloids are literally sprinkled with autoradiographic grains (Fig. 11A) which disappear completely after lipid extraction. These pictures bring to the mind the term intuitively proposed by Wakker (1888) – "elaioplasts" which means - producing fat.

Silver grains and gold grains resulting from autoradiographic and immunogold reactions, respectively are co-localized at the surface of mature lipid bodies at the half unit membrane/core border which clearly indicate their contribution to lipid synthesis contrary to lipid bodies *in statu nascendi* in which lipid synthesis is connected directly with ER (see above).

Recent studies with immunogold reaction and 20 nm colloidal gold coupled with anti-lipase antibodies have shown the presence of lipase near lipid bodies surfaces (Fig. 12) (Kwiatkowska et al., in prep.). Similar lipase localization in *Ricinus communis* was observed by Eastmond (2004) with the use of immunogold method. Lipase is probably responsible for the disappearance of selective labeling of lipotubuloids incubated for 2 h in ^{3}H-palmitic acid and postincubated for 6 h in the non-radioactive medium (Kwiatkowska, 2004). The mature lipid bodies are approximately of the same size (0.1-0.4 µm) during ovary development in spite of active lipid synthesis due to the dynamic balance between lipid synthesis and lipolysis, however the number of lipid bodies in lipotubuloids which grow significantly, increases (Kwiatkowska et al., 2007). Thus the fact that lipid bodies do not enlarge cannot be treated as an unequivocal proof that lipid synthesis does not occur in them.

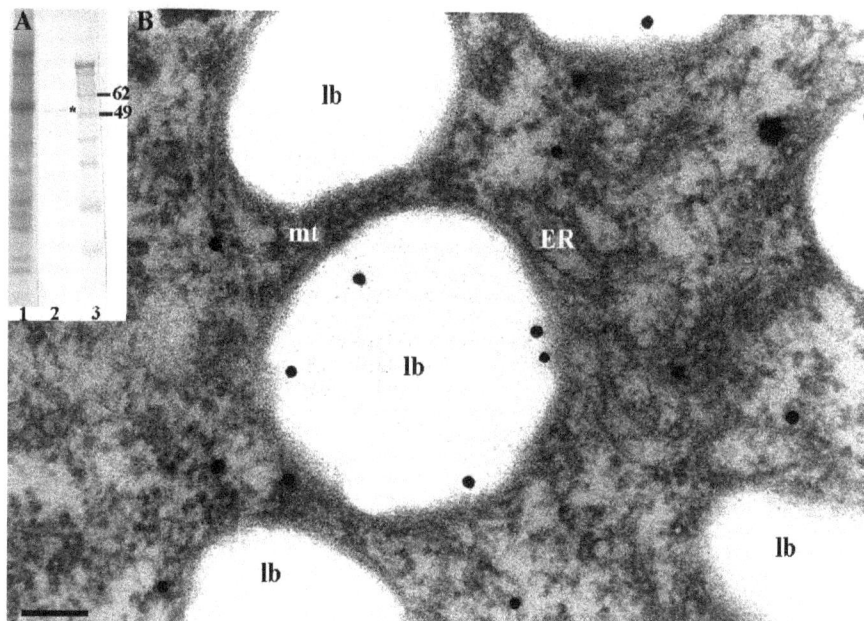

Fig. 12. Immunodetection of lipase in *O. umbellatum* ovary epidermis; A - Western blot analysis; line 1 – SDS-PAGE electrophoretic separation of the ovary epidermis extract; line 2 – Western blotting of the ovary epidermis extract probed with the anti-human lipase antibody; line 3 – molecular mass standards and their weights in kDa; B – lipid bodies with 20 nm gold grains indicating lipase presence; ER – endoplasmic reticulum, lb – lipid bodies, mt – microtubules; bar: 100 nm (Kwiatkowska et al., in prep.).

3. Microtubules and lipotubuloid movement

Microtubules as cytoskeleton elements are mostly involved in movement. Lipotubuloids are characterized with very specific and dynamic movement (Kwiatkowska, 1972a). It consists of sometimes very dynamic rotation with varying speed, direction and axis as well as progressive movement (Fig. 13). Lipotubuloid progressive movement depends on cyclosis, it stops when cytoplasm movement is arrested with dinitrophenol (DNP) which blocks ATP synthesis. Rotation, however, persists for some time after DNP application which suggests that it is autonomous, independent of cytoplasm movement. Also the fact that peripheral speed of the rotating lipotubuloid reaches 31.4 µm/s and is 6.2 times faster than the maximum speed of cytoplasmic motion (Kwiatkowska, 1972a) proves the above suggestion.

The question arises what is the connection between microtubules and lipotubuloid movement. One thing seems certain, microtubules which join lipid bodies create one structure able to move as a unity despite not having its own membrane.

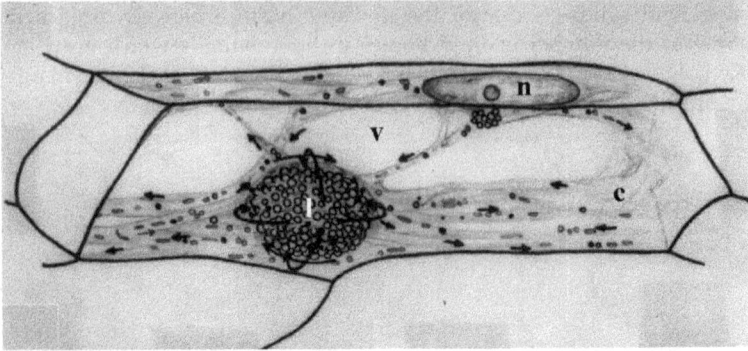

Fig. 13. A scheme of a lipotubuloid in an epidermal cell of *O. umbellatum* stipule which has turned around (several times) within 10-12 s, changing its direction and axis without a change in cellular location; c – cytoplasm, l – lipotubuloid, n – nucleus, v – vacuole, long arrows – the direction of lipotubuloid rotation, short arrows – direction of cytoplasm movement (Kwiatkowska et al., 2009).

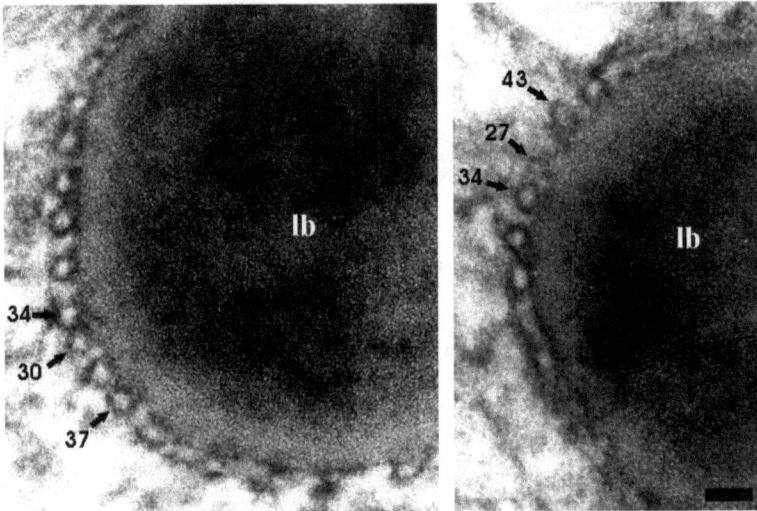

Fig. 14. Fragments of *O. umbellatum* lipid body surrounded with microtubules differing in width; lb – lipid bodies, numbers denote microtubule diameters in nm; bar: 50 nm (Kwiatkowska et al.,2009).

Moreover, it turned out that microtubules of lipotubuloids differ in diameter (Fig. 14), two populations were revealed: wide (43-58 nm) and narrow (24-39 nm). In the lipotubuloids in the ovary epidermis which move less dynamically the number of wide microtubules is smaller (Kwiatkowska et al., 2006) than in the fast-moving lipotubuloids present in stipule (Kwiatkowska et al., 2009). The microtubule diameter depends on the varying number of protofilaments which form them, the bigger the number the greater the diameter (Fig. 15A). Regardless of the above correlation, analyses of microtubule cross-sections revealed that with the same number of filaments (e.g. 10, 11, 12) two microtubule populations were observed both in the control and after DNP removal while under DNP influence only one middle-sized population was present (Fig. 15B). It was also shown that the number of microtubule protofilaments in the control, under DNP influence and after its removal was stable. Analysing wall structure of microtubules varying in size but formed from the same number of protofilaments it was revealed that these changes depended on varying tubulin monomer sizes as well as different distances between them (Fig. 16).

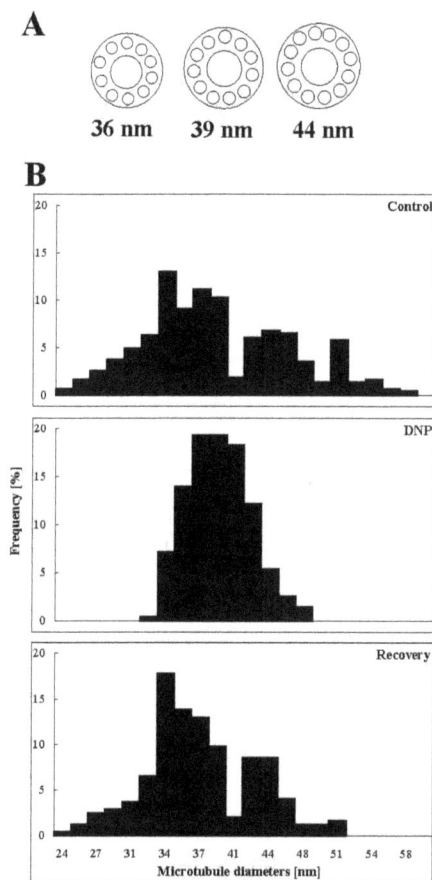

Fig. 15. A – A scheme of microtubules whose width depends on the number of filaments; B - a graph presenting two microtubule populations in the control and after DNP removal, and one microtubule population after DNP application (Kwiatkowska et al., 2009).

In the wider microtubules both these parameters are greater and *vice versa* (Kwiatkowska et al., 2009). All the above proves flexibility of microtubules *in vivo* depending on their functional status. The fixation method used by us makes it possible to "freeze" the microtubule structure in the *in vivo* state due to quick OsO$_4$ penetration as was shown by Omoto & Kung (1980). Other authors observed microtubule flexibility *in vitro* (Nogales et al., 1999; Li et al., 2002; Pampaloni & Florin, 2008).

In *O. umbellatum* lipotubuloids, apart from microtubules, there are also short actin filaments which were observed in ultrastructural pictures (Fig. 3) (Kwiatkowska et al., 2005). A hypothesis has been put forward that interaction of actin filaments with microtubules may determine the transformation of wide microtubules into narrow ones and *vice versa* (Kwiatkowska et al., 2009). Microtubules of varying sizes were observed *in vitro* as a result of tubulin co-sedimentation in the presence of actin and myosin Va (Cao et al., 2004).

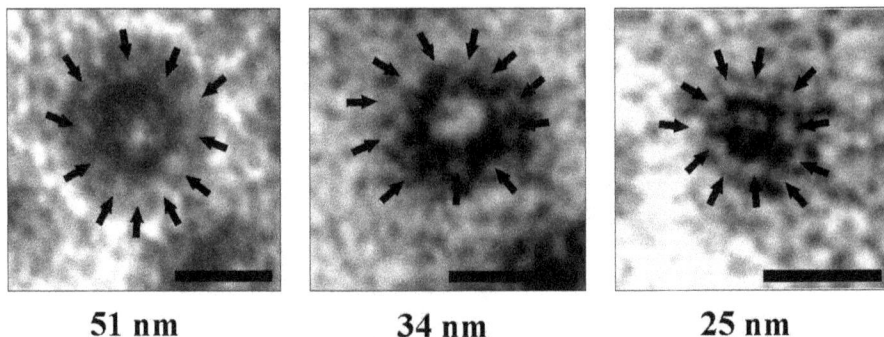

51 nm   **34 nm**   **25 nm**

Fig. 16. Microtubules different in size consisting of the same number of protofilaments; visible differences in monomer sizes (arrows) and in distances between them; bars: 25 nm (Kwiatkowska et al., 2009).

We also suppose that changes in lipotubuloid microtubule sizes might the driving force of their autonomic rotation.

It is worth stressing that the rotary-progressive lipotubuloid movement plays an important role in substance exchange between them and a cell. This is supported by the results concerning the involvement of intracellular motion in spreading various substances in a cell (Verchot-Lubicz & Goldstein, 2010).

4. Microtubules and lipid synthesis in mature lipid bodies

Autoradiographic ultrastructural studies with the use of ^{3}H-palmitic acid showed that incorporation of this precursor into lipids took place at the site of microtubule adhesion to the half unit membrane (Fig. 11B-J) thus a hypothesis has been put forward that these two structures cooperate in lipid synthesis (Kwiatkowska, 2004). It is supported by the fact that after short radioactive incubation microtubules are labeled first while lipid bodies as late as after 2 h (Kwiatkowska et al., 2011b). Thus it can be assumed that microtubules take up lipid precursors, including radioactive particles, and transmit them to the incorporation site.

The immunogold labeling showed that gold grains, indicating the presence of two enzymes: DGAT1 and DGAT2 as well as of phospholipase D also indispensable for lipid synthesis (Andersson et al., 2006), were present at microtubule walls (Fig. 17, 18) (Kwiatkowska et al., 2011b). The results concerning phospholipase D correspond to these of the co-sedimentation assay in which microtubules decorated with phospholipase D were observed (Gardiner et al., 2001; Gardiner et al., 2003; Dhonukshe et al., 2003).

On the basis of autoradiography and immunogold labeling a hypothesis may be put forward that microtubules take an active part in lipid synthesis as transmitters of precursors and enzymes to their respective destinations. Valuable proofs of microtubule involvement in lipid synthesis come from research with their inhibitors. Pacheco et al. (2007) observed that colchicine or taxol similarly blocked lipid body formation, being the reaction to inflammation, in mouse monocytes.

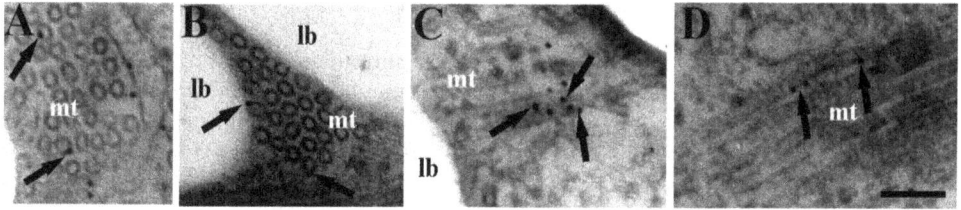

Fig. 17. *O. umbellatum* microtubules in cross (A,B) and longitudinal sections (C,D) with 10 nm gold grains at the surface indicating the presence of DGAT2 (arrows); lb – lipid bodies, mt – microtubules; bar: 100 nm (Kwiatkowska et al., 2011b).

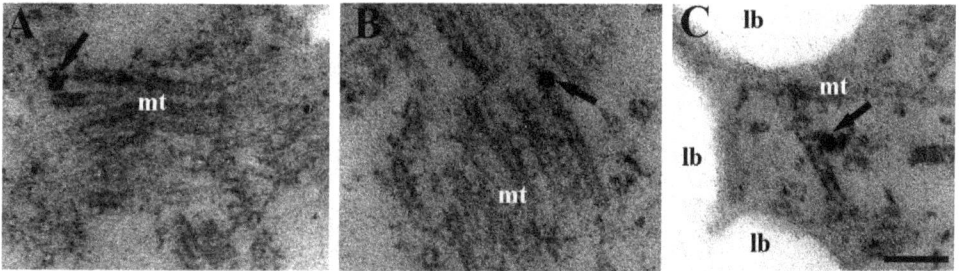

Fig. 18. *O. umbellatum* microtubules in longitudinal sections (A-C) with 20 nm gold grains at the surface indicating the presence of phospholipase D (arrows); lb – lipid bodies, mt – microtubules; bar: 100 nm (Kwiatkowska et al., 2011b).

Recently a similar experiment has been carried out on *O. umbellatum* lipotubuloids, which were incubated for 6 h in propyzamide which is known to induce microtubule degradation (Nakamura et al., 2004; Sedbrook et al., 2004). It turned out that it induced partial microtubule disintegration and changed their structure by forming on their walls dark deposits visible in EM. Most probably they make microtubules lose their transmitting abilities, this leads to the blockade of new lipid synthesis which is reflected by inhibited incorporation of ^{3}H-palmitic acid into lipotubuloids. This seems to be the decisive proof of microtubule role in lipid synthesis (Kwiatkowska et al., 2011b).

We believe that in the case of lipotubuloids of other plants they may also function as transmitters of different substances to lipid bodies, however this issue needs further research.

Up till now there has been no research proving a similar role of microtubules in lipid synthesis in other organisms, although microtubules surrounding single lipid bodies, not organized into lipotubuloids, were observed in *Marchantia paleacea* (Galatis et al., 1978), *Lactuca sativa* (Smith, 1991) and in red alga *Gelidium robustum* (Delivopoulos, 2003). Small sizes and great lability of microtubules probably make observation of more common structural and functional correlation between microtubules and lipid bodies impossible.

As it was mentioned earlier lipotubuloids are rich in ribosomes and rough ER in the form of cisternae and vesicles. No detailed research concerning their functioning was conducted but it easily visible that ribosomes are actively involved in translation as they form numerous polysomes (Fig. 2). Since formation of new lipid bodies and lipid synthesis involve a whole enzymatic system and many regulatory factors active ER and ribosomes in lipotubuloids are indispensable.

Other structures such as mitochondria, microbodies surrounded with single lipid bi-layers (glyoxysomes and peroxysomes) as well as Golgi structures are less numerous and have not been studied in detail so far. It is believed that lipid bodies cooperate with other organelles as "gregarious" organelles (Goodman, 2008). Mitochondria and microbodies are in close contact with lipid bodies. Due to synaptic connections (Binns et al., 2006) there is correlation between release and oxidation of lipid acids resulting from lipolysis (Goodman, 2008; Fujimoto et al., 2008; Guo et al., 2009). Moreover, mitochondria may supply energy and NADPH which make lipid synthesis and lipid bodies biogenesis possible (Walter & Farese, 2009).

Golgi structures which are very dynamic organelles may be involved in microtubule polymerization as many authors believe (Chabin-Brion et al., 2001; Efimov et al., 2007; Kodani & Sutterlin, 2009). On the other hand, COPI (Beller et al., 2008) and COPII complexes (Soni et al., 2009) produced by Golgi structures are evolutionary conserved regulators of lipid homeostasis.

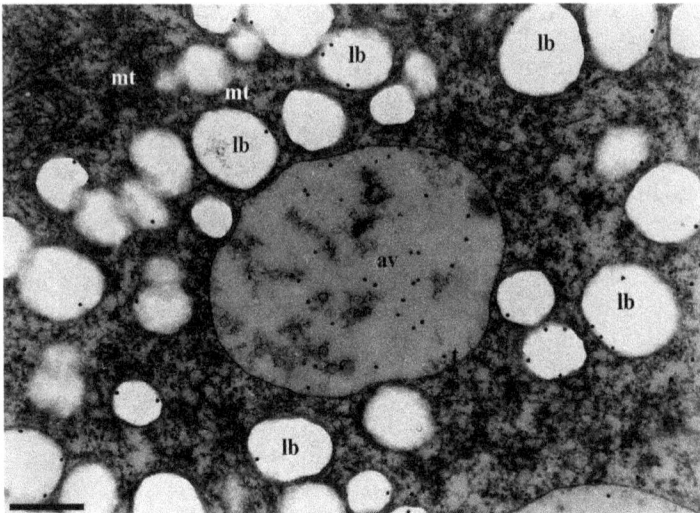

Fig. 19. An autolytic vacuole (av) after immunogold reaction for lipase in *O. umbellatum* lipotubuloid; lb – lipid bodies, mt – microtubules, t – tonoplast; bar: 500 nm (Kwiatkowska et al., in prep.).

At the final stage of lipotubuloid development autolytic vacuoles appear prior to microtubule disappearance leading to disintegration of lipotubuloids into separate lipid bodies (Fig. 6). These vacuoles are surrounded with a tonoplast and contain fragments of membranes and cell structures which is characteristic of autolytic vacuoles. Immunogold labeling revealed in them numerous gold grains indicating the presence of lipase (Fig. 19)

(Kwiatkowska et al., in prep.). The above observation supports earlier results of cytochemical assays revealing lipase and acid phosphatase in lipotubuloids (Kwiatkowska, 1971b). Triggering of autolysis during dynamic metabolism was also observed in animal cells (Dong & Czaja, 2011).

5. Lipotubuloids and cuticle synthesis

It is known that cuticle are produced by epidermis cells which may dedicated more than 50% of their metabolites to this structure (Suh et al., 2005). In the case of *O. umbellatum* lipotubuloids incubated for 2 h in ^{3}H-palmitic acid and postincubated for 6 h in the non-radioactive medium autoradiographic grains first assembled over lipotubuloids become scattered all over a cell (tangential section) (Fig. 20).

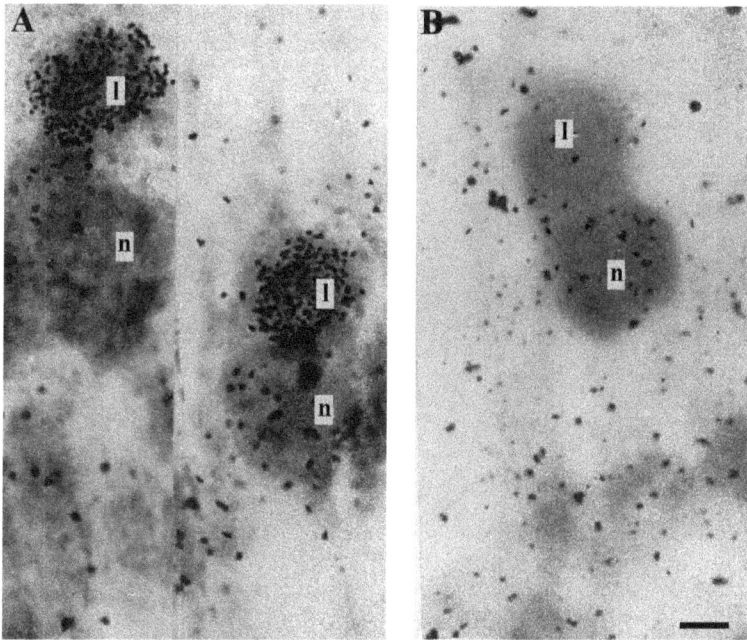

Fig. 20. *O. umbellatum* light microscope autoradiograms; A – silver grains aggregated over lipotubuloids after 2 h incubation in ^{3}H-palmitic acid; B – scattered silver grains after 2 h incubation in ^{3}H-palmitic acid followed by 6 h postincubation in non-radioactive medium - cells in tangential section; l – lipotubuloid, n – nucleus; bar: 10 μm (Kwiatkowska, 1972b).

After postincubation the number of autoradiographic grains falls by about 70% which means that a great amount of lipids was metabolized during 6 h. The remaining autoradiographic grains do not disappear after lipid extraction but their number drops (Tab. 2). Cell radioactive parts insoluble in solvents are visible on the epidermis cross section at the site corresponding to a cuticular layer from Bird's (2008) cuticle scheme (Fig. 21). Thus it seems very probable that there are cutins insoluble in lipid solvents. This part of scattered autoradiographic grains which disappeared after lipid extraction may correspond to waxes which are easily dissolved in organic solvents (Kwiatkowska et al., in prep.). A hypothesis has been put forward that about 30% of lipids from lipotubuloids turn into cuticle. A question arises if this transformation takes place in lipotubuloids or in other cell compartments and the lipids from lipotubuloids are only building blocks. This problem is worth elucidating since many recent results indicate its great importance with regard to economical and biotechnological issues (Heredia et al., 2009; Dominguez et al., 2011). Cutin is the most ubiquitous biopolymer in biosphere (Heredia, 2003). We are planning to take up this question soon.

Labeled area	Incubation in ³H-palmitic acid	Incubation in ³H-palmitic acid after lipid extraction	Incubation in ³H-palmitic acid and 6 h post-incubation in non-radioactive medium	Incubation in ³H-palmitic acid and 6 h post-incubation in non-radioactive medium after lipid extraction
Whole Cell	294 ± 14	4 ± 0.8	102 ± 5	19 ± 0.9
Lipotubuloid	240 ± 9	1 ± 0.5	12 ± 0.4	3 ± 0.8
The rest of cytoplasm and nucleus	54 ± 1.8	3 ± 0.9	90 ± 11	16 ± 1.1

Table 2. Number of silver grains over particular compartments of *O. umbellatum* ovary epidermis cell after incubation in ³H-palmitic acid under different experimental conditions ± SE.

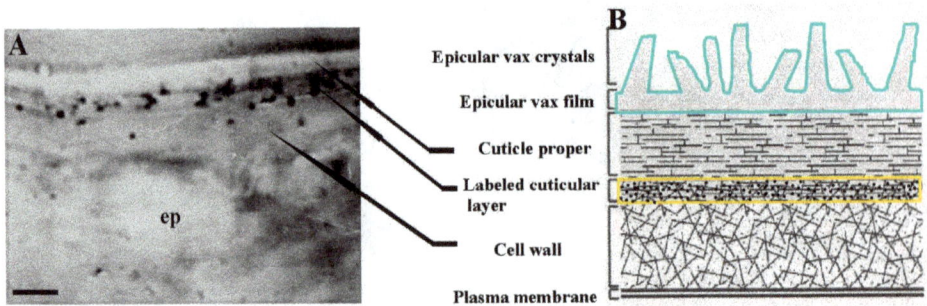

Fig. 21. A - Cross-section of *O. umbellatum* ovary epidermis (ep); autoradiogram after 2 h incubation in ³H-palmitic acid, 6 h postincubation in the non-radioactive medium and after extraction in the lipid solvent; silver grains localized in the cuticular layer (Kwiatkowska et al., in prep.); B - a scheme of cuticle according to Bird (2008 - modified); bar: 5 µm.

6. Conclusions

Lipotubuloids are a very specific, dynamic, complicated set of metabolically active (although seemingly static) lipid bodies containing DGAT and lipase which cooperate with microtubules and other organelles. A lipotubuloid is somewhat independent in a cell which is reflected by its capability for autonomous rotary movement, however, it is closely correlated with the development of the ovary epidermis and cuticle synthesis during seed formation in a fruit. Studies concerning these structures may further elucidate lipid metabolism and their functional relation with a cell and its organelles.

7. Acknowledgements

This work was realized and financed by National Committee of Scientific Research, grant No. NN 303 35 9035.

8. References

Andersson, L., Boström, P., Ericson, J., Rutberg, M., Magnusson, B., Marchesan, D., Rulz, M., Asp, L., Huang, P., Frohan, M.A., Boren, J. & Olofsson, S.O. (2006). PLD1 and ERK2 regulate cytosolic lipid droplet formation. *Journal of Cell Science*, Vol.119, No.11, (June 2006), pp. 2246-225, ISSN 0021-9533

Baud, S. & Lepiniec, L. (2010). Physiological and developmental regulation of seed oil production. *Progress in Lipid Research*, Vol.49, No.3, (July 2010), pp. 235-249, ISSN 0163-7827

Beer, R. (1909). On elaioplasts. *Annals of Botany*, Vol.23, No.1, (January 1909), pp. 63-72, ISSN 0305-7364

Beller, M., Sztalryd, C., Southall, N., Bell, M., Jackle, H., Auld, D.S. & Oliver, B. (2008). COPI complex is regulator of lipid homeostasis. *PLoS Biology*, Vol.6, No.11, (November 2008), pp. 2530-2549, ISSN 1544-9173

Binns, D., Januszewski, T., Chen, Y., Hill J., Markin, V.S., Zhao, Y., Gilpin, C., Chapman, K.D., Anderson, R.G.W. & Goodman, J.M. (2006). An intimate collaboration between peroxisomes and lipid bodies. *Journal of Cell Biology*, Vol.173, No.5, (June 2006), pp. 719-731, ISSN 0021-9525

Biedermann, W. (1920). Der Lipoidgehalt bei Monotropa hypopithys u. Orobanche sporose. *Flora*, Vol.113, pp. 133-154

Bird, D.A. (2008). The role of ABC transporters in cuticular lipid secretion. *Plant Science*, Vol.174, No.6, (June 2008), pp. 563-589, ISSN 0168-9452

Bozza, P.T., Magalhaes, K.G. & Weller, P.F. (2009). Leukocyte lipid bodies – Biogenesis and functions in inflammation. *Biochimica et Biophysica Acta, Molecular and Cell Biology of Lipids*, Vol.1791, No.6, (June 2009), pp. 540-551, ISSN 1388-1981

Buers, I., Robenek, H., Lorkowski, S., Nitsche, Y., Severs, N.J. & Hofnagel, O. (2009). TIP47, a lipid cargo protein involved in macrophage triglyceride metabolism. *Arteriosclerosis, Thrombosis and Vascular Biology*, Vol.29, No.5, (May 2009), pp. 767-773, ISSN 1079-5642

Cao, T.T., Chang, W., Masters, S.E. & Mooseker, M.S. (2004). Myosin-Va binds to and mechanochemically couples microtubules to actin filaments. *Molecular Biology of the Cell*, Vol.15, No.1, (January 2004), pp. 151-161, ISSN 1059-1524

Cermelli, S., Guo, Y., Gross, S.P. & Welte, M.A. (2006). The lipid-droplet proteome reveals that droplets are a protein-storage depot. *Current Biology*, Vol.16, No.18, (September 2006), pp. 1783-1795, ISSN 0960-9822

Chabin-Brion, K., Marceiller, J., Perez, F., Settegrana, C., Drechou, A., Durand, G. & Pous, C. (2001). The Golgi complex is a microtubule-organizing organelle. *Molecular Biology of the Cell*, Vol.12, No.7, (July 2001), pp. 2047-2060, ISSN 1059-1524

Delivopoulos, S.G. (2003). Ultrastructure of cystocarp development in *Gelidium robustum* (Gelidiaceae: Gelidiales: Rhodophyta). *Marine Biology*, Vol.142, No.4, (April 2003), pp. 659-667, ISSN 0025-3162

Dhonukshe, P., Laxalt, A.M., Goedhart, J., Gadella, T.W.J. & Munnik, T. (2003). Phospholipase D activation correlates with microtubule reorganization in living plant cells. *Plant Cell*, Vol.15, No.11, (November 2003), pp. 2666-2679, ISSN 1040-4651

Domínguez, E., Heredia-Guerrero, J.A. & Heredia, A. (2011). The biophysical design of plant cuticles: an overview. *New Phytologist*, Vol.189, No.4, (March 2011), pp. 938-849, ISSN 0028-646X

Dong, H. & Czaja, M.J. (2011). Regulation of lipid droplets by autophagy. Trends in Endocrinology and Matabolism, Vol.22, No.6, (June 2011), pp. 234-240, ISSN 1043-2760

Ducharme, N.A. & Bickel, P.E. (2008). Lipid droplets in lipogenesis and lipolysis. *Endocrinology*, Vol.149, No.3, (March 2008), pp. 942-949, ISSN 0013-7227

Eastmond, P.J. (2004). Cloning and characterization of acid lipase from castor beans. *The Journal of Biological Chemistry*, Vol.279, No.44, (October 2004), pp. 45540-45545, ISSN 0021-9258

Efimov, A., Kharitonov, A., Efimova, N., Loncarek, J., Miller, P.M., Andreyeva, N., Glesson, P., Galjart, N., Maia, A.R.R., MacLeod I.X., Yates, J.R. III, Maiato, H., Khodjakov, Akhmanova, A. & Kaverina, I. (2007). Asymmetric CLASP-dependent nucleation of noncentrosomal microtubules at the *trans*-Golgi network. *Developmental Cell*, Vol.12, No.6, (June 2007), pp. 917-930, ISSN 1534-5807

Faull, A.F. (1935). Elaioplasts in Iris: a morphological study. *Journal of the Arnold Arboretum*, Vol.16, No.2, (April 1935), pp. 225-267, ISSN 0004-26:5

Frey-Wyssling, A., Grieshaber, E. & Muhlethaler, K. (1963). Origin of spherosomes in plant cells. *Journal of Ultrastructure Research*, Vol.8, No.5-6, (June 1963), pp. 506-516, ISSN 0022-5320

Fujimoto, T., Ohasaki, Y., Cheng, J., Suzuki, M. & Shinohara, Y. (2008). Lipid bodies: a classic organelle with new outfits. *Histochemistry of Cell Biology*, Vol.130, No.2, (August 2008), pp. 263-279, ISSN 0948-6143

Fukushima, N., Furuta, D., Hidaka, Y., Moriyama, R. & Tsujiuchi, T. (2009). Post-translational modifications of tubulin in nervous system. *Journal of Neurochemistry*, Vol.109, No.3, (May 2009), pp. 683-693, ISSN 0022-3042

Galatis, B., Apostolakos, P. & Katsaros, C. (1978). Ultrastructural studies on the oil bodies of *Marchantia paleacea* Bert. I. Early stages of oil-body cell differentiation: origination of the oil body. *Canadian Journal of Botany*, Vol.56, No.18, pp. 2252-2267, ISSN 0008-4026

Gardiner, J.C., Harper, J.D.I., Weerakoon, N.D., Colins, D.A., Richtle, S., Gilroy, S., Cyr, R.J. & Mare, J. (2001). A 90-kD phospholoipase D from tobacco binds to microtubulew and plasma membrane. *Plant Cell*, Vol.13, No.9, (September 2001), pp. 2143-2158, ISSN 1040-4651

Gardiner, J.C., Collings, D.A., Harper, J.D.I. & Marc, J. (2003). The effects of phospholipase D-antagonist1-butanol on seedlings development and microtubule organization in *Arabidopsis*. *Plant and Cell Physiology*, Vol.44, No.7, (July 2003), pp. 687-696, ISSN 0032-0781

Goodman, J.M. (2008). The gregarious lipid droplet. *The Journal of Biological Chemistry*, Vol.283,No.42, (October 2009), pp. 28005-28009, ISSN 0021-9258

Górska-Brylass, A. (1962). "Elajoplasts" w ziarnach pyłkowych. *Campanula*. *Acta Societatis Botanicorum Poloniae*, Vol.31, pp. 409-416, ISSN 0001-6977

Guillermond, A. (1922). Sur l'origine et la signification des éleoplastes. *Comptes Rendus des Seances de la Societe Biologie et de ses Filiales*, Vol.86, pp. 437-440

Guo, Y., Cordes, K.R., Farese, R.V. Jr & Walther, T.C. (2009). Lipid droplets at a glance. *Journal of Cell Science*, Vol.122, No.6, (March 2009), pp. 749-752, ISSN 0021-9533

Hammond, J.W., Cai, D. & Verhey, K.J. (2008). Tubulin modifications and their cellular functions. *Current Opinion in Cell Biology*, Vol.20, No.1, (February 2008), pp. 71-76, ISSN 0955-0674

Heredia, A. (2003). Biophysical and biochemical characteristic of cutin, a plant barrier biopolymer. *Biochimica et Biophysica Acta, Molecular and Cell Biology of Lipids*, Vol.1620, No.1-3, (March 2003), pp. 1-7, ISSN 1388-1981

Heredia, A., Heredia-Guerrero, J.A., Domínguez, E. & Benitez J.J. (2009). Cutin synthesis: A slippery paradigm. *Biointerphases*, Vol.4, No.1, (March 2009), pp. P1-P3, ISSN 1559-4106

Hsieh, K. & Huang, A.H.C. (2004). Endoplasmic reticulum, oleosins, and oils in seeds and tapetum cells. *Plant Physiology*, Vol.136, No.3, (November 2004), pp. 3427-3434, ISSN 0032-0889

Huang, A.H.C. (1996). Oleosins and oil bodies in seed and other organs. *Plant Physiology*, Vol.110, No.4, (April 1996), pp. 1055-1061, ISSN 0032-0889

Kodani, A. & Sutterlin, C. (2009). A new function for an old organelle: microtubule nucleation at the Golgi apparatus. *EMBO Journal*, Vol.28, No.8, (April 2009), pp. 995-996, ISSN 0261-4189

Kuerschner, L., Moessinger, C. & Thiele, C. (2008). Imaging of lipid biosynthesis: How a neutral lipid enter lipid droplets. *Traffic*, Vol.9, No.3, (March 2008), pp. 338-352, ISSN 1600-0854

Kwiatkowska, M. (1959). Występowanie elajoplastów w skórce rodzaju *Gentiana* L. *Zeszyty Naukowe UŁ, seria II*, Vol.5, pp. 69-87

Kwiatkowska, M. (1961). Elajoplasty goryczek. cz. II. Obserwacje na materiale utrwalonym, *Acta Societatis Botanicorum Poloniae*, Vol.30, pp. 371-380, ISSN 0001-6977

Kwiatkowska, M. (1963). Elajoplasty dalii. Zeszyty Naukowe UŁ, nauki Matematyczno-Przyrodnicze seria 2, Zeszyt 14, pp. 81-91, (in Polish with French summary)

Kwiatkowska, M. (1966). Investigations on the elaioplasts of *Ornithogalum umbellatum* L. I. morphological, cytochemical and cytoenzymatic observations. *Acta Societatis Botanicorum Poloniae*, Vol.35, pp. 7-16, ISSN 0001-6977

Kwiatkowska, M. (1971a). Fine structure of lipotubuloids (elaioplasts) in *Ornithogalum umbellatum* L. *Acta Societatis Botanicorum Poloniae*, Vol.40, pp. 451-465, ISSN 0001-6977

Kwiatkowska, M. (1971b). Fine structure of lipotubuloids (elaioplasts) in *Ornithogalum umbellatum* in the course of their development. *Acta Societatis Botanicorum Poloniae*, Vol.40, pp. 529-537, ISSN 0001-6977

Kwiatkowska, M. (1972a). Changes in the diameter of microtubules connected with the autonomous rotary motion of the lipotubuloids (elaioplasts). *Protoplasma*, Vol.75, No.4, (December 1972), pp. 345-357, ISSN 0033-183X

Kwiatkowska, M. (1972b). Autoradiographic studies on incorporation of ^{3}H-palmitic acid into lipotubuloids of *Ornithogalum umbellatum* L. *Folia Histochemica et Cytochemica*, Vol.10, No.2, pp. 121-124, ISSN 0239-8508

Kwiatkowska, M. (1973a). Half unit membrane surrounding osmiophilic granules (lipid droplets) of the so-called lipotubuloid in *Ornithogalum*. *Protoploasma*, Vol.77, No.4, (December 1973), pp. 473-476, ISSN 0033-183X

Kwiatkowska, M. (1973b). Poloysaccharides connected with microtubules in the lipotubuloids of *Ornithogalum umbellatum* L. *Histochemie*, Vol.37, No.2, pp. 107-112, ISSN 0018-2222

Kwiatkowska, M. (2004). The incorporation of 3H-palmitic acid into *Ornithogalum umbellatum* lipotubuloids, which are a cytoplasmic domain rich in lipid bodies and microtubules. Light and EM authoradiography. *Acta Societatis Botanicorum Poloniae*, Vol.73, No.3, pp. 181-186, ISSN 0001-6977

Kwiatkowska, M., Popłońska, K. & Stępiński, D. (2005). Actin filaments connected with the microtubules of lipotubuloids, cytoploasmic domains rich in lipid bodies and microtubules. *Protoplasma*, Vol.226, No.3-4, (December 2005), pp. 163-167, ISSN 0033-183X

Kwiatkowska, M., Popłońska, K., Stępiński, D. & Hejnowicz, Z. (2006). Microtubules with different diameter, protofilament number and protofilament spacing in *Ornithogalum umbellatum* ovary epidermis cells. *Folia Histochemica et Cytobiologica*, Vol.44, No.2, pp. 133-138, ISSN 0239-8508

Kwiatkowska, M., Popłońska, K., Kaźmierczak, A., Stępiński, D., Rogala, K. & Polewczyk, K. (2007). Role of DNA endoreduplication, lipotubuloids, and gibberellic acid in epidermal cell growth during fruit development of Ornithogalum umbellatum. *Journal of Experimental Botany*, Vol.58, No.8, (June 2007), pp. 2023-2031, ISSN 0022-0957

Kwiatkowska, M., Popłońska, K. & Stępiński, D. (2009). Diameters of microtubules change during rotation of the lipotubuloids of *Ornithogalum umbellatum* stipule epidermis as result of varying protofilament monomers sizes and distance between them. *Cell Biolology International*, Vol.33, No.12, (December 2009), pp. 1245-1252, ISSN 1065-6995

Kwiatkowska, M., Stępiński, D., Popłońska, K., Wojtczak, A. & Polit, J. (2010). "Elaioplasts" of *Haemanthus albiflos* are true lipotubuloids: cytoplasmic domains rich in lipid bodies entwined by microtubules. *Acta Physiologiae Plantarum*, Vol.32, No.6, (November 2010), pp. 1189-1196, ISSN 0137-5881

Kwiatkowska, M., Stępiński, D., Popłońska, K., Wojtczak, A. & Polit, J.T. (2011a). "Elaioplasts" identified as lipotubuloids in *Althaea rosea, Funkia Sieboldiana* and *Vanilla planifolia* contain lipid bodies connected with microtubules. *Acta Societatis Botanicorum Poloniae*, ISSN 0001-6977, (in press)

Kwiatkowska, M., Popłońska, K., Wojtczak, A., Stępiński, D. & Polit, J.T. (2011b). Lipid body biogenesis and the role of microtubules in lipid synthesis in *Ornithogalum umbellatum* lipotubuloids. *Cell Biology International*, ISSN 1065-6995, (in press)

Kwiatkowska, M., Stępiński, D., Polit, J.T., Popłońska, K. & Wojtczak, A. (2011c). Microtubule heterogeneity of *Ornithogalum umbellatum* ovary epidermal cells: nonstable cortical microtubules and stable lipotubuloid microtubules. *Folia Histochemica et Cytobiologica*, Vol.49, No.2, (July 2011), pp. 285-290, ISSN 0239-8508

Li, H., DeRosier, D., Nicholson, W., Nogales, E. & Downing, K.H. (2002). Microtubule structure at 8 Å resolution. *Structure*, Vol.10, No.10, (October 2002), pp. 1317-1328

Linsbauer, K. (1930). *Die Epidermis* (German), Berlin, Borntraeger, German

Lu, Ch., Napier, J.A., Clemente, T.E. & Cahoon, E.B. (2011). New frontier in oilseed biotechnology meeting the global demand for vegetable oil for food, biofuel and industrial application. *Current Opinion in Biotechnology*, Vol.22, No.2, (April 2011), pp. 252-259, ISSN 0958-1669

Lung, S-C. & Weselake, R.J. (2008). Diacylgycerol acyltransferase: a key mediator of plant triacylglicerool synthesis. *Lipids*, Vol.41, No.12, (December 2008), pp. 1073-1088, ISSN 0024-4201

Luxenburgowa, A. (1928). Recherches cytologiques sur les grain de pollen chez les Malvacees. *Bulletin International l' Academie Polonaire des Sciences Letters ser B*. pp. 363-395

Martin, S. & Parton, R.G. (2005). Caveolin, cholesterol, and lipid bodies. *Seminars in Cell and Developmental Biology* Vol.16, No.2, (April 2005), pp. 163-174, ISSN 1084-9521

Nakamura, M., Naoi, K., Shoji, T. & Hasimoto, T. (2004). Low concentrations of propyzamide and oryzalin alter mikrotubule dynamics in *Arabidopsis* epidermal cells. *Plant and Cell Physiology*, Vol.45, No.9, (September 2004), pp. 1330-1334, ISSN 0032-0781

Nogales, E.A. (1999). Structural view of microtubule dynamics. *Cellular and Molecular Life Sciences*, Vol. 56, No.1-2, (October 1999), pp. 133-142, ISSN 1420-682X

Ohsaki, Y., Cheng, J., Suzuki, M., Sinohara, Y. & Fujimoto, T. (2009). Biogenesis of cytoplasmic lipid droplets. From the lipid ester globule inn the membrane to the visible structure. *Biochimica et Biophysica Acta, Molecular and Cell Biology of Lipids*, Vol. 1791, No.6, (June 2009), pp. 399-407, ISSN 1388-1981

Omoto, C.K. & Kung, C. (1980). Rotation and twist of central-pair microtubules in the cilia of paramecium. *Journal of Cell Biology*, Vol.87, No.1, (October 1980), pp. 33-46, ISSN 0021-9525

Pacheco, P., Vieira-de-Abreu, A., Gomes, R.N., Barbosa-Lima, G., Wermelinger, L.B., May-Monteiro, C.M., Silva, A.R., Bozza, M.T., Castro-Faria-Neto, H.C., Bandeira-Melo, C. & Bozza, P.T. (2007). Monocyte chemoattractant protein-1/CC chemokine ligand 2 controls microtubule-driven biogenesis and leukotriene B4-synthesizing function of macrophage lipid bodies elicited by innate immune response. *Journal of Immunology*, Vol. 179, No.15, (December 2007), pp. 8500-8508, ISSN 0022-1767

Pampaloni, F. & Florin, E.L. (2008). Microtubule architecture: inspiration for novel carbon nanotube-based biomimetic materials. *Trends in Biotechnology*, Vol.26, No.6, (June 2008), pp. 302-310, ISSN 0169-5347

Politis, J. (1911). Sugli elaioplasti nelle mono- e dicotiledoni. *Atti della Academia Nazionale dei Lincei*, Vol.20, No.1, pp. 599-603, ISSN 1120-6330

Raciborski, M. (1895). Elajoplasty liliowatych. Kraków: Akademia umiejętności SII, 6.

Riss, M.M. (1918). Die Antherenhaare von *Cyclanthera pedata* (Schrad.) und einiger anderer Cucurbitaceen. *Flora*, Vol. 111/112, pp. 541-559, ISSN 0367-2530

Robenek, H., Buers, I., Robenek, M.J., Honagel, O., Ruebel, A., Troyer, D. & Severs, N.J. (2011). Topography of lipid droplet-associated proteins: insights from freeze-fracture replica immunogold labeling. *Journal of Lipids*, Vol.2011, Doi:10.1155/2011/409371

Schimper, A.F.W. (1883). Uber die Entwicklung der Chloro-phylllkomer und Farbkorper. *Botanische Zeitung*, Vol.41, pp. 105-114, 121-131

Sedbrook, J.C., Ehrhardt, D.W., Fisher, S.E., Scheible, W-R. & Somerville, C.R. (2004). The Arabidopsis *SKU6/SPIRAL1* gene encodes a plus end-localized microtubule-interacting protein involved in directional cell expansion. *The Plant Cell*, Vol.16, No.6, (June 2004), pp. 1506-1520, ISSN 1040-4651

Siloto, R.M.P., Findlay, K., Lopez-Villalobos, A., Yeung, E.C., Nikiforouk, C.L. & Moloney, M.M. (2006). The accumulation of oleosins determines the size of seed oilbodies in *Arabidopsis*. *The Plant Cell*, Vol.18, No.8, (August 2006), pp. 1961-1974, ISSN 1040-4651

Smith, M.T. (1991). Studies on the anhydrous fixation of dry seeds of lettuce (*Lectuca sativa* L.). *New Phytologist*, Vol. 119, No.4, (December 1991), pp. 575-584, ISSN 0028-646X

Soni, K.G., Mardones, G.A., Sougrat, R., Smirnova, E., Jackson, C.L. & Bonifacino, J.S. (2009). Coatomer-dependent protein delivery to lipid droplets. *Journal of Cell Sciences*, Vol.122, No11, (June 2009), pp. 1834-1841, ISSN 0021-9533

Suh, M.C., Samuels, A.L., Jetter, R., Kunst, L., Pollard, M., Ohlrogge, J., & Beisson, F. (2005). Cuticular lipid composition, surface structure, and gene expression in Arabidopsis stem epidermis. *Plant Physiology*, Vol.139, No.3, (July 2005), pp. 1649-1665, ISSN 0032-0889

Thaler, I. (1956). Studien an „plastidenähnlichen Gebilden" (Elaioplasten und Sterinoplasten). *Protoplasma*, Vol.46, No.1-4, (March 1956), pp. 743-754, ISSN 0033-183X

Tourte, Y. (1964). Observations sur l'infrastructure des èlaïoplastes chez *Haemanthus albiflos*. *Comptes Rendus des Séances de la Société de Biologie et de ses Filiales*, Vol. 158, No.8-9, pp. 1712-1715

Tourte, Y. (1966). Considerations sur la nature, l'origine, et le comportement des elaioplastes chez les Monocotylèdones. *Österreichische botanische Zeitschrift*, Vol. 113, No.3-4, (February 1966), pp. 283-298, ISSN 0029-8948

Verchot-Lubicz, J. & Goldstein, R.E. (2010). Cytoplasmic streaming enables the distribution of molecules and vesicles in large plant cells. *Protoplasma*, Vol.240, No.1-4, (April 2010), pp. 99-107, ISSN 0033-183X

Verhey, K.J. & Gaertig, J. (2007). The tubulin code. *Cell Cycle*, Vol.6, No.17, (September 2007), pp. 2152-2160, ISSN 1538-4101

Wakker, L.C.H. (1888). Studien über die Inhaltskörper der Pflanzenzelle. *Jahrbücher für wissenschaftliche Botanik*, Vol.19, pp. 423-496, ISSN 0091-7451

Walter, T.C. & Farese, R.V. Jr (2009). The life of lipid droplets. *Biochimica et Biophysica Acta, Molecular and Cell Biology of Lipids*, Vol.1791, No.6, (June 2009), pp. 459-466, ISSN 1388-1981

Wałek-Czernecka, A. & Kwiatkowska, M. (1961). Elajoplasty ślazowatych. *Acta Societatis Botanicorum Poloniae*, Vol. 31, pp. 539-543, ISSN 0001-6977

Weber, F. (1955). „Elaïoplasten" fehlen den Schließzellen von *Hosta plantaginea*. *Protoplasma*, Vol. 44, No.4, (December 1955), pp. 460-462, ISSN 0033-183X

Welte, M.A. (2007). Proteins under new management: lipid droplets deliver. *Trends in Cell Biology*, Vol.17, No.8, (August 2007), pp. 363-369, ISSN 0962-8924

Wóycicki, Z. (1929). Sur les cristalloïdes des noyaux et les "eleoplastes" chez *Ornithogalum caudatum*. *Bull Int de l'Acad Polon de Sc et des Lettres*, Vol. 25, pp. 27-39

Zimmermann, A. (1893). Über die Elaioplasten. *Beiträge zur Morphologie und Physiologie der Pflanzenzelle*, Vol.1, pp. 185-197

Yatsu, L.Y., Jacks, T.J. (1972). Spherosome membranes half unit-membranes. *Plant Physiology*, Vol. 49, No.6, (June 1972), pp. 937-943, ISSN 0032-0889

Permissions

The contributors of this book come from diverse backgrounds, making this book a truly international effort. This book will bring forth new frontiers with its revolutionizing research information and detailed analysis of the nascent developments around the world.

We would like to thank Dr. Giuseppe Montanaro and Prof. Bartolomeo Dichio, for lending their expertise to make the book truly unique. They have played a crucial role in the development of this book. Without their invaluable contribution this book wouldn't have been possible. They have made vital efforts to compile up to date information on the varied aspects of this subject to make this book a valuable addition to the collection of many professionals and students.

This book was conceptualized with the vision of imparting up-to-date information and advanced data in this field. To ensure the same, a matchless editorial board was set up. Every individual on the board went through rigorous rounds of assessment to prove their worth. After which they invested a large part of their time researching and compiling the most relevant data for our readers. Conferences and sessions were held from time to time between the editorial board and the contributing authors to present the data in the most comprehensible form. The editorial team has worked tirelessly to provide valuable and valid information to help people across the globe.

Every chapter published in this book has been scrutinized by our experts. Their significance has been extensively debated. The topics covered herein carry significant findings which will fuel the growth of the discipline. They may even be implemented as practical applications or may be referred to as a beginning point for another development. Chapters in this book were first published by InTech; hereby published with permission under the Creative Commons Attribution License or equivalent.

The editorial board has been involved in producing this book since its inception. They have spent rigorous hours researching and exploring the diverse topics which have resulted in the successful publishing of this book. They have passed on their knowledge of decades through this book. To expedite this challenging task, the publisher supported the team at every step. A small team of assistant editors was also appointed to further simplify the editing procedure and attain best results for the readers.

Our editorial team has been hand-picked from every corner of the world. Their multi-ethnicity adds dynamic inputs to the discussions which result in innovative outcomes. These outcomes are then further discussed with the researchers and contributors who give their valuable feedback and opinion regarding the same. The feedback is then collaborated with the researches and they are edited in a comprehensive manner to aid the understanding of the subject.

Apart from the editorial board, the designing team has also invested a significant amount of their time in understanding the subject and creating the most relevant covers. They scrutinized every image to scout for the most suitable representation of the subject and create an appropriate cover for the book.

The publishing team has been involved in this book since its early stages. They were actively engaged in every process, be it collecting the data, connecting with the contributors or procuring relevant information. The team has been an ardent support to the editorial, designing and production team. Their endless efforts to recruit the best for this project, has resulted in the accomplishment of this book. They are a veteran in the field of academics and their pool of knowledge is as vast as their experience in printing. Their expertise and guidance has proved useful at every step. Their uncompromising quality standards have made this book an exceptional effort. Their encouragement from time to time has been an inspiration for everyone.

The publisher and the editorial board hope that this book will prove to be a valuable piece of knowledge for researchers, students, practitioners and scholars across the globe.

List of Contributors

Rebecca S. Lamb
Department of Molecular Genetics, The Ohio State University, Columbus, OH, USA

Miguel Mourato, Rafaela Reis and Luisa Louro Martins
UIQA, Instituto Superior de Agronomia, Technical University of Lisbon, Lisbon, Portugal

Elena Masarovičová and Katarína Kráľová
Comenius University in Bratislava, Slovak Republic

Anna M. De Leonardis, Maria Petrarulo
Pasquale De Vita and Anna M. Mastrangelo CRA-Cereal Research Centre, Foggia, Italy

Geraldo Chavarria
The University of Passo Fundo, Brazil

Henrique Pessoa dos Santos
Embrapa Grape & Wine, Brazil

Eszter Nemeskéri
Research Institute and Model Farms, Center of Agricultural Sciences and Engineering, University of Debrecen, Hungary

Krisztina Molnár and János Nagy
Institute for Land Utilisation, Technology and Regional Development, Center of Agricultural Sciences and Engineering, University of Debrecen, Hungary

Róbert Víg and Attila Dobos
Hungarian Academy of Sciences - University of Debrecen Research Group of Cultivation and Regional Development, Hungary

Giuseppe Tataranni, Bartolomeo Dichio and Cristos Xiloyannis
The University of Basilicata, Italy

Elda B. R. Perotti and Alejandro Pidello
Rosario National University (UNR) - Research Council of UNR, Argentine

Giuseppe Montanaro, Bartolomeo Dichio and Cristos Xiloyannis
Department of Crop Systems, Forestry and Environmental Sciences University of Basilicata, Italy, Turkey

Özgür Çakır, Neslihan Turgut-Kara
Department of Molecular Biology and Genetics, Faculty of Science Istanbul University, Istanbul, Turkey

Şule Arı
Department of Molecular Biology and Genetics, Faculty of Science Istanbul University, Istanbul, Turkey
Research and Application Center for Biotechnology and Genetic Engineering, Istanbul, Turkey

Maharaj Rohanie
University of Trinidad and Tobago, Piarco, Trinidad

Mohammed Ayoub
National Agricultural Marketing Development Corporation, Piarco, Trinidad

Christina K. Kitsaki, Nikos Maragos and Dimitris L. Bouranis
Laboratory of Plant Physiology, Department of Agricultural Biotechnology, Greece Agricultural University of Athens, Athens, Greece

Aiko Amagai
Graduate School of Life Sciences, Tohoku University, Japan

Tomaž Rijavec
Institute of Physical Biology, Slovenia

Qin-Bao Li
Unites States Department of Agriculture – Agricultural Research Service, Gainesville, USA

Marina Dermastia
National Institute of Biology, Slovenia

Prem S. Chourey
Agronomy and Plant Pathology Department University of Florida, Gainesville, FL, USA
Unites States Department of Agriculture – Agricultural Research Service, Gainesville, USA

Zahra Asrar
Department of Biology, Shahid Bahonar University of Kerman, Kerman, Iran

Maria Kwiatkowska, Katarzyna Popłońska, Dariusz Stępiński, Agnieszka Wojtczak, Justyna Teresa Polit and Katarzyna Paszak
University of Łódź, Department of Cytophysiology, Poland

www.ingramcontent.com/pod-product-compliance
Lightning Source LLC
Chambersburg PA
CBHW070712190326
41458CB00004B/952